SATURN

COUPES/SEDANS/WAGONS
1991-98 REPAIR MANUAL

CHILTON'S

**Covers all U.S. and Canadian models of
Saturn SC, SC1, SC2, SL, SL1, SL2, SW1 and SW2**

by Matthew E. Frederick, A.S.E., S.A.E.

CHILTON *Automotive Books*

PUBLISHED BY **HAYNES NORTH AMERICA**, Inc.

Haynes

AUTOMOTIVE
PARTS &
ACCESSORIES
ASSOCIATION   MEMBER

Manufactured in USA
© 1998 Haynes North America, Inc.
ISBN 0-8019-8956-6
Library of Congress Catalog Card No. 97-78116
7890123456 9876543210

Haynes Publishing Group
Sparkford Nr Yeovil
Somerset BA22 7JJ England

Haynes North America, Inc
861 Lawrence Drive
Newbury Park
California 91320 USA

ABCDE
FGHIJ
KL

Contents

Contents

DRIVE TRAIN   **7**

SUSPENSION AND STEERING   **8**

BRAKES   **9**

BODY AND TRIM   **10**

GLOSSARY

MASTER INDEX

SAFETY NOTICE

Proper service and repair procedures are vital to the safe, reliable operation of all motor vehicles, as well as the personal safety of those performing repairs. This manual outlines procedures for servicing and repairing vehicles using safe, effective methods. The procedures contain many NOTES, CAUTIONS and WARNINGS which should be followed, along with standard procedures to eliminate the possibility of personal injury or improper service which could damage the vehicle or compromise its safety.

It is important to note that repair procedures and techniques, tools and parts for servicing motor vehicles, as well as the skill and experience of the individual performing the work vary widely. It is not possible to anticipate all of the conceivable ways or conditions under which vehicles may be serviced, or to provide cautions as to all possible hazards that may result. Standard and accepted safety precautions and equipment should be used when handling toxic or flammable fluids, and safety goggles or other protection should be used during cutting, grinding, chiseling, prying, or any other process that can cause material removal or projectiles.

Some procedures require the use of tools specially designed for a specific purpose. Before substituting another tool or procedure, you must be completely satisfied that neither your personal safety, nor the performance of the vehicle will be endangered.

Although information in this manual is based on industry sources and is complete as possible at the time of publication, the possibility exists that some car manufacturers made later changes which could not be included here. While striving for total accuracy, the authors or publishers cannot assume responsibility for any errors, changes or omissions that may occur in the compilation of this data.

PART NUMBERS

Part numbers listed in this reference are not recommendations by Haynes North America, Inc. for any product brand name. They are references that can be used with interchange manuals and aftermarket supplier catalogs to locate each brand supplier's discrete part number.

SPECIAL TOOLS

Special tools are recommended by the vehicle manufacturer to perform their specific job. Use has been kept to a minimum, but where absolutely necessary, they are referred to in the text by the part number of the tool manufacturer. These tools can be purchased, under the appropriate part number, from your local dealer or regional distributor, or an equivalent tool can be purchased locally from a tool supplier or parts outlet. Before substituting any tool for the one recommended, read the SAFETY NOTICE at the top of this page.

ACKNOWLEDGMENTS

Portions of materials contained herein have been reprinted with the permission of General Motors Corporation, Service Technology Group.

1

GENERAL INFORMATION AND MAINTENANCE

HOW TO USE THIS BOOK

Chilton's Total Car Care manual for the Saturn SC, SL and SW is intended to help you learn more about the inner workings of your vehicle while saving you money on its upkeep and operation.

The beginning of the book will likely be referred to the most, since that is where you will find information for maintenance and tune-up. The other sections deal with the more complex systems of your vehicle. Operating systems from engine through brakes are covered to the extent that the average do-it-yourselfer becomes mechanically involved. This book will not explain such things as rebuilding a differential, for the simple reason that the expertise required and the investment in special tools make this task uneconomical. It will, however, give you detailed instructions to help you change your own brake pads and shoes, replace spark plugs, and perform many more jobs that can save you money, give you personal satisfaction and help you avoid expensive problems.

A secondary purpose of this book is a reference for owners who want to understand their vehicle and/or their mechanics better. In this case, no tools at all are required.

Where to Begin

Before removing any bolts, read through the entire procedure. This will give you the overall view of what tools and supplies will be required. There is nothing more frustrating than having to walk to the bus stop on Monday morning because you were short one bolt on Sunday afternoon. So read ahead and plan ahead. Each operation should be approached logically and all procedures thoroughly understood before attempting any work.

All sections contain adjustments, maintenance, removal and installation procedures, and in some cases, repair or overhaul procedures. When repair is not considered practical, we tell you how to remove the part and then how to install the new or rebuilt replacement. In this way, you at least save the labor costs. Backyard repair of some components is just not practical.

Avoiding Trouble

Many procedures in this book require you to "label and disconnect . . ." a group of lines, hoses or wires. Don't be lulled into thinking you can remember where everything goes—you won't. If you hook up vacuum or fuel lines incorrectly, the vehicle will run poorly, if at all. If you hook up electrical wiring incorrectly, you may instantly learn a very expensive lesson.

You don't need to know the official or engineering name for each hose or line. A piece of masking tape on the hose and a piece on its fitting will allow you to assign your own label such as the letter A or a short name. As long as you remember your own code, the lines can be reconnected by matching similar letters or names. Do remember that tape will dissolve in gasoline or other fluids; if a component is to be washed or cleaned, use another method of identification. A permanent felt-tipped marker can be very handy for marking metal parts. Remove any tape or paper labels after assembly.

Maintenance or Repair?

It's necessary to mention the difference between maintenance and repair. Maintenance includes routine inspections, adjustments, and replacement of parts which show signs of normal wear. Maintenance compensates for wear or deterioration. Repair implies that something has broken or is not working. A need for repair is often caused by lack of maintenance. Example: draining and refilling the automatic transmission fluid is maintenance recommended by the manufacturer at specific mileage intervals. Failure to do this can ruin the transaxle, requiring very expensive repairs. While no maintenance program can prevent items from breaking or wearing out, a general rule can be stated: MAINTENANCE IS CHEAPER THAN REPAIR.

Two basic mechanic's rules should be mentioned here. First, whenever the left side of the vehicle or engine is referred to, it is meant to specify the driver's side. Conversely, the right side of the vehicle means the passenger's side. Second, most screws and bolts are removed by turning counterclockwise, and tightened by turning clockwise.

Safety is always the most important rule. Constantly be aware of the dangers involved in working on an automobile and take the proper precautions. See the information in this section regarding SERVICING YOUR VEHICLE SAFELY and the SAFETY NOTICE on the acknowledgment page.

Avoiding the Most Common Mistakes

Pay attention to the instructions provided. There are 3 common mistakes in mechanical work:

1. Incorrect order of assembly, disassembly or adjustment. When taking something apart or putting it together, performing steps in the wrong order usually just costs you extra time; however, it CAN break something. Read the entire procedure before beginning disassembly. Perform everything in the order in which the instructions say you should, even if you can't immediately see a reason for it. When you're taking apart something that is very intricate, you might want to draw a picture of how it looks when assembled at one point in order to make sure you get everything back in its proper position. We will supply exploded views whenever possible. When making adjustments, perform them in the proper order; often, one adjustment affects another, and you cannot expect even satisfactory results unless each adjustment is made only when it cannot be changed by any other.

2. Overtorquing (or undertorquing). While it is more common for overtorquing to cause damage, undertorquing may allow a fastener to vibrate loose causing serious damage. Especially when dealing with aluminum parts, pay attention to torque specifications and utilize a torque wrench in assembly. If a torque figure is not available, remember that if you are using the right tool to perform the job, you will probably not have to strain yourself to get a fastener tight enough. The pitch of most threads is so slight that the tension you put on the wrench will be multiplied many times in actual force on what you are tightening. A good example of how critical torque is can be seen in the case of spark plug installation, especially where you are putting the plug into an aluminum cylinder head. Too little torque can fail to crush the gasket, causing leakage of combustion gases and consequent overheating of the plug and engine parts. Too much torque can damage the threads or distort the plug, changing the spark gap.

There are many commercial products available for ensuring that fasteners won't come loose, even if they are not torqued just right (a very common brand is Loctite®). If you're worried about getting something together tight enough to hold, but loose enough to avoid mechanical damage during assembly, one of these products might offer substantial insurance. Before choosing a threadlocking compound, read the label on the package and make sure the product is compatible with the materials, fluids, etc. involved.

3. Crossthreading. This occurs when a part such as a bolt is screwed into a nut or casting at the wrong angle and forced. Crossthreading is more likely to occur if access is difficult. It helps to clean and lubricate fasteners, then to start threading with the part to be installed positioned straight in. Then, start the bolt, spark plug, etc. with your fingers. If you encounter resistance, unscrew the part and start over again at a different angle until it can be inserted and turned several times without much effort. Keep in mind that many parts, especially spark plugs, have tapered threads, so that gentle turning will automatically bring the part you're threading to the proper angle, but only if you don't force it or resist a change in angle. Don't put a wrench on the part until it's been tightened a couple of turns by hand. If you suddenly encounter resistance, and the part has not seated fully, don't force it. Pull it back out to make sure it's clean and threading properly.

Always take your time and be patient; once you have some experience, working on your vehicle may well become an enjoyable hobby.

TOOLS AND EQUIPMENT

▶See Figures 1 thru 15

Naturally, without the proper tools and equipment it is impossible to properly service your vehicle. It would also be virtually impossible to catalog every tool that you would need to perform all of the operations in this book. Of course, It would be unwise for the amateur to rush out and buy an expensive set of tools on the theory that he/she may need one or more of them at some time.

The best approach is to proceed slowly, gathering a good quality set of those tools that are used most frequently. Don't be misled by the low cost of bargain tools. It is far better to spend a little more for better quality. Forged wrenches, 6 or 12-point sockets and fine tooth ratchets are by far preferable to their less expensive counterparts. As any good mechanic can tell you, there are few worse experiences than trying to work on a vehicle with bad tools. Your monetary savings will be far outweighed by frustration and mangled knuckles.

Begin accumulating those tools that are used most frequently: those associated with routine maintenance and tune-up. In addition to the normal assortment of screwdrivers and pliers, you should have the following tools:

• Wrenches/sockets and combination open end/box end wrenches in sizes from ⅛-¾ in. or 3mm-19mm (depending on whether your vehicle uses standard or metric fasteners) and a ¹³⁄₁₆ in. or ⅝ in. spark plug socket (depending on plug type).

➡If possible, buy various length socket drive extensions. Universal-joint and wobble extensions can be extremely useful, but be careful when using them, as they can change the amount of torque applied to the socket.

• Jackstands for support.
• Oil filter wrench.
• Spout or funnel for pouring fluids.
• Grease gun for chassis lubrication (unless your vehicle is not equipped with any grease fittings—for details, please refer to information on Fluids and Lubricants found later in this section).
• Hydrometer for checking the battery (unless equipped with a sealed, maintenance-free battery).
• A container for draining oil and other fluids.
• Rags for wiping up the inevitable mess.

In addition to the above items there are several others that are not absolutely necessary, but handy to have around. These include Oil Dry® (or an equivalent oil absorbent gravel—such as cat litter) and the usual supply of lubricants, antifreeze and fluids, although these can be purchased as needed. This is a basic list for routine maintenance, but only your personal needs and desire can accurately determine your list of tools.

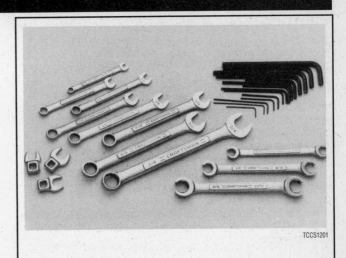

Fig. 2 In addition to ratchets, a good set of wrenches and hex keys will be necessary

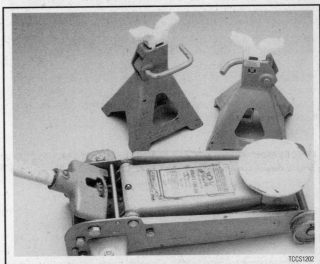

Fig. 3 A hydraulic floor jack and a set of jackstands are essential for lifting and supporting the vehicle

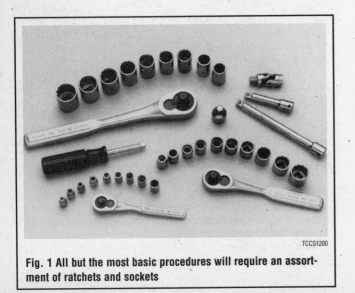

Fig. 1 All but the most basic procedures will require an assortment of ratchets and sockets

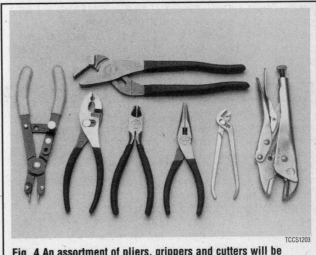

Fig. 4 An assortment of pliers, grippers and cutters will be handy for old rusted parts and stripped bolt heads

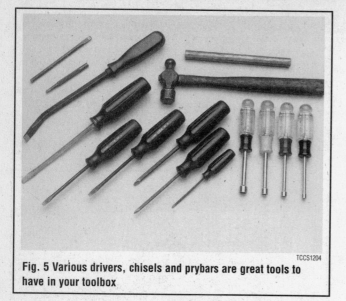

Fig. 5 Various drivers, chisels and prybars are great tools to have in your toolbox

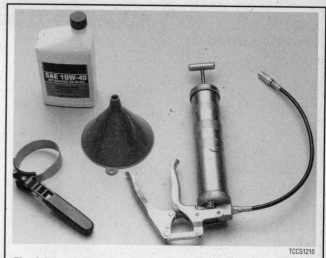

Fig. 8 A few inexpensive lubrication tools will make maintenance easier

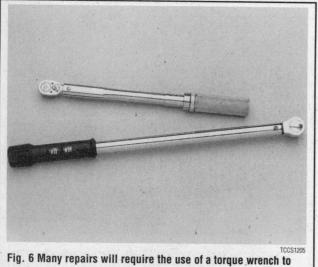

Fig. 6 Many repairs will require the use of a torque wrench to assure the components are properly fastened

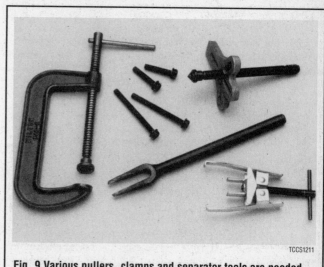

Fig. 9 Various pullers, clamps and separator tools are needed for many larger, more complicated repairs

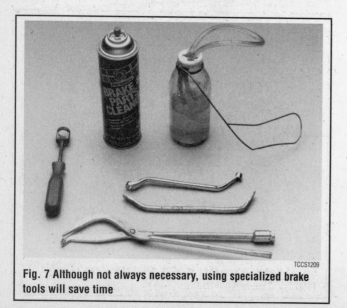

Fig. 7 Although not always necessary, using specialized brake tools will save time

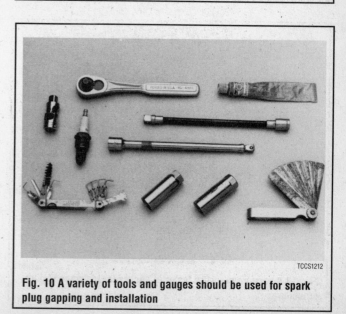

Fig. 10 A variety of tools and gauges should be used for spark plug gapping and installation

After performing a few projects on the vehicle, you'll be amazed at the other tools and non-tools on your workbench. Some useful household items are: a large turkey baster or siphon, empty coffee cans and ice trays (to store parts), ball of twine, electrical tape for wiring, small rolls of colored tape for tagging lines or hoses, markers and pens, a note pad, golf tees (for plugging vacuum lines), metal coat hangers or a roll of mechanics's wire (to hold things out of the way), dental pick or similar long, pointed probe, a strong magnet, and a small mirror (to see into recesses and under manifolds).

A more advanced set of tools, suitable for tune-up work, can be drawn up easily. While the tools are slightly more sophisticated, they need not be

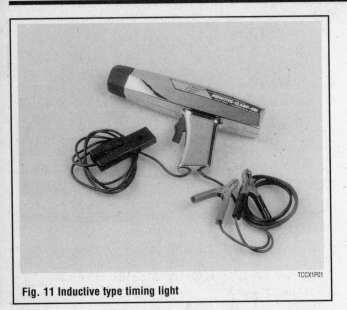

Fig. 11 Inductive type timing light

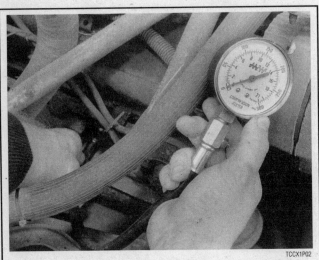

Fig. 12 A screw-in type compression gauge is recommended for compression testing

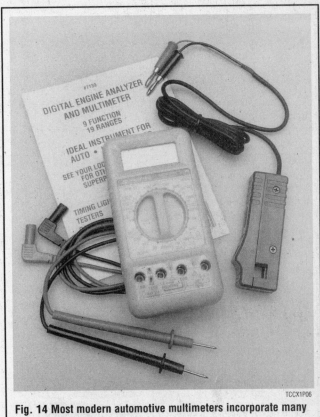

Fig. 14 Most modern automotive multimeters incorporate many helpful features

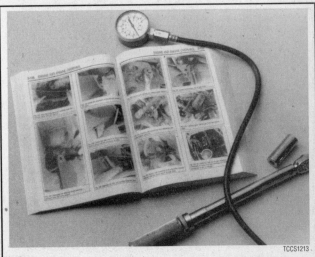

Fig. 13 A vacuum/pressure tester is necessary for many testing procedures

Fig. 15 Proper information is vital, so always have a Chilton Total Car Care manual handy

DIAGNOSTIC TEST EQUIPMENT

Modern vehicles equipped with computer-controlled fuel, emission and ignition systems require modern electronic tools to diagnose problems. Many of these tools are designed solely for the professional mechanic and are too costly and difficult to use for the average do-it-yourselfer. However, various automotive aftermarket companies have introduced products that address the needs of the average home mechanic, providing sophisticated information at affordable cost. Consult your local auto parts store to determine what is available for your vehicle.

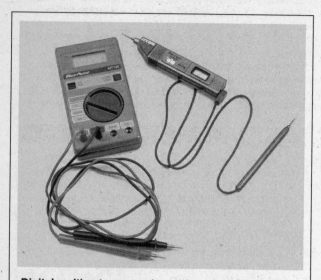

Digital multimeters come in a variety of styles and are a "must-have" for any serious home mechanic. Digital multimeters measure voltage (volts), resistance (ohms) and sometimes current (amperes). These versatile tools are used for checking all types of electrical or electronic components

Trouble code tools allow the home mechanic to extract the "fault code" number from an on-board computer that has sensed a problem (usually indicated by a Check Engine light). Armed with this code, the home mechanic can focus attention on a suspect system or component

Sensor testers perform specific checks on many of the sensors and actuators used on today's computer-controlled vehicles. These testers can check sensors both on or off the vehicle, as well as test the accompanying electrical circuits

Hand-held scanners represent the most sophisticated of all do-it-yourself diagnostic tools. These tools do more than just access computer codes like the code readers above; they provide the user with an actual interface into the vehicle's computer. Comprehensive data on specific makes and models will come with the tool, either built-in or as a separate cartridge

outrageously expensive. There are several inexpensive tach/dwell meters on the market that are every bit as good for the average mechanic as a professional model. Just be sure that it goes to a least 1200–1500 rpm on the tach scale and that it works on 4, 6 and 8-cylinder engines. (If you have one or more vehicles with a diesel engine, a special tachometer is required since diesels don't use spark plug ignition systems). The key to these purchases is to make them with an eye towards adaptability and wide range. A basic list of tune-up tools could include:

- Tach/dwell meter.
- Spark plug wrench and gapping tool.
- Feeler gauges for valve or point adjustment. (Even if your vehicle does not use points or require valve adjustments, a feeler gauge is helpful for many repair/overhaul procedures).

A tachometer/dwell meter will ensure accurate tune-up work on vehicles without electronic ignition. The choice of a timing light should be made carefully. A light which works on the DC current supplied by the vehicle's battery is the best choice; it should have a xenon tube for brightness. On any vehicle with an electronic ignition system, a timing light with an inductive pickup that clamps around the No. 1 spark plug cable is preferred.

In addition to these basic tools, there are several other tools and gauges you may find useful. These include:

- Compression gauge. The screw-in type is slower to use, but eliminates the possibility of a faulty reading due to escaping pressure.
- Manifold vacuum gauge.

- 12V test light.
- A combination volt/ohmmeter
- Induction ammeter. This is used for determining whether or not there is current in a wire. These are handy for use if a wire is broken somewhere in a wiring harness.

As a final note, you will probably find a torque wrench necessary for all but the most basic work. The beam type models are perfectly adequate, although the newer click types (breakaway) are easier to use. The click type torque wrenches tend to be more expensive. Also keep in mind that all types of torque wrenches should be periodically checked and/or recalibrated. You will have to decide for yourself which better fits your purpose.

Special Tools

Normally, the use of special factory tools is avoided for repair procedures, since these are not readily available for the do-it-yourself mechanic. When it is possible to perform the job with more commonly available tools, it will be pointed out, but occasionally, a special tool was designed to perform a specific function and should be used. Before substituting another tool, you should be convinced that neither your safety nor the performance of the vehicle will be compromised.

Special tools can usually be purchased from an automotive parts store or from your dealer. In some cases special tools may be available directly from the tool manufacturer.

SERVICING YOUR VEHICLE SAFELY

▶ **See Figures 16, 17, 18 and 19**

It is virtually impossible to anticipate all of the hazards involved with automotive maintenance and service, but care and common sense will prevent most accidents.

The rules of safety for mechanics range from "don't smoke around gasoline," to "use the proper tool(s) for the job." The trick to avoiding injuries is to develop safe work habits and to take every possible precaution.

Do's

- Do keep a fire extinguisher and first aid kit handy.
- Do wear safety glasses or goggles when cutting, drilling, grinding or prying, even if you have 20–20 vision. If you wear glasses for the sake of vision, wear safety goggles over your regular glasses.
- Do shield your eyes whenever you work around the battery. Batteries contain sulfuric acid. In case of contact with the eyes or skin, flush the area with water or a mixture of water and baking soda, then seek immediate medical attention.

- Do use safety stands (jackstands) for any undervehicle service. Jacks are for raising vehicles; jackstands are for making sure the vehicle stays raised until you want it to come down. Whenever the vehicle is raised, block the wheels remaining on the ground and set the parking brake.
- Do use adequate ventilation when working with any chemicals or hazardous materials. Like carbon monoxide, the asbestos dust resulting from some brake lining wear can be hazardous in sufficient quantities.
- Do disconnect the negative battery cable when working on the electrical system. The secondary ignition system contains EXTREMELY HIGH VOLTAGE. In some cases it can even exceed 50,000 volts.
- Do follow manufacturer's directions whenever working with potentially hazardous materials. Most chemicals and fluids are poisonous if taken internally.
- Do properly maintain your tools. Loose hammerheads, mushroomed punches and chisels, frayed or poorly grounded electrical cords, excessively worn screwdrivers, spread wrenches (open end), cracked sockets, slipping ratchets, or faulty droplight sockets can cause accidents.

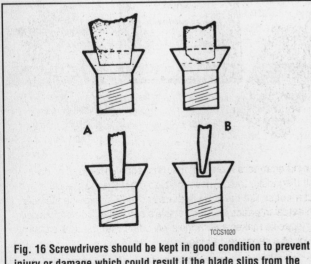

TCCS1020

Fig. 16 Screwdrivers should be kept in good condition to prevent injury or damage which could result if the blade slips from the screw

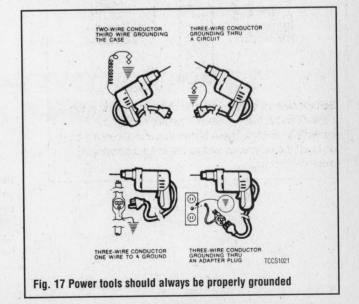

TWO-WIRE CONDUCTOR THIRD WIRE GROUNDING THE CASE

THREE-WIRE CONDUCTOR GROUNDING THRU A CIRCUIT

THREE-WIRE CONDUCTOR ONE WIRE TO A GROUND

THREE-WIRE CONDUCTOR GROUNDING THRU AN ADAPTER PLUG

TCCS1021

Fig. 17 Power tools should always be properly grounded

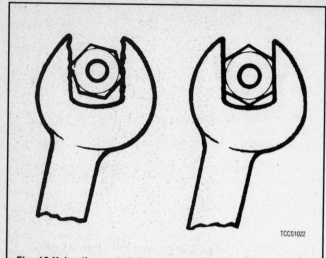

Fig. 18 Using the correct size wrench will help prevent the possibility of rounding off a nut

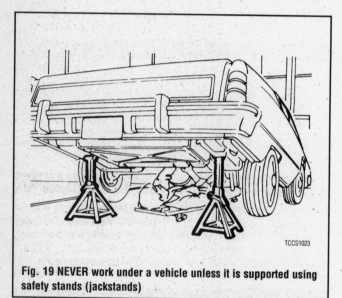

Fig. 19 NEVER work under a vehicle unless it is supported using safety stands (jackstands)

• Likewise, keep your tools clean; a greasy wrench can slip off a bolt head, ruining the bolt and often harming your knuckles in the process.
• Do use the proper size and type of tool for the job at hand. Do select a wrench or socket that fits the nut or bolt. The wrench or socket should sit straight, not cocked.

• Do, when possible, pull on a wrench handle rather than push on it, and adjust your stance to prevent a fall.
• Do be sure that adjustable wrenches are tightly closed on the nut or bolt and pulled so that the force is on the side of the fixed jaw.
• Do strike squarely with a hammer; avoid glancing blows.
• Do set the parking brake and block the drive wheels if the work requires a running engine.

Don'ts

• Don't run the engine in a garage or anywhere else without proper ventilation—EVER! Carbon monoxide is poisonous; it takes a long time to leave the human body and you can build up a deadly supply of it in your system by simply breathing in a little every day. You may not realize you are slowly poisoning yourself. Always use power vents, windows, fans and/or open the garage door.
• Don't work around moving parts while wearing loose clothing. Short sleeves are much safer than long, loose sleeves. Hard-toed shoes with neoprene soles protect your toes and give a better grip on slippery surfaces. Jewelry such as watches, fancy belt buckles, beads or body adornment of any kind is not safe working around a vehicle. Long hair should be tied back under a hat or cap.
• Don't use pockets for toolboxes. A fall or bump can drive a screwdriver deep into your body. Even a rag hanging from your back pocket can wrap around a spinning shaft or fan.
• Don't smoke when working around gasoline, cleaning solvent or other flammable material.
• Don't smoke when working around the battery. When the battery is being charged, it gives off explosive hydrogen gas.
• Don't use gasoline to wash your hands; there are excellent soaps available. Gasoline contains dangerous additives which can enter the body through a cut or through your pores. Gasoline also removes all the natural oils from the skin so that bone dry hands will suck up oil and grease.
• Don't service the air conditioning system unless you are equipped with the necessary tools and training. When liquid or compressed gas refrigerant is released to atmospheric pressure it will absorb heat from whatever it contacts. This will chill or freeze anything it touches. Although refrigerant is normally non-toxic, R-12 becomes a deadly poisonous gas in the presence of an open flame. One good whiff of the vapors from burning refrigerant can be fatal.
• Don't use screwdrivers for anything other than driving screws! A screwdriver used as an prying tool can snap when you least expect it, causing injuries. At the very least, you'll ruin a good screwdriver.
• Don't use a bumper or emergency jack (that little ratchet, scissors, or pantograph jack supplied with the vehicle) for anything other than changing a flat! These jacks are only intended for emergency use out on the road; they are NOT designed as a maintenance tool. If you are serious about maintaining your vehicle yourself, invest in a hydraulic floor jack of at least a 1½ ton capacity, and at least two sturdy jackstands.

FASTENERS, MEASUREMENTS AND CONVERSIONS

Bolts, Nuts and Other Threaded Retainers

⬥ See Figures 20, 21, 22 and 23

Although there are a great variety of fasteners found in the modern car or truck, the most commonly used retainer is the threaded fastener (nuts, bolts, screws, studs, etc). Most threaded retainers may be reused, provided that they are not damaged in use or during the repair. Some retainers (such as stretch bolts or torque prevailing nuts) are designed to deform when tightened or in use and should not be reinstalled.

Whenever possible, we will note any special retainers which should be replaced during a procedure. But you should always inspect the condition of a retainer when it is removed and replace any that show signs of damage. Check all threads for rust or corrosion which can increase the torque necessary to achieve the desired clamp load for which that fastener was

originally selected. Additionally, be sure that the driver surface of the fastener has not been compromised by rounding or other damage. In some cases a driver surface may become only partially rounded, allowing the driver to catch in only one direction. In many of these occurrences, a fastener may be installed and tightened, but the driver would not be able to grip and loosen the fastener again. (This could lead to frustration down the line should that component ever need to be disassembled again).

If you must replace a fastener, whether due to design or damage, you must ALWAYS be sure to use the proper replacement. In all cases, a retainer of the same design, material and strength should be used. Markings on the heads of most bolts will help determine the proper strength of the fastener. The same material, thread and pitch must be selected to assure proper installation and safe operation of the vehicle afterwards.

Thread gauges are available to help measure a bolt or stud's thread. Most automotive and hardware stores keep gauges available to help you

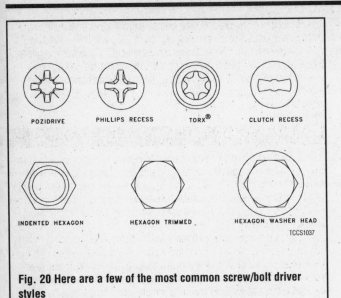

Fig. 20 Here are a few of the most common screw/bolt driver styles

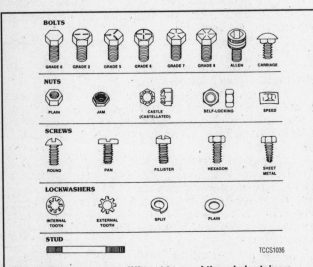

Fig. 21 There are many different types of threaded retainers found on vehicles

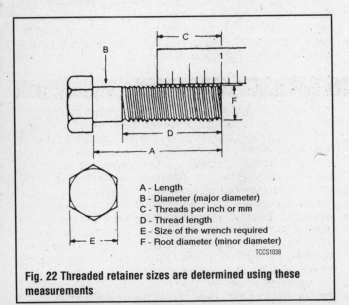

A - Length
B - Diameter (major diameter)
C - Threads per inch or mm
D - Thread length
E - Size of the wrench required
F - Root diameter (minor diameter)

Fig. 22 Threaded retainer sizes are determined using these measurements

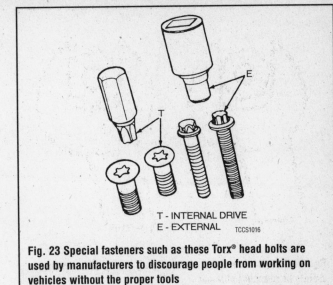

T - INTERNAL DRIVE
E - EXTERNAL

Fig. 23 Special fasteners such as these Torx® head bolts are used by manufacturers to discourage people from working on vehicles without the proper tools

select the proper size. In a pinch, you can use another nut or bolt for a thread gauge. If the bolt you are replacing is not too badly damaged, you can select a match by finding another bolt which will thread in its place. If you find a nut which threads properly onto the damaged bolt, then use that nut to help select the replacement bolt. If however, the bolt you are replacing is so badly damaged (broken or drilled out) that its threads cannot be used as a gauge, you might start by looking for another bolt (from the same assembly or a similar location on your vehicle) which will thread into the damaged bolt's mounting. If so, the other bolt can be used to select a nut; the nut can then be used to select the replacement bolt.

In all cases, be absolutely sure you have selected the proper replacement. Don't be shy, you can always ask the store clerk for help.

✳✳ WARNING

Be aware that when you find a bolt with damaged threads, you may also find the nut or drilled hole it was threaded into has also been damaged. If this is the case, you may have to drill and tap the hole, replace the nut or otherwise repair the threads. NEVER try to force a replacement bolt to fit into the damaged threads.

Torque

Torque is defined as the measurement of resistance to turning or rotating. It tends to twist a body about an axis of rotation. A common example of this would be tightening a threaded retainer such as a nut, bolt or screw. Measuring torque is one of the most common ways to help assure that a threaded retainer has been properly fastened.

When tightening a threaded fastener, torque is applied in three distinct areas, the head, the bearing surface and the clamp load. About 50 percent of the measured torque is used in overcoming bearing friction. This is the friction between the bearing surface of the bolt head, screw head or nut face and the base material or washer (the surface on which the fastener is rotating). Approximately 40 percent of the applied torque is used in overcoming thread friction. This leaves only about 10 percent of the applied torque to develop a useful clamp load (the force which holds a joint together). This means that friction can account for as much as 90 percent of the applied torque on a fastener.

TORQUE WRENCHES

▶ **See Figures 24, 25 and 26**

In most applications, a torque wrench can be used to assure proper installation of a fastener. Torque wrenches come in various designs and most automotive supply stores will carry a variety to suit your needs. A torque wrench should be used any time we supply a specific torque value

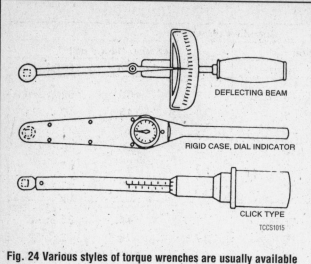

Fig. 24 Various styles of torque wrenches are usually available at your local automotive supply store

for a fastener. A torque wrench can also be used if you are following the general guidelines in the accompanying charts. Keep in mind that because there is no worldwide standardization of fasteners, the charts are a general guideline and should be used with caution. Again, the general rule of "if you are using the right tool for the job, you should not have to strain to tighten a fastener" applies here.

Beam Type

▶ See Figure 27

The beam type torque wrench is one of the most popular types. It consists of a pointer attached to the head that runs the length of the flexible beam (shaft) to a scale located near the handle. As the wrench is pulled, the beam bends and the pointer indicates the torque using the scale.

Click (Breakaway) Type

▶ See Figure 28

Another popular design of torque wrench is the click type. To use the click type wrench you pre-adjust it to a torque setting. Once the torque is reached, the wrench has a reflex signaling feature that causes a momentary breakaway of the torque wrench body, sending an impulse to the operator's hand.

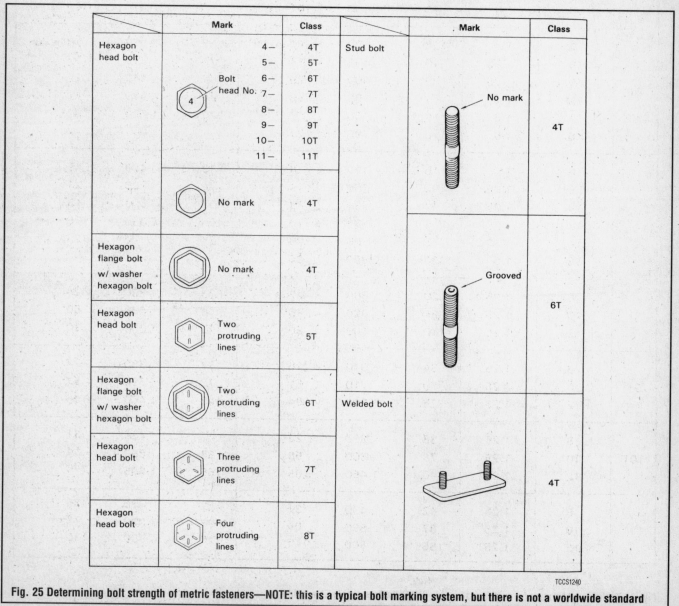

Fig. 25 Determining bolt strength of metric fasteners—NOTE: this is a typical bolt marking system, but there is not a worldwide standard

Class	Diameter mm	Pitch mm	Specified torque					
			Hexagon head bolt			Hexagon flange bolt		
			N·m	kgf·cm	ft·lbf	N·m	kgf·cm	ft·lbf
4T	6	1	5	55	48 in.·lbf	6	60	52 in.·lbf
	8	1.25	12.5	130	9	14	145	10
	10	1.25	26	260	19	29	290	21
	12	1.25	47	480	35	53	540	39
	14	1.5	74	760	55	84	850	61
	16	1.5	115	1,150	83	—	—	—
5T	6	1	6.5	65	56 in.·lbf	7.5	75	65 in.·lbf
	8	1.25	15.5	160	12	17.5	175	13
	10	1.25	32	330	24	36	360	26
	12	1.25	59	600	43	65	670	48
	14	1.5	91	930	67	100	1,050	76
	16	1.5	140	1,400	101	—	—	—
6T	6	1	8	80	69 in.·lbf	9	90	78 in.·lbf
	8	1.25	19	195	14	21	210	15
	10	1.25	39	400	29	44	440	32
	12	1.25	71	730	53	80	810	59
	14	1.5	110	1,100	80	125	1,250	90
	16	1.5	170	1,750	127	—	—	—
7T	6	1	10.5	110	8	12	120	9
	8	1.25	25	260	19	28	290	21
	10	1.25	52	530	38	58	590	43
	12	1.25	95	970	70	105	1,050	76
	14	1.5	145	1,500	108	165	1,700	123
	16	1.5	230	2,300	166	—	—	—
8T	8	1.25	29	300	22	33	330	24
	10	1.25	61	620	45	68	690	50
	12	1.25	110	1,100	80	120	1,250	90
9T	8	1.25	34	340	25	37	380	27
	10	1.25	70	710	51	78	790	57
	12	1.25	125	1,300	94	140	1,450	105
10T	8	1.25	38	390	28	42	430	31
	10	1.25	78	800	58	88	890	64
	12	1.25	140	1,450	105	155	1,600	116
11T	8	1.25	42	430	31	47	480	35
	10	1.25	87	890	64	97	990	72
	12	1.25	155	1,600	116	175	1,800	130

TCCS1241

Fig. 26 Typical bolt torques for metric fasteners—WARNING: use only as a guide

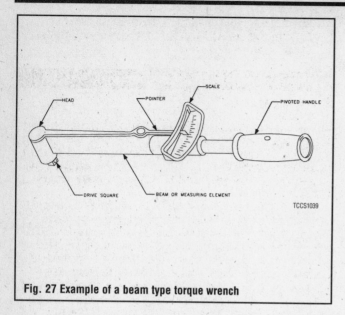

Fig. 27 Example of a beam type torque wrench

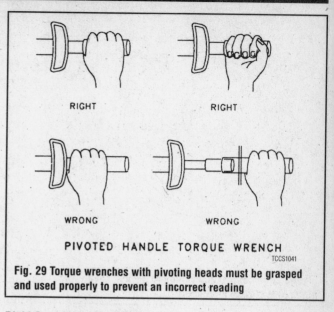

Fig. 29 Torque wrenches with pivoting heads must be grasped and used properly to prevent an incorrect reading

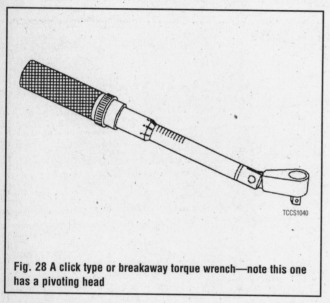

Fig. 28 A click type or breakaway torque wrench—note this one has a pivoting head

Pivot Head Type

▶ See Figures 28 and 29

Some torque wrenches (usually of the click type) may be equipped with a pivot head which can allow it to be used in areas of limited access. BUT, it must be used properly. To hold a pivot head wrench, grasp the handle lightly, and as you pull on the handle, it should be floated on the pivot point. If the handle comes in contact with the yoke extension during the process of pulling, there is a very good chance the torque readings will be inaccurate because this could alter the wrench loading point. The design of the handle is usually such as to make it inconvenient to deliberately misuse the wrench.

➡ It should be mentioned that the use of any U-joint, wobble or extension will have an effect on the torque readings, no matter what type of wrench you are using. For the most accurate readings, install the socket directly on the wrench driver. If necessary, straight extensions (which hold a socket directly under the wrench driver) will have the least effect on the torque reading. Avoid any extension that alters the length of the wrench from the handle to the head/driving point (such as a crow's foot). U-joint or Wobble extensions can greatly affect the readings; avoid their use at all times.

Rigid Case (Direct Reading)

▶ See Figure 30

A rigid case or direct reading torque wrench is equipped with a dial indicator to show torque values. One advantage of these wrenches is that they can be held at any position on the wrench without affecting accuracy. These wrenches are often preferred because they tend to be compact, easy to read and have a great degree of accuracy.

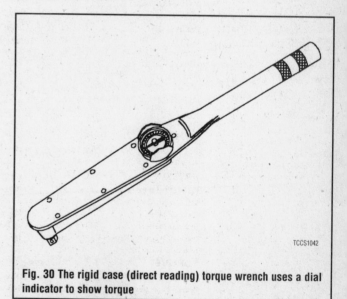

Fig. 30 The rigid case (direct reading) torque wrench uses a dial indicator to show torque

TORQUE ANGLE METERS

▶ See Figure 31

Because the frictional characteristics of each fastener or threaded hole will vary, clamp loads which are based strictly on torque will vary as well. In most applications, this variance is not significant enough to cause worry. But, in certain applications, a manufacturer's engineers may determine that more precise clamp loads are necessary (such is the case with many aluminum cylinder heads). In these cases, a torque angle method of installation would be specified. When installing fasteners which are torque angle tightened, a predetermined seating torque and standard torque wrench are usually used first to remove any compliance from the joint. The fastener is

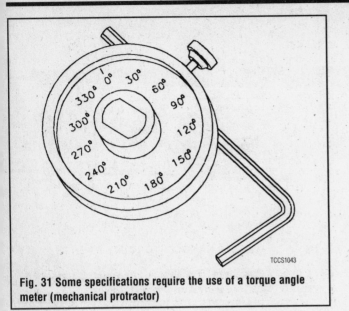

Fig. 31 Some specifications require the use of a torque angle meter (mechanical protractor)

then tightened the specified additional portion of a turn measured in degrees. A torque angle gauge (mechanical protractor) is used for these applications.

Standard and Metric Measurements

♦ See Figure 32

Throughout this manual, specifications are given to help you determine the condition of various components on your vehicle, or to assist you in their installation. Some of the most common measurements include length (in. or cm/mm), torque (ft. lbs., inch lbs. or Nm) and pressure (psi, in. Hg, kPa or mm Hg). In most cases, we strive to provide the proper measurement as determined by the manufacturer's engineers.

Though, in some cases, that value may not be conveniently measured with what is available in your toolbox. Luckily, many of the measuring devices which are available today will have two scales so the Standard or Metric measurements may easily be taken. If any of the various measuring tools which are available to you do not contain the same scale as listed in the specifications, use the accompanying conversion factors to determine the proper value.

The conversion factor chart is used by taking the given specification and multiplying it by the necessary conversion factor. For instance, looking at the first line, if you have a measurement in inches such as "free-play

CONVERSION FACTORS

LENGTH–DISTANCE

Inches (in.)	x 25.4	= Millimeters (mm)	x .0394	= Inches
Feet (ft.)	x .305	= Meters (m)	x 3.281	= Feet
Miles	x 1.609	= Kilometers (km)	x .0621	= Miles

VOLUME

Cubic Inches (in3)	x 16.387	= Cubic Centimeters	x .061	= in3
IMP Pints (IMP pt.)	x .568	= Liters (L)	x 1.76	= IMP pt.
IMP Quarts (IMP qt.)	x 1.137	= Liters (L)	x .88	= IMP qt.
IMP Gallons (IMP gal.)	x 4.546	= Liters (L)	x .22	= IMP gal.
IMP Quarts (IMP qt.)	x 1.201	= US Quarts (US qt.)	x .833	= IMP qt.
IMP Gallons (IMP gal.)	x 1.201	= US Gallons (US gal.)	x .833	= IMP gal.
Fl. Ounces	x 29.573	= Milliliters	x .034	= Ounces
US Pints (US pt.)	x .473	= Liters (L)	x 2.113	= Pints
US Quarts (US qt.)	x .946	= Liters (L)	x 1.057	= Quarts
US Gallons (US gal.)	x 3.785	= Liters (L)	x .264	= Gallons

MASS–WEIGHT

Ounces (oz.)	x 28.35	= Grams (g)	x .035	= Ounces
Pounds (lb.)	x .454	= Kilograms (kg)	x 2.205	= Pounds

PRESSURE

Pounds Per Sq. In. (psi)	x 6.895	= Kilopascals (kPa)	x .145	= psi
Inches of Mercury (Hg)	x .4912	= psi	x 2.036	= Hg
Inches of Mercury (Hg)	x 3.377	= Kilopascals (kPa)	x .2961	= Hg
Inches of Water (H_2O)	x .07355	= Inches of Mercury	x 13.783	= H_2O
Inches of Water (H_2O)	x .03613	= psi	x 27.684	= H_2O
Inches of Water (H_2O)	x .248	= Kilopascals (kPa)	x 4.026	= H_2O

TORQUE

Pounds–Force Inches (in–lb)	x .113	= Newton Meters (N·m)	x 8.85	= in–lb
Pounds–Force Feet (ft–lb)	x 1.356	= Newton Meters (N·m)	x .738	= ft–lb

VELOCITY

Miles Per Hour (MPH)	x 1.609	= Kilometers Per Hour (KPH)	x .621	= MPH

POWER

Horsepower (Hp)	x .745	= Kilowatts	x 1.34	= Horsepower

FUEL CONSUMPTION*

Miles Per Gallon IMP (MPG)	x .354	= Kilometers Per Liter (Km/L)	
Kilometers Per Liter (Km/L)	x 2.352	= IMP MPG	
Miles Per Gallon US (MPG)	x .425	= Kilometers Per Liter (Km/L)	
Kilometers Per Liter (Km/L)	x 2.352	= US MPG	

*It is common to covert from miles per gallon (mpg) to liters/100 kilometers (1/100 km), where mpg (IMP) x 1/100 km = 282 and mpg (US) x 1/100 km = 235.

TEMPERATURE

Degree Fahrenheit (°F)	= (°C x 1.8) + 32
Degree Celsius (°C)	= (°F – 32) x .56

Fig. 32 Standard and metric conversion factors chart

should be 2 in." but your ruler reads only in millimeters, multiply 2 in. by the conversion factor of 25.4 to get the metric equivalent of 50.8mm. Likewise, if the specification was given only in a Metric measurement, for

example in Newton Meters (Nm), then look at the center column first. If the measurement is 100 Nm, multiply it by the conversion factor of 0.738 to get 73.8 ft. lbs.

SERIAL NUMBER IDENTIFICATION

Vehicle

♦ See Figures 33 and 34

The Vehicle Identification Number (VIN) is stamped on a metal plate that is attached to the instrument panel adjacent to the windshield. It can be seen by looking through the lower corner of the windshield on the driver's side of the vehicle. The VIN is also located on various identification stickers throughout the vehicle and on certain metal parts such as the engine or transaxle.

The VIN is a 17-digit combination of numbers and letters. The 1st digit represents the country of manufacture, this is a 1 for all Saturns and means the vehicle was built in the U.S.A. The 2nd and 3rd digits are G8, which represent General Motors and Saturn Corporation. The 4th and 5th digits represent the Saturn car line and the particular sales code of the vehicle. For example, the sales code for a sedan will designate if the car is an SL, SL1 or SL2. The 6th digit identifies the vehicle's body style, in other words, whether it is a coupe, sedan or wagon. The 7th digit represents the safety restraint features with which that particular car was produced. This will mean, "passive restraint seat belts", "passive restraint with a driver's side Supplemental Inflatable Restraint (SIR)" or "active restraints with a driver's and passenger's SIR". The 8th number tells with what 1.9L engine the vehicle is equipped: Single Overhead Camshaft (SOHC) with Throttle Body Injection (TBI), SOHC with Multi-port Fuel Injection (MFI), or Dual Overhead Camshaft (DOHC) with MFI. The 9th digit is a check digit for all vehicles. The 10th digit indicates the model year: M for 1991, N for 1992, P for

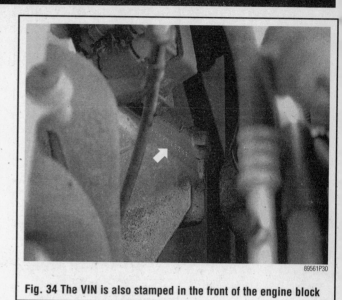

89561P30

Fig. 34 The VIN is also stamped in the front of the engine block

1993, R for 1994, S for 1995, T for 1996, V for 1997, or W for 1998. The 11th digit is a "Z" and indicates that the vehicle was built in the Spring Hill, Tennessee plant. The 12th through 17th digits indicate the production sequence number.

Engine

The engine identification code is located in the VIN at the 8th digit. The VIN can be found on the instrument panel. See the Engine Identification chart for engine VIN codes.

The engine block is also stamped with a 5-digit engine date code on the front side of the block to indicate the day and time of manufacture. The 1st position is a code for hour. The 2nd–4th positions indicate the Julian date (month and day). The 5th position is a code for year.

Transaxle

The transaxle identification code is located on the top of the transmission case to identify the transaxle model and to indicate the day and time of manufacture. The 1st digit is the last number of the year; 1=1991, 2=1992, 3=1993, 4=1994, 5=1995, 6=1996, 7=1997 or 8=1998. The 2nd–4th positions are the 3-digit transaxle code: MP2=base manual 5-speed, MP3=performance manual 5-speed, MP6=base automatic 4-speed, or MP7=performance automatic 4-speed. The 5th position is a 1 for the Spring Hill plant. The 6th–8th digits are the Julian date (month and day), while the 9th digit is a code for hour.

89561P29

Fig. 33 Vehicle Identification Number (VIN) location on the top of the instrument panel

VEHICLE IDENTIFICATION CHART

Engine Code						Model Year	
Code	Liters	Cu. In. (cc)	Cyl.	Fuel Sys.	Eng. Mfg.	Code	Year
7	1.9	116 (1901)	4	MFI	Saturn	M	1991
8	1.9	116 (1901)	4	MFI	Saturn	N	1992
9	1.9	116 (1901)	4	TBI	Saturn	P	1993
MFI - Multi-port Fuel Injection						R	1994
TBI - Throttle Body Injection						S	1995
						T	1996
						V	1997
						W	1998

89561C01

ENGINE IDENTIFICATION

Year	Model	Engine Displacement Liters (cc)	Engine Series (ID/VIN)	Fuel System	No. of Cylinders	Engine Type
1991	Sedan	1.9 (1901)	7	MFI	4	DOHC
	Sedan	1.9 (1901)	9	TBI	4	SOHC
	Coupe	1.9 (1901)	7	MFI	4	DOHC
1992	Sedan	1.9 (1901)	7	MFI	4	DOHC
	Sedan	1.9 (1901)	9	TBI	4	SOHC
	Coupe	1.9 (1901)	7	MFI	4	DOHC
1993	Wagon	1.9 (1901)	7	MFI	4	DOHC
	Wagon	1.9 (1901)	9	TBI	4	SOHC
	Sedan	1.9 (1901)	7	MFI	4	DOHC
	Sedan	1.9 (1901)	9	TBI	4	SOHC
	Coupe	1.9 (1901)	9	TBI	4	SOHC
	Coupe	1.9 (1901)	7	MFI	4	DOHC
1994	Wagon	1.9 (1901)	7	MFI	4	DOHC
	Wagon	1.9 (1901)	9	TBI	4	SOHC
	Sedan	1.9 (1901)	7	MFI	4	DOHC
	Sedan	1.9 (1901)	9	TBI	4	SOHC
	Coupe	1.9 (1901)	9	TBI	4	SOHC
	Coupe	1.9 (1901)	7	MFI	4	DOHC
1995	Wagon	1.9 (1901)	7	MFI	4	DOHC
	Wagon	1.9 (1901)	8	MFI	4	SOHC
	Sedan	1.9 (1901)	7	MFI	4	DOHC
	Sedan	1.9 (1901)	8	MFI	4	SOHC
	Coupe	1.9 (1901)	7	MFI	4	DOHC
	Coupe	1.9 (1901)	8	MFI	4	SOHC
1996	Wagon	1.9 (1901)	7	MFI	4	DOHC
	Wagon	1.9 (1901)	8	MFI	4	SOHC
	Sedan	1.9 (1901)	7	MFI	4	DOHC
	Sedan	1.9 (1901)	8	MFI	4	SOHC
	Coupe	1.9 (1901)	7	MFI	4	DOHC
	Coupe	1.9 (1901)	8	MFI	4	SOHC
1997	Wagon	1.9 (1901)	7	MFI	4	DOHC
	Wagon	1.9 (1901)	8	MFI	4	SOHC
	Sedan	1.9 (1901)	7	MFI	4	DOHC
	Sedan	1.9 (1901)	8	MFI	4	SOHC
	Coupe	1.9 (1901)	7	MFI	4	DOHC
	Coupe	1.9 (1901)	8	MFI	4	SOHC
1998	Wagon	1.9 (1901)	7	MFI	4	DOHC
	Wagon	1.9 (1901)	8	MFI	4	SOHC
	Sedan	1.9 (1901)	7	MFI	4	DOHC
	Sedan	1.9 (1901)	8	MFI	4	SOHC
	Coupe	1.9 (1901)	7	MFI	4	DOHC
	Coupe	1.9 (1901)	8	MFI	4	SOHC

DOHC - Dual Overhead Cam
SOHC - Single Overhead Cam
MFI - Multi-port Fuel Injection
TBI - Throttle Body Injection

89561C02

GENERAL ENGINE SPECIFICATIONS

Year	Engine ID/VIN	Engine Displacement Liters (cc)	Fuel System Type	Net Horsepower @ rpm	Net Torque @ rpm (ft. lbs.)	Bore x Stroke (in.)	Compression Ratio	Oil Pressure @ rpm
1991	7	1.9 (1901)	MFI	124@6000	122@4800	3.23x3.54	9.5:1	36@2000
	9	1.9 (1901)	TBI	85@5000	107@2400	3.23x3.54	9.3:1	36@2000
1992	7	1.9 (1901)	MFI	124@5600	122@4800	3.23x3.54	9.5:1	36@2000
	9	1.9 (1901)	TBI	85@5000	107@2400	3.23x3.54	9.3:1	36@2000
1993	7	1.9 (1901)	MFI	124@5600	122@4800	3.23x3.54	9.5:1	36@2000
	9	1.9 (1901)	TBI	85@5000	107@2400	3.23x3.54	9.3:1	36@2000
1994	7	1.9 (1901)	MFI	124@5600	122@4800	3.23x3.54	9.5:1	36@2000
	9	1.9 (1901)	TBI	85@5000	107@2400	3.23x3.54	9.3:1	36@2000
1995	7	1.9 (1901)	MFI	124@5600	122@4800	3.23x3.54	9.5:1	36@2000
	8	1.9 (1901)	MFI	100@5000	114@2400	3.23x3.54	9.3:1	36@2000
1996	7	1.9 (1901)	MFI	124@5600	122@4800	3.23x3.54	9.5:1	36@2000
	8	1.9 (1901)	MFI	100@5000	114@2400	3.23x3.54	9.3:1	36@2000
1997	7	1.9 (1901)	MFI	124@5600	122@4800	3.23x3.54	9.5:1	36@2000
	8	1.9 (1901)	MFI	100@5000	114@2400	3.23x3.54	9.3:1	36@2000
1998	7	1.9 (1901)	MFI	124@5600	122@4800	3.23x3.54	9.5:1	36@2000
	8	1.9 (1901)	MFI	100@5000	114@2400	3.23x3.54	9.3:1	36@2000

MFI - Multi-port Fuel Injection
TBI - Throttle Body Injection

89561C05

ROUTINE MAINTENANCE AND TUNE-UP

UNDERHOOD MAINTENANCE COMPONENT LOCATIONS

1. Air filter housing
2. Fuel filter (along body frame rail)
3. PCV valve and hose
4. Evaporative canister
5. Battery
6. Accessory drive belt (A/C compressor shown)
7. Upper coolant hose
8. Spark plugs and wires
9. Windshield washer fluid reservoir
10. Engine oil dipstick
11. Engine oil fill cap
12. Transaxle fluid dipstick and fill tube (automatic)
13. Coolant overflow tank
14. Brake fluid reservoir
15. Power steering fluid reservoir
16. Underhood vehicle emissions control information label

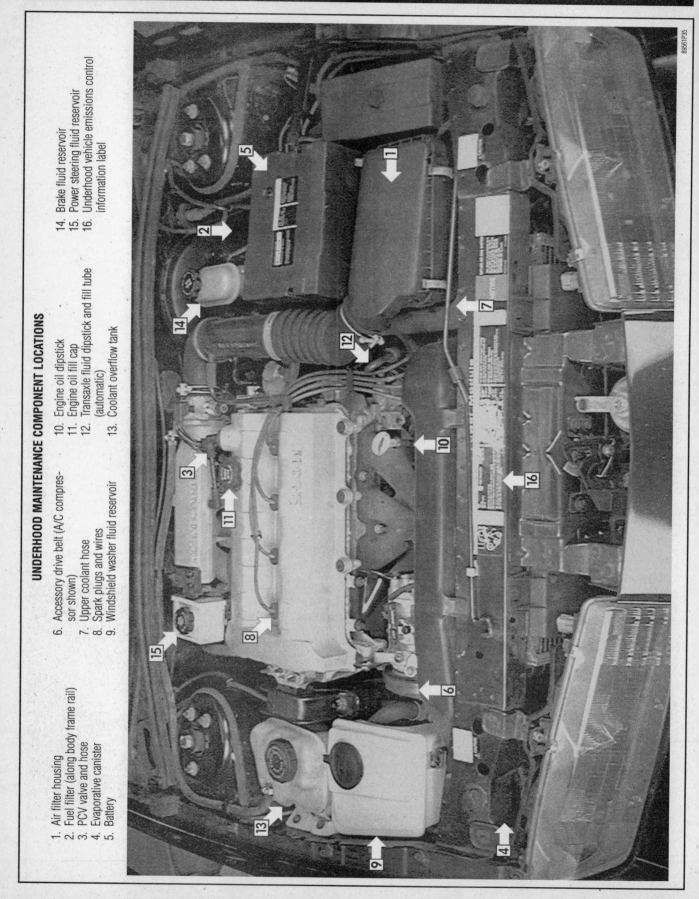

8956IP35

Proper maintenance and tune-up is the key to long and trouble-free vehicle life, and the work can yield its own rewards. Studies have shown that a properly tuned and maintained vehicle can achieve better gas mileage than an out-of-tune vehicle. As a conscientious owner and driver, set aside a Saturday morning, say once a month, to check or replace items which could cause major problems later. Keep your own personal log to jot down which services you performed, how much the parts cost you, the date, and the exact odometer reading at the time. Keep all receipts for such items as engine oil and filters, so that they may be referred to in case of related problems or to determine operating expenses. As a do-it-yourselfer, these receipts are the only proof you have that the required maintenance was performed. In the event of a warranty problem, these receipts will be invaluable.

The literature provided with your vehicle when it was originally delivered includes the factory recommended maintenance schedule. If you no longer have this literature, replacement copies are usually available from the dealer. A maintenance schedule is provided later in this section, in case you do not have the factory literature.

Air Cleaner

The air cleaner is a paper element contained in a plastic housing. It is located on the left front side of the engine compartment on DOHC and 1995–98 SOHC models. It is located behind the engine on 1991–94 SOHC models. The air filter element should be serviced according to the Maintenance Intervals Chart at the end of this section.

➡Check the air filter element more often if the vehicle is operated under severe dusty conditions and replace, as necessary.

REMOVAL & INSTALLATION

1991–94 SOHC Engine

▸ See Figure 36

1. Remove the thumb screws from the top of the cover, then release the 6 fastener clips.
2. Lift the cover up and off of the housing.
3. Remove the air filter element and replace if dirty. If the element is only slightly dusty, it can be cleaned by blowing compressed air through the element from the clean side.
   To install:
4. Wipe all dust from the inside of the housing using a clean rag or cloth.
5. Install the element into the housing, then position the cover over the element.
6. Secure the cover with the fastener clips, then tighten the thumb screws to 35 inch lbs. (4 Nm). Do not overtighten the thumb screws.

DOHC Engine and 1995–98 SOHC Engine

▸ See Figures 37, 38 and 39

1. Release the 2 clips located to the front of the cover.
2. Release the 2 clips located to the rear of the cover.
3. Lift the cover to expose the air filter element, remove the element and replace if dirty. If the element is only slightly dusty, it can be cleaned by blowing compressed air through the element from the clean side.
   To install:
4. Wipe all dust from the inside of the housing using a clean rag or cloth. Inspect the resonator and rubber tubes for damage or cracks.
5. Install the element into the housing, then position the cover over the element.
6. Secure the cover with the fastener clips.

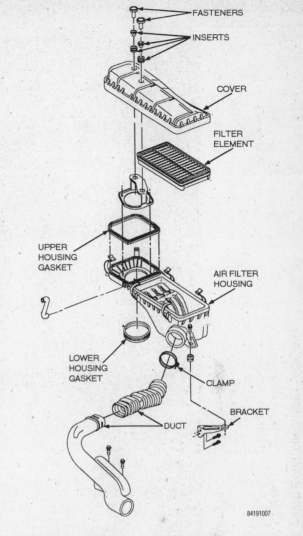

Fig. 36 Exploded view of the air filter components—SOHC engine

Fig. 37 Release the cover retaining clips . . .

Fig. 38 . . . then lift the cover and remove the air filter element

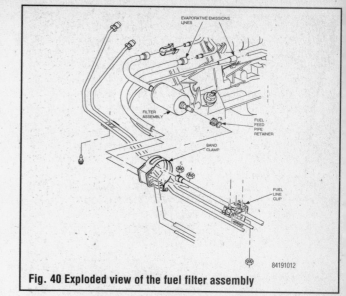

Fig. 40 Exploded view of the fuel filter assembly

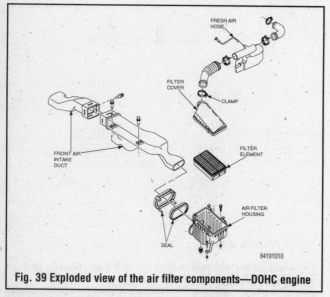

Fig. 39 Exploded view of the air filter components—DOHC engine

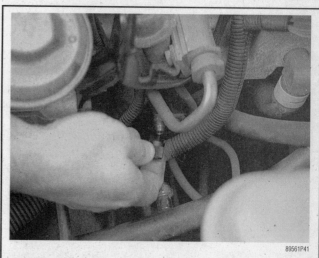

Fig. 41 Unscrew the protective cap from the fuel pressure relief valve

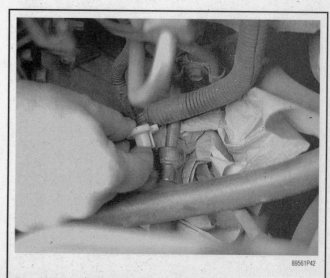

Fig. 42 Use a special fuel line quick-disconnect tool . . .

Fuel Filter

On 1991–97 models, the fuel filter is attached to the vehicle frame in the lower left portion of the engine compartment. On 1998 models, the fuel filter and fuel pressure regulator are one integral component of the new returnless fuel injection system, and is located underneath the vehicle at the forward edge of the left side of the fuel tank. The filter should be serviced according to the Maintenance Intervals Chart at the end of this section.

REMOVAL & INSTALLATION

1991–97 Models

▶ See Figures 40 thru 47

1. Disconnect the negative battery cable from the battery.
2. Remove the fresh air intake hose from the camshaft or rocker cover and remove the air inlet tube.
3. Properly relieve the fuel system pressure as follows:
   a. Remove the air cleaner or air intake duct, as applicable.
   b. Wrap a shop rag around the fuel test port fitting at the lower rear of the engine, remove the cap and connect pressure gauge tool SA9127E or equivalent.

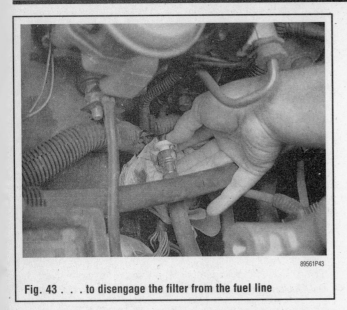

Fig. 43 . . . to disengage the filter from the fuel line

Fig. 44 Be sure to place a rag or paper towel below the fitting to absorb any fuel spillage

Fig. 45 Location of the fuel filter assembly

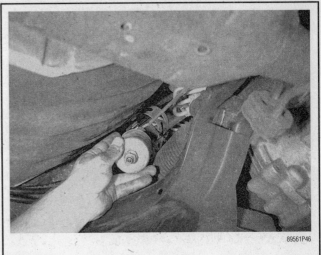

Fig. 46 Remove the fuel filter assembly from beneath the vehicle

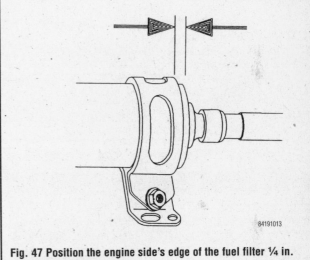

Fig. 47 Position the engine side's edge of the fuel filter ¼ in. (6mm) from the band clamp

   c. Install the bleed hose into an approved container and open the valve to bleed the system pressure. After the system pressure is bled, remove the gauge from the pressure test port and recap it.

   4. Clean the female end of the quick connect fitting by spraying it with penetrating oil, then detach the large underhood fuel line connection located near the intake manifold support brace on the left side of the vehicle using the tool supplied with the replacement filter, SA9157E or equivalent.

   5. Raise and support the vehicle safely.

   6. Disengage the quick connect fitting at the fuel filter inlet (rear of the filter) by pinching the 2 plastic tangs together and pulling on the supply line.

   7. Loosen the fuel filter band clamp nut, but do not completely remove.

   8. Carefully push or pull the filter out of the assembly and discard the filter in an appropriate container.

To install:

   9. If the band clamp was removed, clip the fuel return and vapor lines in place and install 2 new band clamp nuts. Make sure all lines are in place and will not interfere with or be damaged by filter installation, then tighten the bracket nuts to 27 inch lbs. (3 Nm).

   10. Clean the female end of the filter inlet quick connect fitting by holding the line facing downward and spraying penetrating oil up into the fitting. Be careful not to bend or kink the line.

11. If not already installed, insert a new snap lock retainer into the female end of the filter inlet quick connect fitting.

12. Route the filter's nylon outlet line through the band clamp and insert the filter far enough into the band clamp to connect the outlet line to the engine's fuel line attachment. Lubricate the male end of the connector with clean engine oil, snap the connector together and pull on the line to verify proper fitting.

13. Position the filter in the band clamp assembly with the filter's forward edge located ¼ inch (6.35mm) from the front of the band clamp.

14. Lubricate the male end of the fuel supply line with clean engine oil. Snap the line into the fuel filter and pull back to verify that the fitting is secure. Tighten the band clamp nut to 89 inch lbs. (10 Nm).

15. Lower the vehicle and install the air cleaner or intake duct, as applicable. Connect the air inlet tube and fresh air hose.

16. Connect the negative battery cable, then prime the fuel system as follows:

    a. Turn the ignition **ON** for 5 seconds and then **OFF** for 10 seconds.

    b. Repeat the ON/OFF cycle 2 more times.

    c. Crank the engine until it starts.

    d. If it does not start, repeat Steps a–c.

    e. Run the engine and check for leaks.

1998 Models

▶ **See Figures 48 and 49**

1. Disconnect the negative battery cable from the battery.
2. Properly relieve the fuel system pressure as follows:

    a. Remove the air cleaner or air intake duct, as applicable.

    b. Wrap a shop rag around the fuel test port fitting on the fuel supply line at the fuel rail. Remove the protective cap and connect pressure gauge tool SA9127E or equivalent.

    c. Install the bleed hose into an approved container and open the valve to bleed the system pressure. After the system pressure is bled, remove the gauge from the pressure test port and recap it.

3. Raise and support the vehicle safely.
4. Loosen the fuel filter/pressure regulator bracket retaining screws.

✳✳ WARNING

Exercise extreme care when opening the retaining clip. The fuel lines must be retained in this clip; if damaged, the fuel tank assembly must be replaced, since the fitting is not serviced separately.

5. Disengage the quick connect fitting on the left side of the fuel tank by pinching the 2 plastic tangs together and pulling on the supply line.

6. Disengage the EVAP purge line at the 90° quick connect fitting.

7. Slide the outlet of the fuel filter/regulator out of the support on the fuel tank bracket and disengage the fuel feed line at the 90° quick connect fitting.

8. Pivot the fuel filter/regulator down while moving the leg of the mounting bracket out from under the parking brake lines.

➡ **It is not necessary to separate the fuel filter/regulator from the mounting bracket, since both items are serviced as an assembly.**

9. Disengage the fuel feed and return line quick connect fittings on the fuel filter/regulator. Discard the filter in an appropriate container.

To install:

10. Install new fuel line retainers (3) into the female portion of the quick connect fuel line fittings.

11. To ease installation, lubricate the male ends of the fuel filter/regulator with clean engine oil.

12. Install the fuel feed and return lines onto the fuel filter/regulator and snap them closed. Pull back to verify that the fittings are secure.

13. Install the fuel feed, return, and EVAP purge lines into the fuel tank's retaining clip.

14. Slide the leg of the fuel filter/regulator mounting bracket under the parking brake lines and pivot upward.

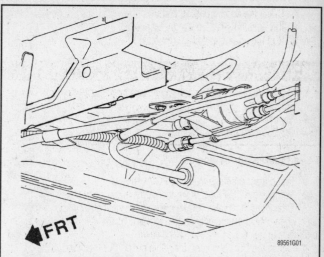

Fig. 48 Fuel filter/pressure regulator location on the vehicle's underbody, at the forward edge of the left side of the fuel tank

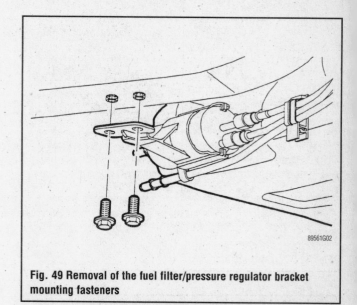

Fig. 49 Removal of the fuel filter/pressure regulator bracket mounting fasteners

✳✳ WARNING

Be sure to route the chassis fuel feed and purge lines above the parking brake cable. The parking brake cable must be firmly secured to the underbody to support the fuel and purge lines.

15. Install the 90° fuel feed line quick connect fitting onto the fuel filter/regulator outlet and snap it closed. Pull back to verify that the fitting is secure.

16. Install the 90° EVAP purge line quick connect fitting to the purge line and snap it closed. Pull back to verify that the fitting is secure.

17. Install the fuel feed outlet pipe of the fuel filter/regulator into the retaining clip on the fuel tank bracket.

18. Install the fuel filter/regulator bracket mounting screws. Tighten the bracket mounting screws to 71 inch lbs. (8 Nm).

19. Lower the vehicle and install the air cleaner or intake duct, as applicable. Connect the air inlet tube and fresh air hose.

20. Connect the negative battery cable, then prime the fuel system as follows:

    a. Turn the ignition **ON** for 5 seconds and then **OFF** for 10 seconds.

    b. Repeat the ON/OFF cycle 2 more times.

c. Crank the engine until it starts.
d. If it does not start, repeat Steps a–c.
e. Run the engine and check for leaks.

PCV Valve

The PCV valve is located in a grommet attached to the top of the rocker or camshaft cover. For crankcase ventilation system testing, refer to Section 4.

REMOVAL & INSTALLATION

▶ **See Figures 50, 51 and 52**

1. Disconnect the hose from the PCV valve.
2. Remove the PCV valve from the camshaft cover grommet.
3. Check the PCV valve for deposits and clogging. The valve should rattle when shaken. If the valve does not rattle, clean the valve with solvent until the plunger is free or replace the valve.
4. Check the PCV hose and the valve cover grommet for clogging and signs of wear or deterioration. Replace as necessary.

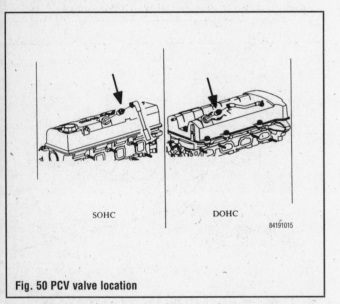

Fig. 50 PCV valve location

Fig. 52 Pull the PCV valve out from the camshaft cover grommet

To install:
5. Connect the PCV hose to the PCV valve.
6. Install the PCV valve in the valve cover grommet.

Evaporative Canister

The vapor or carbon canister is part of the evaporative emission control system. On 1991–97 models, it is located in the right front of the engine compartment and may be accessed by removing the inner wheel liner. On 1998 models, it is located on top of the fuel tank at the left rear corner.

SERVICING

▶ **See Figures 53, 54, 55 and 56**

Servicing the carbon canister is only necessary if it is clogged, cracked or contains liquid fuel, as indicated by odor or excessive weight. For evaporative emission control system testing, removal and installation, refer to Section 4.

Fig. 51 Disconnect the hose from the PCV valve

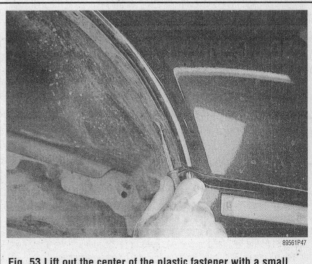

Fig. 53 Lift out the center of the plastic fastener with a small bladed screwdriver

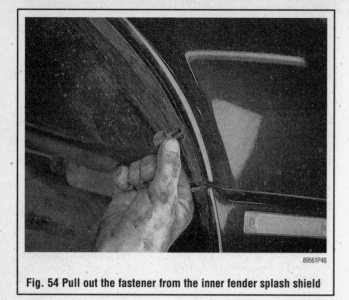

Fig. 54 Pull out the fastener from the inner fender splash shield

Fig. 55 Location of the evaporative canister in the right front of the engine compartment on 1991–97 models

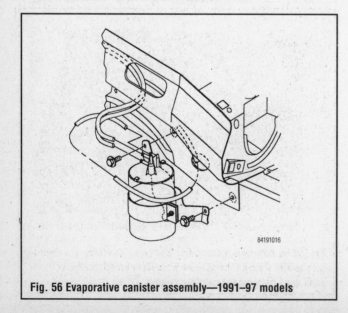

Fig. 56 Evaporative canister assembly—1991–97 models

Battery

PRECAUTIONS

Always use caution when working on or near the battery. Never allow a tool to bridge the gap between the negative and positive battery terminals. Also, be careful not to allow a tool to provide a ground between the positive cable/terminal and any metal component on the vehicle. Either of these conditions will cause a short circuit, leading to sparks and possible personal injury.

Do not smoke, have an open flame or create sparks near a battery; the gases contained in the battery are very explosive and, if ignited, could cause severe injury or death.

All batteries, regardless of type, should be carefully secured by a battery hold-down device. If this is not done, the battery terminals or casing may crack from stress applied to the battery during vehicle operation. A battery which is not secured may allow acid to leak out, making it discharge faster; such leaking corrosive acid can also eat away at components under the hood.

Always visually inspect the battery case for cracks, leakage and corrosion. A white corrosive substance on the battery case or on nearby components would indicate a leaking or cracked battery. If the battery is cracked, it should be replaced immediately.

GENERAL MAINTENANCE

▶ **See Figures 57, 58 and 59**

A battery that is not sealed must be checked periodically for electrolyte level. You cannot add water to a sealed maintenance-free battery (though not all maintenance-free batteries are sealed); however, a sealed battery must also be checked for proper electrolyte level, as indicated by the color of the built-in hydrometer "eye."

Always keep the battery cables and terminals free of corrosion. Check these components about once a year. Refer to the Cables portion of the Battery information, later in this section.

Keep the top of the battery clean, as a film of dirt can help completely discharge a battery that is not used for long periods. A solution of baking soda and water may be used for cleaning, but be careful to flush this off with clear water. DO NOT let any of the solution into the filler holes. Baking soda neutralizes battery acid and will de-activate a battery cell.

Batteries in vehicles which are not operated on a regular basis can fall victim to parasitic loads (small current drains which are constantly drawing current from the battery). Normal parasitic loads may drain a battery on a vehicle that is in storage and not used for 6–8 weeks. Vehicles that have additional accessories such as a cellular phone, an alarm system or other devices that increase parasitic load may discharge a battery sooner. If the

Fig. 57 A typical location for the built-in hydrometer on maintenance-free batteries

Fig. 58 Brush on a solution of baking soda and water to clean the tray

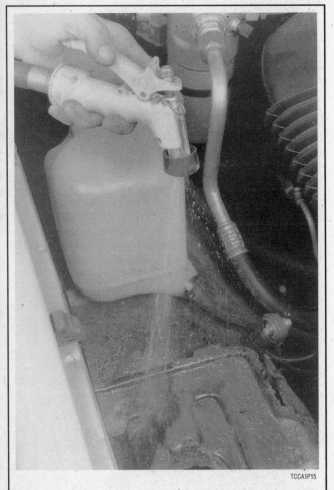

Fig. 59 After cleaning the tray thoroughly, wash it off with some water

vehicle is to be stored for 6–8 weeks in a secure area and the alarm system, if present, is not necessary, the negative battery cable should be disconnected at the onset of storage to protect the battery charge.

Remember that constantly discharging and recharging will shorten battery life. Take care not to allow a battery to be needlessly discharged.

When the battery is eventually removed for replacement or other reason, it is a good idea and opportunity to check the condition of the battery tray. Clear it of any debris, and check it for soundness (the battery tray can be cleaned with a baking soda and water solution).

BATTERY FLUID

Check the battery electrolyte level at least once a month, or more often in hot weather or during periods of extended vehicle operation. On non-sealed batteries, the level can be checked either through the case on translucent batteries or by removing the cell caps on opaque-cased types. The electrolyte level in each cell should be kept filled to the split ring inside each cell, or the line marked on the outside of the case.

If the level is low, add only distilled water through the opening until the level is correct. Each cell is separate from the others, so each must be checked and filled individually. Distilled water should be used, because the chemicals and minerals found in most drinking water are harmful to the battery and could significantly shorten its life.

If water is added in freezing weather, the vehicle should be driven several miles to allow the water to mix with the electrolyte. Otherwise, the battery could freeze.

Although some maintenance-free batteries have removable cell caps for access to the electrolyte, the electrolyte condition and level on all sealed maintenance-free batteries must be checked using the built-in hydrometer "eye." The exact type of eye varies between battery manufacturers, but most apply a sticker to the battery itself explaining the possible readings. When in doubt, refer to the battery manufacturer's instructions to interpret battery condition using the built-in hydrometer.

➡ **Although the readings from built-in hydrometers found in sealed batteries may vary, a green eye usually indicates a properly charged battery with sufficient fluid level. A dark eye is normally an indicator of a battery with sufficient fluid, but one which may be low in charge. And a light or yellow eye is usually an indication that electrolyte supply has dropped below the necessary level for battery (and hydrometer) operation. In this last case, sealed batteries with an insufficient electrolyte level must usually be discarded.**

Checking the Specific Gravity

▶ **See Figures 60, 61, 62 and 63**

A hydrometer is required to check the specific gravity on all batteries that are not maintenance-free. On batteries that are maintenance-free, the specific gravity is checked by observing the built-in hydrometer "eye" on the top of the battery case. Check with your battery's manufacturer for proper interpretation of its built-in hydrometer readings.

Fig. 60 On non-maintenance-free batteries, the fluid level can be checked through the case on translucent models; the cell caps must be removed on other models

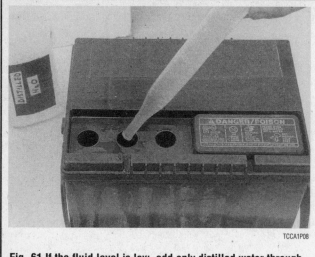

Fig. 61 If the fluid level is low, add only distilled water through the opening until the level is correct

Fig. 62 Check the specific gravity of the battery's electrolyte with a hydrometer

❊❊ CAUTION

Battery electrolyte contains sulfuric acid. If you should splash any on your skin or in your eyes, flush the affected area with plenty of clear water. If it lands in your eyes, get medical help immediately.

The fluid (sulfuric acid solution) contained in the battery cells will tell you many things about the condition of the battery. Because the cell plates must be kept submerged below the fluid level in order to operate, maintaining the fluid level is extremely important. And, because the specific gravity of the acid is an indication of electrical charge, testing the fluid can be an aid in determining if the battery must be replaced. A battery in a vehicle with a properly operating charging system should require little maintenance, but careful, periodic inspection should reveal problems before they leave you stranded.

As stated earlier, the specific gravity of a battery's electrolyte level can be used as an indication of battery charge. At least once a year, check the specific gravity of the battery. It should be between 1.20 and 1.26 on the gravity scale. Most auto supply stores carry a variety of inexpensive battery testing hydrometers. These can be used on any non-sealed battery to test the specific gravity in each cell.

The battery testing hydrometer has a squeeze bulb at one end and a nozzle at the other. Battery electrolyte is sucked into the hydrometer until the float is lifted from its seat. The specific gravity is then read by noting the position of the float. If gravity is low in one or more cells, the battery should be slowly charged and checked again to see if the gravity has come up. Generally, if after charging, the specific gravity between any two cells varies more than 50 points (0.50), the battery should be replaced, as it can no longer produce sufficient voltage to guarantee proper operation.

CABLES

◗ See Figures 64 thru 69

Once a year (or as necessary), the battery terminals and the cable clamps should be cleaned. Loosen the clamps and remove the cables, negative cable first. On batteries with posts on top, the use of a puller specially made for this purpose is recommended. These are inexpensive and available in most auto parts stores. Side terminal battery cables are secured with a small bolt.

Clean the cable clamps and the battery terminal with a wire brush, until all corrosion, grease, etc., is removed and the metal is shiny. It is especially important to clean the inside of the clamp thoroughly (an old knife is useful here), since a small deposit of foreign material or oxidation there will prevent a sound electrical connection and inhibit either starting or charging. Special tools are available for cleaning these parts, one type for conventional top post batteries and another type for side terminal batteries. It is also a good idea to apply some dielectric grease to the terminal, as this will aid in the prevention of corrosion.

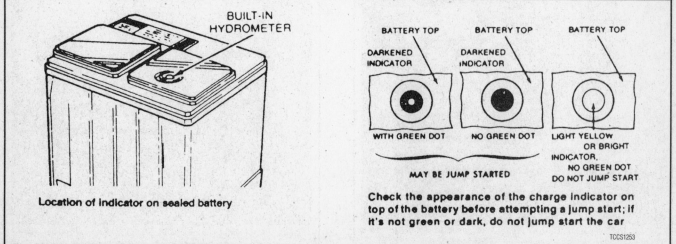

Fig. 63 A typical sealed (maintenance-free) battery with a built-in hydrometer—NOTE that the hydrometer eye may vary between battery manufacturers; always refer to the battery's label

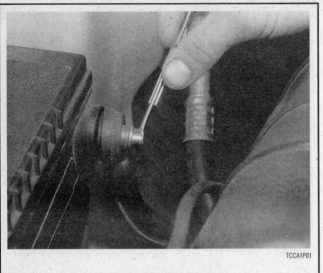

Fig. 64 Loosen the battery cable retaining nut . . .

Fig. 65 . . . then disconnect the cable from the battery

Fig. 66 A wire brush may be used to clean any corrosion or foreign material from the cable

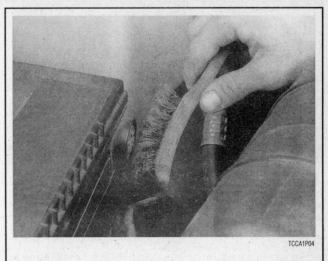

Fig. 67 The wire brush can also be used to remove any corrosion or dirt from the battery terminal

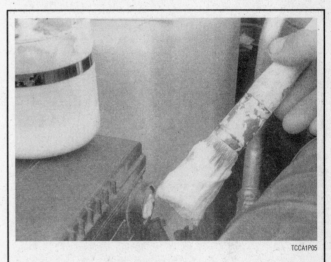

Fig. 68 The battery terminal can also be cleaned using a solution of baking soda and water

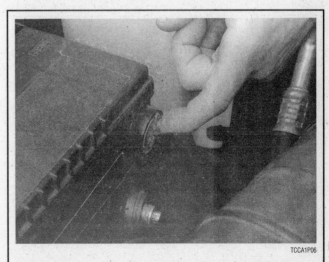

Fig. 69 Before connecting the cables, it's a good idea to coat the terminals with a small amount of dielectric grease

After the clamps and terminals are clean, reinstall the cables, negative cable last; DO NOT hammer the clamps onto battery posts. Tighten the clamps securely, but do not distort them. Give the clamps and terminals a thin external coating of grease after installation, to retard corrosion.

Check the cables at the same time that the terminals are cleaned. If the cable insulation is cracked or broken, or if the ends are frayed, the cable should be replaced with a new cable of the same length and gauge.

CHARGING

✱✱ CAUTION

The chemical reaction which takes place in all batteries generates explosive hydrogen gas. A spark can cause the battery to explode and splash acid. To avoid serious personal injury, be sure there is proper ventilation and take appropriate fire safety precautions when connecting, disconnecting, or charging a battery and when using jumper cables.

A battery should be charged at a slow rate to keep the plates inside from getting too hot. However, if some maintenance-free batteries are allowed to discharge until they are almost "dead," they may have to be charged at a high rate to bring them back to "life." Always follow the charger manufacturer's instructions on charging the battery.

REPLACEMENT

When it becomes necessary to replace the battery, select one with an amperage rating equal to or greater than the battery originally installed. Deterioration and just plain aging of the battery cables, starter motor, and associated wires makes the battery's job harder in successive years. The slow increase in electrical resistance over time makes it prudent to install a new battery with a greater capacity than the old.

Serpentine Drive Belt

INSPECTION

▶ **See Figures 70, 71, 72, 73 and 74**

Inspect the belt for signs of glazing or cracking. A glazed belt will be perfectly smooth from slippage, while a good belt will have a slight texture of fabric visible. Cracks will usually start at the inner edge of the belt and run outward. All worn or damaged drive belts should be replaced immedi-

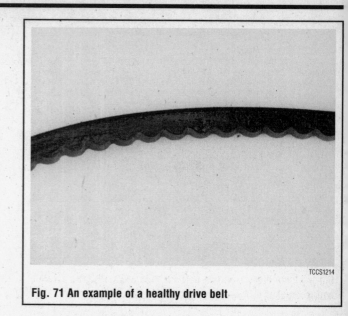
TCCS1214

Fig. 71 An example of a healthy drive belt

TCCS1215

Fig. 72 Deep cracks in this belt will cause flex, building up heat that will eventually lead to belt failure

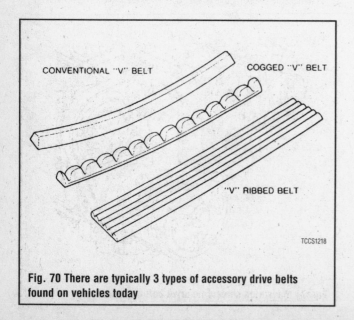

CONVENTIONAL "V" BELT

COGGED "V" BELT

"V" RIBBED BELT

TCCS1218

Fig. 70 There are typically 3 types of accessory drive belts found on vehicles today

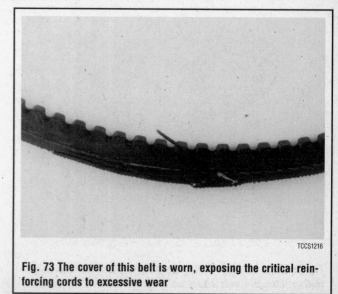
TCCS1216

Fig. 73 The cover of this belt is worn, exposing the critical reinforcing cords to excessive wear

Fig. 74 Installing too wide a belt can result in serious belt wear and/or breakage

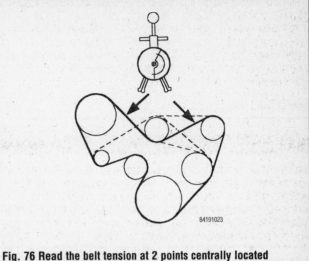

Fig. 76 Read the belt tension at 2 points centrally located between the indicated pulleys

ately. It is best to replace the drive belt as a preventive maintenance measure, during this service operation.

Belt Tension

♦ See Figures 75 and 76

The belt is automatically adjusted using a spring loaded tensioner. If belt slippage is suspected or unusual belt noises occurs, the following procedure should be used to determine if the tensioner or belt is at fault.

1. Start the engine and allow it to warm up to normal operating temperature with all accessories (such as the A/C) turned **ON;** this should take approximately 10 minutes. If the vehicle is equipped with power steering, turn the steering wheel to the left and right several times during warm-up to make sure a proper load is placed on the belt.
2. Turn the engine and accessories **OFF.**
3. Using a 14mm or 9⁄16 in. wrench, depress the tensioner arm until the belt becomes loosened, then slowly allow the tensioner to return to position and apply tension to the belt. Do not allow the tensioner to snap against the belt.
4. Inspect the markings located on the tensioner arm. If the marks on the arm fall outside the operating range, the drive belt must be replaced.

➡**Inspection of the belt using a tension gauge must be performed with the upper engine mount removed.**

5. Using tool SA9181-NE or an equivalent calibrated belt tension gauge, measure and note the tension readings at the 2 points located centrally between 2 belt pulleys as shown in the illustration. The dotted lines represent other possible paths which your belt may follow, depending on the engine and accessories with which your vehicle is equipped.
6. Repeat the measurements from the previous step 2 more times, then calculate the average result for each test location. The readings should be 50–65 lbs. (22.7–29.2 kg) for new belts or a minimum of 45 lbs. (20.4 kg) for used belts. If the readings are out of this range or do not meet the minimum and the drive belt passed the test in Step 4, the drive belt tensioner must be replaced.

Belt Alignment

♦ See Figure 77

1. Measure the distance from the front machined surface of the cylinder block to the inboard edge of the belt at the location as shown.
2. The distance should be 1.102–1.220 in. (28–31mm).
3. If the distance does not fall within the specification, check the following:
- Check that the belt is properly located within the pulley grooves.
- Check the drive belt for wear at the edges and, if necessary, replace a worn belt.
- Make sure the belt pulleys are not bent or damaged.

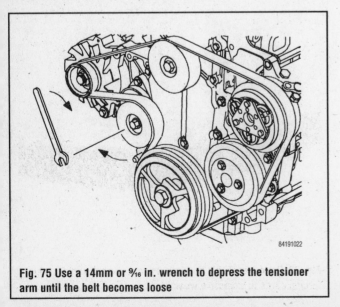

Fig. 75 Use a 14mm or 9⁄16 in. wrench to depress the tensioner arm until the belt becomes loose

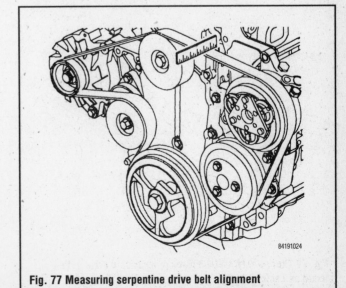

Fig. 77 Measuring serpentine drive belt alignment

• Make sure the accessory assemblies have proper shaft and bearing end-play.

• Check the pulley hubs for proper installation on their shafts.

REMOVAL & INSTALLATION

Belt

▶ **See Figures 78 and 79**

1. Loosen the drive belt by depressing the tensioner arm using a 14mm or 9/16 in. wrench.

2. Remove the belt from the idler or air conditioning compressor, as applicable.

3. Remove the drive belt from the vehicle.

To install:

4. Install the belt around the pulleys, except for the front cover idler or air conditioning compressor.

5. Depress the tensioner arm and slip the belt over the idler or air conditioning compressor pulley. Make sure the belt ribs are properly aligned on the pulleys.

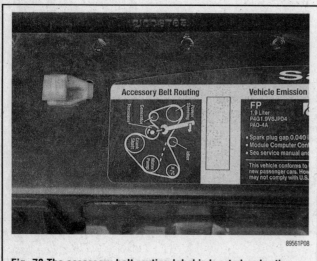

Fig. 78 The accessory belt routing label is located under the hood

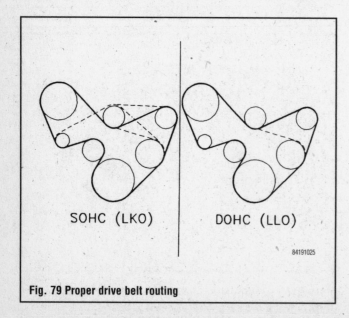

Fig. 79 Proper drive belt routing

6. If the tensioner idler pulley retaining bolt is loose, remove the bolt and apply Loctite® 242 or equivalent to the bolt threads. Install the bolt and tighten to 22 ft. lbs. (30 Nm).

Belt Tensioner

▶ **See Figures 80 and 81**

1. On 1992 and later models equipped with the torque axis mount system, position a 1 in. x 1 in. x 2 in. long wooden block between the torque strut and cradle in order to support the mount. This will allow removal and installation of the right upper mount without the necessity of lifting or jacking the powertrain.

2. On 1992 and later models, remove the 3 engine mount-to-front cover nuts and the 2 mount-to-midrail bracket nuts, then remove the upper mount. This will allow the powertrain to rest on the wooden block positioned earlier.

3. Remove the serpentine drive belt.

4. Check the position of the engine to the right side midrails; it may be possible to remove the tensioner without removing the power steering pump. If this is preferable, raise and support the vehicle safely, then remove the right tire and splash shield.

5. If it was determined that tensioner removal is necessary, remove the power steering pump and bracket from the engine.

6. Remove the upper and lower fasteners attaching the tensioner to the engine.

7. Remove the tensioner assembly from the vehicle. If necessary, the engine may be moved slightly toward the driver's side by prying on the motor mount and frame rail. Do not pry against the aluminum engine and accessories or damage may occur.

To install:

8. Install the tensioner using the upper and lower fasteners, then tighten the fasteners to 22 ft. lbs. (30 Nm).

9. If removed, install the power steering pump assembly.

10. If removed, install the right tire and splash shield, then remove the support and lower the vehicle.

11. Install the accessory drive belt, making sure it is properly aligned on the pulleys.

12. On 1992 and later vehicles equipped with the torque axis mount system, position the upper mount and install the 3 mount to front cover nuts. Install the 2 mount-to-midrail bracket nuts, then tighten the 5 mount fasteners slowly and evenly to 52 ft. lbs. (70 Nm). Remove the wooden block from the engine cradle.

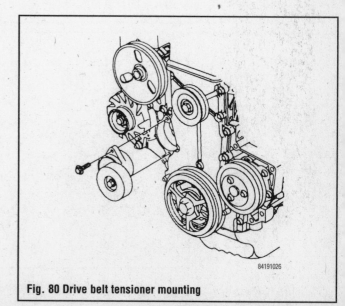

Fig. 80 Drive belt tensioner mounting

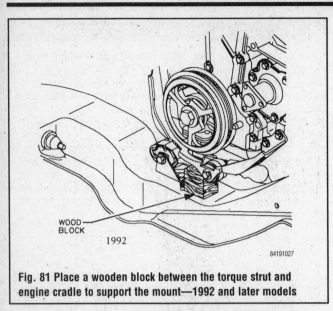

Fig. 81 Place a wooden block between the torque strut and engine cradle to support the mount—1992 and later models

Fig. 83 A hose clamp that is too tight can cause older hoses to separate and tear on either side of the clamp

Timing Chain

All Saturn models utilize a timing chain driven, interference-type, non-freewheeling engine. If the timing chain breaks, the valves in the cylinder head may strike the pistons, causing potentially serious (also time-consuming and expensive) engine damage. Refer to Section 3 for information on servicing the timing chain.

➡ **Despite the potential for serious engine damage, the manufacturer does not give a recommendation for timing chain replacement as a preventive measure. Such replacement, however, may be prudent on high mileage engines.**

Hoses

INSPECTION

▶ **See Figures 82, 83, 84 and 85**

Upper and lower radiator hoses, along with the heater hoses, should be checked for deterioration, leaks and loose hose clamps at least every 15,000 miles (24,000 km). It is also wise to check the hoses periodically

Fig. 84 A soft spongy hose (identifiable by the swollen section) will eventually burst and should be replaced

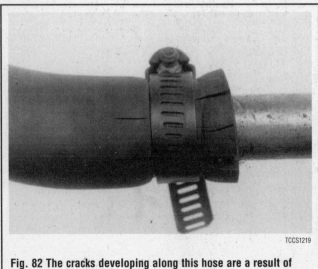

Fig. 82 The cracks developing along this hose are a result of age-related hardening

Fig. 85 Hoses are likely to deteriorate from the inside if the cooling system is not periodically flushed

in early spring and at the beginning of fall or winter, when you are performing other maintenance. A quick visual inspection could discover a weakened hose which might have left you stranded if it had remained unrepaired.

Whenever you are checking the hoses, make sure the engine and cooling system are cold. Visually inspect for cracking, rotting or collapsed hoses, and replace as necessary. Run your hand along the length of the hose. If a weak or swollen spot is noted when squeezing the hose wall, the hose should be replaced.

REMOVAL & INSTALLATION

▶ **See Figures 86 and 87**

❊❊ CAUTION

Disconnect the negative battery cable or fan motor wiring harness connector before replacing any radiator/heater hose. The fan may come ON, under certain circumstances, even though the ignition is OFF.

1. Remove the surge tank pressure cap.

❊❊ CAUTION

Never remove the pressure cap while the engine is running or personal injury from scalding hot coolant or steam may result. If possible, wait until the engine has cooled to remove the pressure cap. If this is not possible, wrap a thick cloth around the pressure cap and turn it slowly to the first stop. Step back while the pressure is released from the cooling system. When you are sure all the pressure has been released, press down on the cap with the cloth, then turn the cap and remove it.

➡**If removing just the upper radiator hose, only a little coolant must be drained. To remove hoses positioned lower on the engine, such as a lower radiator hose, the entire cooling system must be emptied.**

2. Position a suitable container under the radiator and block, then drain the coolant by opening the radiator drain and removing the engine drain plug. The engine drain plug is a bolt which can be removed with a ratchet and socket or wrench. To open the radiator drain, grasp the plug firmly and turn counterclockwise until loosened, then use a small prybar to dislodge the plug and large rubber gasket. Make sure the plug is fully loosened before prying to prevent plug or radiator damage.

❊❊ CAUTION

When draining the coolant, keep in mind that cats and dogs are attracted by ethylene glycol antifreeze, and are quite likely to drink any that is left in an uncovered container or in puddles on the ground. This will prove fatal in sufficient quantity. Always drain the coolant into a sealable container. Coolant may be reused unless it is contaminated or several years old.

3. Loosen the hose clamps at each end of the hose requiring replacement. Clamps are usually either of the spring tension type (which require pliers to squeeze the tabs and loosen) or of the screw tension type (which require screw or hex drivers to loosen). Pull the clamps back on the hose away from the connection.

4. Twist, pull and slide the hose off the fitting, taking care not to damage the neck of the component from which the hose is being removed.

➡**If the hose is stuck at the connection, do not try to insert a screwdriver or other sharp tool under the hose end in an effort to free it, as the connection and/or hose may become damaged. Heater connections especially may be easily damaged by such a procedure. If the hose is to be replaced, use a single-edged razor blade to make a slice along the portion of the hose which is stuck on the connection, perpendicular to the end of the hose. Do not cut deep, so as to prevent damaging the connection. The hose can then be peeled from the connection and discarded.**

5. Clean both hose mounting connections. Inspect the condition of the hose clamps and replace them, if necessary.

To install:

6. Dip the ends of the new hose into clean engine coolant to ease installation.

7. Slide the hose clamps over the replacement hose and slide the hose ends over the connections into position.

8. Position and secure the clamps at least 0.12 in. (3mm) from the ends of the hose. Make sure they are located beyond the raised bead of the connector.

9. Close the radiator or engine drains and properly refill the cooling system with the clean drained engine coolant or a suitable mixture of ethylene glycol coolant and water.

10. If available, install a pressure tester and check for leaks. If a pressure tester is not available, run the engine until normal operating temperature is reached (allowing the system to naturally pressurize), then check for leaks.

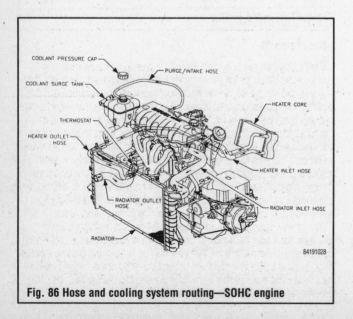

Fig. 86 Hose and cooling system routing—SOHC engine

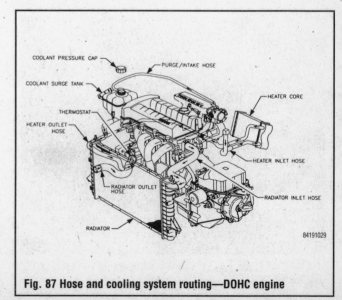

Fig. 87 Hose and cooling system routing—DOHC engine

✳✳ CAUTION

If you are checking for leaks with the system at normal operating temperature, BE EXTREMELY CAREFUL not to touch any moving or hot engine parts. Once operating temperature has been reached, shut the engine OFF, and check for leaks around the hose fittings and connections which were removed earlier.

CV-Boots

INSPECTION

▶ See Figures 88 and 89

The CV (Constant Velocity) boots should be checked for damage each time the oil is changed and any other time the vehicle is raised for service. These boots keep water, grime, dirt and other damaging matter from entering the CV-joints. Any of these could cause early CV-joint failure, which can be expensive to repair. Heavy grease thrown around the inside of the front wheel(s) and on the brake caliper can be an indication of a torn boot.

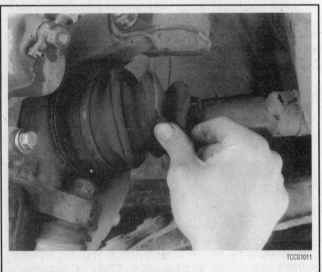

Fig. 88 CV-boots must be inspected periodically for damage

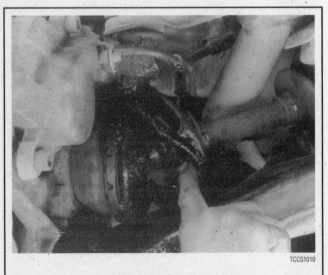

Fig. 89 A torn boot should be replaced immediately

Thoroughly check the boots for missing clamps and tears. If the boot is damaged, it should be replaced immediately. Please refer to Section 7 for procedures.

Spark Plugs

▶ See Figure 90

A typical spark plug consists of a metal shell surrounding a ceramic insulator. A metal electrode extends downward through the center of the insulator and protrudes a small distance. Located at the end of the plug and attached to the side of the outer metal shell is the side electrode. The side electrode bends in at a 90° angle so that its tip is just past and parallel to the tip of the center electrode. The distance between these two electrodes (measured in thousandths of an inch or hundredths of a millimeter) is called the spark plug gap.

The spark plug does not produce a spark, but instead provides a gap across which the current can arc. The coil produces anywhere from 20,000 to 50,000 volts (depending on the type and application) which travels through the wires to the spark plugs. The current passes along the center electrode and jumps the gap to the side electrode, and in doing so, ignites the air/fuel mixture in the combustion chamber.

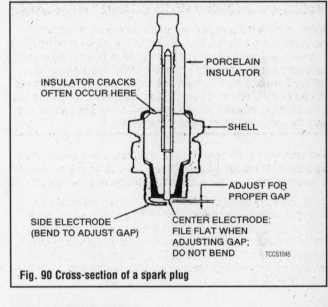

Fig. 90 Cross-section of a spark plug

SPARK PLUG HEAT RANGE

▶ See Figure 91

Spark plug heat range is the ability of the plug to dissipate heat. The longer the insulator (or the farther it extends into the engine), the hotter the plug will operate; the shorter the insulator (the closer the electrode is to the block's cooling passages), the cooler it will operate. A plug that absorbs little heat and remains too cool will quickly accumulate deposits of oil and carbon, since it is not hot enough to burn them off. This leads to plug fouling and, consequently, to misfiring. A plug that absorbs too much heat will have no deposits but, due to the excessive heat, the electrodes will burn away quickly and might possibly lead to pre-ignition or other ignition problems. Pre-ignition takes place when plug tips get so hot that they glow sufficiently to ignite the air/fuel mixture before the actual spark occurs. This early ignition will usually cause a pinging during low speeds and heavy loads.

The general rule of thumb for choosing the correct heat range when picking a spark plug is: if most of your driving is long distance, high speed travel, use a colder plug; if most of your driving is stop and go, use a hotter plug. Original equipment plugs are generally a good compromise between the 2 styles and most people never have the need to change their plugs from the factory-recommended heat range.

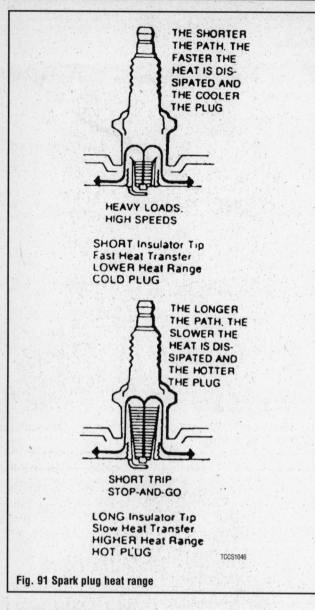

THE SHORTER THE PATH, THE FASTER THE HEAT IS DISSIPATED AND THE COOLER THE PLUG

HEAVY LOADS,
HIGH SPEEDS

SHORT Insulator Tip
Fast Heat Transfer
LOWER Heat Range
COLD PLUG

THE LONGER THE PATH, THE SLOWER THE HEAT IS DISSIPATED AND THE HOTTER THE PLUG

SHORT TRIP
STOP-AND-GO

LONG Insulator Tip
Slow Heat Transfer
HIGHER Heat Range
HOT PLUG

TCCS1046

Fig. 91 Spark plug heat range

REMOVAL & INSTALLATION

▶ See Figures 92, 93 and 94

A set of spark plugs usually requires replacement after about 20,000–30,000 miles (32,000–48,000 km), depending on your style of driving. In normal operation, plug gap increases about 0.001 in. (0.025mm) for every 2500 miles (4000 km). As the gap increases, the plug's voltage requirement also increases. It requires a greater voltage to jump the wider gap and about two to three times as much voltage to fire the plug at high speeds than at idle. The improved air/fuel ratio control of modern fuel injection, combined with the higher voltage output of modern ignition systems, will often allow an engine to run significantly longer on a set of standard spark plugs, but keep in mind that efficiency will drop as the gap widens (along with fuel economy and power).

When you're removing spark plugs, work on one at a time. Don't start by removing the plug wires all at once, because, unless you number them, they may become mixed up. Take a minute before you begin and number the wires with tape.

1. Disconnect the negative battery cable and, if the vehicle has been run recently, allow the engine to thoroughly cool.

2. Carefully twist the spark plug wire boot to loosen it, then pull upward and remove the boot from the plug. Be sure to pull on the boot and not on

89561P36

Fig. 92 Carefully twist the spark plug wire boot to loosen it, then pull upward and remove the boot from the plug

89561P37

Fig. 93 Using a spark plug socket that is equipped with a rubber insert to properly hold the plug, turn the spark plug counterclockwise to loosen

89561P38

Fig. 94 Remove the spark plug by lifting it out of the bore

the wire, otherwise the connector located inside the boot may become separated.

3. Using compressed air, blow any water or debris from the spark plug well to assure that no harmful contaminants are allowed to enter the combustion chamber when the spark plug is removed. If compressed air is not available, use a rag or a brush to clean the area.

➡ **Remove the spark plugs when the engine is cold, if possible, to prevent damage to the threads. If removal of the plugs is difficult, apply a few drops of penetrating oil or silicone spray to the area around the base of the plug, and allow it a few minutes to work.**

4. Using a spark plug socket that is equipped with a rubber insert to properly hold the plug, turn the spark plug counterclockwise to loosen and remove the spark plug from the bore.

✳ WARNING

Be sure not to use a flexible extension on the socket. Use of a flexible extension may allow a shear force to be applied to the plug. A shear force could break the plug off in the cylinder head, leading to costly and frustrating repairs.

To install:

5. Inspect the spark plug boot for tears or damage. If a damaged boot is found, the spark plug wire must be replaced.

6. Using a wire feeler gauge, check and adjust the spark plug gap. When using a gauge, the proper size should pass between the electrodes with a slight drag. The next larger size should not be able to pass while the next smaller size should pass freely.

7. Carefully thread the plug into the bore by hand. If resistance is felt before the plug is almost completely threaded, back the plug out and begin threading again. In small, hard to reach areas, an old spark plug wire and boot could be used as a threading tool. The boot will hold the plug while you twist the end of the wire and the wire is supple enough to twist before it would allow the plug to crossthread.

✳ WARNING

Do not use the spark plug socket to thread the plugs. Always carefully thread the plug by hand or using an old plug wire to prevent the possibility of crossthreading and damaging the cylinder head bore.

8. Carefully tighten the spark plug. If the plug you are installing is equipped with a crush washer, seat the plug, then tighten about ¼ turn to crush the washer. If you are installing a tapered seat plug, tighten the plug to specifications provided by the vehicle or plug manufacturer.

9. Apply a small amount of silicone dielectric compound to the end of the spark plug lead or inside the spark plug boot to prevent sticking, then install the boot to the spark plug and push until it clicks into place. The click may be felt or heard, then gently pull back on the boot to assure proper contact.

INSPECTION & GAPPING

▸ **See Figures 95 thru 105**

Check the plugs for deposits and wear. If they are not going to be replaced, clean the plugs thoroughly. Remember that any kind of deposit will decrease the efficiency of the plug. Plugs can be cleaned on a spark plug cleaning machine, which can sometimes be found in service stations, or you can do an acceptable job of cleaning with a stiff brush. If the plugs are cleaned, the electrodes must be filed flat. Use an ignition points file, not an emery board or the like, which will leave deposits. The electrodes must be filed perfectly flat with sharp edges; rounded edges reduce the spark plug voltage by as much as 50%.

Check spark plug gap before installation. The ground electrode (the

Fig. 95 A normally worn spark plug should have light tan or gray deposits on the firing tip

Fig. 96 A carbon fouled plug, identified by soft, sooty, black deposits, may indicate an improperly tuned vehicle. Check the air cleaner, ignition components and engine control system

L-shaped one connected to the body of the plug) must be parallel to the center electrode and the specified size wire gauge (please refer to the Tune-Up Specifications chart for details) must pass between the electrodes with a slight drag.

➡ **NEVER adjust the gap on a used platinum type spark plug.**

Always check the gap on new plugs, as they are not always set correctly at the factory. Do not use a flat feeler gauge when measuring the gap on a used plug, because the reading may be inaccurate. A round-wire type gapping tool is the best way to check the gap. The correct gauge should pass through the electrode gap with a slight drag. If you're in doubt, try one size

TCCS1212

Fig. 97 A variety of tools and gauges are needed for spark plug service

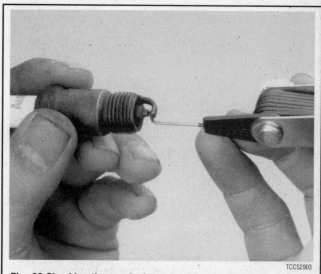

TCCS2903

Fig. 99 Checking the spark plug gap with a feeler gauge

TCCS2137

Fig. 98 A physically damaged spark plug may be evidence of severe detonation in that cylinder. Watch that cylinder carefully between services, as a continued detonation will not only damage the plug, but could also damage the engine

TCCS2138

Fig. 100 An oil fouled spark plug indicates an engine with worn piston rings and/or bad valve seals allowing excessive oil to enter the chamber

smaller and one larger. The smaller gauge should go through easily, while the larger one shouldn't go through at all. Wire gapping tools usually have a bending tool attached. Use that to adjust the side electrode until the proper distance is obtained. Absolutely NEVER attempt to bend the center electrode. Also, be careful not to bend the side electrode too far or too often, as it may weaken and break off within the engine, requiring removal of the cylinder head to retrieve it.

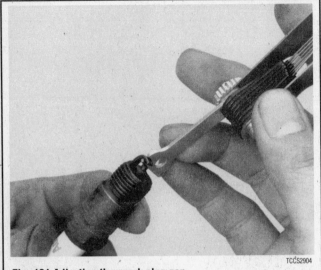

Fig. 101 Adjusting the spark plug gap

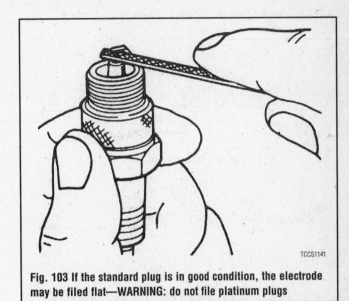

Fig. 103 If the standard plug is in good condition, the electrode may be filed flat—WARNING: do not file platinum plugs

Fig. 102 This spark plug has been left in the engine too long, as evidenced by the extreme gap. Plugs with such an extreme gap can cause misfiring and stumbling accompanied by a noticeable lack of power

Fig. 104 A bridged or almost bridged spark plug, identified by a build-up between the electrodes, caused by excessive carbon or oil build-up on the plug

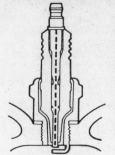

Tracking Arc
High voltage arcs between a fouling deposit on the insulator tip and spark plug shell. This ignites the fuel/air mixture at some point along the insulator tip, retarding the ignition timing which causes a power and fuel loss.

Wide Gap
Spark plug electrodes are worn so that the high voltage charge cannot arc across the electrodes. Improper gapping of electrodes on new or "cleaned" spark plugs could cause a similar condition. Fuel remains unburned and a power loss results.

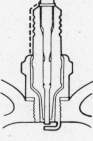

Flashover
A damaged spark plug boot, along with dirt and moisture, could permit the high voltage charge to short over the insulator to the spark plug shell or the engine. AC's buttress insulator design helps prevent high voltage flashover.

Fouled Spark Plug
Deposits that have formed on the insulator tip may become conductive and provide a "shunt" path to the shell. This prevents the high voltage from arcing between the electrodes. A power and fuel loss is the result.

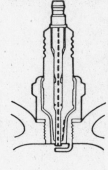

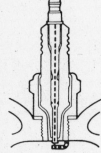

Bridged Electrodes
Fouling deposits between the electrodes "ground out" the high voltage needed to fire the spark plug. The arc between the electrodes does not occur and the fuel air mixture is not ignited. This causes a power loss and exhausting of raw fuel.

Cracked Insulator
A crack in the spark plug insulator could cause the high voltage charge to "ground out." Here, the spark does not jump the electrode gap and the fuel air mixture is not ignited. This causes a power loss and raw fuel is exhausted.

TCCS2001

Fig. 105 Used spark plugs which show damage may indicate engine problems

Spark Plug Wires

TESTING

▶ **See Figure 106**

At every tune-up/inspection, visually check the spark plug cables for burns, cuts, or breaks in the insulation. Check the boots and the nipples on the coil. Replace any damaged wiring.

Every 50,000 miles (80,000 km) or 60 months, the resistance of the wires should be checked with an ohmmeter. Wires with excessive resistance will cause misfiring, and may make the engine difficult to start in damp weather.

To check resistance, disengage the spark plug wire from the spark plug end and the ignition coil terminal at the opposite end. Connect one lead of an ohmmeter to one end of the wire and the other lead to the opposite end. Wire resistance should read below 12,000 ohms.

REMOVAL & INSTALLATION

When you're removing spark plug wires, remove and install one at a time. Don't start by removing the plug wires all at once, because, unless

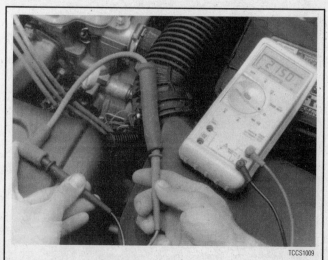

TCCS1009

Fig. 106 Checking individual plug wire resistance with a digital ohmmeter

you number them, they may become mixed up. Take a minute before you begin and number the wires with tape.

1. Disconnect the negative battery cable and, if the vehicle has been run recently, allow the engine to thoroughly cool.

2. Carefully twist the spark plug wire boot to loosen it, then pull upward and remove the boot from the plug. Be sure to pull on the boot and not on the wire, otherwise the connector located inside the boot may become separated.

To install:

3. Apply a small amount of silicone dielectric compound to the end of the spark plug lead and inside the spark plug boot to prevent sticking, then install the boot to the spark plug and push until it clicks into place. The click may be felt or heard, then gently pull back on the boot to assure proper contact.

4. Connect the negative battery cable.

Ignition Timing

GENERAL INFORMATION

There is no conventional distributor for the DIS system. Instead, timing is controlled by the Powertrain Control Module (PCM) and/or the DIS ignition module. The PCM has the ability to advance or retard ignition timing for optimal engine performance. No timing adjustments are necessary or possible.

Valve Lash

All Saturn engines utilize hydraulic valve lifters. No valve lash adjustments are necessary or possible.

Idle Speed and Mixture Adjustments

IDLE SPEED

▶ See Figures 107, 108, 109, 110 and 111

The idle stop screw controls the minimum idle speed of the engine, from which the PCM will raise idle speed as necessary for operating condition and load. The stop screw is preset at the factory and requires no periodic adjustments. Adjustments should be performed ONLY when the throttle body has been replaced and/or proper idle speed cannot be obtained. Be aware that improper adjustment of the minimum idle speed could result in false PCM trouble codes, idle instability and problems with shifting of the automatic transaxle.

The engine should be at normal operating temperature, the A/C and cooling fans should be OFF when making adjustments.

1. Because residue accumulation can affect idle speed, clean the throttle body bore before making any adjustments. Use a clean rag and a carburetor cleaner that does not contain methyl ethyl ketone. Take extreme care not to scratch or damage the throttle body bore or valve. Then check the idle speed to be sure adjustment is necessary. Proper idle speeds are as follows:
   - SOHC with manual or automatic transaxle in N—700–800 rpm
   - SOHC with automatic transaxle in D—600–700 rpm
   - SOHC with automatic transaxle in D and A/C ON—725–825 rpm
   - DOHC with manual transaxle in N—800–900 rpm
   - DOHC with automatic transaxle in D—700–800 rpm

2. Block the wheels and apply the parking brake.

3. Connect IAC tester SA9195E or equivalent to the IAC valve at the throttle body, a suitable scan tool to the diagnostic connector under the dash, or the equivalent of either to the appropriate location. Use the tool to bottom the IAC pintle in the throttle. If a scan tool was used, disconnect the IAC valve electrical connector after the pintle is bottomed to make sure it remains bottomed.

4. Remove the idle stop screw plug or cover. For SOHC TBI engines only, remove the plug by piercing it with an awl and applying leverage. On DOHC and SOHC MFI engines, remove the idle stop screw cover.

5. Insert the IAC air plug in the throttle body; use tool SA9196E for TBI or tool SA9106E for MFI or an equivalent plug.

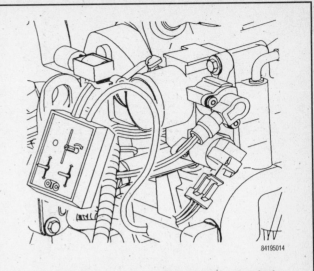

Fig. 107 Connect the IAC valve tester to the valve terminal

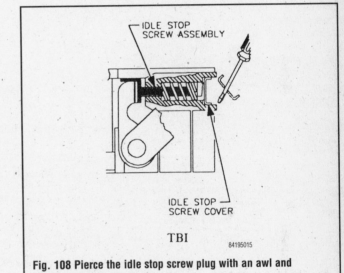

Fig. 108 Pierce the idle stop screw plug with an awl and remove—SOHC TBI engine

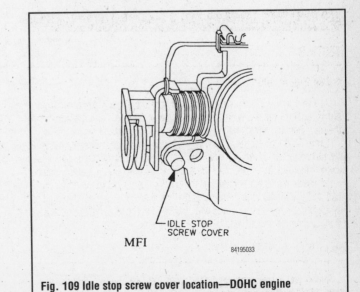

Fig. 109 Idle stop screw cover location—DOHC engine

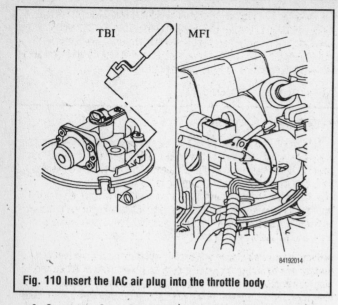

Fig. 110 Insert the IAC air plug into the throttle body

Fig. 111 Connecting the Saturn PDT or an equivalent scan tool to the ALDL

6. Connect the Saturn Portable Diagnostic Tool (PDT) or equivalent scan tool to the Assembly Line Diagnostic Link (ALDL) located under the dash, start the engine and check the minimum idle speed. Minimum idle speed should be 450–650 rpm for all engines.

7. If not within specification, adjust the idle screw to obtain a minimum idle speed of 500–600 rpm.

8. Turn the ignition **OFF** and reconnect the IAC electrical connector.

9. Using the scan tool, check the TPS voltage. Do not replace the TPS unless its reading is not between 0.35–0.70 volts.

10. Remove the IAC air plug and install the idle stop plug or cover.

11. Start the engine and check for proper idle operation.

12. Shut the engine **OFF** and remove the scan tool.

IDLE MIXTURE

The air/fuel mixture is controlled by the Powertrain Control Module (PCM) during all modes of operation; therefore, no idle mixture adjustment is necessary or possible. For more information, refer to Section 5 of this manual.

GASOLINE ENGINE TUNE-UP SPECIFICATIONS

Year	Engine ID/VIN	Engine Displacement Liters (cc)	Spark Plugs Gap (in.)	Ignition Timing (deg.) MT	Ignition Timing (deg.) AT	Fuel Pump (psi)	Idle Speed (rpm) MT	Idle Speed (rpm) AT	Valve Clearance In.	Valve Clearance Ex.
1991	7	1.9 (1901)	0.040	①	①	46	800-900 ②	700-800 ②	HYD	HYD
	9	1.9 (1901)	0.040	①	①	38	700-800 ②	600-700 ②	HYD	HYD
1992	7	1.9 (1901)	0.040	①	①	46	800-900 ②	700-800 ②	HYD	HYD
	9	1.9 (1901)	0.040	①	①	38	700-800 ②	600-700 ②	HYD	HYD
1993	7	1.9 (1901)	0.040	①	①	46	800-900 ②	700-800 ②	HYD	HYD
	9	1.9 (1901)	0.040	①	①	38	700-800 ②	600-700 ②	HYD	HYD
1994	7	1.9 (1901)	0.040	①	①	46	800-900 ②	700-800 ②	HYD	HYD
	9	1.9 (1901)	0.040	①	①	38	700-800 ②	600-700 ②	HYD	HYD
1995	7	1.9 (1901)	0.040	①	①	46	800-900 ②	700-800 ②	HYD	HYD
	8	1.9 (1901)	0.040	①	①	46	700-800 ②	600-700 ②	HYD	HYD
1996	7	1.9 (1901)	0.040	①	①	46	800-900 ②	700-800 ②	HYD	HYD
	8	1.9 (1901)	0.040	①	①	46	700-800 ②	600-700 ②	HYD	HYD
1997	7	1.9 (1901)	0.040	①	①	46	800-900 ②	700-800 ②	HYD	HYD
	8	1.9 (1901)	0.040	①	①	46	700-800 ②	600-700 ②	HYD	HYD
1998	7	1.9 (1901)	0.040	①	①	40-44	800-900 ②	700-800 ②	HYD	HYD
	8	1.9 (1901)	0.040	①	①	40-44	700-800 ②	600-700 ②	HYD	HYD

NOTE: The Vehicle Emission Control Information label often reflects specification changes made during production. Label figures must be used if they differ from those in this chart.

HYD Hydraulic

① These engines are equipped with Distributorless Ignition System (DIS), therefore the ignition timing is not adjustable.

② Manual speed with transaxle gear in N
Automatic speed with transaxle gear in D

89561C03

Air Conditioning System

SYSTEM SERVICE & REPAIR

➡️**It is recommended that the A/C system be serviced by an EPA Section 609 certified automotive technician utilizing a refrigerant recovery/recycling machine.**

The do-it-yourselfer should not service his/her own vehicle's A/C system for many reasons, including legal concerns, personal injury, environmental damage and cost. Following are some of the reasons why you may decide not to service your own vehicle's A/C system.

According to the U.S. Clean Air Act, it is a federal crime to service or repair (involving the refrigerant) a Motor Vehicle Air Conditioning (MVAC) system for money without being EPA certified. It is also illegal to vent R-12 and R-134a refrigerants into the atmosphere. Selling or distributing A/C system refrigerant (in a container which contains less than 20 pounds of refrigerant) to any person who is not EPA 609 certified is also not allowed by law.

State and/or local laws may be more strict than the federal regulations, so be sure to check with your state and/or local authorities for further information. For further federal information on the legality of servicing your A/C system, call the EPA Stratospheric Ozone Hotline.

➡️**Federal law dictates that a fine of up to $25,000 may be levelled on people convicted of venting refrigerant into the atmosphere. Additionally, the EPA may pay up to $10,000 for information or services leading to a criminal conviction of the violation of these laws.**

When servicing an A/C system you run the risk of handling or coming in contact with refrigerant, which may result in skin or eye irritation or frostbite. Although low in toxicity (due to chemical stability), inhalation of concentrated refrigerant fumes is dangerous and can result in death; cases of fatal cardiac arrhythmia have been reported in people accidentally subjected to high levels of refrigerant. Some early symptoms include loss of concentration and drowsiness.

➡️**Generally, the limit for exposure is lower for R-134a than it is for R-12. Exceptional care must be practiced when handling R-134a.**

Also, refrigerants can decompose at high temperatures (near gas heaters or open flame), which may result in hydrofluoric acid, hydrochloric acid and phosgene (a fatal nerve gas).

R-12 refrigerant can damage the environment because it is a Chlorofluorocarbon (CFC), which has been proven to add to ozone layer depletion, leading to increasing levels of UV radiation. UV radiation has been linked with an increase in skin cancer, suppression of the human immune system, an increase in cataracts, damage to crops, damage to aquatic organisms, an increase in ground-level ozone, and increased global warming. R-134a refrigerant is a greenhouse gas which, if allowed to vent into the atmosphere, will contribute to global warming (the Greenhouse Effect).

It is usually more economically feasible to have a certified MVAC automotive technician perform A/C system service on your vehicle. Some possible reasons for this are as follows:

• While it is illegal to service an A/C system without the proper equipment, the home mechanic would have to purchase an expensive refrigerant recovery/recycling machine to service his/her own vehicle.

• Since only a certified person may purchase refrigerant—according to the Clean Air Act, there are specific restrictions on selling or distributing A/C system refrigerant—it is legally impossible (unless certified) for the home mechanic to service his/her own vehicle. Procuring refrigerant in an illegal fashion exposes one to the risk of paying a $25,000 fine to the EPA.

R-12 Refrigerant Conversion

If your vehicle still uses R-12 refrigerant, one way to save A/C system costs down the road is to investigate the possibility of having your system converted to R-134a. The older R-12 systems can be easily converted to R-134a refrigerant by a certified automotive technician by installing a few new components and changing the system oil.

The cost of R-12 is steadily rising and will continue to increase, because it is no longer imported or manufactured in the United States. Therefore, it is often possible to have an R-12 system converted to R-134a and recharged for less than it would cost to just charge the system with R-12.

If you are interested in having your system converted, contact local automotive service stations for more details and information.

PREVENTIVE MAINTENANCE

▸ **See Figures 112 and 113**

Although the A/C system should not be serviced by the do-it-yourselfer, preventive maintenance can be practiced and A/C system inspections can be performed to help maintain the efficiency of the vehicle's A/C system. For preventive maintenance, perform the following:

• The easiest and most important preventive maintenance for your A/C system is to be sure that it is used on a regular basis. Running the system for five minutes each month (no matter what the season) will help ensure that the seals and all internal components remain lubricated.

➡️**Some newer vehicles automatically operate the A/C system compressor whenever the windshield defroster is activated. When running, the compressor lubricates the A/C system components; therefore, the A/C system would not need to be operated each month.**

TCCS1233

Fig. 112 A coolant tester can be used to determine the freezing and boiling levels of the coolant in your vehicle

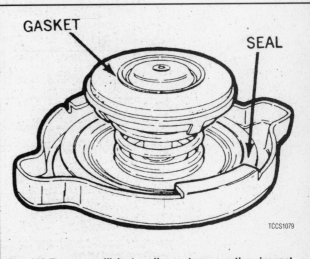

GASKET

SEAL

TCCS1079

Fig. 113 To ensure efficient cooling system operation, inspect the radiator cap gasket and seal

• In order to prevent heater core freeze-up during A/C operation, it is necessary to maintain proper antifreeze protection. Use a hand-held coolant tester (hydrometer) to periodically check the condition of the antifreeze in your engine's cooling system.

➡**Antifreeze should not be used longer than the manufacturer specifies.**

• For efficient operation of an air conditioned vehicle's cooling system, the radiator cap should have a holding pressure which meets manufacturer's specifications. A cap which fails to hold these pressures should be replaced.

• Any obstruction of or damage to the condenser configuration will restrict air flow which is essential to its efficient operation. It is, therefore, a good rule to keep this unit clean and in proper physical shape.

➡**Bug screens which are mounted in front of the condenser (unless they are original equipment) are regarded as obstructions.**

• The condensation drain tube expels any water which accumulates on the bottom of the evaporator housing into the engine compartment. If this tube is obstructed, the air conditioning performance can be restricted and condensation buildup can spill over onto the vehicle's floor.

SYSTEM INSPECTION

▶ **See Figure 114**

Although the A/C system should not be serviced by the do-it-yourselfer, preventive maintenance can be practiced and A/C system inspections can be performed to help maintain the efficiency of the vehicle's A/C system. For A/C system inspection, perform the following:

The easiest and often most important check for the air conditioning system consists of a visual inspection of the system components. Visually inspect the air conditioning system for refrigerant leaks, damaged compressor clutch, abnormal compressor drive belt tension and/or condition, plugged evaporator drain tube, blocked condenser fins, disconnected or broken wires, blown fuses, corroded connections and poor insulation.

A refrigerant leak will usually appear as an oily residue at the leakage point in the system. The oily residue soon picks up dust or dirt particles from the surrounding air and appears greasy. Through time, this will build up and appear to be a heavy dirt-impregnated grease.

For a thorough visual and operational inspection, check the following:

• Check the surface of the radiator and condenser for dirt, leaves or other material which might block air flow.

• Check for kinks in hoses and lines. Check the system for leaks.

• Make sure the drive belt is properly tensioned. When the air conditioning is operating, make sure the drive belt is free of noise or slippage.

• Make sure the blower motor operates at all appropriate positions, then check for distribution of the air from all outlets with the blower on **HIGH** or **MAX**.

➡**Keep in mind that under conditions of high humidity, air discharged from the A/C vents may not feel as cold as expected, even if the system is working properly. This is because vaporized moisture in humid air retains heat more effectively than dry air, thereby making humid air more difficult to cool.**

• Make sure the air passage selection lever is operating correctly. Start the engine and warm it to normal operating temperature, then make sure the temperature selection lever is operating correctly.

Windshield Wipers

ELEMENT (REFILL) CARE AND REPLACEMENT

▶ **See Figures 115 thru 124**

For maximum effectiveness and longest element life, the windshield and wiper blades should be kept clean. Dirt, tree sap, road tar and so on will cause streaking, smearing and blade deterioration if left on the glass. It is

Fig. 115 Bosch® wiper blade and fit kit

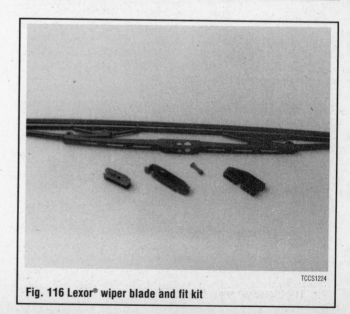

Fig. 116 Lexor® wiper blade and fit kit

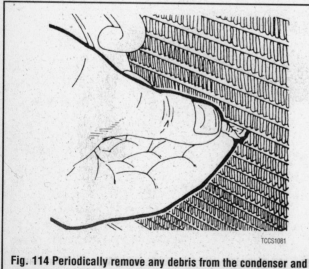

Fig. 114 Periodically remove any debris from the condenser and radiator fins

Fig. 117 Pylon® wiper blade and adaptor

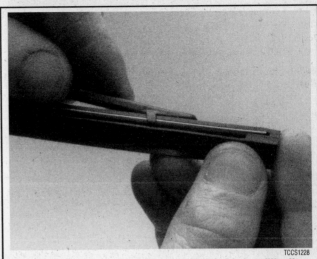

Fig. 120 To remove and install a Lexor® wiper blade refill, slip out the old insert and slide in a new one

Fig. 118 Trico® wiper blade and fit kit

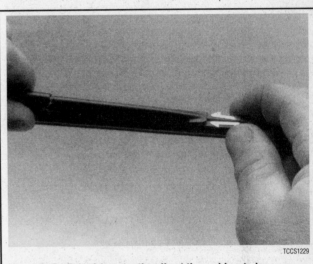

Fig. 121 On Pylon® inserts, the clip at the end has to be removed prior to sliding the insert off

Fig. 119 Tripledge® wiper blade and fit kit

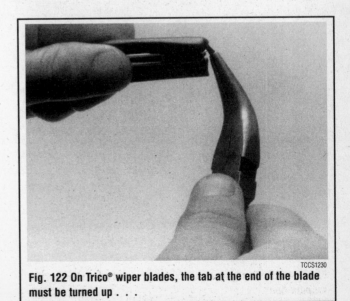

Fig. 122 On Trico® wiper blades, the tab at the end of the blade must be turned up . . .

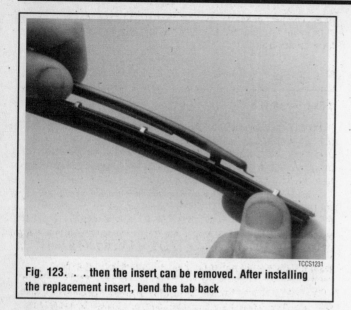

Fig. 123. . . then the insert can be removed. After installing the replacement insert, bend the tab back

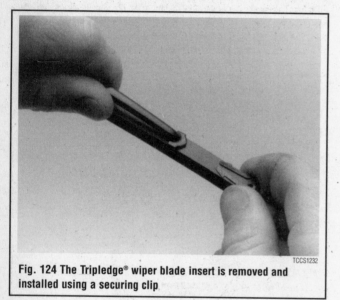

Fig. 124 The Tripledge® wiper blade insert is removed and installed using a securing clip

advisable to wash the windshield carefully with a commercial glass cleaner at least once a month. Wipe off the rubber blades with the wet rag afterwards. Do not attempt to move wipers across the windshield by hand; damage to the motor and drive mechanism will result.

To inspect and/or replace the wiper blade elements, place the wiper switch in the **LOW** speed position and the ignition switch in the **ACC** position. When the wiper blades are approximately vertical on the windshield, turn the ignition switch to **OFF.**

Examine the wiper blade elements. If they are found to be cracked, broken or torn, they should be replaced immediately. Replacement intervals will vary with usage, although ozone deterioration usually limits element life to about one year. If the wiper pattern is smeared or streaked, or if the blade chatters across the glass, the elements should be replaced. It is easiest and most sensible to replace the elements in pairs.

If your vehicle is equipped with aftermarket blades, there are several different types of refills and your vehicle might have any kind. Aftermarket blades and arms rarely use the exact same type blade or refill as the original equipment. Here are some typical aftermarket blades; not all may be available for your vehicle:

The Anco® type uses a release button that is pushed down to allow the refill to slide out of the yoke jaws. The new refill slides back into the frame and locks in place.

Some Trico® refills are removed by locating where the metal backing strip or the refill is wider. Insert a small screwdriver blade between the frame and metal backing strip. Press down to release the refill from the retaining tab.

Other types of Trico® refills have two metal tabs which are unlocked by squeezing them together. The rubber filler can then be withdrawn from the frame jaws. A new refill is installed by inserting the refill into the front frame jaws and sliding it rearward to engage the remaining frame jaws. There are usually four jaws; be certain when installing that the refill is engaged in all of them. At the end of its travel, the tabs will lock into place on the front jaws of the wiper blade frame.

Another type of refill is made from polycarbonate. The refill has a simple locking device at one end which flexes downward out of the groove into which the jaws of the holder fit, allowing easy release. By sliding the new refill through all the jaws and pushing through the slight resistance when it reaches the end of its travel, the refill will lock into position.

To replace the Tridon® refill, it is necessary to remove the wiper blade. This refill has a plastic backing strip with a notch about 1 in. (25mm) from the end. Hold the blade (frame) on a hard surface so that the frame is tightly bowed. Grip the tip of the backing strip and pull up while twisting counterclockwise. The backing strip will snap out of the retaining tab. Do this for the remaining tabs until the refill is free of the blade. The length of these refills is molded into the end and they should be replaced with identical types.

Regardless of the type of refill used, be sure to follow the part manufacturer's instructions closely. Make sure that all of the frame jaws are engaged as the refill is pushed into place and locked. If the metal blade holder and frame are allowed to touch the glass during wiper operation, the glass will be scratched.

Tires and Wheels

Common sense and good driving habits will afford maximum tire life. Fast starts, sudden stops and hard cornering are tough on tires and will shorten their useful life span. Make sure that you don't overload the vehicle or run with incorrect pressure in the tires. Both of these practices will increase tread wear.

➡For optimum tire life, keep the tires properly inflated, rotate them often and have the wheel alignment checked periodically.

Inspect your tires frequently. Be especially careful to watch for bubbles in the tread or sidewall, deep cuts or underinflation. Replace any tires with bubbles in the sidewall. If cuts are so deep that they penetrate to the cords, discard the tire. Any cut in the sidewall of a radial tire renders it unsafe. Also look for uneven tread wear patterns that may indicate the front end is out of alignment or that the tires are out of balance.

TIRE ROTATION

▶ See Figures 125, 126 and 127

Tires must be rotated periodically to equalize wear patterns that vary with a tire's position on the vehicle. Tires will also wear in an uneven way as the front steering/suspension system wears to the point where the alignment should be reset.

Rotating the tires will ensure maximum life for the tires as a set, so you will not have to discard a tire early due to wear on only part of the tread. Regular rotation is required to equalize wear.

When rotating "unidirectional tires," make sure that they always roll in the same direction. This means that a tire used on the left side of the vehicle must not be switched to the right side and vice-versa. Such tires should only be rotated front-to-rear or rear-to-front, while always remaining on the same side of the vehicle. These tires are marked on the sidewall as to the direction of rotation; observe the marks when reinstalling the tire(s).

Some styled or "mag" wheels may have different offsets front to rear. In these cases, the rear wheels must not be used up front and vice-versa. Furthermore, if these wheels are equipped with unidirectional tires, they cannot be rotated unless the tire is remounted for the proper direction of rotation.

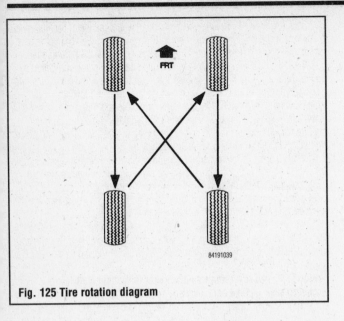

Fig. 125 Tire rotation diagram

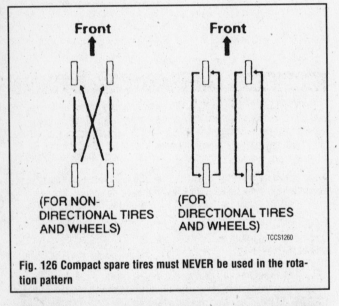

Front   Front

(FOR NON-DIRECTIONAL TIRES AND WHEELS)   (FOR DIRECTIONAL TIRES AND WHEELS)

Fig. 126 Compact spare tires must NEVER be used in the rotation pattern

Fig. 127 Unidirectional tires are identifiable by sidewall arrows and/or the word "rotation"

➡ The compact or space-saver spare is strictly for emergency use. It must never be included in the tire rotation or placed on the vehicle for everyday use.

TIRE DESIGN

♦ See Figure 128

For maximum satisfaction, tires should be used in sets of four. Mixing of different types (radial, bias-belted, fiberglass belted) must be avoided. In most cases, the vehicle manufacturer has designated a type of tire on which the vehicle will perform best. Your first choice when replacing tires should be to use the same type of tire that the manufacturer recommends.

When radial tires are used, tire sizes and wheel diameters should be selected to maintain ground clearance and tire load capacity equivalent to the original specified tire. Radial tires should always be used in sets of four.

> ❊❊ **CAUTION**
>
> **Radial tires should never be used on only the front axle.**

When selecting tires, pay attention to the original size as marked on the tire. Most tires are described using an industry size code sometimes referred to as P-Metric. This allows the exact identification of the tire specifications, regardless of the manufacturer. If selecting a different tire size or brand, remember to check the installed tire for any sign of interference with the body or suspension while the vehicle is stopping, turning sharply or heavily loaded.

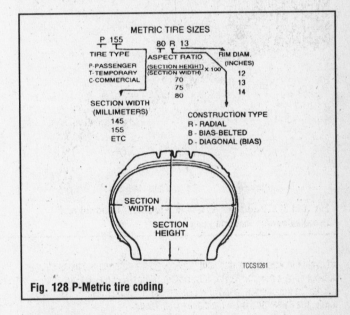

Fig. 128 P-Metric tire coding

Snow Tires

Good radial tires can produce a big advantage in slippery weather, but in snow, a street radial tire does not have sufficient tread to provide traction and control. The small grooves of a street tire quickly pack with snow and the tire behaves like a billiard ball on a marble floor. The more open, chunky tread of a snow tire will self-clean as the tire turns, providing much better grip on snowy surfaces.

To satisfy municipalities requiring snow tires during weather emergencies, most snow tires carry either a Mud and Snow (M + S) designation after the tire size stamped on the sidewall, or the designation "all-season." In general, no change in tire size is necessary when buying snow tires.

Most manufacturers strongly recommend the use of 4 snow tires on their vehicles for reasons of stability. If snow tires are fitted only to the drive wheels, the opposite end of the vehicle may become very unstable when braking or turning on slippery surfaces. This instability can lead to unpleasant endings if the driver can't counteract the slide in time.

Note that snow tires, whether 2 or 4, will affect vehicle handling in all non-snow situations. The stiffer, heavier snow tires will noticeably change the turning and braking characteristics of the vehicle. Once the snow tires are installed, you must re-learn the behavior of the vehicle and drive accordingly.

➡ **Consider buying extra wheels on which to mount the snow tires. Once done, the "snow wheels" can be installed and removed as needed. This eliminates the potential damage to tires or wheels from seasonal removal and installation. Even if your vehicle has styled wheels, see if inexpensive steel wheels are available. Although the look of the vehicle will change, the expensive wheels will be protected from salt, curb hits and pothole damage.**

TIRE STORAGE

If they are mounted on wheels, store the tires at proper inflation pressure. All tires should be kept in a cool, dry place. If they are stored in the garage or basement, do not let them stand on a concrete floor; set them on strips of wood, a mat or a large stack of newspaper. Keeping them away from direct moisture is of paramount importance. Tires should not be stored upright, but in a flat position.

INFLATION & INSPECTION

▶ **See Figures 129 thru 136**

The importance of proper tire inflation cannot be overemphasized. A tire employs air as part of its structure. It is designed around the supporting strength of the air at a specified pressure. For this reason, improper inflation drastically reduces the tire's ability to perform as intended. A tire will lose some air in day-to-day use; having to add a few pounds of air periodically is not necessarily a sign of a leaking tire.

Two items should be a permanent fixture in every glove compartment: an accurate tire pressure gauge and a tread depth gauge. Check the tire pressure (including the spare) regularly with a pocket type gauge. Too often, the gauge on the end of the air hose at your corner garage is not accurate because it suffers too much abuse. Always check tire pressure when the tires are cold, as pressure increases with temperature. If you must move the vehicle to check the tire inflation, do not drive more than a mile before checking. A cold tire is generally one that has not been driven for more than three hours.

A plate or sticker is normally provided somewhere in the vehicle (door post, hood, tailgate or trunk lid) which shows the proper pressure for the tires. Never counteract excessive pressure build-up by bleeding off air pressure (letting some air out). This will cause the tire to run hotter and wear quicker.

Fig. 130 Tires with deep cuts, or cuts which show bulging should be replaced immediately

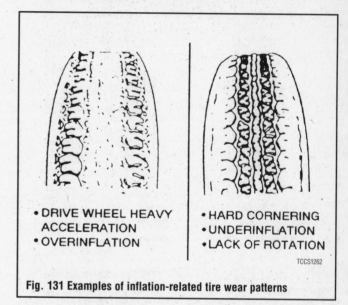

- DRIVE WHEEL HEAVY ACCELERATION
- OVERINFLATION

- HARD CORNERING
- UNDERINFLATION
- LACK OF ROTATION

Fig. 131 Examples of inflation-related tire wear patterns

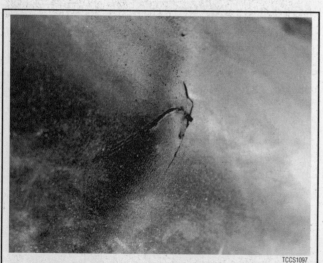

Fig. 129 Tires should be checked frequently for any sign of puncture or damage

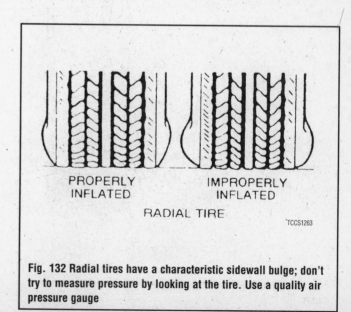

PROPERLY INFLATED

IMPROPERLY INFLATED

RADIAL TIRE

Fig. 132 Radial tires have a characteristic sidewall bulge; don't try to measure pressure by looking at the tire. Use a quality air pressure gauge

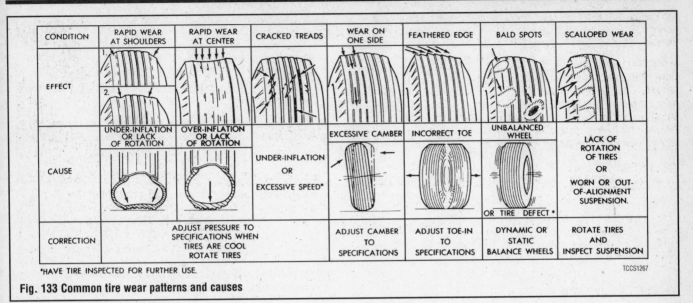

CONDITION	RAPID WEAR AT SHOULDERS	RAPID WEAR AT CENTER	CRACKED TREADS	WEAR ON ONE SIDE	FEATHERED EDGE	BALD SPOTS	SCALLOPED WEAR
EFFECT							
CAUSE	UNDER-INFLATION OR LACK OF ROTATION	OVER-INFLATION OR LACK OF ROTATION	UNDER-INFLATION OR EXCESSIVE SPEED*	EXCESSIVE CAMBER	INCORRECT TOE	UNBALANCED WHEEL OR TIRE DEFECT *	LACK OF ROTATION OF TIRES OR WORN OR OUT-OF-ALIGNMENT SUSPENSION.
CORRECTION	ADJUST PRESSURE TO SPECIFICATIONS WHEN TIRES ARE COOL ROTATE TIRES			ADJUST CAMBER TO SPECIFICATIONS	ADJUST TOE-IN TO SPECIFICATIONS	DYNAMIC OR STATIC BALANCE WHEELS	ROTATE TIRES AND INSPECT SUSPENSION

*HAVE TIRE INSPECTED FOR FURTHER USE.

TCCS1267

Fig. 133 Common tire wear patterns and causes

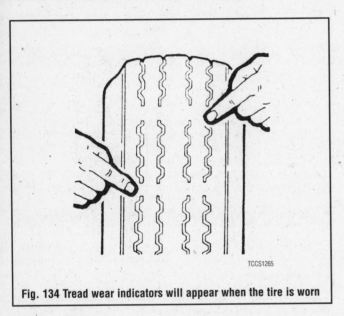

TCCS1265

Fig. 134 Tread wear indicators will appear when the tire is worn

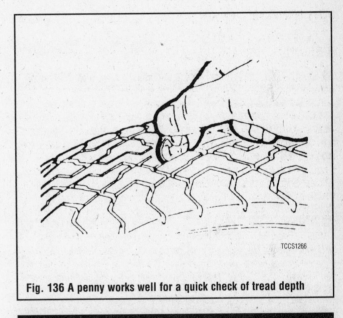

TCCS1266

Fig. 136 A penny works well for a quick check of tread depth

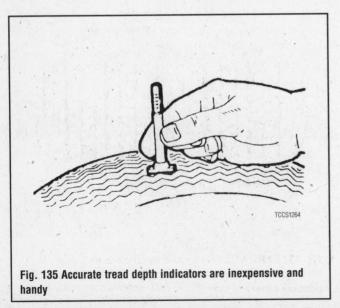

TCCS1264

Fig. 135 Accurate tread depth indicators are inexpensive and handy

> ※ **CAUTION**
>
> **Never exceed the maximum tire pressure embossed on the tire! This is the pressure to be used when the tire is at maximum loading, but it is rarely the correct pressure for everyday driving. Consult the owner's manual or the tire pressure sticker for the correct tire pressure.**

Once you've maintained the correct tire pressures for several weeks, you'll be familiar with the vehicle's braking and handling personality. Slight adjustments in tire pressures can fine-tune these characteristics, but never change the cold pressure specification by more than 2 psi. A slightly softer tire pressure will give a softer ride but also yield lower fuel mileage. A slightly harder tire will give crisper dry road handling but can cause skidding on wet surfaces. Unless you're fully attuned to the vehicle, stick to the recommended inflation pressures.

All tires made since 1968 have built-in tread wear indicator bars that show up as ½ in. (13mm) wide smooth bands across the tire when only ¹⁄₁₆ in. (1.5mm) of tread remains. The appearance of tread wear indicators means that the tires should be replaced. In fact, many states have laws prohibiting the use of tires with less than this amount of tread.

You can check your own tread depth with an inexpensive gauge or by using a Lincoln head penny. Slip the Lincoln penny (with Lincoln's head upside-down) into several tread grooves. If you can see the top of Lincoln's head in 2 adjacent grooves, the tire has less than 1/16 in. (1.5mm) tread left and should be replaced. You can measure snow tires in the same manner by using the "tails" side of the Lincoln penny. If you can see the top of the Lincoln memorial, it's time to replace the snow tire(s).

CARE OF SPECIAL WHEELS

If you have invested money in magnesium, aluminum alloy or sport wheels, special precautions should be taken to make sure your investment is not wasted and that your special wheels look good for the life of the vehicle.

Special wheels are easily damaged and/or scratched. Occasionally check the rims for cracking, impact damage or air leaks. If any of these are found, replace the wheel. But in order to prevent this type of damage and the costly replacement of a special wheel, observe the following precautions:

• Use extra care not to damage the wheels during removal, installation, balancing, etc. After removal of the wheels from the vehicle, place them on a mat or other protective surface. If they are to be stored for any length of time, support them on strips of wood. Never store tires and wheels upright; the tread may develop flat spots.

• When driving, watch for hazards; it doesn't take much to crack a wheel.

• When washing, use a mild soap or non-abrasive dish detergent (keeping in mind that detergent tends to remove wax). Avoid cleansers with abrasives or the use of hard brushes. There are many cleaners and polishes for special wheels.

• If possible, remove the wheels during the winter. Salt and sand used for snow removal can severely damage the finish of a wheel.

• Make certain the recommended lug nut torque is never exceeded or the wheel may crack. Never use snow chains on special wheels; severe scratching will occur.

Maintenance Light

RESETTING

On 1991–93 vehicles equipped with an automatic transaxle, the Powertrain Control Module (PCM) stores a parameter called "oil life left percent." This is used by the computer to illuminate the SERVICE ENGINE SOON light when the percent reaches a minimum level, and to inform a technician that the transmission fluid is in need of a change. If a suitable scan tool is available, connect it and reset the parameter to 100 percent by accessing the "special test function". If you do not have access to a suitable scan tool as described in Section 4 of this manual, the vehicle should be taken to a dealer or shop that has the appropriate tool. Simply tell them that the transmission oil has just been changed and the oil life PCM parameter needs to be reset.

FLUIDS AND LUBRICANTS

Fluid Disposal

Used fluids such as engine oil, transmission fluid, antifreeze and brake fluid are hazardous wastes and must be disposed of properly. Before draining any fluids, consult with the local authorities; in many areas, waste oil, antifreeze, etc. is being accepted as a part of recycling programs. A number of service stations and auto parts stores are also accepting waste fluids for recycling.

Be sure of the recycling center's policies before draining any fluids, as many will not accept different fluids that have been mixed together, such as oil and antifreeze.

Fuel and Engine Oil Recommendations

♦ **See Figure 137**

All Saturns are equipped with a catalytic converter, necessitating the use of unleaded gasoline. The use of leaded gasoline will damage the catalytic converter. Both the SOHC and DOHC engines are designed to use unleaded gasoline with a minimum octane rating of 87, which usually means regular unleaded.

Oil must be selected with regard to the anticipated temperatures during the period before the next oil change. Using the chart, select the oil viscosity for the lowest expected temperature and you will be assured of easy cold starting and sufficient engine protection. The oil you pour into your engine should have an American Petroleum Institute (API) designation of SG marked on the container. For maximum fuel economy benefits, use an oil with the Roman Numeral II next to the words Energy Conserving in the API Service Symbol.

Engine

OIL LEVEL CHECK

♦ **See Figures 138, 139 and 140**

Check the engine oil level every time you fill the fuel tank. Make sure the oil level is between the **FULL** and L (or **ADD**) marks. The engine and oil

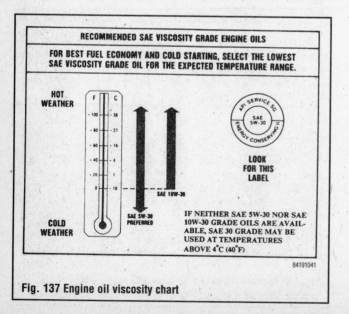

Fig. 137 Engine oil viscosity chart

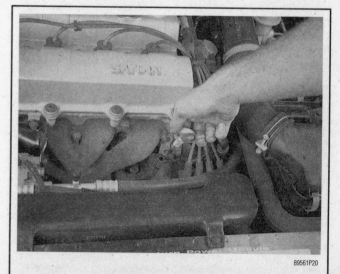

Fig. 138 Remove the engine oil dipstick from the guide tube

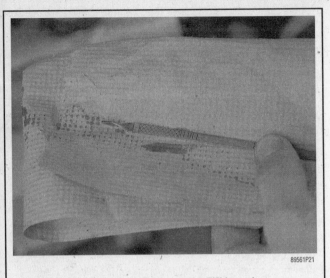

Fig. 139 Make sure the oil level is at the FULL mark

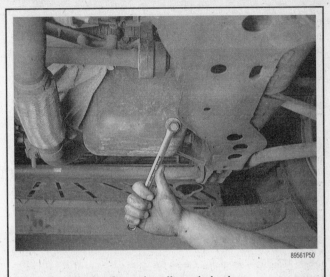

Fig. 141 Loosening the engine oil pan drain plug

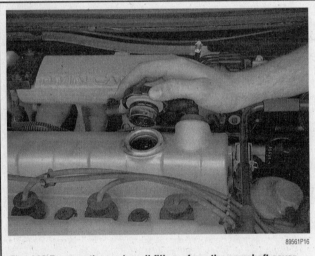

Fig. 140 Remove the engine oil fill cap from the camshaft cover and pour in the correct amount of engine oil—DOHC engine shown

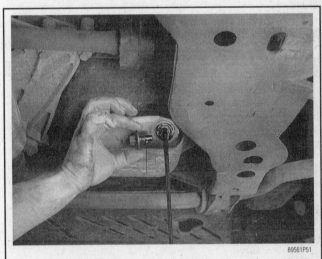

Fig. 142 Remove the plug, allowing the oil to drain into a suitable container

must be warm and the vehicle parked on level ground to get an accurate reading. Always allow a few minutes after turning the engine OFF for the oil to drain back into the pan before checking, or an inaccurate reading will result. Check the engine oil level as follows:

1. Open the hood and locate the engine oil dipstick.

2. If the engine is hot, you may want to wrap a rag around the dipstick handle before removing it.

3. Remove the dipstick and wipe it with a clean, lint-free rag, then reinsert it into the dipstick tube. Make sure it is inserted all the way to avoid an inaccurate reading.

4. Pull out the dipstick and note the oil level. It should be between the marks, as stated above.

5. If the oil level is below the lower mark, replace the dipstick and add fresh oil to bring the level within the proper range. Do not overfill.

6. Recheck the oil level and close the hood.

OIL & FILTER CHANGE

◆ See Figures 141, 142, 143, 144 and 145

The engine oil and oil filter should be changed together, at the recommended interval on the Maintenance Intervals chart. The oil should be changed more frequently if the vehicle is being operated in very dusty

Fig. 143 After loosening the oil filter with an oil filter wrench, turn the filter counterclockwise

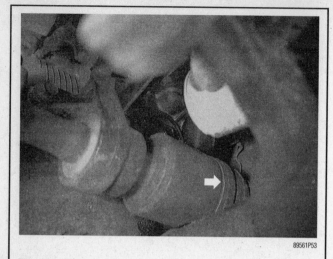

Fig. 144 When removing the oil filter, be careful not to spill any oil on the CV-boot

Fig. 145 Before installing a new oil filter, lightly coat the rubber gasket with clean oil

areas. Before draining the oil, make sure the engine is at operating temperature. Hot oil will hold more impurities in suspension and will flow better, allowing the removal of more oil and dirt.

As noted earlier, used oil has been classified as a hazardous waste and must be disposed of properly. Before draining any oil from the engine crankcase, make sure you are aware of the proper disposal procedures for your area.

Change the oil and filter as follows:

1. Run the engine until it reaches the normal operating temperature, then shut off the engine and remove the oil filler cap.

2. Raise and safely support the front of the car. If possible, make sure the engine drain plug is lower than the rest of the oil pan. Position a drain pan beneath the plug.

✳✳ CAUTION

The EPA warns that prolonged contact with used engine oil may cause a number of skin disorders, including cancer! You should make every effort to minimize your exposure to used engine oil. Protective gloves should be worn when changing the oil. Wash your hands and any other exposed skin areas as soon as possible after exposure to used engine oil. Soap and water, or water-less hand cleaner, should be used.

3. Wipe the drain plug and the surrounding area clean. Loosen the drain plug with a socket or box wrench, then remove it by hand using a rag to shield your fingers from the heat. Push in on the plug as you turn it out, so that no oil escapes until the plug is completely removed.

4. Allow the oil to drain into the pan. Be careful if the engine is at operating temperature, as the oil is hot enough to burn you.

5. Clean and install the drain plug, making sure that the gasket is still on the plug. Use a new drain plug gasket whenever possible, but if the gasket is damaged, a new gasket MUST be installed. Tighten the drain plug to 27 ft. lbs. (37 Nm).

6. The oil filter is on the rear right side of the block protruding towards the firewall. Slide the drain pan under the oil filter. Slip an oil filter wrench onto the filter and turn counterclockwise to loosen the filter, allowing much of the oil to neatly drain into the container by way of the oil pan drip deflector. Do not remove the filter immediately after loosening it or oil will drip onto the axle shaft boot.

7. Wrap a rag around the filter and unscrew it the rest of the way by hand. Be careful of hot oil which may run down the side of the filter.

➡When the oil filter is removed, make sure the old filter gasket has also been removed from the engine or a proper seal will not be achieved with the new filter. More than a few people have installed the replacement filter with the old gasket still in place and have wound up with a garage floor full of clean engine oil.

8. If the filter's case separates from the base plate, a 10mm socket or wrench may be used to remove the base plate and threaded adapter from the engine. Separate the 2 parts, clean the adapter and block threads, then install the adapter and tighten to 22 ft. lbs. (30 Nm).

9. Clean the oil filter adapter on the engine with a clean rag. Wipe the oil pan drip deflector and drip rail dry prior to installing a new oil filter.

10. Coat the rubber gasket on the replacement filter with clean engine oil. Place the filter in position on the adapter fitting and screw it on by hand until resistance is felt. Using a standard oil filter tool, tighten the filter an additional ¾–1 turn.

11. Pull the drain pan from under the vehicle, remove the supports and lower the vehicle to the ground.

12. Fill the crankcase with the proper type and quantity of engine oil right away. If the engine is started and run without oil in the crankcase, serious damage may occur almost immediately.

13. Run the engine and check for leaks. Stop the engine and check the oil level.

➡Remember to properly dispose of (recycle) the old engine oil.

Manual Transaxle

FLUID RECOMMENDATIONS

Saturn Transaxle Fluid (P/N 21005966) is recommended. If not available, DEXRON®IIE or DEXRON®III transmission/transaxle fluid is also acceptable.

LEVEL CHECK

▶ **See Figures 146, 147 and 148**

1. The manual transaxle fluid should be checked **COLD**, with the car parked on a level surface.

2. If necessary, move the car to a level spot and allow the powertrain to cool.

3. Reach below the brake master cylinder, located at the firewall on the left side of the engine compartment, to the top of the transaxle. Grasp the hinged dipstick lever and pull upwards to withdraw the dipstick. The dipstick hinge will automatically straighten and release the expansion plug at the bottom of the hinge, so the stick may be pulled from the transaxle.

4. Note the fluid level, it should be at the **FULL** line. If necessary, add

Fig. 146 Withdraw the manual transaxle dipstick with the hinge in the upright position

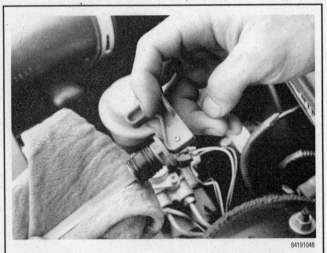
Fig. 147 When the dipstick is installed, position the hinge in the locked position to seal the transaxle

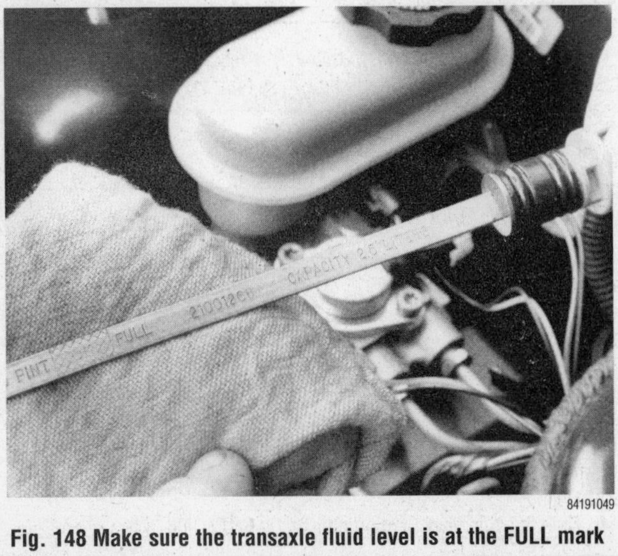

fluid to bring the amount up to the proper level. Be careful not to overfill the transaxle.

5. Insert the dipstick to the top of the transaxle case and push the hinge back to the locked position. This will expand the rubber plug below the hinge to lock and seal the transaxle.

DRAIN & REFILL

♦ **See Figure 149**

The manual transaxle fluid should be changed when the vehicle reaches 6000 miles (10,000 km). No periodic changing should be necessary after the initial change.

1. Start the engine and allow it to warm to normal operating temperature, then shut the engine **OFF**.

2. If possible, raise and safely support the vehicle so that it is level. If not, support the vehicle so the drain plug is the lowest point of the transaxle case.

3. Position a drain pan under the plug.

4. Clean the area around the drain plug, then remove the plug using a socket or box wrench. Push in on the plug as you turn it out, so that no fluid escapes until the plug is completely removed. Allow the fluid to drain into the pan.

Fig. 149 Manual transaxle drain plug location

5. Clean the drain plug, then lubricate the plug using clean transaxle fluid. Wipe off the excess fluid and install the plug using a new washer (1991–early 93 models) or integral rubber seal (late 1993–98 models). On 1991–early 93 models, tighten the drain plug to 40 ft. lbs. (55 Nm). On late 1993–98 models, tighten the drain plug to 22 ft. lbs. (30 Nm).

6. Slide the drain pan out from under the vehicle, then remove the supports and carefully lower the vehicle to the ground.

7. Withdraw the transaxle dipstick and position a suitable long necked funnel into the top of the transaxle. Fill the transaxle to the proper level using the recommended fluid.

8. Check the fluid level, then install and lock the transaxle dipstick.

Automatic Transaxle

FLUID RECOMMENDATIONS

Saturn Transaxle Fluid (P/N 21005966) is recommended. If not available, DEXRON®IIE or DEXRON®III transmission/transaxle fluid is also acceptable.

Fig. 148 Make sure the transaxle fluid level is at the FULL mark

LEVEL CHECK

▶ **See Figure 150**

1. Start the engine and operate the vehicle for 15 minutes or until the normal operating temperature is reached. An incorrect level reading will be obtained if the vehicle is operated under the following conditions immediately before checking the dipstick:
   - In an ambient temperature of 90°F (32°C) or above
   - At sustained highway speeds
   - In heavy city traffic, during hot weather
   - When used as a towing vehicle
2. Park the vehicle on a level surface and apply the parking brake.
3. Move the gear selector through all gears, then position the selector in **P**.
4. With all accessories turned **OFF**, allow the vehicle to idle for 3 minutes.
5. Withdraw the dipstick and check for proper fluid level as compared with the illustration. Check the fluid color and condition; fluid should be smooth, transparent and red. If the fluid is not transparent or if it is a dark brown, the fluid is contaminated or overheated and must be replaced.
6. If necessary, change the fluid or add fluid to bring the amount to the proper level.

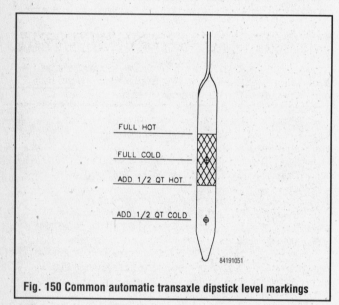

Fig. 150 Common automatic transaxle dipstick level markings

FLUID & FILTER CHANGE

▶ **See Figures 151 thru 157**

The automatic transaxle fluid and filter should be changed when the vehicle reaches 30,000 miles (50,000 km). Although the manufacturer recommends changing the filter with only every other fluid change thereafter, we at Chilton recommend that the fluid and filter be changed together, at the recommended interval in the Maintenance Intervals chart. A few dollars more spent on a filter today may offer greater protection against transaxle wear and repair in the future.

Change the fluid and filter as follows:

1. Warm the fluid to the normal operating temperature of 190–200°F (88–93°C). It should take approximately 15 miles of highway driving to reach this temperature.
2. If possible, raise and safely support the vehicle so that it is level. If not, support the vehicle so the drain plug is at the lowest point of the transaxle case.
3. Position a drain pan beneath the plug.
4. Clean the area around the drain plug, then remove the plug using a socket or box wrench. Push in on the plug as you turn it out, so that no fluid escapes until the plug is completely removed. Allow the fluid to completely drain into the pan; it may take 5 minutes or longer.

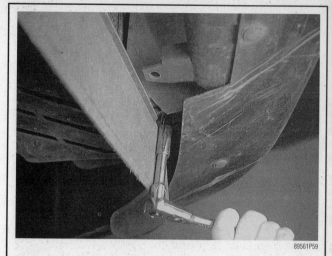

Fig. 151 Loosen the front splash shield/air dam mounting fasteners

Fig. 152 Remove the front splash shield/air dam assembly from the vehicle

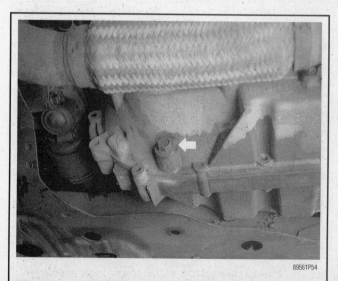

Fig. 153 Location of automatic transaxle fluid drain plug

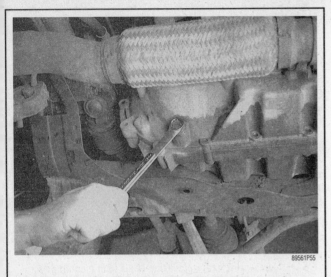

Fig. 154 Loosen the drain plug with an appropriate wrench

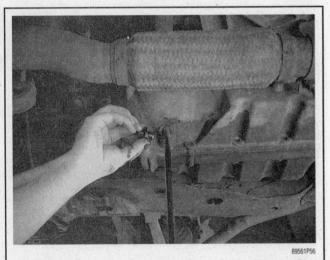

Fig. 155 Remove the plug and drain the transaxle oil into a suitable container

Fig. 156 After loosening it with a strap-type wrench, remove the transaxle oil filter by turning counterclockwise

Fig. 157 Removal of the transaxle oil filter may be easier from above the engine cradle crossmember

✳✳ CAUTION

The transaxle fluid was heated to operating temperature for maximum flow. Use a rag and protective gloves to make sure you are not burned by the extremely hot fluid.

5. Clean the drain plug, then lubricate the plug using clean transaxle fluid. Wipe off the excess fluid and install the plug using a new washer (1991–early 93 models) or integral rubber seal (late 1993–98 models). On 1991–early 93 models, tighten the drain plug to 33 ft. lbs. (45 Nm). On late 1993–98 models, tighten the drain plug to 22 ft. lbs. (30 Nm).

6. Position the drain pan beneath the transaxle fluid filter.

7. Remove the air deflector/shield mounting fasteners and assembly from beneath the front of the vehicle to allow access to the transaxle spin-on fluid pressure filter.

8. Use a nylon strap-type filter wrench to loosen the pressure filter, then remove the filter from the transaxle. Make sure the filter seal was removed from the transaxle with the filter. If equipped, remove the magnet from the old filter and install it on the new filter.

9. Lubricate the new filter seal using clean transaxle fluid, then install the filter by hand and tighten until the seal makes contact. Tighten the filter 1 additional turn by hand or, if necessary, using the nylon strap filter wrench. Do not use any tool that might scratch, dent or damage the filter. A filter that is scratched, dented or damaged must be immediately replaced.

10. Hold the air deflector and shield assembly in proper position and install the mounting fasteners.

11. Slide the drain pan out from under the vehicle, then remove the supports and carefully lower the vehicle to the ground.

12. Using a funnel, slowly refill the transaxle with the recommended type and quantity of fluid. Check the fluid level cold.

13. Start the engine and check for leaks, then check the level hot and add fluid as necessary to achieve the proper level.

14. On 1991–93 vehicles, the Powertrain Control Module (PCM) stores a parameter called "oil life left percent." This is used by the computer to illuminate the SERVICE ENGINE SOON light when the percent reaches a minimum level and to inform a technician that the fluid is in need of a change. If a suitable scan tool is available, connect it and reset the parameter to 100 percent by accessing the "special test function". If you do not have access to a suitable scan tool as described in Section 4 of this manual, the vehicle should be taken to a dealer or shop that has the appropriate tool. Simply tell them that the transmission oil has just been changed and the oil life PCM parameter needs to be reset.

Cooling System

▶ See Figures 158 and 159

Check the cooling system at the intervals specified in the applicable Maintenance Intervals chart at the end of this section.

If necessary, hose clamps should be checked and soft or cracked hoses replaced. Damp spots or accumulations of rust or dye near hoses, the water pump or other areas indicate areas of possible leakage. Check the surge tank cap for a worn or cracked gasket. If the cap doesn't seal properly, fluid will be lost and the engine will overheat. A worn cap should be replaced with a new one. The surge tank should be free of rust and the coolant should be free from oil. If oil is found in the coolant, the engine thermostat will not function correctly, therefore the system must be flushed and filled with fresh coolant.

Periodically clean any debris such as leaves, paper, insects, etc. from the radiator fins. Pick the large pieces off by hand. The smaller pieces can be washed away with water pressure from a hose.

Carefully straighten any bent radiator fins with a pair of needlenose pliers. Be careful—the fins are very soft. Don't wiggle the fins back and forth too much. Straighten them once and try not to move them again.

FLUID RECOMMENDATIONS

The recommended fluid for most Saturn vehicles is a 50/50 mixture of a non-phosphate ethylene glycol antifreeze and water for year-round use. Use a good quality antifreeze with rust and other corrosion inhibitors, along with acid neutralizers.

➡**While regular ethylene glycol antifreeze is green in color, some 1996 and later Saturn vehicle cooling systems are filled with DEX-COOL® silicate-free antifreeze, which is orange colored and good for 5 years or 100,000 miles (161,030 km). To verify the type of engine coolant, check the top of the coolant reservoir cap for a green dot or an orange dot. The color of the dot should be the same as the color of the coolant. Do not mix DEX-COOL® antifreeze with regular ethylene glycol antifreeze.**

LEVEL CHECK

▶ See Figures 160 and 161

With the engine cold, check the surge tank (reservoir) located against the center of the right fender in the engine compartment. The coolant

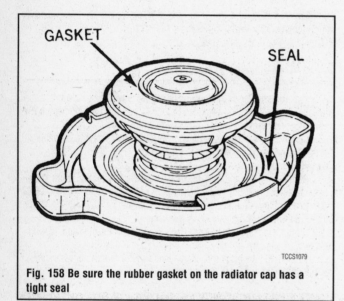

Fig. 158 Be sure the rubber gasket on the radiator cap has a tight seal

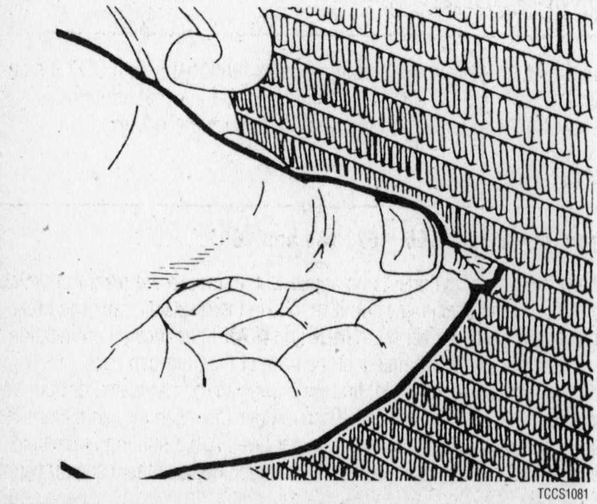

TCCS1079

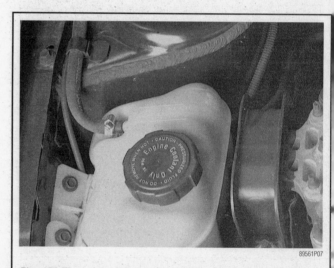

Fig. 160 The surge tank is located on the right fender. Do not remove the cap when the engine is hot

89561P07

Fig. 159 Periodically remove all debris from the radiator fins

TCCS1081

Fig. 161 With the engine cold, fill up the coolant surge tank until the coolant level is at the FULL COLD line

89561P02

should be at the **FULL COLD** level. If it is low, check for leaks and repair as necessary. Add a suitable coolant mixture to bring the coolant to the proper level.

DRAIN & REFILL

▶ See Figures 162, 163 and 164

1. Make sure the engine is cool and that the vehicle is parked on a level surface, then remove the surge tank pressure cap.
2. Position a large drain pan under the radiator drain plug on the right side of the radiator and engine drain plug under the right front of the block.
3. Open the radiator drain and remove the engine drain plug, then allow the coolant to drain from the system.
4. Close the radiator plug and install the engine plug to the cylinder block. Tighten the engine drain plug to 26 ft. lbs. (35 Nm).
5. Fill the system through the surge tank using a mixture of water and the recommended type of antifreeze.
6. Start the engine and check for leaks.

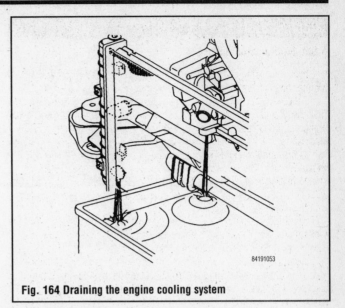

Fig. 164 Draining the engine cooling system

7. After the engine has run for 2 or 3 minutes, add coolant as necessary to bring the coolant level to the surge tank **FULL COLD** line.
8. Install the surge tank pressure cap.

FLUSHING & CLEANING THE SYSTEM

1. Prepare a flushing solution consisting of 2 ounces of Calgon®, or an equivalent automatic dishwasher detergent to 1 gallon (3.8L) of clean water.
2. Properly drain the engine cooling system.
3. Remove the thermostat to permit the flushing solution to circulate through the entire cooling system.
4. Fill the cooling system with the flushing solution.
5. Run the engine for 5 minutes, then drain the flushing solution into a clean container.
6. Repeat Steps 4 and 5.
7. Fill the cooling system with clean water.
8. Run the engine for 5 minutes, then drain the water from the cooling system.
9. Properly fill the engine cooling system with a mixture of water and the recommended type of antifreeze.

Brake Master Cylinder

FLUID RECOMMENDATIONS

The brake master cylinder requires brake fluid that meets DOT 3 standards. DO NOT use DOT 5 silicone fluid or any fluid that contains a mineral or paraffin base oil, or system damage could occur.

LEVEL CHECK

▶ See Figures 165, 166, 167, 168 and 169

The brake master cylinder is equipped with a translucent reservoir which enables fluid level checking without removing the reservoir cap. The brake fluid level should be between the **MIN** and **MAX** lines located on the side of the reservoir. The **MAX** line is at the base of the filler cap neck.

If it is necessary to add fluid, first wipe away any accumulated dirt or grease from the reservoir and cap. Then remove the reservoir cap by twisting it counterclockwise. Add fluid to the proper level. Avoid spilling brake fluid on any painted surface, as it will harm the finish. Replace the reservoir cap.

Fig. 162 Coolant drain plug located at the lower right side of the radiator

Fig. 163 Loosen the engine drain plug located under the right front of the engine block

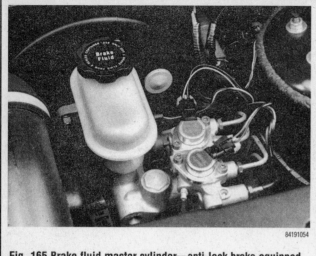

Fig. 165 Brake fluid master cylinder—anti-lock brake equipped model shown

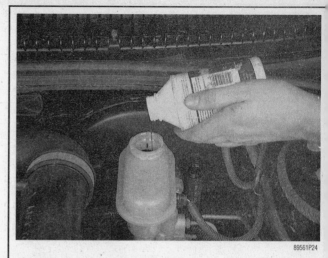

Fig. 168 Add brake fluid to the proper level. Be careful not to spill any fluid on painted surfaces

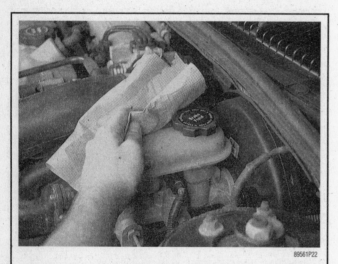

Fig. 166 With a paper towel or rag, wipe away any accumulated dirt or grease from the reservoir and cap

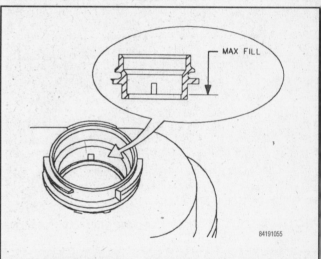

Fig. 169 The brake master cylinder MAX fill line is at the base of the filler cap neck

Clutch Master Cylinder

FLUID RECOMMENDATIONS

The clutch master cylinder requires brake fluid that meets DOT 3 standards. DO NOT use DOT 5 silicone fluid or any fluid that contains a mineral or paraffin base oil, or system damage could occur.

LEVEL CHECK

▶ See Figures 170 and 171

The clutch master cylinder is located beneath the brake master cylinder and slightly more toward the outside of the vehicle. The filler cap has a built-in dipstick and may be withdrawn to determine the fluid level.

If it is necessary to add fluid or withdraw the cap, first wipe away any accumulated dirt or grease from the reservoir and cap. Then remove the reservoir cap by twisting it counterclockwise. Add fluid to the proper level, as indicated on the dipstick. Avoid spilling brake fluid on any painted surface, as it will harm the finish. Replace the reservoir cap.

Fig. 167 Remove the fluid reservoir cap by turning counterclockwise

Fig. 170 Twist and remove the cap to check the clutch master cylinder fluid level

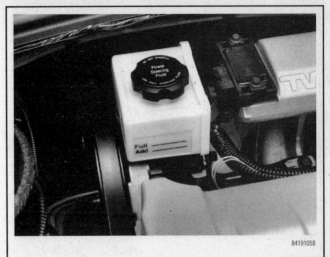

Fig. 172 Power steering fluid reservoir with FULL and ADD lines shown

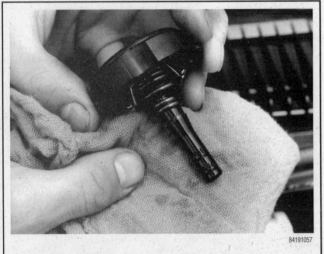

Fig. 171 The clutch master cylinder fluid level should be between the FULL and ADD lines

Fig. 173 Unscrew the power steering fluid cap from the reservoir

Power Steering Pump

FLUID RECOMMENDATIONS

GM power steering fluid specification 9985010 or equivalent must be used. Failure to use a fluid which meets these specifications may cause damage and fluid leaks.

LEVEL CHECK

▶ See Figures 172, 173 and 174

The power steering system is equipped with a translucent reservoir, mounted on the power steering pump, which enables fluid level checking without removing the reservoir cap. The power steering fluid level should be between the **FULL** and **ADD** lines located on the side of the reservoir. The fluid level markings on the reservoir match the **FULL** and **ADD** indicators on the dipstick. To check the power steering fluid level using the reservoir cap/dipstick, perform the following steps:

1. Center the front wheels to the straight-ahead position and turn the engine **OFF**.

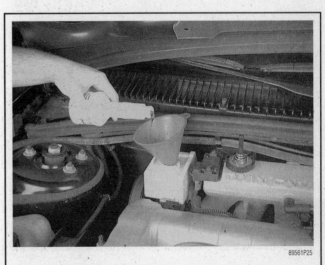

Fig. 174 Pour power steering fluid into the reservoir until it reaches the FULL line

2. Allow a sufficient amount of time for the engine to cool. Also, be sure to clean the reservoir cap and surrounding area with a rag before removing the cap.

3. Unscrew the cap from the power steering fluid reservoir and wipe the dipstick using a clean rag.

4. Place the cap back on the reservoir and tighten it completely.

5. Remove the cap from the reservoir again and check the fluid level, as indicated on the dipstick. The fluid level should read between the **FULL** and **ADD** lines on the dipstick. If necessary, add the proper fluid to reach the **FULL** mark.

Chassis Greasing

The steering and suspension joints on your car are sealed at the factory and do not require any form of periodic lubrication.

At least annually, lubricate the transaxle shift linkage with GM-6031M or an equivalent chassis grease.

Also annually, the parking brake cable guides, underbody contact points and linkage should be lubricated with GM-6031M or an equivalent chassis grease.

Body Lubrication and Maintenance

At least once a year, use a multi-purpose grease to lubricate the body door hinges, including the hood, fuel door and trunk or liftgate hinges and latches. The glove box, console doors and folding seat hardware should also be lightly lubricated. Be careful not to stain the interior fabrics with the grease.

The door and window weatherstripping should be lubricated with silicone lubricant. Flush the underbody using plain water to remove any corrosive materials picked up from the road and used for ice, snow or dust control. Make sure you thoroughly clean areas where mud and dirt may collect. If necessary, loosen sediment packed in closed areas before flushing.

To preserve the appearance of your car, it should be washed periodically with a mild soap or detergent and water solution. A liquid dishwashing detergent that DOES NOT contain any abrasives may be useful in loosening dirt or grease from the vehicle, while not endangering the finish. Only wash the vehicle when the body feels cool and the vehicle is in the shade. Rinse the entire vehicle with cold water, then wash and rinse one panel at a time, beginning with the roof and upper areas. After washing is complete, rinse the vehicle one final time and dry with a soft cloth or chamois. Air drying a vehicle, by driving at highway speeds for a few miles, may quickly and easily dry a vehicle without the need for excessive elbow grease.

TRAILER TOWING

General Recommendations

Your vehicle was primarily designed to carry passengers and cargo. It is important to remember that towing a trailer will place additional loads on your vehicle's engine, drive train, steering, braking and other systems. However, if you decide to tow a trailer, using the prior equipment is a must.

Local laws may require specific equipment such as trailer brakes or fender mounted mirrors. Check with local authorities.

Trailer Weight

The weight of the trailer is the most important factor. A good weight-to-horsepower ratio is about 35:1, that is 35 lbs. of Gross Combined Weight (GCW) for every horsepower your engine develops. Multiply the engine's rated horsepower by 35 and subtract the weight of the vehicle passengers and luggage. The number remaining is the approximate ideal maximum weight you should tow, although a numerically higher axle ratio can help compensate for heavier weight.

Hitch (Tongue) Weight

▶ **See Figure 176**

Calculate the hitch weight in order to select a proper hitch. The weight of the hitch is usually 9–11% of the trailer gross weight and should be mea-

Periodic waxing will remove harmful deposits from the vehicles surface and protect the finish. If the finish has dulled due to age or neglect, polishing may be necessary to restore the original gloss.

There are many specialized products available at your local auto parts store to care for the appearance of painted metal surfaces, plastic, chrome, wheels and tires as well as the interior upholstery and carpeting. Be sure to follow the manufacturer's instructions before using them.

Wheel Bearings

REPACKING

▶ **See Figure 175**

All Saturn wheel bearing assemblies are an integral component of the wheel hub assembly. They are sealed at the factory and do not require any form of periodic lubrication. Refer to Section 8 for information on servicing the wheel bearing/hub assembly.

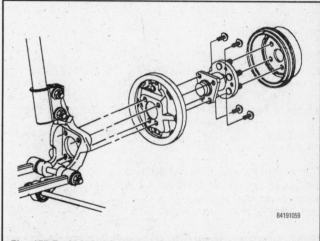

84191059

Fig. 175 Exploded view of the rear bearing/hub mounting—rear drum brake vehicles

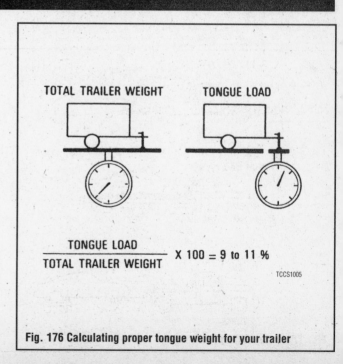

TCCS1005

Fig. 176 Calculating proper tongue weight for your trailer

sured with the trailer loaded. Hitches fall into various categories: those that mount on the frame and rear bumper, the bolt-on type, or the weld-on distribution type used for larger trailers. Axle-mounted or clamp-on bumper hitches should never be used.

Check the gross weight rating of your trailer. Tongue weight is usually figured as 10% of gross trailer weight. Therefore, a trailer with a maximum gross weight of 2000 lbs. will have a maximum tongue weight of 200 lbs. Class I trailers fall into this category. Class II trailers are those with a gross weight rating of 2000–3000 lbs., while Class III trailers fall into the 3500–6000 lb. category. Class IV trailers are those over 6000 lbs. and are for use with fifth wheel trucks, only.

When you've determined the hitch that you'll need, follow the manufacturer's installation instructions exactly, especially when it comes to fastener torques. The hitch will be subjected to a lot of stress and good hitches come with hardened bolts. Never substitute an inferior bolt for a hardened bolt.

Cooling

ENGINE

Overflow Tank

One of the most common, if not THE most common, problem associated with trailer towing is engine overheating. If you have a cooling system without an expansion tank, you'll definitely need to get an aftermarket expansion tank kit, preferably one with at least a 2 quart capacity. These kits are easily installed on the radiator's overflow hose, and come with a pressure cap designed for expansion tanks.

Oil Cooler

Aftermarket engine oil coolers are helpful for prolonging engine oil life and reducing overall engine temperatures. Both of these factors increase engine life. While not absolutely necessary in towing Class I and some Class II trailers, they are recommended for heavier Class II and all Class III

towing. Engine oil cooler systems usually consist of an adapter, screwed on in place of the oil filter, a remote filter mounting and a multi-tube, finned heat exchanger, which is mounted in front of the radiator or air conditioning condenser.

TRANSAXLE

An automatic transaxle is usually recommended for trailer towing. Modern automatics have proven reliable and, of course, easy to operate in trailer towing. The increased load of a trailer, however, causes an increase in the temperature of the automatic transaxle fluid. Heat is the worst enemy of an automatic transaxle. As the temperature of the fluid increases, the life of the fluid decreases.

It is essential, therefore, that you install an automatic transaxle cooler. The cooler, which consists of a multi-tube, finned heat exchanger, is usually installed in front of the radiator or air conditioning compressor, and hooked in-line with the transaxle cooler tank inlet line. Follow the cooler manufacturer's installation instructions.

Select a cooler of a sufficient capacity, based upon the combined gross weights of the vehicle and trailer.

Cooler manufacturers recommend that you use an aftermarket cooler in addition to, and not instead of, the present cooling tank in your radiator. If you prefer to use it in place of the radiator cooling tank, get a cooler at least two sizes larger than normally necessary.

➡**A transaxle cooler can, sometimes, cause slow or harsh shifting in the transaxle during cold weather, until the fluid has a chance to come up to normal operating temperature. Some coolers can be purchased with, or retrofitted with, a temperature bypass valve, which will allow fluid flow through the cooler only when the fluid has reached a certain operating temperature.**

Handling A Trailer

Towing a trailer with ease and safety requires a certain amount of experience. It's a good idea to learn the feel of a trailer by practicing turning, stopping and backing in an open area such as an empty parking lot.

TOWING THE VEHICLE

◆ **See Figure 177**

When towing is required, the vehicle should be flat bedded or towed with the front wheels off of the ground on a wheel lift, to prevent damage to the transaxle. DO NOT allow your vehicle to be towed by a sling type tow truck,

if it is at all avoidable. If it is necessary to tow the vehicle from the rear, a wheel dolly should be placed under the front tires.

Regardless of whether the vehicle is equipped with a manual transaxle, push starting the vehicle IS NOT RECOMMENDED under any circumstance.

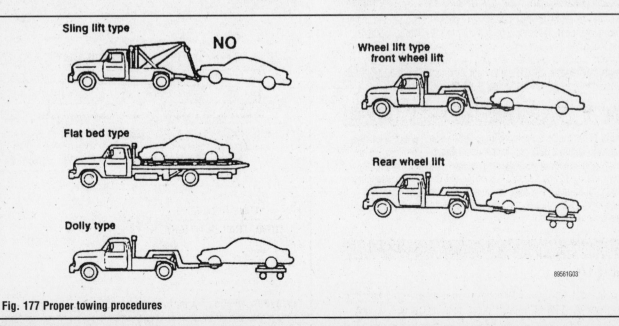

Fig. 177 Proper towing procedures

JUMP STARTING A DEAD BATTERY

▶ **See Figure 178**

Whenever a vehicle is jump started, precautions must be followed in order to prevent the possibility of personal injury. Remember that batteries contain a small amount of explosive hydrogen gas which is a by-product of battery charging. Sparks should always be avoided when working around batteries, especially when attaching jumper cables. To minimize the possibility of accidental sparks, follow the procedure carefully.

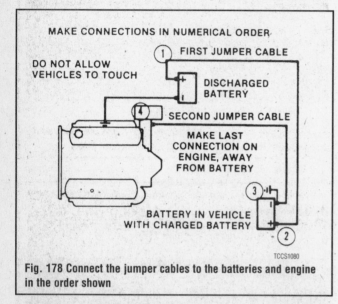

MAKE CONNECTIONS IN NUMERICAL ORDER

① FIRST JUMPER CABLE

DO NOT ALLOW VEHICLES TO TOUCH

DISCHARGED BATTERY

④ SECOND JUMPER CABLE

MAKE LAST CONNECTION ON ENGINE, AWAY FROM BATTERY

③

BATTERY IN VEHICLE WITH CHARGED BATTERY

②

TCCS1080

Fig. 178 Connect the jumper cables to the batteries and engine in the order shown

✳✳ WARNING

NEVER hook the batteries up in a series circuit or the entire electrical system will go up in smoke, including the starter!

The batteries are connected in a parallel circuit (positive terminal to positive terminal, negative terminal to engine ground). Hooking the batteries up in a parallel circuit increases battery cranking power without increasing total battery voltage output. Output remains at 12 volts. On the other hand, hooking two 12 volt batteries up in a series circuit (positive terminal to negative terminal, and other positive terminal to engine ground) increases total battery output to 24 volts (12 volts plus 12 volts).

Jump Starting Precautions

• Be sure that both batteries are of the same voltage. Vehicles covered by this manual and most vehicles on the road today utilize a 12 volt charging system.
• Be sure that both batteries are of the same polarity (have the same terminal grounded); in most cases, it is NEGATIVE.
• Be sure that the vehicles are not touching or a short circuit could occur.
• On serviceable batteries, be sure the vent cap holes are not obstructed.
• Do not smoke or allow sparks anywhere near the batteries.
• In cold weather, make sure the battery electrolyte is not frozen. This can occur more readily in a battery that has been in a state of discharge.
• Do not allow electrolyte to contact your skin or clothing.

Jump Starting Procedure

1. Make sure that the voltages of the 2 batteries are the same. Most batteries and charging systems are of the 12 volt variety.
2. Pull the jumping vehicle (with the good battery) into a position so the jumper cables can reach the discharged battery and that vehicle's engine. Make sure that the vehicles do NOT touch.
3. Place the transmissions/transaxles of both vehicles in **Neutral** (MT) or **P** (AT), as applicable, then firmly set their parking brakes.

➡**If necessary for safety reasons, the hazard lights on both vehicles may be operated throughout the entire procedure without significantly increasing the difficulty of jumping the dead battery.**

4. Turn all lights and accessories **OFF** on both vehicles. Make sure the ignition switches on both vehicles are turned to the **OFF** position.
5. Cover the battery cell caps with a rag, but do not cover the terminals.
6. Make sure the terminals on both batteries are clean and free of corrosion, or proper electrical connection will be impeded. If necessary, clean the battery terminals before proceeding.
7. Identify the positive (+) and negative (-) terminals on both batteries.
8. Connect the first jumper cable to the positive (+) terminal of the dead battery, then connect the other end of that cable to the positive (+) terminal of the booster (good) battery.
9. Connect one end of the other jumper cable to the negative (-) terminal on the booster battery and the final cable clamp to an engine bolt head, alternator bracket or other solid, metallic point on the engine with the dead battery. Try to pick a ground on the engine that is positioned away from the battery, in order to minimize the possibility of the 2 clamps touching should one loosen during the procedure. DO NOT connect this clamp to the negative (-) terminal of the bad battery.

✳✳ CAUTION

Be very careful to keep the jumper cables away from moving parts (cooling fan, belts, etc.) on both engines.

10. Check to make sure that the cables are routed away from any moving parts, then start the donor vehicle's engine. Run the engine at moderate speed for several minutes to allow the dead battery a chance to receive some initial charge.
11. With the donor vehicle's engine still running slightly above idle, try to start the vehicle with the dead battery. Crank the engine for no more than 10 seconds at a time and let the starter cool for at least 20 seconds between tries. If the vehicle does not start in 3 tries, it is likely that something else is also wrong or that the battery needs additional time to charge.
12. Once the vehicle is started, allow it to run at idle for a few seconds to make sure that it is operating properly.
13. Turn ON the headlights, heater blower and, if equipped, the rear defroster of both vehicles in order to reduce the severity of voltage spikes and subsequent risk of damage to the vehicles' electrical systems when the cables are disconnected. This step is especially important to any vehicle equipped with a computer control module.
14. Carefully disconnect the cables in the reverse order of connection. Start with the negative cable that is attached to the engine ground, then the negative cable on the donor battery. Disconnect the positive cable from the donor battery and finally, disconnect the positive cable from the formerly dead battery. Be careful when disconnecting the cables from the positive terminals not to allow the alligator clips to touch any metal on either vehicle, or a short and sparks will occur.

JACKING

▶ See Figures 179, 180, 181 and 182

Your vehicle was supplied with a jack for emergency road repairs. This jack is fine for changing a flat tire or other short term procedures not requiring you to go beneath the vehicle. If it is used in an emergency situation, carefully follow the instructions provided either with the jack or in your owner's manual. Do not attempt to use the jack on any portions of the vehicle other than specified by the vehicle manufacturer. Always block the diagonally opposite wheel when using a jack.

A more convenient way of jacking is the use of a hydraulic garage or floor jack. You may use the floor jack to support the vehicle on the side members at the front or rear, the center of the rear crossmember assembly, or the trailing engine cradle. The engine cradle has been coated with a special finish to protect it.

Never place the jack under the radiator, engine or transmission components. Severe and expensive damage will result when the jack is raised. Additionally, never jack under the floorpan or bodywork; the metal will deform.

Fig. 181 Raise the rear of the vehicle with a hydraulic jack positioned under the center member of the rear suspension

Fig. 179 Raise the front of the vehicle with a hydraulic jack and position the jackstand under the frame rail

Fig. 182 Support the rear of the vehicle with a jackstand positioned squarely under each outer bodyside pinch weld, just ahead of the rear wheels

Fig. 180 Support the front of the vehicle with a jackstand positioned squarely under each frame rail

Whenever you plan to work under the vehicle, you must support it on jackstands or ramps. Never use cinder blocks or stacks of wood to support the vehicle, even if you're only going to be under it for a few minutes. Never crawl under the vehicle when it is supported only by the tire-changing jack or other floor jack.

➡**Always position a block of wood or small rubber pad on top of the jack or jackstand to protect the lifting point's finish when lifting or supporting the vehicle.**

Small hydraulic, screw or scissors jacks are satisfactory for raising the vehicle. Drive-on trestles or ramps are also a handy and safe way to both raise and support the vehicle. Be careful though, some ramps may be too steep to drive your vehicle onto without scraping the front bottom panels. Never support the vehicle by any suspension member (unless specifically instructed to do so by a repair manual) or by an underbody panel.

Jacking Precautions

The following safety points cannot be overemphasized:
• Always block the opposite wheel or wheels to keep the vehicle from rolling off the jack.
• Before raising the front of the vehicle, firmly apply the parking brake.

• When the drive wheels are to remain on the ground, leave the vehicle in gear to help prevent it from rolling.
• Always use jackstands to support the vehicle when you are working underneath. Place the stands beneath the vehicle's jacking brackets. Before climbing underneath, rock the vehicle a bit to make sure it is firmly supported.

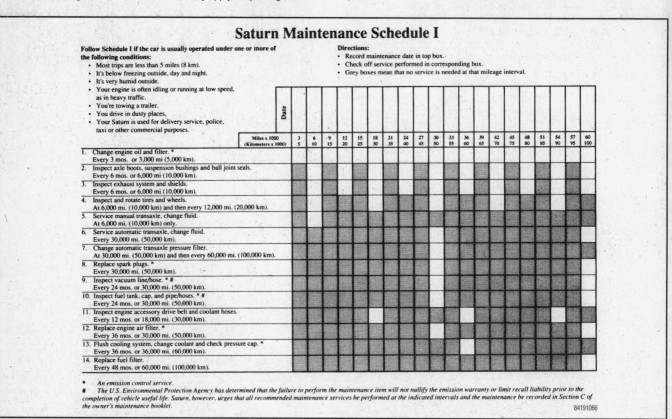

Saturn Maintenance Schedule I

Follow Schedule I if the car is usually operated under one or more of the following conditions:
• Most trips are less than 5 miles (8 km).
• It's below freezing outside, day and night.
• It's very humid outside.
• Your engine is often idling or running at low speed, as in heavy traffic.
• You're towing a trailer.
• You drive in dusty places.
• Your Saturn is used for delivery service, police, taxi or other commercial purposes.

Directions:
• Record maintenance date in top box.
• Check off service performed in corresponding box.
• Grey boxes mean that no service is needed at that mileage interval.

Miles x 1000 (Kilometers x 1000): 3/5, 6/10, 9/15, 12/20, 15/25, 18/30, 21/35, 24/40, 27/45, 30/50, 33/55, 36/60, 39/65, 42/70, 45/75, 48/80, 51/85, 54/90, 57/95, 60/100

1. Change engine oil and filter. * — Every 3 mos. or 3,000 mi (5,000 km).
2. Inspect axle boots, suspension bushings and ball joint seals. — Every 6 mos. or 6,000 mi (10,000 km).
3. Inspect exhaust system and shields. — Every 6 mos. or 6,000 mi (10,000 km).
4. Inspect and rotate tires and wheels. — At 6,000 mi. (10,000 km) and then every 12,000 mi. (20,000 km).
5. Service manual transaxle, change fluid. — At 6,000 mi. (10,000 km) only.
6. Service automatic transaxle, change fluid. — Every 30,000 mi. (50,000 km).
7. Change automatic transaxle pressure filter. — At 30,000 mi. (50,000 km) and then every 60,000 mi. (100,000 km).
8. Replace spark plugs. * — Every 30,000 mi. (50,000 km).
9. Inspect vacuum line/hose. * # — Every 24 mos. or 30,000 mi. (50,000 km).
10. Inspect fuel tank, cap, and pipe/hoses. * # — Every 24 mos. or 30,000 mi. (50,000 km).
11. Inspect engine accessory drive belt and coolant hoses. — Every 12 mos. or 18,000 mi. (30,000 km).
12. Replace engine air filter. * — Every 36 mos. or 30,000 mi. (50,000 km).
13. Flush cooling system, change coolant and check pressure cap. * — Every 36 mos. or 36,000 mi. (60,000 km).
14. Replace fuel filter. — Every 48 mos. or 60,000 mi. (100,000 km).

* An emission control service.
\# The U.S. Environmental Protection Agency has determined that the failure to perform the maintenance item will not nullify the emission warranty or limit recall liability prior to the completion of vehicle useful life. Saturn, however, urges that all recommended maintenance services be performed at the indicated intervals and the maintenance be recorded in Section C of the owner's maintenance booklet.

84191066

Saturn Maintenance Schedule II

Schedule II should only be followed if none of the conditions for Schedule I apply.
• If trips are more than 5 miles (8 km), and more than half of these miles include non-stop highway driving.

Directions:
• Record maintenance date in top box.
• Check off service performed in corresponding box.
• Grey boxes mean that no service is needed at that mileage interval.

Miles x 1000 (Kilometers x 1000): 3/5, 6/10, 9/15, 12/20, 15/25, 18/30, 21/35, 24/40, 27/45, 30/50, 33/55, 36/60, 39/65, 42/70, 45/75, 48/80, 51/85, 54/90, 57/95, 60/100

1. Change engine oil and filter. * — Every 6 mos. or 6,000 mi (10,000 km).
2. Inspect axle boots, suspension bushings and ball joint seals. — Every 6 mos. or 6,000 mi (10,000 km).
3. Inspect exhaust system and shields. — Every 6 mos. or 6,000 mi (10,000 km).
4. Inspect and rotate tires and wheels. — At 6,000 mi. (10,000 km) and then every 12,000 mi. (20,000 km).
5. Service manual transaxle, change fluid. — At 6,000 mi. (10,000 km) only.
6. Service automatic transaxle, change fluid. — Every 30,000 mi. (50,000 km).
7. Change automatic transaxle pressure filter. — At 30,000 mi. (50,000 km) and then every 60,000 mi. (100,000 km).
8. Replace spark plugs. * — Every 30,000 mi. (50,000 km).
9. Inspect vacuum line/hose. * # — Every 24 mos. or 30,000 mi. (50,000 km).
10. Inspect fuel tank, cap, and pipe/hoses. * # — Every 24 mos. or 30,000 mi. (50,000 km).
11. Inspect engine accessory drive belt and coolant hoses. — Every 12 mos. or 18,000 mi. (30,000 km).
12. Replace engine air filter. * — Every 36 mos. or 30,000 mi. (50,000 km).
13. Flush cooling system, change coolant and check pressure cap. * — Every 36 mos. or 36,000 mi. (60,000 km).
14. Replace fuel filter. — Every 48 mos. or 60,000 mi. (100,000 km).

* An emission control service.
\# The U.S. Environmental Protection Agency has determined that the failure to perform the maintenance item will not nullify the emission warranty or limit recall liability prior to the completion of vehicle useful life. Saturn, however, urges that all recommended maintenance services be performed at the indicated intervals and the maintenance be recorded in Section C of the owner's maintenance booklet.

84191067

CAPACITIES

Year	Model	Engine ID/VIN	Engine Displacement Liters (cc)	Engine Oil with Filter	Transmission (pts.)			Drive Axle (pts.)	Fuel Tank (gal.)	Cooling System (qts.)
					4-Spd	5-Spd	Auto.			
1991	Sedan	7	1.9 (1901)	4.0	—	5.2	8.4 ①	—	12.8	7.0
	Sedan	9	1.9 (1901)	4.0	—	5.2	8.4 ①	—	12.8	7.0
	Coupe	7	1.9 (1901)	4.0	—	5.2	8.4 ①	—	12.8	7.0
1992	Sedan	7	1.9 (1901)	4.0	—	5.2	8.4 ①	—	12.8	7.0
	Sedan	9	1.9 (1901)	4.0	—	5.2	8.4 ①	—	12.8	7.0
	Coupe	7	1.9 (1901)	4.0	—	5.2	8.4 ①	—	12.8	7.0
1993	Wagon	7	1.9 (1901)	4.0	—	5.2	8.4 ①	—	12.8	7.0
	Wagon	9	1.9 (1901)	4.0	—	5.2	8.4 ①	—	12.8	7.0
	Sedan	7	1.9 (1901)	4.0	—	5.2	8.4 ①	—	12.8	7.0
	Sedan	9	1.9 (1901)	4.0	—	5.2	8.4 ①	—	12.8	7.0
	Coupe	7	1.9 (1901)	4.0	—	5.2	8.4 ①	—	12.8	7.0
	Coupe	9	1.9 (1901)	4.0	—	5.2	8.4 ①	—	12.8	7.0
1994	Wagon	7	1.9 (1901)	4.0	—	5.2	8.4 ①	—	12.8	7.0
	Wagon	9	1.9 (1901)	4.0	—	5.2	8.4 ①	—	12.8	7.0
	Sedan	7	1.9 (1901)	4.0	—	5.2	8.4 ①	—	12.8	7.0
	Sedan	9	1.9 (1901)	4.0	—	5.2	8.4 ①	—	12.8	7.0
	Coupe	7	1.9 (1901)	4.0	—	5.2	8.4 ①	—	12.8	7.0
	Coupe	9	1.9 (1901)	4.0	—	5.2	8.4 ①	—	12.8	7.0
1995	Wagon	7	1.9 (1901)	4.0	—	5.2	8.4 ①	—	12.8	7.0
	Wagon	8	1.9 (1901)	4.0	—	5.2	8.4 ①	—	12.8	7.0
	Sedan	7	1.9 (1901)	4.0	—	5.2	8.4 ①	—	12.8	7.0
	Sedan	8	1.9 (1901)	4.0	—	5.2	8.4 ①	—	12.8	7.0
	Coupe	7	1.9 (1901)	4.0	—	5.2	8.4 ①	—	12.8	7.0
	Coupe	8	1.9 (1901)	4.0	—	5.2	8.4 ①	—	12.8	7.0
1996	Wagon	7	1.9 (1901)	4.0	—	5.2	8.4 ①	—	12.8	7.0
	Wagon	8	1.9 (1901)	4.0	—	5.2	8.4 ①	—	12.8	7.0
	Sedan	7	1.9 (1901)	4.0	—	5.2	8.4 ①	—	12.8	7.0
	Sedan	8	1.9 (1901)	4.0	—	5.2	8.4 ①	—	12.8	7.0
	Coupe	7	1.9 (1901)	4.0	—	5.2	8.4 ①	—	12.8	7.0
	Coupe	8	1.9 (1901)	4.0	—	5.2	8.4 ①	—	12.8	7.0
1997	Wagon	7	1.9 (1901)	4.0	—	5.2	8.4 ①	—	12.2	7.0
	Wagon	8	1.9 (1901)	4.0	—	5.2	8.4 ①	—	12.2	7.0
	Sedan	7	1.9 (1901)	4.0	—	5.2	8.4 ①	—	12.2	7.0
	Sedan	8	1.9 (1901)	4.0	—	5.2	8.4 ①	—	12.2	7.0
	Coupe	7	1.9 (1901)	4.0	—	5.2	8.4 ①	—	12.2	7.0
	Coupe	8	1.9 (1901)	4.0	—	5.2	8.4 ①	—	12.2	7.0
1998	Wagon	7	1.9 (1901)	4.0	—	5.2	8.4 ①	—	12.1	7.0
	Wagon	8	1.9 (1901)	4.0	—	5.2	8.4 ①	—	12.1	7.0
	Sedan	7	1.9 (1901)	4.0	—	5.2	8.4 ①	—	12.1	7.0
	Sedan	8	1.9 (1901)	4.0	—	5.2	8.4 ①	—	12.1	7.0
	Coupe	7	1.9 (1901)	4.0	—	5.2	8.4 ①	—	12.1	7.0
	Coupe	8	1.9 (1901)	4.0	—	5.2	8.4 ①	—	12.1	7.0

89561C04

① Overhaul capacity 14.8 pts.

ENGLISH TO METRIC CONVERSION: MASS (WEIGHT)

Current **mass** measurement is expressed in pounds and ounces (lbs. & ozs.). The metric unit of mass (or weight) is the kilogram (kg). Even although this table does not show conversion of masses (weights) larger than 15 lbs, it is easy to calculate larger units by following the data immediately below.

To convert ounces (oz.) to grams (g): multiply th number of ozs. by 28
To convert grams (g) to ounces (oz.): multiply the number of grams by .035

To convert pounds (lbs.) to kilograms (kg): multiply the number of lbs. by .45
To convert kilograms (kg) to pounds (lbs.): multiply the number of kilograms by 2.2

lbs	kg	lbs	kg	oz	kg	oz	kg
0.1	0.04	0.9	0.41	0.1	0.003	0.9	0.024
0.2	0.09	1	0.4	0.2	0.005	1	0.03
0.3	0.14	2	0.9	0.3	0.008	2	0.06
0.4	0.18	3	1.4	0.4	0.011	3	0.08
0.5	0.23	4	1.8	0.5	0.014	4	0.11
0.6	0.27	5	2.3	0.6	0.017	5	0.14
0.7	0.32	10	4.5	0.7	0.020	10	0.28
0.8	0.36	15	6.8	0.8	0.023	15	0.42

ENGLISH TO METRIC CONVERSION: TEMPERATURE

To convert Fahrenheit (°F) to Celsius (°C): take number of °F and subtract 32; multiply result by 5; divide result by 9

To convert Celsius (°C) to Fahrenheit (°F): take number of °C and multiply by 9; divide result by 5; add 32 to total

Fahrenheit (F)		Celsius (C)		Fahrenheit (F)		Celsius (C)		Fahrenheit (F)		Celsius (C)	
°F	°C	°C	°F	°F	°C	°C	°F	°F	°C	°C	°F
−40	−40	−38	−36.4	80	26.7	18	64.4	215	101.7	80	176
−35	−37.2	−36	−32.8	85	29.4	20	68	220	104.4	85	185
−30	−34.4	−34	−29.2	90	32.2	22	71.6	225	107.2	90	194
−25	−31.7	−32	−25.6	95	35.0	24	75.2	230	110.0	95	202
−20	−28.9	−30	−22	100	37.8	26	78.8	235	112.8	100	212
−15	−26.1	−28	−18.4	105	40.6	28	82.4	240	115.6	105	221
−10	−23.3	−26	−14.8	110	43.3	30	86	245	118.3	110	230
−5	−20.6	−24	−11.2	115	46.1	32	89.6	250	121.1	115	239
0	−17.8	−22	−7.6	120	48.9	34	93.2	255	123.9	120	248
1	−17.2	−20	−4	125	51.7	36	96.8	260	126.6	125	257
2	−16.7	−18	−0.4	130	54.4	38	100.4	265	129.4	130	266
3	−16.1	−16	3.2	135	57.2	40	104	270	132.2	135	275
4	−15.6	−14	6.8	140	60.0	42	107.6	275	135.0	140	284
5	−15.0	−12	10.4	145	62.8	44	112.2	280	137.8	145	293
10	−12.2	−10	14	150	65.6	46	114.8	285	140.6	150	302
15	−9.4	−8	17.6	155	68.3	48	118.4	290	143.3	155	311
20	−6.7	−6	21.2	160	71.1	50	122	295	146.1	160	320
25	−3.9	−4	24.8	165	73.9	52	125.6	300	148.9	165	329
30	−1.1	−2	28.4	170	76.7	54	129.2	305	151.7	170	338
35	1.7	0	32	175	79.4	56	132.8	310	154.4	175	347
40	4.4	2	35.6	180	82.2	58	136.4	315	157.2	180	356
45	7.2	4	39.2	185	85.0	60	140	320	160.0	185	365
50	10.0	6	42.8	190	87.8	62	143.6	325	162.8	190	374
55	12.8	8	46.4	195	90.6	64	147.2	330	165.6	195	383
60	15.6	10	50	200	93.3	66	150.8	335	168.3	200	392
65	18.3	12	53.6	205	96.1	68	154.4	340	171.1	205	401
70	21.1	14	57.2	210	98.9	70	158	345	173.9	210	410
75	23.9	16	60.8	212	100.0	75	167	350	176.7	215	414

TCCS1C01

ENGLISH TO METRIC CONVERSION: LENGTH

To convert inches (ins.) to millimeters (mm): multiply number of inches by 25.4

To convert millimeters (mm) to inches (ins.): multiply number of millimeters by .04

Inches		Decimals	Milli-meters	Inches to millimeters inches	mm	Inches		Decimals	Milli-meters	Inches to millimeters inches	mm
	1/64	0.051625	0.3969	0.0001	0.00254		33/64	0.515625	13.0969	0.6	15.24
	1/32	0.03125	0.7937	0.0002	0.00508	17/32		0.53125	13.4937	0.7	17.78
	3/64	0.046875	1.1906	0.0003	0.00762		35/64	0.546875	13.8906	0.8	20.32
1/16		0.0625	1.5875	0.0004	0.01016	9/16		0.5625	14.2875	0.9	22.86
	5/64	0.078125	1.9844	0.0005	0.01270		37/64	0.578125	14.6844	1	25.4
	3/32	0.09375	2.3812	0.0006	0.01524	19/32		0.59375	15.0812	2	50.8
	7/64	0.109375	2.7781	0.0007	0.01778		39/64	0.609375	15.4781	3	76.2
1/8		0.125	3.1750	0.0008	0.02032	5/8		0.625	15.8750	4	101.6
	9/64	0.140625	3.5719	0.0009	0.02286		41/64	0.640625	16.2719	.5	127.0
	5/32	0.15625	3.9687	0.001	0.0254	21/32		0.65625	16.6687	6	152.4
	11/64	0.171875	4.3656	0.002	0.0508		43/64	0.671875	17.0656	7	177.8
3/16		0.1875	4.7625	0.003	0.0762	11/16		0.6875	17.4625	8	203.2
	13/64	0.203125	5.1594	0.004	0.1016		45/64	0.703125	17.8594	9	228.6
	7/32	0.21875	5.5562	0.005	0.1270	23/32		0.71875	18.2562	10	254.0
	15/64	0.234375	5.9531	0.006	0.1524		47/64	0.734375	18.6531	11	279.4
1/4		0.25	6.3500	0.007	0.1778	3/4		0.75	19.0500	12	304.8
	17/64	0.265625	6.7469	0.008	0.2032		49/64	0.765625	19.4469	13	330.2
	9/32	0.28125	7.1437	0.009	0.2286	25/32		0.78125	19.8437	14	355.6
	19/64	0.296875	7.5406	0.01	0.254		51/64	0.796875	20.2406	15	381.0
5/16		0.3125	7.9375	0.02	0.508	13/16		0.8125	20.6375	16	406.4
	21/64	0.328125	8.3344	0.03	0.762		53/64	0.828125	21.0344	17	431.8
	11/32	0.34375	8.7312	0.04	1.016	27/32		0.84375	21.4312	18	457.2
	23/64	0.359375	9.1281	.05	1.270		55/64	0.859375	21.8281	19	482.6
3/8		0.375	9.5250	0.06	1.524	7/8		0.875	22.2250	20	508.0
	25/64	0.390625	9.9219	0.07	1.778		57/64	0.890625	22.6219	21	533.4
	13/32	0.40625	10.3187	0.08	2.032	29/32		0.90625	23.0187	22	558.8
	27/64	0.421875	10.7156	0.09	2.286		59/64	0.921875	23.4156	23	584.2
7/16		0.4375	11.1125	0.1	2.54	15/16		0.9375	23.8125	24	609.6
	29/64	0.453125	11.5094	0.2	5.08		61/64	0.953125	24.2094	25	635.0
	15/32	0.46875	11.9062	0.3	7.62	31/32		0.96875	24.6062	26	660.4
	31/64	0.484375	12.3031	0.4	10.16		63/64	0.984375	25.0031	27	690.6
1/2		0.5	12.7000	0.5	12.70						

ENGLISH TO METRIC CONVERSION: TORQUE

To convert foot-pounds (ft. lbs.) to Newton-meters: multiply the number of ft. lbs. by 1.3

To convert inch-pounds (in. lbs.) to Newton-meters: multiply the number of in. lbs. by .11

in lbs	N·m	in lbs	N·m	in lbs	N·m	in lbs	N·m	in lbs	N·m	in lbs	N·m
0.1	0.01	1	0.11	10	1.13	19	2.15	28	3.16		
0.2	0.02	2	0.23	11	1.24	20	2.26	29	3.28		
0.3	0.03	3	0.34	12	1.36	21	2.37	30	3.39		
0.4	0.04	4	0.45	13	1.47	22	2.49	31	3.50		
0.5	0.06	5	0.56	14	1.58	23	2.60	32	3.62		
0.6	0.07	6	0.68	15	1.70	24	2.71	33	3.73		
0.7	0.08	7	0.78	16	1.81	25	2.82	34	3.84		
0.8	0.09	8	0.90	17	1.92	26	2.94	35	3.95		
0.9	0.10	9	1.02	18	2.03	27	3.05	36	4.0		

ENGLISH TO METRIC CONVERSION: TORQUE

Torque is now expressed as either foot-pounds (ft./lbs.) or inch-pounds (in./lbs.). The metric measurement unit for torque is the Newton-meter (Nm). This unit—the Nm—will be used for all SI metric torque references, both the present ft./lbs. and in./lbs.

ft lbs	N-m	ft lbs	N-m	ft lbs	N-m	ft lbs	N-m
0.1	0.1	33	44.7	74	100.3	115	155.9
0.2	0.3	34	46.1	75	101.7	116	157.3
0.3	0.4	35	47.4	76	103.0	117	158.6
0.4	0.5	36	48.8	77	104.4	118	160.0
0.5	0.7	37	50.7	78	105.8	119	161.3
0.6	0.8	38	51.5	79	107.1	120	162.7
0.7	1.0	39	52.9	80	108.5	121	164.0
0.8	1.1	40	54.2	81	109.8	122	165.4
0.9	1.2	41	55.6	82	111.2	123	166.8
1	1.3	42	56.9	83	112.5	124	168.1
2	2.7	43	58.3	84	113.9	125	169.5
3	4.1	44	59.7	85	115.2	126	170.8
4	5.4	45	61.0	86	116.6	127	172.2
5	6.8	46	62.4	87	118.0	128	173.5
6	8.1	47	63.7	88	119.3	129	174.9
7	9.5	48	65.1	89	120.7	130	176.2
8	10.8	49	66.4	90	122.0	131	177.6
9	12.2	50	67.8	91	123.4	132	179.0
10	13.6	51	69.2	92	124.7	133	180.3
11	14.9	52	70.5	93	126.1	134	181.7
12	16.3	53	71.9	94	127.4	135	183.0
13	17.6	54	73.2	95	128.8	136	184.4
14	18.9	55	74.6	96	130.2	137	185.7
15	20.3	56	75.9	97	131.5	138	187.1
16	21.7	57	77.3	98	132.9	139	188.5
17	23.0	58	78.6	99	134.2	140	189.8
18	24.4	59	80.0	100	135.6	141	191.2
19	25.8	60	81.4	101	136.9	142	192.5
20	27.1	61	82.7	102	138.3	143	193.9
21	28.5	62	84.1	103	139.6	144	195.2
22	29.8	63	85.4	104	141.0	145	196.6
23	31.2	64	86.8	105	142.4	146	198.0
24	32.5	65	88.1	106	143.7	147	199.3
25	33.9	66	89.5	107	145.1	148	200.7
26	35.2	67	90.8	108	146.4	149	202.0
27	36.6	68	92.2	109	147.8	150	203.4
28	38.0	69	93.6	110	149.1	151	204.7
29	39.3	70	94.9	111	150.5	152	206.1
30	40.7	71	96.3	112	151.8	153	207.4
31	42.0	72	97.6	113	153.2	154	208.8
32	43.4	73	99.0	114	154.6	155	210.2

TCCS1C03

ENGLISH TO METRIC CONVERSION: FORCE

Force is presently measured in pounds (lbs.). This type of measurement is used to measure spring pressure, specifically how many pounds it takes to compress a spring. Our present force unit (the pound) will be replaced in SI metric measurements by the Newton (N). This term will eventually see use in specifications for electric motor brush spring pressures, valve spring pressures, etc.

To convert pounds (lbs.) to Newton (N): multiply the number of lbs. by 4.45

lbs	N	lbs	N	lbs	N	oz	N
0.01	0.04	21	93.4	59	262.4	1	0.3
0.02	0.09	22	97.9	60	266.9	2	0.6
0.03	0.13	23	102.3	61	271.3	3	0.8
0.04	0.18	24	106.8	62	275.8	4	1.1
0.05	0.22	25	111.2	63	280.2	5	1.4
0.06	0.27	26	115.6	64	284.6	6	1.7
0.07	0.31	27	120.1	65	289.1	7	2.0
0.08	0.36	28	124.6	66	293.6	8	2.2
0.09	0.40	29	129.0	67	298.0	9	2.5
0.1	0.4	30	133.4	68	302.5	10	2.8
0.2	0.9	31	137.9	69	306.9	11	3.1
0.3	1.3	32	142.3	70	311.4	12	3.3
0.4	1.8	33	146.8	71	315.8	13	3.6
0.5	2.2	34	151.2	72	320.3	14	3.9
0.6	2.7	35	155.7	73	324.7	15	4.2
0.7	3.1	36	160.1	74	329.2	16	4.4
0.8	3.6	37	164.6	75	333.6	17	4.7
0.9	4.0	38	169.0	76	338.1	18	5.0
1	4.4	39	173.5	77	342.5	19	5.3
2	8.9	40	177.9	78	347.0	20	5.6
3	13.4	41	182.4	79	351.4	21	5.8
4	17.8	42	186.8	80	355.9	22	6.1
5	22.2	43	191.3	81	360.3	23	6.4
6	26.7	44	195.7	82	364.8	24	6.7
7	31.1	45	200.2	83	369.2	25	7.0
8	35.6	46	204.6	84	373.6	26	7.2
9	40.0	47	209.1	85	378.1	27	7.5
10	44.5	48	213.5	86	382.6	28	7.8
11	48.9	49	218.0	87	387.0	29	8.1
12	53.4	50	224.4	88	391.4	30	8.3
13	57.8	51	226.9	89	395.9	31	8.6
14	62.3	52	231.3	90	400.3	32	8.9
15	66.7	53	235.8	91	404.8	33	9.2
16	71.2	54	240.2	92	409.2	34	9.4
17	75.6	55	244.6	93	413.7	35	9.7
18	80.1	56	249.1	94	418.1	36	10.0
19	84.5	57	253.6	95	422.6	37	10.3
20	89.0	58	258.0	96	427.0	38	10.6

TCCS1C04

ENGLISH TO METRIC CONVERSION: LIQUID CAPACITY

Liquid or fluid capacity is presently expressed as pints, quarts or gallons, or a combination of all of these. In the metric system the liter (l) will become the basic unit. Fractions of a liter would be expressed as deciliters, centiliters, or most frequently (and commonly) as milliliters.

To convert pints (pts.) to liters (l): multiply the number of pints by .47
To convert liters (l) to pints (pts.): multiply the number of liters by 2.1
To convert quarts (qts.) to liters (l): multiply the number of quarts by .95

To convert liters (l) to quarts (qts.): multiply the number of liters by 1.06
To convert gallons (gals.) to liters (l): multiply the number of gallons by 3.8
To convert liters (l) to gallons (gals.): multiply the number of liters by .26

gals	liters	qts	liters	pts	liters
0.1	0.38	0.1	0.10	0.1	0.05
0.2	0.76	0.2	0.19	0.2	0.10
0.3	1.1	0.3	0.28	0.3	0.14
0.4	1.5	0.4	0.38	0.4	0.19
0.5	1.9	0.5	0.47	0.5	0.24
0.6	2.3	0.6	0.57	0.6	0.28
0.7	2.6	0.7	0.66	0.7	0.33
0.8	3.0	0.8	0.76	0.8	0.38
0.9	3.4	0.9	0.85	0.9	0.43
1	3.8	1	1.0	1	0.5
2	7.6	2	1.9	2	1.0
3	11.4	3	2.8	3	1.4
4	15.1	4	3.8	4	1.9
5	18.9	5	4.7	5	2.4
6	22.7	6	5.7	6	2.8
7	26.5	7	6.6	7	3.3
8	30.3	8	7.6	8	3.8
9	34.1	9	8.5	9	4.3
10	37.8	10	9.5	10	4.7
11	41.6	11	10.4	11	5.2
12	45.4	12	11.4	12	5.7
13	49.2	13	12.3	13	6.2
14	53.0	14	13.2	14	6.6
15	56.8	15	14.2	15	7.1
16	60.6	16	15.1	16	7.6
17	64.3	17	16.1	17	8.0
18	68.1	18	17.0	18	8.5
19	71.9	19	18.0	19	9.0
20	75.7	20	18.9	20	9.5
21	79.5	21	19.9	21	9.9
22	83.2	22	20.8	22	10.4
23	87.0	23	21.8	23	10.9
24	90.8	24	22.7	24	11.4
25	94.6	25	23.6	25	11.8
26	98.4	26	24.6	26	12.3
27	102.2	27	25.5	27	12.8
28	106.0	28	26.5	28	13.2
29	110.0	29	27.4	29	13.7
30	113.5	30	28.4	30	14.2

TCCS1C05

ENGLISH TO METRIC CONVERSION: PRESSURE

The basic unit of pressure measurement used today is expressed as pounds per square inch (psi). The metric unit for psi will be the kilopascal (kPa). This will apply to either fluid pressure or air pressure, and will be frequently seen in tire pressure readings, oil pressure specifications, fuel pump pressure, etc.

To convert pounds per square inch (psi) to kilopascals (kPa): multiply the number of psi by 6.89

Psi	kPa	Psi	kPa	Psi	kPa	Psi	kPa
0.1	0.7	37	255.1	82	565.4	127	875.6
0.2	1.4	38	262.0	83	572.3	128	882.5
0.3	2.1	39	268.9	84	579.2	129	889.4
0.4	2.8	40	275.8	85	586.0	130	896.3
0.5	3.4	41	282.7	86	592.9	131	903.2
0.6	4.1	42	289.6	87	599.8	132	910.1
0.7	4.8	43	296.5	88	606.7	133	917.0
0.8	5.5	44	303.4	89	613.6	134	923.9
0.9	6.2	45	310.3	90	620.5	135	930.8
1	6.9	46	317.2	91	627.4	136	937.7
2	13.8	47	324.0	92	634.3	137	944.6
3	20.7	48	331.0	93	641.2	138	951.5
4	27.6	49	337.8	94	648.1	139	958.4
5	34.5	50	344.7	95	655.0	140	965.2
6	41.4	51	351.6	96	661.9	141	972.2
7	48.3	52	358.5	97	668.8	142	979.0
8	55.2	53	365.4	98	675.7	143	985.9
9	62.1	54	372.3	99	682.6	144	992.8
10	69.0	55	379.2	100	689.5	145	999.7
11	75.8	56	386.1	101	696.4	146	1006.6
12	82.7	57	393.0	102	703.3	147	1013.5
13	89.6	58	399.9	103	710.2	148	1020.4
14	96.5	59	406.8	104	717.0	149	1027.3
15	103.4	60	413.7	105	723.9	150	1034.2
16	110.3	61	420.6	106	730.8	151	1041.1
17	117.2	62	427.5	107	737.7	152	1048.0
18	124.1	63	434.4	108	744.6	153	1054.9
19	131.0	64	441.3	109	751.5	154	1061.8
20	137.9	65	448.2	110	758.4	155	1068.7
21	144.8	66	455.0	111	765.3	156	1075.6
22	151.7	67	461.9	112	772.2	157	1082.5
23	158.6	68	468.8	113	779.1	158	1089.4
24	165.5	69	475.7	114	786.0	159	1096.3
25	172.4	70	482.6	115	792.9	160	1103.2
26	179.3	71	489.5	116	799.8	161	1110.0
27	186.2	72	496.4	117	806.7	162	1116.9
28	193.0	73	503.3	118	813.6	163	1123.8
29	200.0	74	510.2	119	820.5	164	1130.7
30	206.8	75	517.1	120	827.4	165	1137.6
31	213.7	76	524.0	121	834.3	166	1144.5
32	220.6	77	530.9	122	841.2	167	1151.4
33	227.5	78	537.8	123	848.0	168	1158.3
34	234.4	79	544.7	124	854.9	169	1165.2
35	241.3	80	551.6	125	861.8	170	1172.1
36	248.2	81	558.5	126	868.7	171	1179.0

TCCS1C06

ENGLISH TO METRIC CONVERSION: PRESSURE

The basic unit of pressure measurement used today is expressed as pounds per square inch (psi). The metric unit for psi will be the kilopascal (kPa). This will apply to either fluid pressure or air pressure, and will be frequently seen in tire pressure readings, oil pressure specifications, fuel pump pressure, etc.

To convert pounds per square inch (psi) to kilopascals (kPa): multiply the number of psi by 6.89

Psi	kPa	Psi	kPa	Psi	kPa	Psi	kPa
172	1185.9	216	1489.3	260	1792.6	304	2096.0
173	1192.8	217	1496.2	261	1799.5	305	2102.9
174	1199.7	218	1503.1	262	1806.4	306	2109.8
175	1206.6	219	1510.0	263	1813.3	307	2116.7
176	1213.5	220	1516.8	264	1820.2	308	2123.6
177	1220.4	221	1523.7	265	1827.1	309	2130.5
178	1227.3	222	1530.6	266	1834.0	310	2137.4
179	1234.2	223	1537.5	267	1840.9	311	2144.3
180	1241.0	224	1544.4	268	1847.8	312	2151.2
181	1247.9	225	1551.3	269	1854.7	313	2158.1
182	1254.8	226	1558.2	270	1861.6	314	2164.9
183	1261.7	227	1565.1	271	1868.5	315	2171.8
184	1268.6	228	1572.0	272	1875.4	316	2178.7
185	1275.5	229	1578.9	273	1882.3	317	2185.6
186	1282.4	230	1585.8	274	1889.2	318	2192.5
187	1289.3	231	1592.7	275	1896.1	319	2199.4
188	1296.2	232	1599.6	276	1903.0	320	2206.3
189	1303.1	233	1606.5	277	1909.8	321	2213.2
190	1310.0	234	1613.4	278	1916.7	322	2220.1
191	1316.9	235	1620.3	279	1923.6	323	2227.0
192	1323.8	236	1627.2	280	1930.5	324	2233.9
193	1330.7	237	1634.1	281	1937.4	325	2240.8
194	1337.6	238	1641.0	282	1944.3	326	2247.7
195	1344.5	239	1647.8	283	1951.2	327	2254.6
196	1351.4	240	1654.7	284	1958.1	328	2261.5
197	1358.3	241	1661.6	285	1965.0	329	2268.4
198	1365.2	242	1668.5	286	1971.9	330	2275.3
199	1372.0	243	1675.4	287	1978.8	331	2282.2
200	1378.9	244	1682.3	288	1985.7	332	2289.1
201	1385.8	245	1689.2	289	1992.6	333	2295.9
202	1392.7	246	1696.1	290	1999.5	334	2302.8
203	1399.6	247	1703.0	291	2006.4	335	2309.7
204	1406.5	248	1709.9	292	2013.3	336	2316.6
205	1413.4	249	1716.8	293	2020.2	337	2323.5
206	1420.3	250	1723.7	294	2027.1	338	2330.4
207	1427.2	251	1730.6	295	2034.0	339	2337.3
208	1434.1	252	1737.5	296	2040.8	240	2344.2
209	1441.0	253	1744.4	297	2047.7	341	2351.1
210	1447.9	254	1751.3	298	2054.6	342	2358.0
211	1454.8	255	1758.2	299	2061.5	343	2364.9
212	1461.7	256	1765.1	300	2068.4	344	2371.8
213	1468.7	257	1772.0	301	2075.3	345	2378.7
214	1475.5	258	1778.8	302	2082.2	346	2385.6
215	1482.4	259	1785.7	303	2089.1	347	2392.5

TCCS1C07

2

ENGINE ELECTRICAL

DISTRIBUTORLESS IGNITION SYSTEM

General Information

The Distributorless Ignition System (DIS) for all Saturn engines is an electronic system designed to provide spark for air/fuel combustion in response to timing commands from the Powertrain Control Module (PCM). System components include the PCM and the ignition module, which contains 2 coil packs. Each coil pack is made up of 2 spark towers. Spark plug wires deliver voltage from the towers to the spark plugs located in the cylinder head bores. The DIS module receives inputs from the Crankshaft Position Sensor (CPS) to monitor engine position and rotation. The module provides output signals, based on the CPS signal, which are used to drive the instrument cluster tachometer, and are also used by the PCM to determine engine timing.

When the ignition is switched to the **ON** (or RUN) position, battery voltage is applied to the DIS module, but no spark occurs because the CPS sensor shows no engine rotation. When the engine begins to rotate and reference signals are received, the PCM allows the DIS to fire when the first cylinder reaches Top Dead Center (TDC), the end of the compression stroke.

During normal engine operation, the PCM will control spark timing advance or retard according to various sensor inputs. This is called the Electronic Spark Timing (EST) mode. While operating in the EST mode, the PCM will vary engine timing for optimum performance. PCM control of the ignition timing will continue unless a problem occurs and the BYPASS mode is entered, during which the DIS module will determine engine timing.

If the vehicle stalls, the engine will cease rotation, thus ending CPS reference pulses. Should this occur, the DIS will discharge any charged coils and halt spark plug firing within 500 milliseconds after the last CPS reference pulse. The DIS will not fire any plugs again until engine rotation resumes.

Diagnosis and Testing

Before beginning any diagnosis and testing procedures, visually inspect the components of the ignition system and check for the following:
- Discharged battery
- Damaged or loose connections
- Damaged electrical insulation
- Poor coil and spark plug connections
- Ignition module connections
- Blown fuses
- Damaged vacuum hoses
- Damaged spark plugs

Check the spark plug wires and boots for signs of poor insulation that could cause cross firing. Make sure the battery is fully charged and that all accessories are off during diagnosis and testing. Make sure the idle speed is within specification.

You will need a good quality digital volt-ohmmeter and a spark tester to check the ignition system. A spark tester resembles a spark plug without threads and the side electrode removed. Do not attempt to use a modified spark plug.

SECONDARY SPARK TEST

When it is suspected that a spark plug may not be firing, the simplest test consists of a visual inspection as noted above. If no loose or damaged wires can be found, each coil-to-plug wire should be checked for spark.

1. Twist the boot of the spark plug wire to loosen it, then carefully pull upwards and disconnect the wire from the plug.
2. Connect the spark plug wire to Spark Tester tool SA9199Z or equivalent.
3. Connect the spark tester to a suitable engine ground.
4. Crank the engine and watch for a strong spark across the tester gap.
5. If the spark is missing or weak, check the resistance of the coil-to-plug wire; it should be less than 12,000 ohms. Replace any wire with too high a resistance and repeat the test. If the wire resistance is good, but the spark is missing or weak, proceed to the DIS MODULE SPARK TEST.
6. If no spark can be found for all of the wires, proceed to the NO SPARK TEST.

NO SPARK TEST

▶ See Figures 1 and 2

1. If no spark can be found at any spark plug or coil, you must first determine if a cranking rpm signal is present from the Crankshaft Position Sensor (CPS). This can be accomplished in various ways:

Connect a scan tool as described in Section 4 and watch for a CPS signal as the engine is cranked.

Watch the tachometer for any signs of motion as the engine is cranked.

Disengage the 6-pin connector from the DIS ignition module and connect a voltmeter to terminal **E** (the second terminal from the end with the WHITE wire) and watch for AC voltage as the engine is cranked.

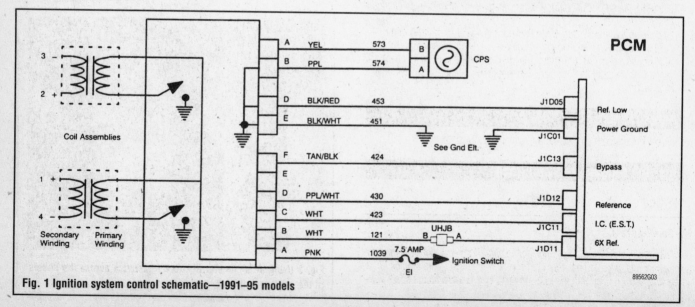

Fig. 1 Ignition system control schematic—1991–95 models

89562G03

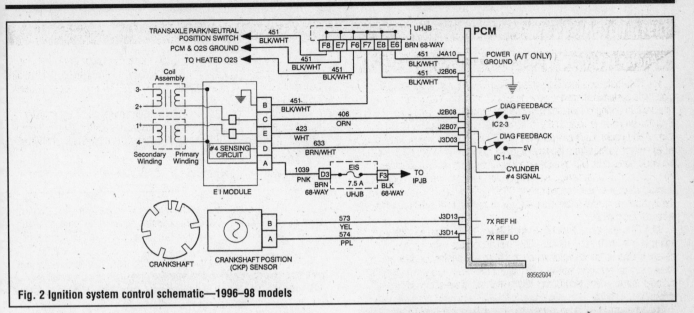

Fig. 2 Ignition system control schematic—1996–98 models

2. If it has been determined that cranking rpm is present, proceed as follows:

    a. Disengage the 6-pin connector from the DIS module, then turn the ignition key **ON** and measure the voltage to ground at terminal **A** (the ignition switch circuit). If battery voltage is not present, check for a blown fuse or an open or shorted circuit.

    b. Disengage the 5-pin connector from the DIS module, then check the resistance of terminal **E** to ground. If resistance is above 200 ohms, there is an open in the circuit. If the resistance is below 200 ohms, the problem is a loose terminal or a faulty DIS module.

3. If it has been determined that no cranking rpm is present, proceed as follows:

    a. Disengage the 5-pin connector from the DIS module, then measure the resistance between terminals **A** and **B** of the CPS. Sensor resistance should be 700–900 ohms.

    b. If CPS and circuit resistance is less than specification, remove the sensor from the engine and check the resistance directly; check to see if the sensor is still magnetized and check the continuity of the sensor wiring. Replace a sensor that does not have the proper resistance and/or is demagnetized, or repair open/shorted wires.

    c. If CPS resistance was within specification, set the voltmeter to the AC volt scale and connect the meter probes to terminals **A** and **B** of the DIS 5-pin connector. Crank the engine and watch for voltage. If while cranking, the sensor is putting out less than 200 millivolts, replace the CPS.

    d. With the ignition key **ON**, check for voltage at terminal **A** of the DIS module 6-pin connector. If no voltage is present, check for a blown fuse or an open/short in the circuit wiring. If voltage is present, check for loose terminals; if loose terminals are not the problem, the DIS module is at fault.

Adjustments

For information on all tune-up adjustments, refer to Section 1.

Ignition Module and Coil Packs

TESTING

▶ **See Figure 3**

When it is suspected that one or more spark plugs may not be firing and the spark plug wires do not seem to be at fault, this test offers a quick check to see if the ignition module and coil packs are producing spark voltage.

✳✳ CAUTION

The DIS system produces extremely high voltages. Never disconnect or touch a system component while the engine is running or the key is in the RUN position. Furthermore, it is always safest to disconnect the negative battery cable before servicing any system components.

1. Turn the ignition to the **OFF** position, then tag and disconnect all 4 spark plug wires from the ignition module towers.

2. Make sure that you are not touching the vehicle, then have an assistant crank the engine for a few seconds. Sparks should appear alternately between each pair of ignition coil towers while the engine is cranking.

➡ **Do not crank the engine for more than a few seconds at a time or starter damage may occur. Crank the engine for a few seconds, then release the ignition key and pause for an equal amount of time before cranking the vehicle again.**

3. If sparks appear between the 2 towers, the DIS module, Crankshaft Position Sensor (CPS) and the ignition coil packs are known to be good.

4. If a single coil pack is suspect from an initial spark test, or no sparks appeared between its 2 towers, switch the coil pack to the opposite position

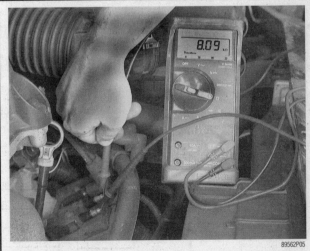

Fig. 3 Using an ohmmeter to check resistance across the towers of the ignition coil pack

on the ignition module (the coil packs are interchangeable) and check to see if the problem follows it. If the problem follows, the coil pack must be replaced. If the formerly good coil pack does not produce spark when switched with the suspect coil on the module, and the suspect coil works in the new position, then the ignition module is bad and must be replaced. Also, using an ohmmeter, check the resistance across the suspect coil's towers; it should be 7,000–10,000 ohms.

5. If no spark can be found across any tower, proceed with the NO SPARK TEST.

REMOVAL & INSTALLATION

♦ **See Figures 4, 5, 6 and 7**

1. Properly disable the SIR system, if equipped, and disconnect the negative battery cable.
2. Label and disconnect the spark plug wires from the DIS ignition module located on the front side of the bell housing.
3. Unfasten the electrical connectors from the ignition module.
4. Remove the 4 retaining bolts and the DIS unit.
5. If necessary, the coils may be removed from the unit at this time by using a pair of needlenose pliers to squeeze the retaining tabs while pulling the coils upward.

To install:

6. If removed, install the coils to the ignition module by aligning the coils with the module terminals, then carefully guiding the coils into position.
7. Run a 6 x 1.0mm tap through the module mounting holes on the bell housing to remove any remaining thread sealant residue and verify that the module and bell housing mating surfaces are clean and free from grit or dirt.
8. Always use new module mounting bolts. Install the ignition module/coil assembly using the new bolts equipped with factory applied thread sealant and tighten the bolts to 61 inch lbs. (7 Nm) using a torque wrench. Be careful when tightening the mounting bolts, and verify that each bolt head is properly seated on the module unit when tightened. If a bolt is not properly seated when the torque reaches the proper specification, remove the bolt and clean the bore threads again using the tap.
9. Attach the electrical connectors and spark plug wires to the module unit. Make sure the spark plug wires are properly connected to avoid engine damage.
10. Connect the negative battery cable and, if equipped, properly enable the SIR system.
11. Start the engine and check operation.

Crankshaft Position Sensor (CPS)

For information on servicing the Crankshaft Position Sensor, refer to Electronic Engine Controls in Section 4.

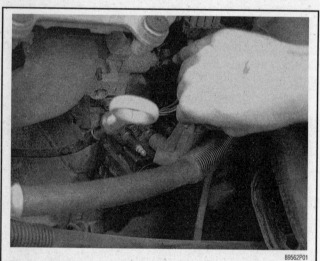

Fig. 4 Label and disconnect each spark plug wire from the ignition coil pack

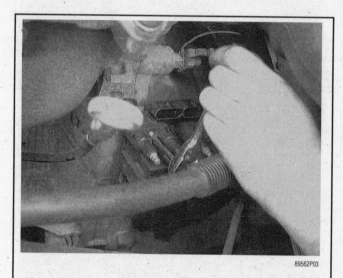

Fig. 6 Loosen and remove the ignition coil pack mounting bolts

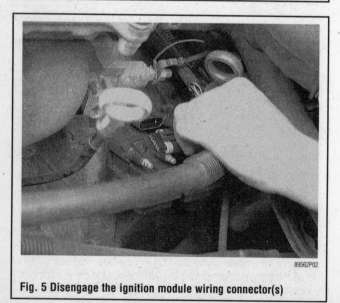

Fig. 5 Disengage the ignition module wiring connector(s)

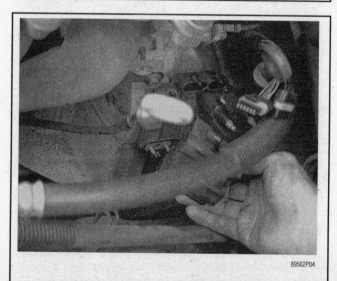

Fig. 7 Remove each ignition coil pack from the vehicle

FIRING ORDERS

▶ **See Figure 8**

To maintain the correct engine firing order, it is imperative that you label all spark plug wires before disconnecting any of them. An engine with a Distributorless Ignition System (DIS) will run poorly if the spark plug wires are replaced incorrectly.

➡ **To avoid confusion, remove and tag the spark plug wires one at a time, for replacement.**

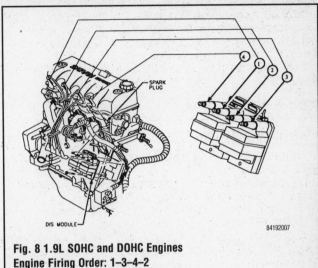

**Fig. 8 1.9L SOHC and DOHC Engines
Engine Firing Order: 1–3–4–2
Distributorless Ignition System**

84192007

CHARGING SYSTEM

General Information

The battery and charging sub-system consists of a battery and an alternator with a built-in voltage regulator. The alternator is a lightweight, high performance component that provides electrical power for charging the battery and operating the accessories while the engine is running.

The alternator is constructed essentially of a rotor mounted on bearings in two end frames, a stator assembly, six silicon diodes, an internally mounted voltage regulator, and an integral fluid intrusion shield. The alternator develops AC voltages which are converted to DC current by way of a rectifier circuit.

The charging system also includes the Powertrain Control Module (PCM) software required to provide idle boost during low voltage conditions. The battery supplies current to the starter motor to crank the engine. After the engine starts, the charging system recharges the battery and supplies power to a variety of electrical devices in the vehicle. The PCM monitors battery voltage and will boost idle speed if the voltage drops below a calibrated amount.

Alternator Precautions

To avoid damage to the alternator unit, several precautions must be observed.
• If the battery is removed for any reason, make sure it is reconnected with the correct polarity. Reversing the battery connections may result in damage to the one-way rectifiers.
• When utilizing a booster battery as a starting aid, always connect the positive terminals to each other, and the negative terminal from the booster battery to a good engine ground on the vehicle being started.
• Never use a fast charger as a booster to start vehicles.
• Disconnect the battery cables when charging the battery with a fast charger.
• Never attempt to polarize the alternator.
• Do not use test lights of more than 12 volts when checking diode continuity.
• Do not short across or ground any of the alternator terminals.
• The polarity of the battery, alternator and regulator must be matched and considered before making any electrical connections within the system.
• Never separate the alternator on an open circuit. Make sure all connections within the circuit are clean and tight.

• Disconnect the battery ground terminal when performing any service on electrical components.
• Disconnect the battery if arc welding is to be done on the vehicle.

Alternator

TESTING

Whenever troubleshooting the charging system, always check for obvious problems before proceeding, such as loose belts, bad electrical or battery connections, blown fuses, broken wires, etc. The battery must be in good condition and in the proper state of charge.

When performing the following tests, inspect the serpentine accessory drive belt to make sure it is properly tensioned. Also, make sure the ignition is **OFF** before connecting or disconnecting test equipment.

System Function Test

1. Turn the ignition key to the **OFF** position and verify that the charge lamp (which looks like a battery and is located in the left lamp grouping of the instrument cluster) is not illuminated.
2. Turn the ignition key to the **ON** position and verify that the charge lamp illuminates. If the lamp does not illuminate, check that the battery contains the proper voltage (and charge if necessary), then make sure the alternator and battery connections are clean and tight.
3. Start the engine and verify that the charge lamp extinguishes and remains unlit as long as the engine is running.

Carbon Pile Load Alternator Test

▶ **See Figure 9**

1. Make sure the battery is properly charged (on original type batteries, a green eye should be visible through the built-in hydrometer) and that all battery and alternator connections are clean and tight.
2. Install the red tester cable to the positive battery terminal and the black cable to the negative battery terminal.
3. Clamp the ammeter connection to one of the battery terminals.
4. Start the engine, then run at 2000 rpm and observe the voltmeter; if the voltage is uncontrolled or exceeds 16 volts, the alternator is bad.

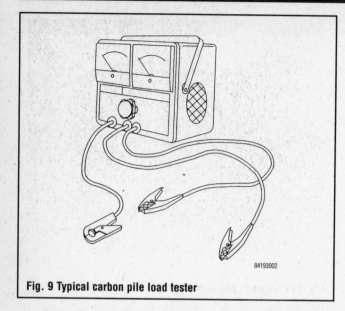

Fig. 9 Typical carbon pile load tester

5. If the voltage is below 16 volts, turn the carbon pile load tester **ON** and, while maintaining 2000 rpm, adjust the tester to obtain a maximum current reading on the ammeter. Do not allow the voltage to fall below 13 volts.

6. If the output current is greater than 60 amps or within 15 amps of the rated alternator voltage, the alternator is good.

REMOVAL & INSTALLATION

▶ **See Figures 10 thru 19**

1. Disconnect the negative battery cable.
2. Remove the serpentine drive belt.
3. Raise and support the vehicle safely.
4. Remove the right front wheel from the vehicle.
5. Remove the right wheel inner fender splash shield.
6. If equipped, remove the alternator dust shield attaching bolt and unclip the shield from the alternator. When removing the dust shield, be careful not to damage the rubber boot over the battery's positive terminal.
7. Disengage the alternator field wire connector.
8. Remove the retaining nut and wire from the positive (B+) terminal on the back of the alternator, using alternator output stud wrench SA9401C or equivalent, as a back-up wrench to prevent the stud from rotating. If the positive terminal stud rotates, the plastic insulator will crack or break, causing alternator failure.
9. Remove the upper and lower alternator attaching bolts.
10. Remove the alternator from the vehicle through the wheel well opening.

To install:

11. Position the alternator in the vehicle and install the lower attaching bolt.
12. Install the upper attaching bolt and tighten both bolts to 24 ft. lbs. (32 Nm).
13. Engage the alternator field wire connector.
14. Install the B+ terminal wire. The battery terminal wire should be placed on the back of the alternator between a 10 and 11 o'clock position.
15. Be sure to use the alternator output stud wrench SA9401C to prevent the terminal stud from rotating, then tighten the alternator positive terminal nut to 89 inch lbs. (10 Nm).
16. Install the alternator dust shield and tighten the fastener bolt to 89 inch lbs. (10 Nm).

Fig. 10 Compress the belt tensioner and disengage the serpentine belt, then remove it from the top of the engine compartment

17. Install the serpentine drive belt.
18. Install the wheel well splash shield and right front wheel. Tighten the wheel lugs in a criss-cross pattern to 103 ft. lbs. (140 Nm).
19. Remove the supports and lower the vehicle.
20. Connect the negative battery cable.

Fig. 11 After removing the front wheel, remove the inner fender splash shield

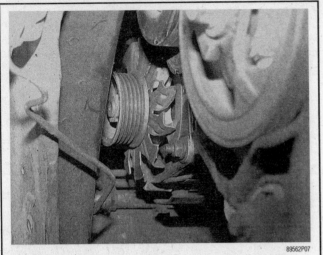

Fig. 12 View of the alternator location with the inner fender splash shield removed

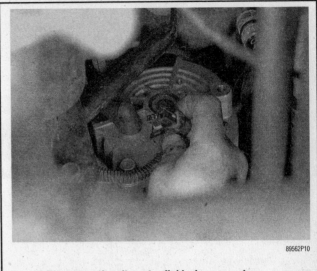

Fig. 15 Disengage the alternator field wire connector

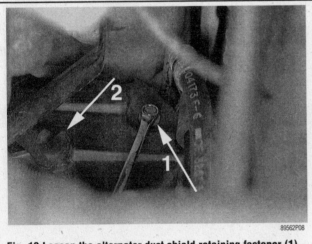

Fig. 13 Loosen the alternator dust shield retaining fastener (1), being careful not to damage the dust boot over the positive battery terminal (2)

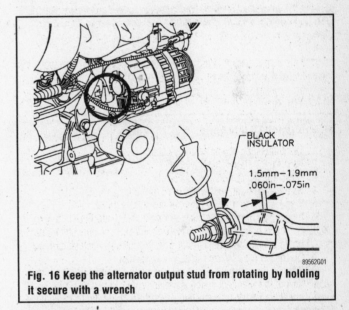

Fig. 16 Keep the alternator output stud from rotating by holding it secure with a wrench

Fig. 14 Remove the alternator dust shield from beneath the vehicle

Fig. 17 If it is more accessible, remove the upper alternator attaching bolt from the top of the engine compartment

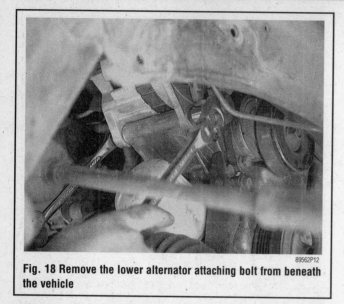

Fig. 18 Remove the lower alternator attaching bolt from beneath the vehicle

89562P12

89562P14

Fig. 19 Remove the alternator through the wheel well opening

Regulator

REMOVAL & INSTALLATION

The electronic voltage regulator is contained within the alternator and, like other internal alternator components, cannot be serviced separately.

STARTING SYSTEM

General Information

The starting system is designed to start the engine. The battery provides power to the starter motor to crank the engine when starting. The system consists of a starter motor with a solenoid, a starter relay, a transaxle mounted park/neutral switch (automatic transaxle) or clutch pedal actuated switch (manual transaxle), a crank contact in the ignition switch, and associated wiring.

With the ignition switch in the **START** position, the starter relay enabled and the transaxle in the **P** (Park) or **N** (Neutral) position, power flows to the small terminal of the starter solenoid. Power through the pull-in winding also flows through the starter motor. The motor rotates slowly at first so the gears mesh and the solenoid contacts close. This allows full battery power to flow directly to the cranking motor. With full battery voltage at the M terminal, the motor armature turns and cranks the engine.

When the ignition switch is released from the **START** position, power to the small terminal is cut off. The battery continues to supply power to the cranking motor through the pull-in and hold-in windings. The current through the windings flows in opposing directions so it creates opposite magnetic fields. The magnetic fields cancel each other out, releasing the driver mechanism. The shift lever returns to its at-rest position and the starter solenoid contacts open, cutting off power to the starter motor.

Starter

TESTING

▶ See Figure 20

Before testing the starting system, perform a preliminary visual check. Inspect the wiring and connections for the battery cable and the starter solenoid, then check the ignition switch for proper operation. When the switch is turned **ON**, the instrument cluster lights should illuminate and accessories such as the radio may be operated. Check the instrument panel junction block for blown fuses. The ignition fuse "IGN3" must be intact.

Clutch Start Safety Switch

On vehicles equipped with a manual transaxle, a clutch safety switch is mounted on the clutch pedal bracket to prevent starting the vehicle unless the pedal is depressed. If the switch is suspected, check it as follows:

1. Detach the electrical connector from the switch and position a digital volt-ohmmeter to test resistance through the switch.
2. The circuit should be open (infinite resistance) when the clutch pedal is in the clutch engaged or upper position.
3. Fully depress the clutch pedal and check resistance; if the switch is good, resistance should be below 2 ohms.

Neutral Start Safety Switch

On vehicles equipped with an automatic transaxle, a neutral safety switch is mounted on the rear side of the transaxle bell housing to prevent starting the vehicle unless the shifter is in **NEUTRAL** or **PARK**. If the switch is suspected, check it as follows:

1. Detach the electrical connector from the neutral safety switch that contains only 2-pins. These should be terminals **A** and **B**.
2. With the transaxle in **REVERSE**, **3RD** or **2ND**, measure the resistance of the switch across terminals **A** and **B**; the circuit should be open (infinite resistance).
3. With the transaxle in **PARK** or **NEUTRAL**, the resistance across the two switch terminals should be less than 2 ohms.

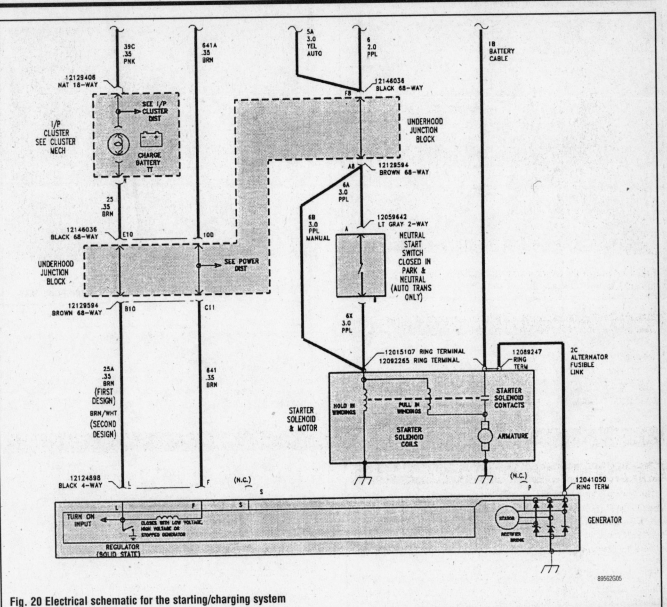

Fig. 20 Electrical schematic for the starting/charging system

REMOVAL & INSTALLATION

▶ **See Figures 21 thru 29**

1. Disconnect the negative battery cable.
2. Raise and support the vehicle safely.
3. If equipped, remove the starter shield pin by pulling on it with pliers, then lift and release the shield from the solenoid.
4. Spray the solenoid electrical connection nuts and studs with penetrating oil, and allow it to seep in for a minute to loosen the connections. While the oil is soaking in, tag the solenoid electrical connectors for identification during assembly.

➡ **It is very important that the solenoid electrical connection nuts and studs are sprayed with penetrating oil prior to removal to avoid damage to the solenoid end cap. A cracked cap will allow debris and moisture to enter and corrode the solenoid contacts.**

5. Carefully loosen the bolts and detach the starter electrical connectors. Position the wires to the side and out of the way.
6. Remove the lower starter mounting bolt.
7. Using an obstruction wrench, remove the upper starter mounting bolt. If the upper starter mounting bolt cannot be reached from below the vehicle, it may be accessible from above, through the intake manifold support bracket opening.
8. If equipped, remove the bolt attaching the starter rear support bracket to the vehicle. Rotate the starter until the bracket misses the axle shaft support bracket.
9. Carefully support the starter, pulling it rearward and toward the left side of the vehicle to remove it.

To install:

10. If equipped on the old starter, install the rear starter support bracket to the new starter. Tighten the bracket nuts to 7 ft. lbs. (9 Nm).
11. Guide the starter into the bell housing and rotate the assembly until the lower bolt hole in the starter aligns.

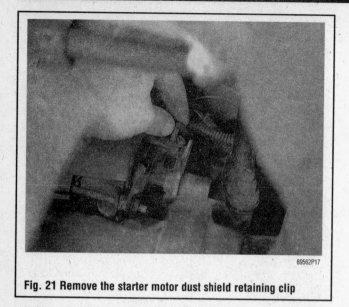

Fig. 21 Remove the starter motor dust shield retaining clip

Fig. 22 Lift off and remove the starter motor dust shield from beneath the vehicle

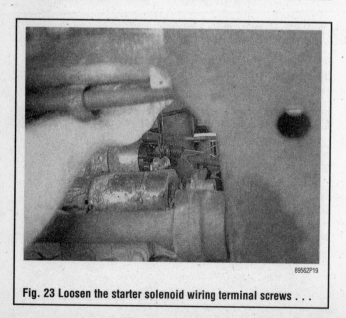

Fig. 23 Loosen the starter solenoid wiring terminal screws . . .

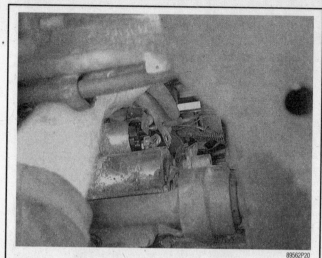

Fig. 24 . . . then label and disengage the wiring harnesses from the starter solenoid, carefully positioning them out of the way

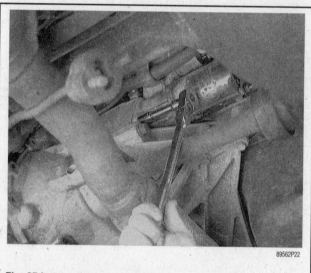

Fig. 25 Loosen the lower starter motor mounting bolt . . .

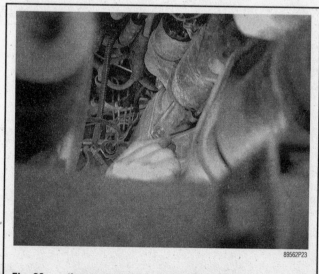

Fig. 26 . . . then remove the bolt from the vehicle

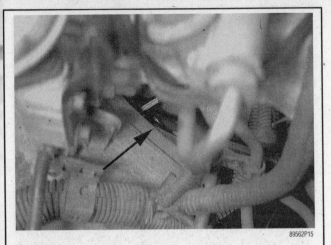

Fig. 27 If accessibility is limited from below the vehicle, use the opening in the intake manifold support bracket to reach the upper starter bolt

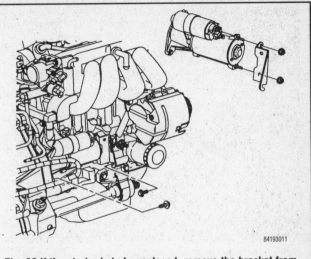

Fig. 29 If the starter is to be replaced, remove the bracket from the old starter

Fig. 28 Remove the rear mounting bracket bolt, then remove the starter from the vehicle

12. Verify that the bracket is properly aligned and loosely install the bracket bolt, then loosely install the starter mounting bolts. It may be necessary to raise or lower the vehicle for access to both the upper and lower bolts, but do not tighten any bolts until all of the mounting bolts have been started.

13. Tighten both upper and lower starter assembly attachment bolts to 27 ft. lbs. (37 Nm).

14. Reconnect the electrical wires and install the nuts. Be careful not to overtighten the nuts and crack the solenoid end cap. Route the wiring cable so it will not contact any rough surfaces. Tighten the starter positive terminal nut to 89 inch lbs. (10 Nm) and the solenoid terminal nut to 44 inch lbs. (5 Nm).

15. If equipped, install the starter shield and push pin, being careful that the pin is positioned for possible future removal.

16. Lower the vehicle.

17. Connect the negative battery cable and verify operation.

STARTER SOLENOID REPLACEMENT

The Saturn original equipment starter and solenoid are a single unit, and must be replaced as an assembly if the unit fails.

SENDING UNITS

➡ **This section describes the operating principles of sending units, warning lights and gauges. For sensors which provide information to the Powertrain Control Module (PCM),** *and function as electronic engine controls***, refer to Section 4 of this manual.**

Instrument panels contain a number of indicating devices (gauges and warning lights). These devices are composed of two separate components. One is the sending unit, mounted on the engine or other remote part of the vehicle, and the other is the actual gauge or light in the instrument panel.

Several types of sending units exist; however, most can be characterized as being either a pressure type or a resistance type. Pressure type sending units convert liquid pressure into an electrical signal which is sent to the gauge. Resistance type sending units are most often used to measure temperature and use variable resistance to control the current flow back to the indicating device. Both types of sending units are connected in series by a wire to the battery (through the ignition switch). When the ignition is turned **ON**, current flows from the battery through the indicating device and on to the sending unit.

Fuel Level Sender

OPERATION

This sending unit is a rheostat type, meaning that the resistance (or signal) changes according to the fuel level inside the tank. On some models, the Powertrain Control Module (PCM) monitors the resistance of the fuel level sending unit to calculate the amount of fuel in the tank, and supplies voltage to the sending unit through a network of resistors and a return to ground. On other models, the PCM is *not* part of the circuit.

TESTING

▶ **See Figures 30, 31, 32 and 33**

➡The fuel level sending unit can be most accurately diagnosed while still in the vehicle with a full or empty fuel tank.

1. Be sure that the ignition is turned **OFF**. Open the trunk lid or liftgate and turn back the carpet against the left side.
2. Unplug the in-line fuel connector.
3. With the ohmmeter probes placed at connector pins B and C of the sending unit side connector, check the resistance and compare it to the figures in the appropriate chart.
4. If the resistance reading(s) do not coincide with the figures in the chart, replace the fuel level sending unit.
5. Attach the in-line fuel connector and reposition the carpet.

REMOVAL & INSTALLATION

The fuel level sending unit is mounted on the fuel pump module assembly, which is located in the fuel tank. If the sending unit requires replacement, the fuel tank must be removed. Refer to Section 5 for Fuel Tank removal and installation.

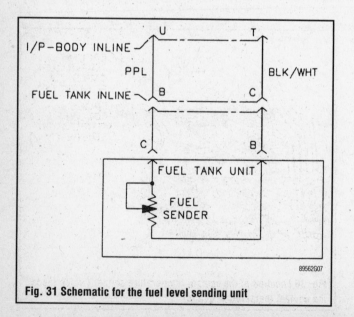

Fig. 30 Location of the in-line fuel connector

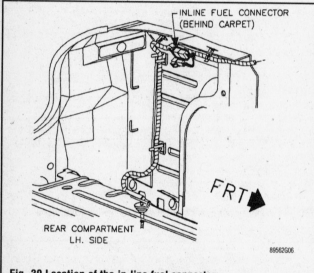

Fig. 31 Schematic for the fuel level sending unit

Level Indicator	Fuel Quantity Gallons	Fuel Quantity Liters	Ohms (Range)
Full/Full Stop	11.8	44.7	33.6-37.6
3/4	9.1	34.4	65.1-72.1
1/2	6.5	24.6	94.0-105.1
1/4	3.8	14.4	130.0-155.0
"E" Range	2.0	7.6	199.0-240.0
Empty Stop (Below "E")	1.4	2.3	245.0-253.0

89562G08

Fig. 32 Fuel level sending unit resistance chart for 1991–96 models

GAUGE READING	APPROXIMATE OHMS RESISTANCE	APPROXIMATE FUEL
<EMPTY	677 OHMS	1.0 GAL.
1/8	195 OHMS	1.6 GALS.
1/4	163 OHMS	3.1 GALS.
1/2	113 OHMS	6.2 GALS.
3/4	68 OHMS	8.8 GALS
>FULL	18 OHMS	12.2 GALS.

89562G09

Fig. 33 Fuel level sending unit resistance chart for 1997–98 models

Coolant Temperature Sender

➡On 1991–95 Saturn models, a coolant temperature sender, as well as a coolant temperature sensor, are used. Be careful not to confuse the single-wire coolant temperature sender used by the instrument cluster temperature gauge with the 2-wire *sensor* used by the PCM. On 1996–98 Saturn models, only one 2-wire sensor is used to perform both functions. For information on the Coolant Temperature Sensor, refer to Section 4.

TESTING

▶ **See Figures 34 and 35**

1. Remove the temperature sender from the engine.
2. Position the coolant temperature sender in such a way that the metal shaft (opposite end from the electrical connector) is situated in a container of water. Make sure that the electrical connector is not submerged and that only the tip of the sending unit's body is in the water.
3. Heat the container of water at a medium rate. While the water is warming, continue to measure the resistance of the terminal and the metal body of the sender:

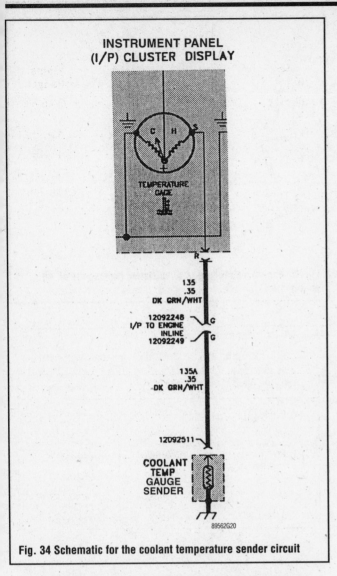

INSTRUMENT PANEL
(I/P) CLUSTER DISPLAY

Fig. 34 Schematic for the coolant temperature sender circuit

DEGREES (°C)	DEGREES (°F)	SENSOR RESISTANCE (OHMS)
−40	−40	77k – 109k
−29	−20	39k – 53k
−18	0	21k – 27k
−7	20	11k – 15k
4	40	6.6k – 8.4k
16	60	3.9k – 4.5k
27	80	2.4k – 2.7k
38	100	1.5k – 1.7k
49	120	.98k – 1.1k
60	140	650 – 730
72	160	430 – 480
83	180	302 – 334
94	200	215 – 235
105	220	159 – 172
120	248	104 – 113
140	284	63 – 68

84194026

Fig. 35 Sender resistance chart for the coolant temperature sender

a. As the water warms up, the resistance exhibited by the ohmmeter goes down in a steady manner: the temperature sender is good.

b. As the water warms up, the resistance does not change or changes in erratic jumps: the sender is bad; replace it with a new one.

4. Install the good unit or a new coolant temperature sender into the engine, then connect the negative battery cable.

REMOVAL & INSTALLATION

▶ See Figure 36

The sender is threaded into the coolant passage at the rear of the cylinder head.

1. Turn the ignition to the **OFF** position.
2. Disconnect the negative battery cable.
3. Position a clean container under the engine or radiator plug, and drain the engine coolant to a level below the sender.
4. Using your hand, a pair of pliers, or with the aid of a small prying tool, gently squeeze the sides of the sender's electrical connector, then remove the connector from the sender. Do not pull on the wire.
5. Using an appropriate sized deep well socket, remove the sender from the engine.

To install:

6. Apply a coat of Loctite® 242 or an equivalent threadlock to the coolant temperature sender threads, then install the sender and tighten to 71 inch lbs. (8 Nm).

7. Attach the sender's electrical connector. Push in until a click is heard and pull back slightly to confirm a positive engagement.
8. Connect the negative battery cable.
9. Refill the engine cooling system.

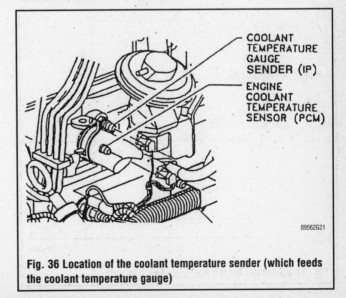

COOLANT
TEMPERATURE
GAUGE
SENDER (IP)

ENGINE
COOLANT
TEMPERATURE
SENSOR (PCM)

89562G21

Fig. 36 Location of the coolant temperature sender (which feeds the coolant temperature gauge)

Oil Pressure Sender

TESTING

1. To test the normally closed oil pressure sender, disengage the electrical connector and measure the resistance between the switch terminal (terminal for the wire to the warning lamp) and the metal housing. The ohmmeter should read 0 ohms.

2. To test the sending unit, measure the resistance between the sender terminal and the metal housing. The ohmmeter should read an open circuit (infinite resistance).

3. Start the engine.

4. Once again, test each terminal against the metal housing:

a. The oil pressure switch terminal-to-housing circuit should read an open circuit, if there is oil pressure present.

b. The sending unit-to-housing circuit should read between 15–80 ohms, depending on the engine speed, oil temperature and oil viscosity.

5. To test the oil pressure sender only, rev the engine and watch the ohms reading, which should fluctuate slightly (within the range of 15–80 ohms) as rpm increases.

6. If the above results were not obtained, replace the sending unit/switch with a new one.

REMOVAL & INSTALLATION

▶ **See Figure 37**

The oil pressure sender is located in the rear center of the block, facing the firewall. The sender is slightly below and toward the passenger's side of the vehicle from the knock sensor.

1. Disconnect the negative battery cable.
2. Detach the sender electrical connector.

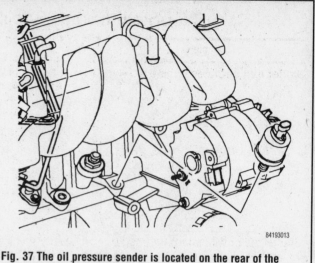

84193013

Fig. 37 The oil pressure sender is located on the rear of the engine block, between the alternator and the knock sensor

3. Using a 3 inch long wobble drive extension and a ¾ inch (19mm) crow's foot, carefully loosen and remove the sender from the engine block.

➡ **If it is too difficult to access the sender from the engine compartment, raise and support the vehicle safely, then loosen and remove the sender from underneath the vehicle.**

To install:

4. Install the oil pressure sender in the engine block and tighten to 26 ft. lbs. (35 Nm).
5. Attach the oil pressure sender's electrical connector.
6. If raised, remove the supports and carefully lower the vehicle.
7. Connect the negative battery cable.

Troubleshooting Basic Starting System Problems

Problem	Cause	Solution
Starter motor rotates engine slowly	• Battery charge low or battery defective	• Charge or replace battery
	• Defective circuit between battery and starter motor	• Clean and tighten, or replace cables
	• Low load current	• Bench-test starter motor. Inspect for worn brushes and weak brush springs.
	• High load current	• Bench-test starter motor. Check engine for friction, drag or coolant in cylinders. Check ring gear-to-pinion gear clearance.
Starter motor will not rotate engine	• Battery charge low or battery defective	• Charge or replace battery
	• Faulty solenoid	• Check solenoid ground. Repair or replace as necessary.
	• Damaged drive pinion gear or ring gear	• Replace damaged gear(s)
	• Starter motor engagement weak	• Bench-test starter motor
	• Starter motor rotates slowly with high load current	• Inspect drive yoke pull-down and point gap, check for worn end bushings, check ring gear clearance
	• Engine seized	• Repair engine
Starter motor drive will not engage (solenoid known to be good)	• Defective contact point assembly	• Repair or replace contact point assembly
	• Inadequate contact point assembly ground	• Repair connection at ground screw
	• Defective hold-in coil	• Replace field winding assembly
Starter motor drive will not disengage	• Starter motor loose on flywheel housing	• Tighten mounting bolts
	• Worn drive end busing	• Replace bushing
	• Damaged ring gear teeth	• Replace ring gear or driveplate
	• Drive yoke return spring broken or missing	• Replace spring
Starter motor drive disengages prematurely	• Weak drive assembly thrust spring	• Replace drive mechanism
	• Hold-in coil defective	• Replace field winding assembly
Low load current	• Worn brushes	• Replace brushes
	• Weak brush springs	• Replace springs

TCCS2C01

Troubleshooting Basic Charging System Problems

Problem	Cause	Solution
Noisy alternator	• Loose mountings • Loose drive pulley • Worn bearings • Brush noise • Internal circuits shorted (High pitched whine)	• Tighten mounting bolts • Tighten pulley • Replace alternator • Replace alternator • Replace alternator
Squeal when starting engine or accelerating	• Glazed or loose belt	• Replace or adjust belt
Indicator light remains on or ammeter indicates discharge (engine running)	• Broken belt • Broken or disconnected wires • Internal alternator problems • Defective voltage regulator	• Install belt • Repair or connect wiring • Replace alternator • Replace voltage regulator/alternator
Car light bulbs continually burn out—battery needs water continually	• Alternator/regulator overcharging	• Replace voltage regulator/alternator
Car lights flare on acceleration	• Battery low • Internal alternator/regulator problems	• Charge or replace battery • Replace alternator/regulator
Low voltage output (alternator light flickers continually or ammeter needle wanders)	• Loose or worn belt • Dirty or corroded connections • Internal alternator/regulator problems	• Replace or adjust belt • Clean or replace connections • Replace alternator/regulator

TCCS2C02

3

ENGINE AND ENGINE OVERHAUL

1.9L SOHC (1901cc) ENGINE SPECIFICATIONS

Description	English Specifications	Metric Specifications
Type	In-line 4	
Displacement	116 cu. in.	1901cc
Number of Cylinders	4	
Bore	3.2 in.	82mm
Stroke	3.5 in.	90mm
Compression Ratio	9.3:1	
Firing Order	1-3-4-2	
Cylinder Block		
Top side of main bearing bore-to-deck	7.4511 in.	189.26mm
Longitudinal deck warpage	0.0031-0.0040 in.	0.08-0.10mm
Transverse deck warpage	0.0009-0.0020 in.	0.024-0.050mm
Cylinder bore		
Diameter		
No. 1, 2 and 3	3.2280-3.2297 in.	81.990-82.035mm
No.4	3.2283-3.2301 in.	82.000-82.045mm
Re-boring limit	0.0157 in.	0.40mm
Out-of-round limit	0.0004-0.0020 in.	0.01-0.05mm
Taper limit	0.0004-0.0020 in.	0.01-0.05mm
Camshaft		
Lobe height (intake and exhaust)	0.2520-0.2556 in.	6.400-6.493mm
Journal diameter	1.7470-1.7490 in.	44.375-44.424mm
Journal-to-bearing clearance	0.0020-0.0054 in.	0.051-0.138mm
End-play	0.0028-0.0098 in.	0.070-0.250mm
Camshaft run-out	0.0020-0.0028 in.	0.05-0.07mm
Cylinder Head		
Longitudinal deck warpage	0.0028-0.0040 in.	0.07-0.10mm
Transverse deck warpage	0.0012-0.0020 in.	0.03-0.05mm
Lifter bore diameter	0.8434-0.8445 in.	21.422-21.450mm
Lifter diameter	0.8417-0.8427 in.	21.380-21.405mm
Lifter-to-bore clearance	0.0007-0.0028 in.	0.017-0.070mm
Valve guide bore diameter	0.4460-0.4480 in.	11.337-11.368mm
Valve stem diameter		
Intake	0.273-0.274 in.	6.935-6.962mm
Exhaust	0.272-0.274 in.	6.915-6.950mm
Valve stem-to-guide clearance		
Intake	0.0010-0.0040 in.	0.025-0.115mm
Exhaust	0.0015-0.0050 in.	0.037-0.135mm
Valve face angle	44.75-45.5°	45.0-45.5°
Valve seat angle	44.5-45.5°	44.5-45.5°
Valve seat width		
Intake	0.0394-0.0630 in.	1.0-1.6mm
Exhaust	0.0512-0.0750 in.	1.3-1.9mm

89563C01

1.9L SOHC (1901cc) ENGINE SPECIFICATIONS

Description	English Specifications	Metric Specifications
Cylinder Head (cont.)		
Valve spring pressure		
Minimum	196 lbs. @ 1.28 in.	870N @ 30mm
Maximum	211 lbs. @ 1.28 in.	983N @ 30mm
Valve spring free-length	1.8898-1.9134 in.	48.001-48.600mm
Rocker arm shaft diameter	0.620-0.625 in.	15.860-15.875mm
Piston and Connecting Rod		
Piston diameter	3.2266-3.2277 in.	81.956-81.984mm
Piston-to-bore clearance		
No. 1,2 and 3	0.0002-0.0028 in.	0.006-0.070mm
No. 4	0.0006-0.0028 in.	0.016-0.070mm
Piston pin diameter	0.7676-0.7677 in.	19.496-19.500mm
Piston pin-to-piston clearance	0.0001-0.0004 in.	0.002-0.011mm
Ring groove clearance		
Top ring	0.0016-0.0035 in.	0.040-0.090mm
Second ring	0.0012-0.0031 in.	0.030-0.080mm
Ring end-gap (1991-94)		
Top	0.0098-0.0236 in.	0.25-0.60mm
Second	0.0098-0.0236 in.	0.25-0.60mm
Oil control	0.0098-0.0551 in.	0.25-1.40mm
Ring end-gap (1995-98)		
Top	0.0098-0.0197 in.	0.25-0.50mm
Second	0.0098-0.0236 in.	0.25-0.60mm
Oil control	0.0098-0.0551 in.	0.25-1.40mm
Connecting rod alignment		
Twist	0.0110 in. per 3.94 in.	0.28mm per 100mm
Bend	0.0126 in. per 3.94 in.	0.32mm per 100mm
Connecting rod side-play on crankshaft	0.0065-0.0185 in.	0.165-0.470mm
Crankshaft		
Main journal diameter	2.2437-2.2444 in.	56.990-57.007mm
Main journal out-of-round limit	0.0004 in.	0.01mm
Main bearing-to-crankshaft oil clearance	0.0002-0.0024 in.	0.006-0.060mm
Connecting rod journal diameter	1.8500-1.8508 in.	46.99-47.01mm
Connecting rod journal out-of-round limit	0.0004 in.	0.01mm
Connecting rod bearing oil clearance	0.00039-0.00290 in.	0.010-0.073mm
Crankshaft run-out	0.0020 in.	0.05mm
Crankshaft end-play	0.0020-0.0098 in.	0.05-0.25mm
Oil Pump		
Tip clearance	0.006 in. max.	0.15mm max.
End-to-end clearance	0.0016-0.0050 in.	0.040-0.128 mm
Body-to-side clearance	0.006-0.011 in.	0.150-0.277mm

89563C02

1.9L DOHC (1901cc) ENGINE SPECIFICATIONS

Description	English Specifications	Metric Specifications
Type	In-line 4	
Displacement	116 cu. in.	1901cc
Number of Cylinders	4	
Bore	3.2 in.	82mm
Stroke	3.5 in.	90mm
Compression Ratio	9.5:1	
Firing Order	1-3-4-2	
Cylinder Block		
Top side of main bearing bore-to-deck	7.4511 in.	189.26mm
Longitudinal deck warpage	0.0031-0.0040 in.	0.08-0.10mm
Transverse deck warpage	0.0009-0.0020 in.	0.024-0.050mm
Cylinder bore		
Diameter		
No. 1, 2 and 3	3.2280-3.2297 in.	81.990-82.035mm
No.4	3.2283-3.2301 in.	82.000-82.045mm
Re-boring limit	0.0157 in.	0.40mm
Out-of-round limit	0.0004-0.0020 in.	0.01-0.05mm
Taper limit	0.0004-0.0020 in.	0.01-0.05mm
Camshaft		
Lobe height		
Intake	0.3510-0.3559 in.	8.91-9.04mm
Exhaust	0.3390-0.3441 in.	8.61-8.74mm
Journal diameter	1.1390-1.1406 in.	28.925-28.970mm
Journal-to-bearing clearance	0.0012-0.0050 in.	0.030-0.125mm
End-play	0.002-0.010 in.	0.05-0.25mm
Camshaft run-out	0.0020-0.0040 in.	0.05-0.10mm
Cylinder Head		
Longitudinal deck warpage	0.0028-0.0040 in.	0.07-0.10mm
Transverse deck warpage	0.0012-0.0020 in.	0.03-0.05mm
Lifter bore diameter	1.2992-1.3000 in.	33.00-33.03mm
Lifter diameter	1.2970-1.2982 in.	32.947-32.975mm
Lifter-to-bore clearance	0.001-0.003 in.	0.025-0.083mm
Valve guide bore diameter	0.4460-0.4475 in.	11.337-11.368mm
Valve stem diameter		
Intake	0.273-0.274 in.	6.935-6.962mm
Exhaust	0.2720-0.2736 in.	6.919-6.950mm
Valve stem-to-guide clearance		
Intake	0.0010-0.0044 in.	0.025-0.113mm
Exhaust	0.0015-0.0050 in.	0.037-0.131mm
Valve face angle	44.5-45.5°	44.5-45.5°
Valve seat angle	44.5-45.5°	44.5-45.5°
Valve seat width		
Intake	0.031-0.0653 in.	0.79-1.66mm
Exhaust	0.041-0.0756 in.	1.05-1.92mm
Valve spring pressure		
Minimum	157 lbs. @ 0.984 in.	700N @ 25mm
Maximum	180 lbs. @ 0.984 in.	802N @ 25mm

89563C03

1.9L DOHC (1901cc) ENGINE SPECIFICATIONS

Description	English Specifications	Metric Specifications
Cylinder Head (cont.)		
Valve spring free-length	1.61 in.	40.84mm
Piston and Connecting Rod		
Piston diameter	3.2266-3.2277 in.	81.956-81.984mm
Piston-to-bore clearance		
No. 1,2 and 3	0.0002-0.0028 in.	0.006-0.070mm
No. 4	0.0006-0.0028 in.	0.016-0.070mm
Piston pin diameter	0.7676-0.7677 in.	19.496-19.500mm
Piston pin-to-piston clearance	0.0001-0.0004 in.	0.002-0.011mm
Ring groove clearance		
Top ring	0.0016-0.0035 in.	0.040-0.090mm
Second ring	0.0012-0.0031 in.	0.030-0.080mm
Ring end-gap (1991-94)		
Top	0.0098-0.0236 in.	0.25-0.60mm
Second	0.0098-0.0236 in.	0.25-0.60mm
Oil control	0.0098-0.0551 in.	0.25-1.40mm
Ring end-gap (1995-98)		
Top	0.0098-0.0197 in.	0.25-0.50mm
Second	0.0098-0.0236 in.	0.25-0.60mm
Oil control	0.0098-0.0551 in.	0.25-1.40mm
Connecting rod alignment		
Twist	0.0110 in. per 3.94 in.	0.28mm per 100mm
Bend	0.0126 in. per 3.94 in.	0.32mm per 100mm
Connecting rod side-play on crankshaft	0.0065-0.0185 in.	0.165-0.470mm
Crankshaft		
Main journal diameter	2.2437-2.2444 in.	56.990-57.007mm
Main journal out-of-round limit	0.0004 in.	0.01mm
Main bearing-to-crankshaft oil clearance	0.0002-0.0024 in.	0.006-0.060mm
Connecting rod journal diameter	1.8500-1.8508 in.	46.99-47.01mm
Connecting rod journal out-of-round limit	0.0004 in.	0.01mm
Connecting rod bearing oil clearance	0.00039-0.00290 in.	0.010-0.073mm
Crankshaft run-out	0.0020 in.	0.05mm
Crankshaft end-play	0.0020-0.0098 in.	0.05-0.25mm
Oil Pump		
Tip clearance	0.006 in. max.	0.15mm max.
End-to-end clearance	0.0016-0.0050 in.	0.040-0.128 mm
Body-to-side clearance	0.006-0.011 in.	0.150-0.277mm

89563C04

ENGINE MECHANICAL

Engine

REMOVAL & INSTALLATION

In the process of removing the engine, you will come across a number of steps which call for the removal of a separate component or system, such as "disconnect the exhaust system" or "remove the radiator." In most instances, a detailed removal procedure can be found elsewhere in this manual.

It is virtually impossible to list each individual wire and hose which must be disconnected, simply because so many different model and engine combinations have been manufactured. Careful observation and common sense are the best possible approaches to any repair procedure.

Removal and installation of the engine can be made easier if you follow these basic points:

- If you have to drain any of the fluids, use a suitable container.
- Always tag any wires or hoses and, if possible, the components they came from before disconnecting them.
- Because there are so many bolts and fasteners involved, store and label the retainers from components separately in muffin pans, jars or coffee cans. This will prevent confusion during installation.
- After unbolting the transaxle, always make sure it is properly supported.
- If it is necessary to disconnect the air conditioning system, have this service performed by a qualified technician using a recovery/recycling station. If the system does not have to be disconnected, unbolt the compressor and set it aside.
- When unbolting the engine mounts, always make sure the engine is properly supported. When removing/installing the engine, make sure that any lifting devices are properly attached to the engine. It is recommended that if your engine is supplied with lifting hooks, your lifting apparatus be attached to them.
- Remove the engine from its compartment slowly, checking that no hoses, wires or other components are still connected.
- After the engine is clear of the compartment, place it on an engine stand or workbench.
- After the engine has been removed, you can perform a partial or full teardown of the engine using the procedures outlined in this manual.

Factory Recommended Procedure

▶ See Figures 1, 2, 3, 4 and 5

➡The manufacturer recommends that the engine and transaxle be removed as a complete assembly. Disconnect the cradle and lower the entire powertrain assembly, instead of lifting it out of the vehicle. Both the SOHC and DOHC engines are removed or installed in the same manner.

1. Properly disable the SIR system, if equipped. Disconnect the negative and then the positive battery cable.
2. Drain the engine coolant from the radiator and engine block into a suitable clean container.

✳✳ CAUTION

When draining the coolant, keep in mind that cats and dogs are attracted by ethylene glycol antifreeze, and are quite likely to drink any that is left in an uncovered container or in puddles on the ground. This will prove fatal in sufficient quantity. Always drain the coolant into a sealable container. Coolant may be reused unless it is contaminated or several years old.

3. Properly relieve the fuel system pressure as directed in Section 5 of this manual.
4. Remove the complete air cleaner/intake duct assembly, as applicable.

5. Disconnect and label the following electrical plugs and vacuum lines, as applicable:
   a. The coolant temperature sensor(s).
   b. Oxygen sensor and clip at the transaxle front mount bracket.
   c. Idle Air Control (IAC) valve.
   d. The ignition coil module connector(s).
   e. Throttle Position Sensor (TPS).
   f. Manifold Absolute Pressure (MAP) sensor.
   g. Exhaust Gas Recirculation (EGR) solenoid.
   h. Brake booster vacuum hose from the booster or intake manifold.
   i. Disconnect the 2 ground connectors from the transaxle attachment studs at the rear side of the cylinder block.
   j. Fuel injector electrical connectors.
6. Disconnect the following automatic or manual transaxle electrical connectors. If access is difficult, wait until the vehicle is safely raised and supported, then unplug the connections from underneath the vehicle:
   a. The 3 neutral safety switch connectors.
   b. Valve body actuator connection.
   c. Turbine speed sensor.
   d. Temperature sensor.
   e. For manual transaxles only, the reverse light switch.
   f. The 2 gear shift (PRNDL switch) connectors.
7. Disconnect the accelerator cable assembly.
8. Separate the fuel supply and return lines at their connectors. On 1991–94 models, utilize fuel line disconnect service tool SA9157E or equivalent. Plug the lines to prevent fuel contamination or loss. The lines may be tied to the master cylinder lines to help prevent fuel spillage and to keep them out of the way.
9. Disconnect the upper radiator hose and the cylinder head outlet and the de-aeration hose at the engine.
10. Remove the serpentine drive belt and if equipped, remove the air conditioning compressor from its brackets and with the hoses attached. Support the compressor from the front crossbar.

➡**It is not necessary to discharge the A/C compressor during engine removal, but be careful not to kink, damage or rupture the refrigerant lines.**

11. If equipped, disconnect the automatic transaxle cooler lines at the transaxle by pinching the plastic connector tabs and carefully pulling back on the lines. Plug the openings to prevent fluid loss or contamination.
12. Disconnect the automatic transaxle shifter cable or the manual shifter cables from the transaxle.
13. If equipped with a manual transaxle, remove the 2 hydraulic slave cylinder retaining nuts from the clutch housing studs, then slide the cylinder and bracket assembly from the studs. Rotate the clutch actuator ¼ turn counterclockwise while pushing toward the housing to disengage the bayonet connector and remove it from the clutch housing. Support the clutch hydraulic system to the battery tray; being sure not to kink or pinch the hydraulic lines.
14. Using a length of an appropriate wire, tie the radiator, condenser and fan to the front crossbar. Route the wire around the 2 fan shroud supports and the crossbar.
15. Raise and support the front end of the vehicle safely.
16. Remove the front wheels and remove the fasteners connecting the side and front fender shields to the cradle.
17. Remove the brake caliper bracket attaching bolts (2 on each side) and hang the caliper assemblies from the shock tower springs using wire. Do not hang the assembly by the brake hose or damage to the brake hydraulic system may occur. The springs and shocks will remain with the body when the powertrain is lowered.
18. Disconnect the struts from the knuckles on each side of the vehicle (2 bolts per side). This will allow the knuckle and hub assembly to remain with the powertrain cradle when lowered. The stabilizer bar will remain attached to the cradle and the lower control arms.

19. Using a suitable hose clamp tool, disconnect the lower radiator and heater return hoses from the engine. Disconnect the heater inlet hose at the front of the dash or the engine.

20. Disconnect the steering shaft and pressure switch connectors at the gear, as applicable.

21. Disconnect the front exhaust pipe at the manifold, catalytic converter and powertrain stiffening bracket.

22. For 1991 vehicles, remove the powertrain stiffening bracket bolts, except the 3 bolts holding the torque resistor bracket to the transaxle.

23. Remove the flywheel cover and torque converter bolts, if equipped.

24. Remove the alternator and starter shields.

25. Label and disconnect all the remaining electrical and vacuum connectors from the following components:

    a. Unplug the connectors from the starter solenoid and the battery feed.

    b. The alternator field and battery feed connectors.

    c. Oil pressure sensor.

    d. Knock sensor.

    e. Crankshaft position sensor.

    f. If equipped, the EVO solenoid.

    g. Vehicle speed sensor.

    h. Canister purge solenoid.

    i. Powertrain Control Module (PCM) and Oxygen sensor.

    j. If equipped, the ABS wheel sensor connector grounds.

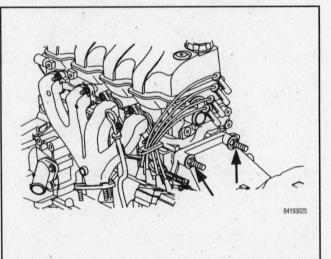

Fig. 1 Unfasten the ground connectors from the transaxle attachment studs

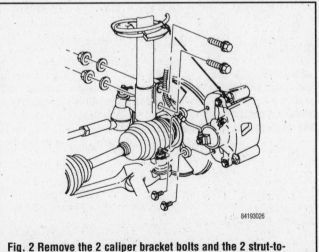

Fig. 2 Remove the 2 caliper bracket bolts and the 2 strut-to-knuckle attaching bolts in order to free the knuckle and hub assembly

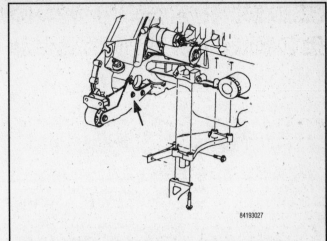

Fig. 3 For 1991 vehicles, remove the powertrain stiffening bracket, but do not remove the 3 bolts attaching the torque restrictor bracket to the transaxle

26. Unclip the brake lines from the rear side of the cradle.

27. Carefully remove the electrical harness from the engine and transaxle, then lay the electrical harness on top of the underhood junction block and battery cover.

28. For 1992–98 vehicles with a torque axis mount system, place a 1 in. x 1 in. x 2 in. long block of wood between the torque strut and cradle to ease removal and installation of the torque engine mount. Remove the 3 right side upper engine torque axis to front cover nuts and the 2 mount to midrail bracket nuts, allowing the powertrain to rest on the block of wood.

➡**Placing a block of wood under the torque axis mount prior to removing the upper mount will allow the engine to rest on the wood, thus preventing the engine from shifting. If the engine is not to be removed from the cradle, this will allow you to install the engine and the upper mount without jacking or raising the engine.**

29. Place a powertrain support dolly under the cradle. Use two 4 in. x 4 in. x 36 in. pieces of wood to support the cradle on the dolly.

30. Remove the 2 right side front engine mount torque strut bracket-to-cradle nuts.

31. Remove the 4 cradle attaching bolts and carefully lower the complete powertrain assembly from the vehicle. Verify that all necessary components are disconnected and free before complete removal.

32. Attach the 2 washers, located between the cradle and body, to the cradle. They must be repositioned and installed during cradle installation.

33. Disconnect the spark plug wires at the ignition module.

34. If applicable, remove the power steering pump and bracket. Support the assembly, in an upright position, from the cradle or the steering gear.

35. Install a suitable engine lifting device to the service support brackets.

36. Remove the front mount assembly and disconnect the motion restrictor bracket, if applicable.

37. Place a ½ in. x 1 in. x 3 in. block of wood under the axle shaft and remove the starter support bracket bolt, intake manifold support brace (on DOHC engines), and 3 axle shaft bracket support bolts. Allow the bracket to rotate rearward. Lift the engine slightly for clearance, as necessary.

38. For 1991 vehicles, remove the front engine mount assembly and disconnect the motion restrictor cable used with the DOHC engine and a manual transaxle.

39. For 1992–98 vehicles with a torque axis mount system, place a 4 in. x 4 in. x 6 in. long block of wood under the transaxle housing for support.

40. For 1992–98 vehicles, remove the engine strut bracket and torque strut from the front of the engine as an assembly. Lift the engine slightly as necessary for removal.

41. Remove the 4 transaxle attaching bolts/studs and separate the assembly. Manual transaxles will require the engine to be moved about 4 in. (100mm) forward in the cradle to disengage the input shaft.

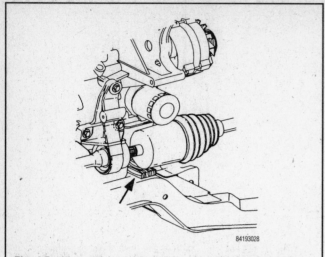

Fig. 4 Position a ½ in. x 1 in. x 3 in. block of wood under the axle shaft for support

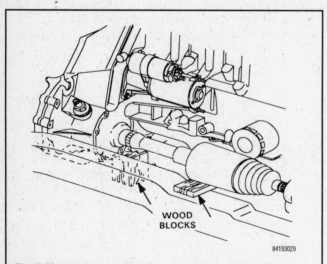

Fig. 5 Place a 4 in. x 4 in. x 6 in. long block of wood under the transaxle housing for support

42. Carefully lift the engine off the cradle.

To install:

43. Installation of the automatic transaxle assembly is the reverse of removal. However, please note the following important steps:

44. If installing a manual transaxle, align the yellow dot on the clutch pressure plate near the mark on the flywheel. Use clutch alignment tool SA9145T or equivalent to align the disk and input shaft, then tighten the pressure plate bolts to 19 ft. lbs. (25 Nm).

45. If installing an automatic transaxle, the yellow dot on the torque converter must be in the 6 o'clock position when the first flexplate to torque converter bolt is tightened.

46. Tighten the lower transaxle-to-engine bolts to 96 ft. lbs. (130 Nm) and the upper transaxle-to-engine bolts to 66 ft. lbs. (90 Nm) and the stiffening bracket-to-powertrain fasteners to 40 ft. lbs. (54 Nm).

47. Tighten the front engine mount-to-engine bolts to 41 ft. lbs. (55 Nm). For the 1991 DOHC engine with manual transaxle, tighten the motion restrictor bracket to 40 ft. lbs. (54 Nm).

48. For 1991 vehicles, tighten the front engine mount-to-cradle nut to 52 ft. lbs. (70 Nm).

49. For 1992–98 vehicles, tighten the engine mount torque strut-to-cradle bracket fasteners to 52 ft. lbs. (70 Nm). Hand-tighten the cradle fasteners, but do not torque until the upper midrail mount is installed.

50. Tighten the axle shaft bracket fasteners to 41 ft. lbs. (55 Nm) and starter bracket to 80 inch lbs. (9 Nm).

51. Carefully lift the powertrain and cradle into position. Make sure the radiator grommets are correctly aligned and that the 2 washers are reinstalled between the cradle and body at each rear cradle attachment position. If necessary, use two ³⁄₁₆ in. x 18 in. long guide pins in the forward cradle holes (located next to the attaching holes) to help align the cradle. Tighten the cradle to body fasteners to 151 ft. lbs. (205 Nm).

52. Tighten the steering shaft U-joint bolt to 35 ft. lbs. (47 Nm).

53. Tighten the starter solenoid connector to 44 inch lbs. (5 Nm).

54. Tighten the alternator and starter battery connectors to 89 inch lbs. (10 Nm).

55. Install the following components, if previously removed:
   a. Tighten the oil pressure sensor to 26 ft. lbs. (35 Nm).
   b. Tighten the knock sensor to 133 inch lbs. (15 Nm).
   c. Tighten the crankshaft position sensor mounting fastener(s) to 80 inch lbs. (9 Nm).
   d. Tighten canister purge valve mounting screw to 22 ft. lbs. (30 Nm).

56. Connect the wiring harness PCM ground and tighten to 89 inch lbs. (10 Nm).

57. Connect the wiring harness to the transaxle case/engine block and tighten to 18 ft. lbs. (25 Nm).

58. Tighten the engine stiffening bracket bolts to 35 ft. lbs. (47 Nm).

59. Tighten the exhaust pipe-to-intake manifold fasteners to 23 ft. lbs. (31 Nm), the pipe-to-stiffener bracket fasteners to 35 ft. lbs. (47 Nm), the pipe-to-support bracket fasteners to 23 ft. lbs. (31 Nm) and the pipe-to-catalytic converter fasteners to 18 ft. lbs. (25 Nm).

60. Tighten the cylinder block drain plug to 27 ft. lbs. (36 Nm), then close the radiator drain.

61. Tighten the steering knuckle-to-strut attachment bolts to 148 ft. lbs. (200 Nm).

62. Tighten the brake caliper assembly bolts to 81 ft. lbs. (110 Nm).

63. If equipped with a manual transaxle, tighten the hydraulic clutch slave cylinder fasteners to 19 ft. lbs. (25 Nm).

64. Tighten the A/C compressor-to-front bracket bolts to 40 ft. lbs. (54 Nm) and the compressor-to-rear bracket bolts to 22 ft. lbs. (30 Nm).

65. For 1992–98 vehicles equipped with a torque axis mount system, tighten the 2 engine mounts-to-midrail bracket nuts to 37 ft. lbs. (50 Nm). Tighten the torque axis mount-to-front cover nuts uniformly to 37 ft. lbs. (50 Nm).

66. For 1992–98 vehicles, tighten the strut bracket-to-cradle nuts to 37 ft. lbs. (50 Nm).

67. If equipped with automatic transaxle, tighten the torque converter-to-flexplate bolts to 52 ft. lbs. (70 Nm). Tighten the dust cover fasteners to 89 inch lbs. (10 Nm).

68. Tighten the wheel lugs, in a crisscross pattern to 103 ft. lbs. (140 Nm).

69. Lower the vehicle to the ground.

70. Apply a drop of clean engine oil to the male ends of the fuel line connectors and attach the line quick connect fittings. Make sure the lines are not kinked or damaged.

➡**Check the upper cooling module grommets for binding or misalignment. The module retaining pins must be centered in the grommets supported by the brackets. If the grommets are pinched, loosen the brackets and reposition them. It is extremely important that the cooling module be able to move freely.**

71. Thoroughly inspect the transaxle and engine compartment area to be sure that all wires, hoses and lines have been connected. Also inspect to make sure that all components and fasteners have been properly installed.

72. Connect the positive, then the negative battery cables.

73. Fill the engine cooling system, then check all engine and transaxle fluids, add or fill as necessary.

74. Prime the fuel system by cycling the ignition **ON** for 5 seconds and **OFF** for 10 seconds a few times without cranking the engine. Start the engine and check for leaks.

75. Perform a short road test and check the engine again for leaks. Make sure the cooling system is filled to the surge tank FULL COLD line.

Alternative Procedure

The following procedure applies only to Saturn vehicles equipped with an automatic transaxle. For vehicles equipped with a manual transaxle, the engine should be removed following the factory recommended procedure.

➡️**It may be possible to remove the engine on manual transaxle-equipped vehicles in a similar manner, but the engine will most likely not clear the input shaft of the transaxle; in this case, the complete engine/transaxle assembly must be removed. In order to do that, you must remove any components which may inhibit the transaxle, such as the brake master cylinder and booster, air cleaner housing, battery tray, etc.**

1. Properly disable the SIR system, if equipped. Disconnect the negative and then the positive battery cable.
2. Remove the hood. Refer to Section 10.
3. Drain the engine coolant from the radiator and engine block into a suitable clean container.

❉❉ CAUTION

When draining the coolant, keep in mind that cats and dogs are attracted by ethylene glycol antifreeze, and are quite likely to drink any that is left in an uncovered container or in puddles on the ground. This will prove fatal in sufficient quantity. Always drain the coolant into a sealable container. Coolant may be reused unless it is contaminated or several years old.

4. Remove the radiator and fan assembly. Refer to the procedures in this section.
5. Properly relieve the fuel system pressure, as described in Section 5 of this manual.
6. Remove the complete air cleaner/intake duct assembly, as applicable.
7. Detach and label the following electrical connections and vacuum lines, as applicable:
   a. Coolant temperature sensor(s).
   b. Oxygen sensor and clip at the transaxle front mount bracket.
   c. Idle Air Control (IAC) valve.
   d. Ignition coil module connector(s).
   e. Throttle Position Sensor (TPS).
   f. Manifold Absolute Pressure (MAP) sensor.
   g. Exhaust Gas Recirculation (EGR) solenoid.
   h. Brake booster vacuum hose from the booster or intake manifold.
   i. 2 ground connectors from the transaxle attachment studs at the rear of the cylinder block.
   j. Fuel injector electrical connectors.
8. Disconnect the accelerator cable assembly.
9. Separate the fuel supply and return lines at their connectors. On 1991–94 models, utilize fuel line disconnect service tool SA9157E or equivalent. Plug the lines to prevent fuel contamination or loss. The lines may be tied to the master cylinder lines to help prevent fuel spillage and to keep them out of the way.
10. Disconnect the upper radiator hose and cylinder head outlet, and the de-aeration hose at the engine.
11. Remove the serpentine drive belt and, if equipped, remove the air conditioning compressor from its brackets with the hoses attached. Support the compressor from the front crossbar.

➡️**It is not necessary to discharge the A/C compressor during engine removal, but be careful not to kink, damage or rupture the refrigerant lines.**

12. Raise and support the front end of the vehicle safely.
13. Remove the passenger side front wheel and remove the fasteners connecting the side and front fender shields to the cradle.
14. Remove the crankshaft pulley.
15. Using a suitable hose clamp tool, remove the lower radiator hose and disconnect the heater return hoses from the engine. Disconnect the heater inlet hose at the front of the dashboard or engine.

16. Disconnect the front exhaust pipe at the manifold, catalytic converter and powertrain stiffening bracket.
17. Remove the powertrain stiffening bracket bolts. On 1991 vehicles, do not remove the 3 bolts holding the torque resistor bracket to the transaxle.
18. Remove the flywheel cover and torque converter bolts.
19. Remove the alternator and starter, as described in Section 2.
20. Label and disconnect all the remaining electrical and vacuum connectors from the following components:
   a. Oil pressure sensor.
   b. Knock sensor.
   c. Crankshaft position sensor.
   d. EVO solenoid (if equipped).
   e. Canister purge solenoid.
   f. Powertrain Control Module (PCM) and oxygen sensor.
   g. ABS wheel sensor connector grounds (if equipped and necessary).
21. Carefully position the electrical harness away from the engine and transaxle.
22. For 1992–98 vehicles with a torque axis mount system, place a 1 in. x 1 in. x 2 in. long block of wood between the torque strut and cradle to ease removal and installation of the torque engine mount. Remove the 3 right side upper engine torque axis-to-front cover nuts and the 2 mount-to-midrail bracket nuts, allowing the powertrain to rest on the block of wood.

➡️**Placing a block of wood under the torque axis mount prior to removing the upper mount will allow the engine to rest on the wood, thus preventing the engine from shifting. If the engine is not to be removed from the cradle, this will allow you to install the engine and the upper mount without jacking or raising the engine.**

23. If applicable, remove the power steering pump and bracket. Support the assembly out of the way, in an upright position.
24. Place a ½ in. x 1 in. x 3 in. block of wood under the axle shaft and remove the starter support bracket bolt, intake manifold support brace (on DOHC engines), and 3 axle shaft bracket support bolts. Allow the bracket to rotate rearward. Lift the engine slightly for clearance, as necessary.
25. Attach a suitable lifting device to the engine service support brackets and carefully raise the hoist to relieve the engine weight from its mounts.
26. Remove the 2 right side front engine mount torque strut bracket-to-cradle nuts.
27. Disconnect the spark plug wires at the ignition module.
28. Support the transaxle assembly with a floor jack.
29. Remove the front mount assembly and disconnect the motion restrictor bracket, if applicable.
30. For 1992–98 vehicles with a torque axis mount system, place a 4 in. x 4 in. x 6 in. long block of wood under the transaxle housing for support.
31. For 1992–98 vehicles, remove the engine strut bracket and torque strut from the front of the engine as an assembly. Lift the engine slightly as necessary for removal.
32. Remove the 4 transaxle attaching bolts/studs and separate the transaxle assembly.
33. Carefully lift the engine from the vehicle.
   To install:
34. Installation of the engine is the reverse of the removal procedure. However, please note the following steps:
35. Before lowering the engine into the vehicle, ensure that the torque converter in the transaxle is fully seated.
36. Position one of the bolt holes on the torque converter facing straight down. Also, rotate the engine so as to position one of the flywheel-to-torque converter bolt holes facing straight down.
37. When mating the engine to the transaxle, ensure that the locating dowels are fully seated.
38. Install the engine-to-transaxle bolts. Then, install the motor mounts. Tighten all fasteners to the proper torque specifications. Refer to the specification charts in this section.
39. After all hoses, lines, and connections are installed, fill the engine and radiator with the proper type and quantity of fluids, then check the transaxle and power steering fluid levels. Adjust as necessary.

Rocker Arm (Camshaft/Valve) Cover

REMOVAL & INSTALLATION

SOHC Engine

♦ See Figures 6, 7 and 8

1. Disconnect the negative battery cable. Remove the air cleaner assembly and air inlet duct.
2. Disconnect the PCV valve, fresh air hose and any remaining vacuum hoses or electrical connectors from the valve cover.
3. Remove the 2 fasteners and washers from the silicone insulators located in the top of the valve cover.
4. Disconnect the de-aeration line from the cylinder head outlet (located at one end of the cover) and from the support bracket. Have a rag handy to wipe up the small amount of coolant which may escape from the line. The MAP sensor (1991–94 models only) and de-aeration line may remain attached to the cover.
5. Carefully remove the valve cover and gasket from the engine. If the cover does not come off easily, tap the sides gently with a rubber mallet to loosen the cover. Do not pry against the mating surfaces or damage may occur.

To install:

6. Inspect the silicone insulators located in the valve cover fastener bores and replace if cracked or deteriorated.
7. Inspect the valve cover gasket which is bonded to the cover flange. If it is damaged, or if replacement is necessary to assure a proper seal, remove the gasket from the cover and clean the old RTV from the flange surface using a chlorinated solvent and a wire brush. Most carburetor cleaners or brake clean solvents should be appropriate.
8. If the gasket is to be replaced, apply a thin 0.08 in. (2mm) bead of RTV to the cover flange. Firmly press the flat side of the gasket into the flange and install the cover while the RTV is still wet to assure a proper bond. RTV should only be placed between the cover flange and gasket, but NOT onto the cylinder head mating surface, so that the gasket will separate from the head easily and may be reused.
9. Apply a small 0.2 in. (5mm) bead of RTV to the cylinder head T-joints where the head meets the front cover to assure a good seal. This is the only spot where RTV will be placed between the cylinder head and the cover gasket.
10. Install the valve cover and gasket, then install the washers and fasteners. Tighten the fasteners to 22 ft. lbs. (30 Nm). Connect the de-aeration line to the support bracket and to the cylinder head.

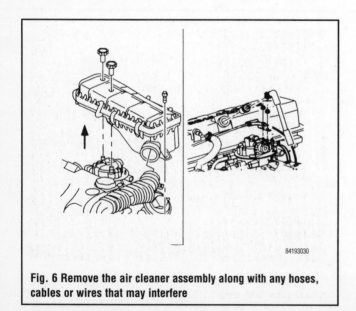

Fig. 6 Remove the air cleaner assembly along with any hoses, cables or wires that may interfere

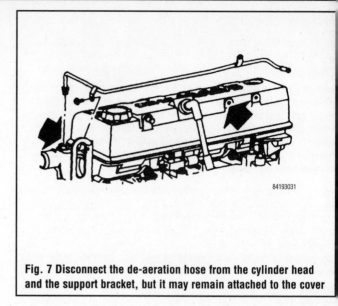

Fig. 7 Disconnect the de-aeration hose from the cylinder head and the support bracket, but it may remain attached to the cover

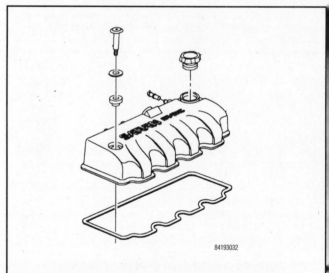

Fig. 8 Exploded view of the SOHC valve cover assembly

11. Connect the PCV valve, fresh air hose and any remaining vacuum hoses or electrical connectors to the valve cover.
12. Install the air cleaner assembly and air inlet duct, then carefully tighten the fasteners. Connect the negative battery cable.

DOHC Engine

♦ See Figures 9 thru 16

1. Disconnect the negative battery cable.
2. Disconnect the fresh air hose from the cover fitting and the PCV valve from the cover grommet.
3. On 1991–94 models, disconnect the vacuum hose and electrical connector from the EGR solenoid located at the end of the camshaft cover, near the EGR valve and the engine lifting bracket.

➡On 1991–94 models, if it is preferred or necessary, the bolt may be removed from the EGR solenoid to disconnect the solenoid from the camshaft cover.

4. The original spark plug wires are numbered, if the numbers are illegible or if the wires have been replaced with non-labeled parts, tag the wires for installation purposes. Disconnect the spark plug wires from the plugs.

5. Remove the fasteners, washers and insulators, then remove the camshaft cover and gasket from the engine. If the cover is stuck, tap the sides gently using a rubber mallet. Do not damage the aluminum cylinder head by prying the cover loose.

To install:

6. Inspect the silicone insulators through which the camshaft cover fasteners are installed and replace if cracked or deteriorated.

7. Inspect the camshaft cover gasket which is bonded to the cover flange. If it is damaged, or if replacement is necessary to assure a proper seal, remove the gasket from the cover and clean the old RTV from the flange surface using a chlorinated solvent and a wire brush. Most carburetor cleaners or brake clean solvents should be appropriate.

8. If the gasket is to be replaced, apply a thin 0.08 in. (2mm) bead of RTV to the cover flange. Firmly press the flat side of the gasket into the flange and install the cover while the RTV is still wet to assure a proper bond. RTV should only be placed between the cover flange and gasket, but NOT onto the cylinder head mating surface, so that the gasket will separate from the head easily and may be reused.

9. Apply a small 0.2 in. (5mm) bead of RTV to the cylinder head T-joints where the head meets the front cover to assure a good seal. This is the only spot where RTV will be placed between the cylinder head and the cover gasket.

Fig. 11 Lift the camshaft cover from the top of the engine. If the cover is stuck, tap it with a rubber mallet

Fig. 9 Loosen the camshaft cover retaining fasteners with a Torx® driver

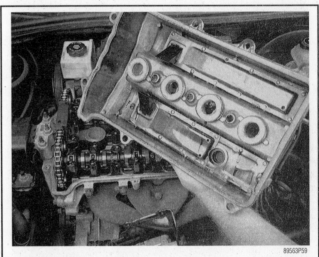

Fig. 12 Examine the camshaft cover inside and out for cracks or damage

Fig. 10 Remove the camshaft cover fasteners, washers and insulators. Examine for cracks or deterioration and replace, if necessary

Fig. 13 Examine the camshaft cover gasket material for signs of deterioration, and replace if necessary

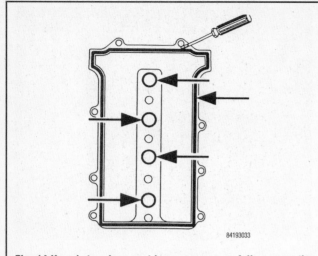

Fig. 14 If gasket replacement is necessary, carefully remove the old gasket from the camshaft cover surfaces

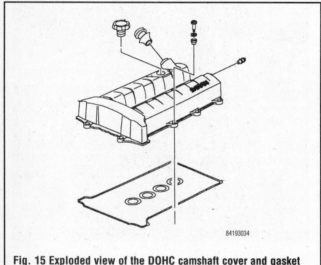

Fig. 15 Exploded view of the DOHC camshaft cover and gasket assembly

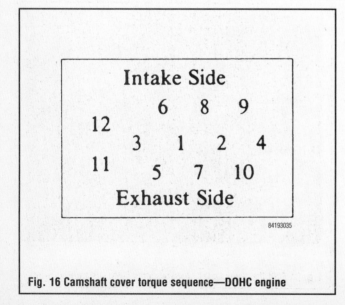

Fig. 16 Camshaft cover torque sequence—DOHC engine

10. Install the camshaft cover and gasket using the washers and fasteners. Tighten the fasteners uniformly and in the proper sequence to 89 inch lbs. (10 Nm).

11. Connect the wires to the spark plugs as labeled.

12. If equipped, connect the electrical plug and vacuum hose to the EGR solenoid valve.

13. Install the PCV valve to the grommet and the fresh air hose to the fitting.

14. Connect the negative battery cable.

Rocker Arms/Shafts

Only the SOHC engine has a cylinder head that uses rocker arms in its valve train.

REMOVAL & INSTALLATION

▶ **See Figures 17 and 18**

1. Disconnect the negative battery cable.

2. Remove the rocker arm cover, then inspect the cover silicone isolators for cracks or deterioration and replace as necessary.

3. Uniformly remove the rocker arm assembly bolts, then carefully remove the 2 shaft and rocker arm assemblies. The shafts may be unsnapped from the lifter guideplates leaving the lifters and plates in the cylinder head. If this cannot be accomplished, remove the guideplates and lifters, but make sure to reposition the lifters in the same bores and guideplates during installation.

4. If necessary, remove the rocker arms from the shafts.

To install:

5. If removed, oil the shafts and install the rocker arms into position on the shafts.

6. Snap one end of each lifter guide plate retaining spring onto the rocker arm shaft between the No. 1–No. 2 and the No. 3–No. 4 cylinder rocker arms.

7. Install the rocker arm shaft assemblies. To prevent valve or piston damage, be sure the rocker arm tangs are squarely seated on the lifter plungers and the retaining springs are positioned in the guide plate slots.

➡**If difficulty is encountered aligning the rocker arms on the valves and lifters, use a flat piece of wood, cardboard or an extension bar of suitable length on top of the shafts and rocker arms to hold both assemblies in position.**

8. Tighten the 5 rocker arm bolts on each shaft in a uniform sequence to 18 ft. lbs. (25 Nm). Verify the proper position and seating of all rocker components.

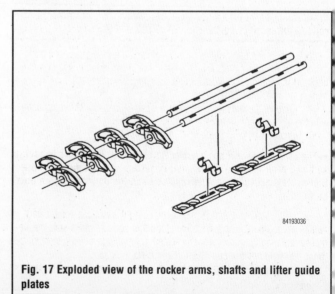

Fig. 17 Exploded view of the rocker arms, shafts and lifter guide plates

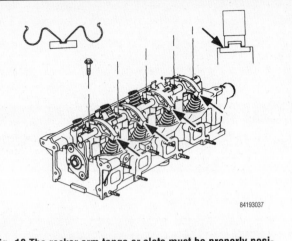

Fig. 18 The rocker arm tangs or slots must be properly positioned on the plungers and the springs must be positioned on the guide plate slots to prevent valve or piston damage

9. Apply a small drop of RTV to each cylinder head and front cover T-joint. Inspect the rocker arm cover gasket and replace if necessary. Install the gasket and rocker arm cover, then tighten the fasteners uniformly to 22 ft. lbs. (30 Nm). Make sure all cover hoses and components are reconnected after installation.

10. Connect the negative battery cable, start the engine and check for leaks.

Thermostat

REMOVAL & INSTALLATION

▸ **See Figures 19 thru 24**

1. Make sure the engine is cold, and position a suitable clean drainpan under the radiator and engine drains. Open the radiator drain and remove the engine drain plug located at the front right of the engine, draining the coolant into the container.

❋❋ CAUTION

When draining the coolant, keep in mind that cats and dogs are attracted by ethylene glycol antifreeze, and are quite likely to drink any that is left in an uncovered container or in puddles on the ground. This will prove fatal in sufficient quantity. Always drain the coolant into a sealable container. Coolant may be reused unless it is contaminated or several years old.

2. Disconnect the lower radiator hose from the thermostat housing.

3. Remove the 2 bolts from the water inlet housing, then remove the housing and thermostat assembly. Remove and discard the O-ring from the mating surface.

4. Remove the thermostat from the housing using the tool provided with the replacement thermostat element.

To install:

➡ **The thermostat will not function correctly if it has come in contact with oil. If oil is found in the cooling system at any time, both the thermostat cartridge must be replaced and the cooling system must be flushed.**

5. Install the replacement thermostat using the tool provided, being careful not to damage or scratch the aluminum surface. Make sure the element's retaining tangs are properly seated in the 2 legs and the element piston is correctly positioned in the inlet housing.

6. Install a new O-ring and position the housing to the engine. Tighten the retaining bolts to 22 ft. lbs. (30 Nm).

Fig. 19 Loosen the thermostat housing bolts . . .

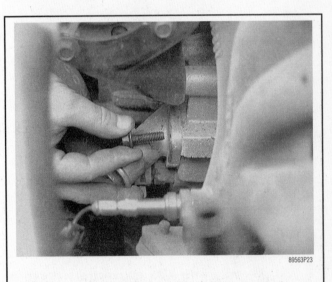

Fig. 20 . . . then remove each bolt from the engine block

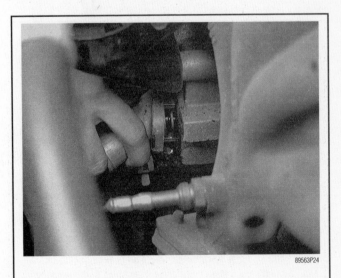

Fig. 21 Pull the thermostat housing away from the engine block

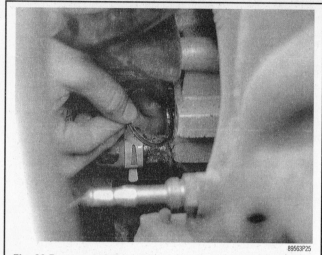

Fig. 22 Remove and discard the thermostat housing O-ring gasket

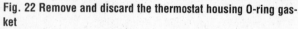

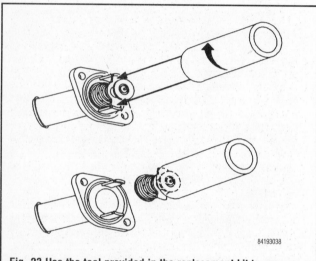

Fig. 23 Use the tool provided in the replacement kit to remove the old thermostat from the housing

Fig. 24 Unlike most thermostats which sit in the housing or engine block, the Saturn thermostat is secured to the housing by its own spring tension

7. Close the radiator drain plug and install the cylinder block drain plug. Tighten the block plug to 26 ft. lbs. (35 Nm).

8. Install the hose to the inlet housing.

9. Connect the negative battery cable and properly fill the engine cooling system.

10. Run the engine and check for leaks. Make sure the cooling system is filled to the proper level.

Intake Manifold

REMOVAL & INSTALLATION

SOHC Engine

▶ See Figures 25, 26 and 27

1. Disconnect the negative battery cable and drain the engine coolant from the radiator and engine block, into a suitable container.

2. Remove the air cleaner, disconnecting the fresh air tube at the valve cover. Remove the PCV tube and hose.

3. Properly relieve the fuel system pressure at the test port.

4. Remove the retaining screw from the fuel line retaining bracket.

5. On 1991–94 models, disconnect the fuel supply and return lines at the close-coupled connectors using service tool SA9157E or equivalent. On 1995–98 models, disconnect the fuel line(s) at the quick-connect fitting(s) by compressing the two tangs of the retainer and pulling.

6. On 1991–94 models, using a backup wrench, disconnect the steel fuel supply and return lines from the throttle body unit.

7. Plug the line(s) to prevent fuel contamination or loss.

8. Disconnect the throttle cable from the throttle body, then remove the throttle cable bracket attaching nuts and position the assembly aside.

9. Label and disconnect the wiring from the intake manifold, throttle body and valve cover components, as follows:

   a. The fuel injector.
   b. Idle Air Control (IAC) valve.
   c. Throttle Position Sensor (TPS).
   d. Exhaust Gas Recirculation (EGR) valve.
   e. Manifold Absolute Pressure (MAP) sensor.

10. Remove the wiring tube and lay the harness away from the manifold onto the Under Hood Junction Block (UHJB).

11. Label and disconnect all vacuum hoses from the throttle body unit. Disconnect the vacuum line from the brake booster.

12. Disconnect the heater hose from the intake manifold and the de-aeration line fitting at the cylinder head coolant outlet. Remove the 2 clamps and lay the line onto the coolant bottle.

13. Remove the intake manifold support bracket bolt located next to the starter. If necessary, the bolt can be removed from below the vehicle.

14. Remove the serpentine drive belt, then remove the power steering pump and support the pump next to the right side dash panel sufficiently away from the intake manifold and cylinder head.

15. Remove the manifold retaining nuts, then remove the manifold and throttle body assembly. It may be much easier to access the lower manifold nuts from under the vehicle. Remove and discard the old gasket from the mating surfaces.

To install:

16. Thoroughly clean all gasket mating surfaces. Be careful not to damage or score the aluminum surface. If replaced, use Loctite® 290 or equivalent to seal the new PCV valve inlet tube into the manifold.

17. Position the new gasket, then install the manifold and retaining nuts. Tighten the nuts in sequence to 22 ft. lbs. (30 Nm).

18. Install the power steering pump and tighten the fasteners to 27 ft. lbs. (38 Nm).

19. Install the serpentine drive belt.

20. Connect the coolant hose, the de-aeration line and clamps, then install the manifold support bracket bolt. Tighten the bolt to 22 ft. lbs. (30 Nm).

21. Lubricate the male ends of the fuel lines with a few drops of clean engine oil, then connect the fuel supply and return lines.

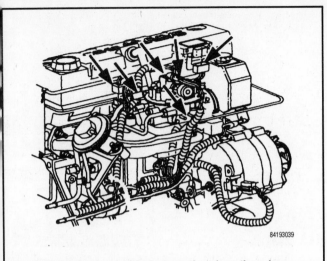

Fig. 25 Unfasten the electrical connections from these locations—1991–94 SOHC engine

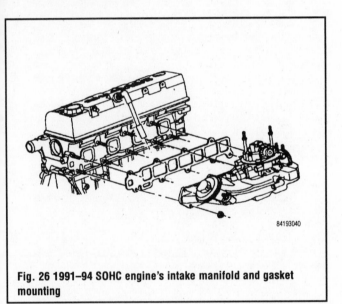

Fig. 26 1991–94 SOHC engine's intake manifold and gasket mounting

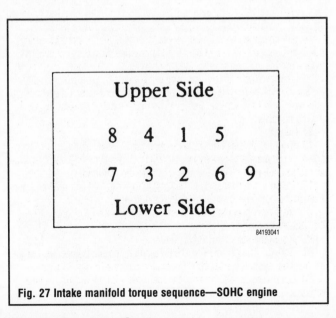

Fig. 27 Intake manifold torque sequence—SOHC engine

22. On 1991–94 models connect the fuel lines to the throttle body unit and tighten the fittings to 19 ft. lbs. (25 Nm).

23. Secure the fuel line(s) in the retaining bracket and tighten the mounting screw to 36 inch lbs. (4 Nm).

24. Reposition the wiring harness and connect the wiring and vacuum hoses to their original locations. The harness leads to the TPS and EGR solenoid must be routed between the intake manifold runners.

25. Inspect the air cleaner/throttle body unit gasket and replace, if necessary. Install the air cleaner and fresh air tubes, then install the PCV valve hose.

26. Connect the negative battery cable, close the radiator drain and install the engine block drain plug. Tighten the block drain plug to 26 ft. lbs. (35 Nm). Fill the engine cooling system.

27. Prime the fuel system by cycling the ignition switch **ON** for 5 seconds, then **OFF** for 10 seconds and repeating 2 times. Start the engine and check for leaks.

DOHC Engine

▶ **See Figures 28 thru 45**

1. Disconnect the negative battery cable and drain the engine coolant from the radiator and engine block, into a suitable container.

2. Remove the air inlet tube, disconnecting the fresh air tube at the camshaft cover. Lift the resonator upward to disengage it from the engine service support bracket. Remove the PCV valve and hose.

3. Properly relieve the fuel system pressure at the test port.

4. Remove the retaining screw from the fuel line retaining bracket.

5. On 1991–94 models, disconnect the fuel supply and return lines at the close-coupled connectors using service tool SA9157E or equivalent. On 1995–98 models, disconnect the fuel line(s) at the quick-connect fitting(s) by compressing the two tangs of the retainer and pulling.

6. Plug the line(s) to prevent fuel contamination or loss.

7. Disconnect the throttle cable from the throttle body, then remove the throttle cable bracket attaching nuts and position the assembly aside.

8. Label and disconnect the wiring from the intake manifold and surrounding components, as follows:

   a. The fuel injectors.
   b. Idle Air Control (IAC) valve.
   c. Throttle Position Sensor (TPS).
   d. Manifold Absolute Pressure (MAP) sensor.

9. Disconnect the heater and de-aeration hoses from the intake manifold outlets. The heater hose may be removed at the firewall, if necessary. Disconnect the EGR solenoid vacuum hose.

10. Position the wiring harness over the brake master cylinder, then remove the intake manifold support bracket bolt attached to the manifold next to the brake master cylinder.

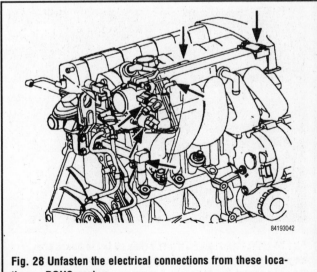

Fig. 28 Unfasten the electrical connections from these locations—DOHC engine

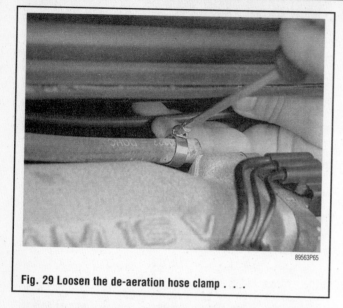

Fig. 29 Loosen the de-aeration hose clamp . . .

Fig. 32 Remove the intake manifold support bracket bolt at the manifold, near the master cylinder

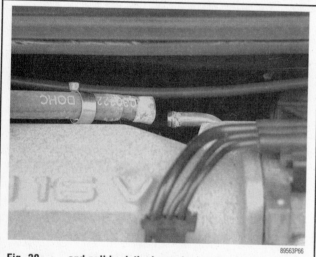

Fig. 30 . . . and pull back the hose clamp, then disconnect the hose from the intake manifold

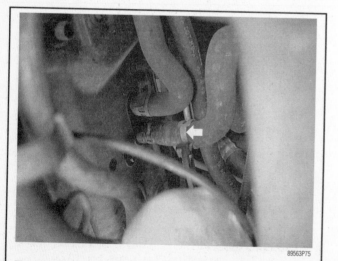

Fig. 31 Disconnect the heater hose at the firewall. It may be easier to access the hose from underneath the vehicle

11. Remove the serpentine drive belt. Remove the power steering pump assembly with the support bracket, then remove the upper pump bracket attachment bolts and position the pump away from the manifold and cylinder head, near the right dash panel.

12. Remove the 3 upper intake manifold attachment nuts, then raise and support the front of the vehicle safely.

13. Remove the lower power steering unit support bracket. Remove the intake manifold support bracket bolt located next to the alternator, then the lower bracket bolt and remove from the vehicle.

14. Disconnect the canister purge solenoid and brake booster vacuum hoses.

15. Remove the intake manifold attaching stud, remove the supports and lower the vehicle.

16. Remove the remaining fasteners and the intake manifold assembly, then remove and discard the old gasket.

To install:

17. Thoroughly clean the gasket mating surfaces. Be careful not to score or damage the aluminum sealing surfaces. If installing a new coolant de-aeration tube elbow into the manifold use Loctite® 290 or equivalent to achieve a proper seal.

18. Position the new gasket, then install the intake manifold and retaining nuts. Tighten the nuts in sequence to 22 ft. lbs. (30 Nm).

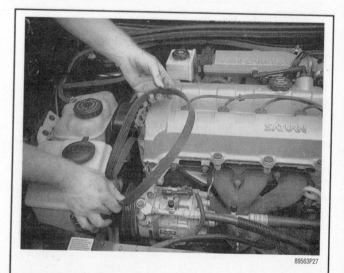

Fig. 33 Remove the serpentine drive belt

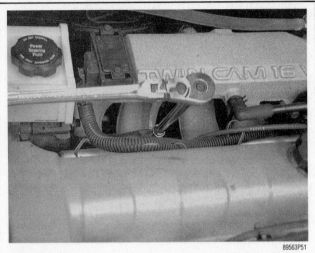

Fig. 34 Remove the power steering pump-to-upper intake manifold bracket bolt

Fig. 35 Access the power steering pump front bracket attaching bolts through the pulley holes

Fig. 36 Remove the upper intake manifold attaching nuts

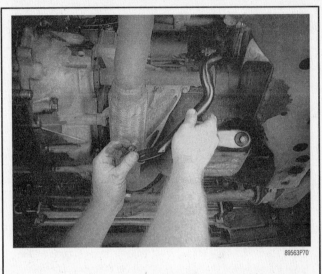

Fig. 37 Remove the power steering pump's lower bracket brace

Fig. 38 Remove the intake manifold support bracket's top mounting bolt, located behind the alternator . . .

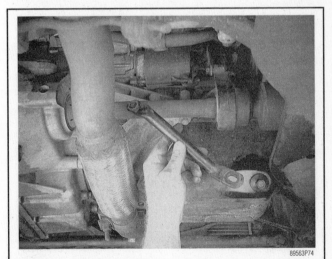

Fig. 39 . . . then, after removing the lower bolt, remove the support bracket from the vehicle

Fig. 40 Disconnect the brake power booster vacuum hose at the nipple on the intake manifold, between the manifold and firewall

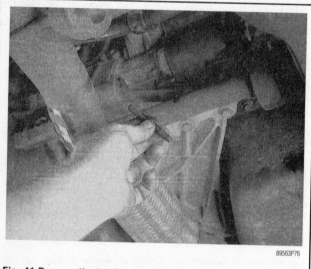

Fig. 41 Remove the intake manifold attaching stud

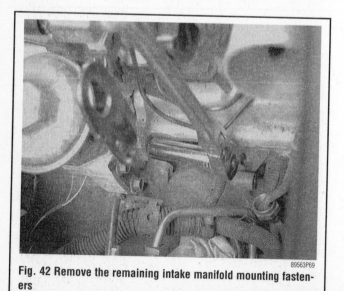

Fig. 42 Remove the remaining intake manifold mounting fasteners

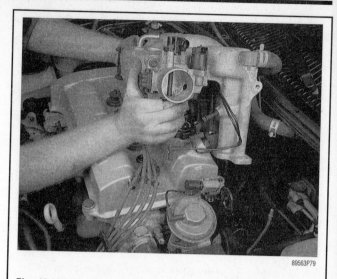

Fig. 43 Lift the intake manifold up and out of the vehicle

Fig. 44 Remove and discard the intake manifold gasket

Upper Side

5   2   3

7   4   1   6

Lower Side

Fig. 45 Intake manifold torque sequence—DOHC engine

19. Install the power steering pump and brackets. Tighten the fasteners to 28 ft. lbs. (38 Nm).

20. Install the serpentine drive belt making sure the belt is properly aligned on the pulleys.

21. Connect the heater hose and de-aeration line to the manifold.

22. Position the manifold support brackets and install the bolts. Tighten the right block bolt to 41 ft. lbs. (55 Nm), then tighten the left block bolt and the support bracket to intake manifold bolts to 22 ft. lbs. (30 Nm).

23. Lubricate the male ends of the fuel lines with a few drops of clean engine oil, then connect the fuel supply and return lines.

24. Secure the fuel line(s) in the retaining bracket and tighten the mounting screw to 36 inch lbs. (4 Nm).

25. Connect the throttle cable to the throttle body and install the support bracket. Tighten the bracket retaining bolts to 19 ft. lbs. (25 Nm). Verify that the cable locking tangs are fully engaged when assembled.

26. Position the wiring harness and connect all electrical connectors and vacuum hoses in their original locations.

27. Install the PCV hose, the air inlet tube and resonator.

28. Connect the negative battery cable, close the radiator drain and install the engine block drain plug. Tighten the block drain plug to 26 ft. lbs. (35 Nm). Fill the engine cooling system.

29. Prime the fuel system by cycling the ignition switch **ON** for 5 seconds, then **OFF** for 10 seconds and repeating 2 times. Start the engine and check for leaks.

Exhaust Manifold

REMOVAL & INSTALLATION

▶ **See Figures 46 thru 56**

1. Disconnect the negative battery cable, then raise and support the vehicle safely using suitable jackstands.

2. If equipped, remove the two front exhaust pipe to engine support bracket mounting fasteners.

3. Remove the pipe-to-manifold nuts and lower the pipe.

4. Remove the supports and lower the vehicle.

➡ **When performing the next step, do NOT disconnect the refrigerant lines.**

5. If equipped, remove the air conditioning compressor and bracket from the engine, which will first require the removal of the serpentine belt, then position them aside.

Fig. 47 Remove the pipe-to-manifold nuts in order to separate the flanges

Fig. 48 Pull down slightly to separate the front pipe from the intake manifold

Fig. 46 Remove the front exhaust pipe-to-engine support bracket mounting fasteners

Fig. 49 Remove the A/C compressor mounting bracket bolts at the front of the engine . . .

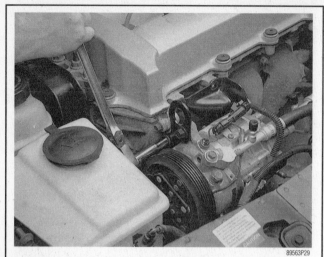

Fig. 50 . . . and on the side, then move the compressor away from the engine

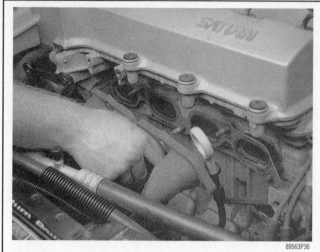

Fig. 53 . . . and separate the exhaust manifold from the cylinder head

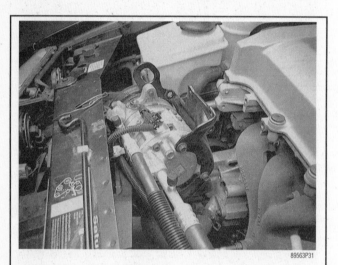

Fig. 51 Using a heavy wire hanger, or equivalent, secure the A/C compressor to the front engine compartment crossmember

Fig. 54 Remove and discard the exhaust manifold-to-cylinder head gasket. Be sure to clean the gasket mating surfaces

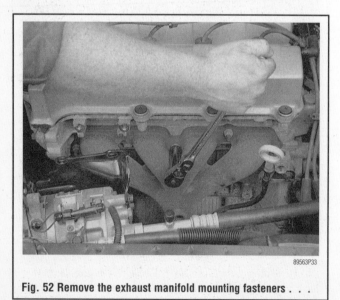

Fig. 52 Remove the exhaust manifold mounting fasteners . . .

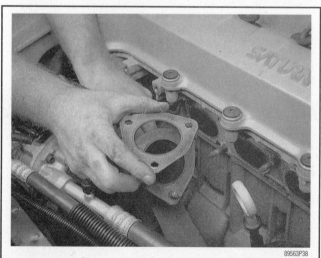

Fig. 55 Remove and discard the front pipe-to-exhaust manifold flange gasket

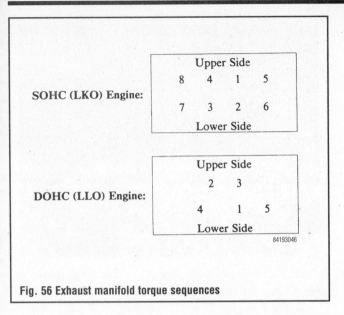

SOHC (LKO) Engine:	Upper Side			
	8	4	1	5
	7	3	2	6
	Lower Side			

DOHC (LLO) Engine:	Upper Side		
		2	3
	4	1	5
	Lower Side		

84193046

Fig. 56 Exhaust manifold torque sequences

6. Disconnect the oxygen sensor connector. If necessary, use a 19mm, 6-point crow's foot wrench to remove the oxygen sensor from the manifold.

7. Remove the manifold retaining nuts and remove the manifold from the cylinder head. Remove and discard the old gaskets from the mating surfaces.

To install:

8. Thoroughly clean the gasket mating surfaces, being careful not to score or damage the aluminum surface.

9. Install the new gasket with the smooth side facing the manifold, then install the manifold and attaching nuts. Tighten the nuts in sequence to 16 ft. lbs. (22 Nm) for the SOHC engine or to 23 ft. lbs. (31 Nm) for the DOHC engine.

10. If replacing the oxygen sensor, coat the threads with nickel based anti-seize compound and tighten to 33 ft. lbs. (45 Nm). Connect the oxygen sensor electrical connector.

11. Install the air conditioning compressor and brackets. Tighten all fasteners except the front bracket-to-compressor fasteners to 19 ft. lbs. (25 Nm). Tighten the front bracket-to-compressor fasteners to 40 ft. lbs. (54 Nm).

12. Raise and support the vehicle safely using jackstands, then install a new gasket onto the studs between the pipe and manifold.

13. Connect the pipe and manifold, then tighten the fasteners in a crosswise pattern to 23 ft. lbs. (31 Nm).

14. If necessary, position the exhaust pipe to engine support bracket in place and install the two mounting fasteners. Tighten the mounting fasteners to 23 ft. lbs. (31 Nm).

15. Lower the vehicle.

16. Connect the negative battery cable, start the engine and check for exhaust leaks.

Radiator

On 1991–93 Saturn models, the radiator drain plug is threaded into a removable drain housing. The housing may be replaced at any point after the coolant is drained from the system. To remove the housing, use a pair of pliers to pinch the tabs closed, then pull the housing straight back and out of the radiator. Squeeze the tabs of the replacement to begin inserting it into the bore, then carefully snap it into position.

REMOVAL & INSTALLATION

▶ See Figures 57 thru 76

1. Disconnect the negative battery cable. Drain the engine coolant from the radiator and block drains into a suitable clean container.

❋❋ CAUTION

When draining the coolant, keep in mind that cats and dogs are attracted by ethylene glycol antifreeze, and are quite likely to drink any that is left in an uncovered container or in puddles on the ground. This will prove fatal in sufficient quantity. Always drain the coolant into a sealable container. Coolant may be reused unless it is contaminated or several years old.

89563P01

Fig. 57 Disconnect the intake air temperature sensor, then remove the air intake duct

89563P02

Fig. 58 Loosen the upper coolant hose clamp and pull it back, away from the radiator . . .

Fig. 59 . . . then disconnect the upper coolant hose from the radiator

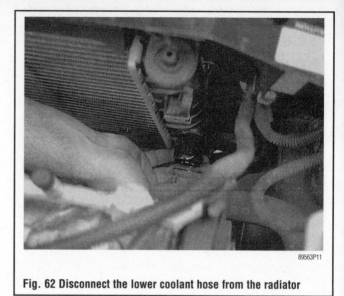

Fig. 62 Disconnect the lower coolant hose from the radiator

Fig. 60 If equipped with an automatic transaxle, loosen the cooler line fitting at the radiator . . .

2. Remove the air intake ducts and disconnect the temperature sensor connector, then remove the air cleaner housing.

3. Loosen the clamp and disconnect the upper hose from the radiator. If equipped with an automatic transaxle, disconnect the upper transaxle fluid cooler line. Plug the openings to prevent transaxle fluid contamination or loss.

4. Remove the electric cooling fan assembly.

5. Loosen the clamp and disconnect lower hose from the radiator.

6. Raise and support the front of the vehicle safely using suitable jack-stands.

7. Disconnect the fasteners, the carefully lower the splash shield from the vehicle to gain access below the radiator.

8. If equipped with an automatic transaxle, disconnect and plug the lower transaxle cooler line.

9. Remove the 2 condenser bracket-to-radiator bolts from either side of the radiator. Wire the condenser to the frame assembly to keep it in place, then remove the supports and carefully lower the vehicle.

10. Remove the upper radiator nuts and brackets. On air conditioning equipped vehicles, remove the upper radiator seal.

11. Carefully lift the radiator from the vehicle. If necessary, squeeze the drain housing tabs and withdraw it from the bottom of the radiator.

Fig. 61 . . . then disconnect the transaxle cooler line from the radiator. Plug the line to prevent fluid loss

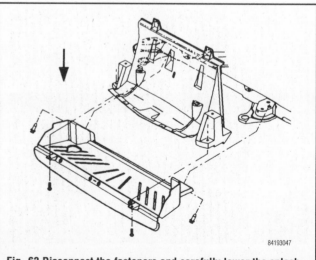

Fig. 63 Disconnect the fasteners and carefully lower the splash shield from the vehicle

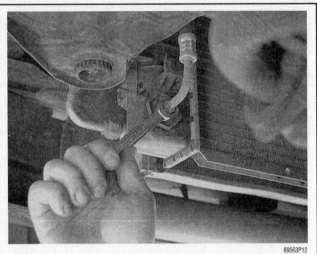

Fig. 64 If equipped with an automatic transaxle, loosen the lower cooler line fitting at the radiator . . .

Fig. 67 Remove the upper radiator mounting nuts and brackets

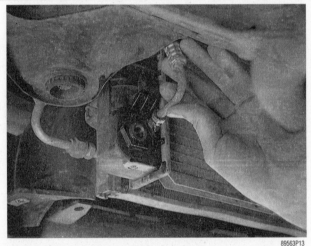

Fig. 65 . . . then disconnect the lower cooler line from the radiator. Plug the line to prevent fluid loss

Fig. 68 If replacing the radiator, the rubber mounting brackets must be installed on the new radiator

Fig. 66 Remove the 2 condenser bracket-to-radiator bolts

Fig. 69 If equipped with A/C, remove the upper radiator seal

Fig. 70 Carefully lift the radiator up and out of the vehicle

Fig. 71 A dirt covered radiator, such as this one, should be sprayed with a garden hose

Fig. 72 Be careful not to lose the lower radiator mounting grommets

To install:

12. If installing a new radiator on 1991–93 models, make sure the drain housing is in position. If the replacement did not come with a drain housing, insert the old drain housing into the radiator and press until it snaps into position. Be careful not to press the housing through the hole and into the radiator tank.

13. Install the radiator into the vehicle. Install the upper seal, if applicable, then install the brackets and retaining nuts. Tighten the retaining nuts to 89 inch lbs. (10 Nm). Be sure the L-shaped brackets do not pinch the radiator locating pins and that the radiator moves freely in the grommets.

14. Raise the front of the vehicle and safely support using jackstands.

15. Install the condenser bracket bolts, then if applicable, remove the plugs and install the lower automatic transaxle cooler line. Tighten the condenser bracket bolts to 53 inch lbs. (6 Nm) and the lower transaxle oil cooler line to 20 ft. lbs. (27 Nm).

16. Install the splash shield, remove the supports and lower the vehicle.

17. Install the lower radiator hose with the clamp tangs positioned at 11 o'clock.

18. Install the cooling fan assembly.

19. For 1991 vehicles with an automatic transaxle, connect the upper transaxle cooler line and tighten. For 1992–98 vehicles with an automatic transaxle, connect the upper transaxle cooler line at a 35 degree angle

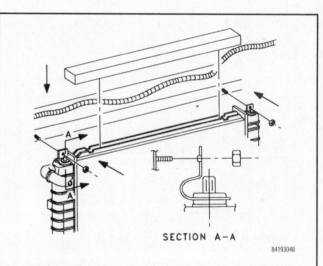

SECTION A–A

Fig. 73 Make sure the L-shaped brackets do not pinch the radiator locating pins

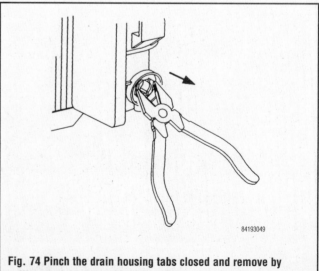

Fig. 74 Pinch the drain housing tabs closed and remove by pulling straight out—1991–93 models

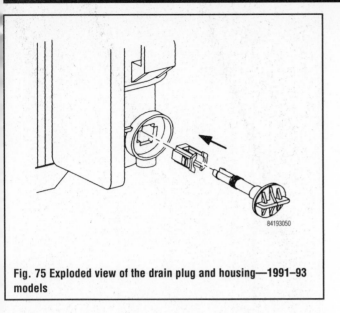

Fig. 75 Exploded view of the drain plug and housing—1991-93 models

Fig. 77 Remove the top, right side fan motor assembly mounting bolt . . .

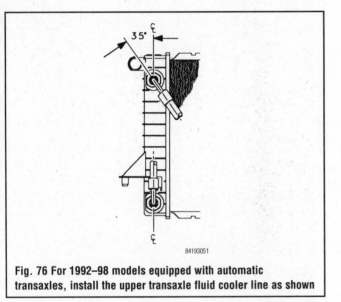

Fig. 76 For 1992-98 models equipped with automatic transaxles, install the upper transaxle fluid cooler line as shown

Fig. 78 . . . then remove the top, left side fan motor assembly mounting bolt

inward from vertical and hold while tightening. Tighten the lower transaxle oil cooler line to 20 ft. lbs. (27 Nm).

20. Install the upper radiator hose with the clamp tangs at 1 o'clock. Tighten the hose clamp to 31 inch lbs. (3.5 Nm).

21. Install the air cleaner housing.

22. Install the intake air ducts and connect the air temperature sensor plug.

23. Close the radiator drain plug and install the cylinder block drain plug. Tighten the block plug to 26 ft. lbs. (35 Nm).

24. Connect the negative battery cable and properly fill the engine cooling system.

25. Start and run the engine to check for coolant leaks.

Electric Cooling Fan

REMOVAL & INSTALLATION

▶ **See Figures 77, 78, 79 and 80**

1. Disconnect the negative battery cable.

2. Remove the air cleaner/intake duct assembly and unplug temperature sensor connector.

Fig. 79 Lift the fan assembly out of the engine compartment enough . . .

Fig. 80 . . . to disengage the fan motor connector and remove the assembly from the vehicle

3. Unplug the wiring harness from the fan motor electrical connector.

4. Remove the top fan motor assembly bolt(s).

5. If equipped with air conditioning and an automatic transaxle, it may be necessary to loosen the top automatic transaxle cooler line and position it aside for clearance.

6. Lift the fan assembly off the lower mounting brackets. Move the assembly to the left and rotate counterclockwise lifting the right side up past the radiator hose, then remove the assembly from the vehicle.

7. If necessary, remove the fan blade and motor from the shroud:

a. Hold the fan and remove the left-hand threaded nut, then pull the fan from the motor shaft.

b. Remove the motor mounting screws.

c. Separate the motor from the shroud.

To install:

8. If removed, install the motor and fan blade. Tighten the left-hand threaded fan nut to 27–44 inch lbs. (3–5 Nm).

9. Install the assembly with the lower left corner 1st. Rotate the assembly clockwise to place the lower left mount under the lower radiator hose and position the assembly onto the mounting brackets.

10. Install and tighten the upper retaining bolt(s) to 53 inch lbs. (6 Nm).

11. If disconnected, install and tighten the automatic transaxle fluid cooler line. Tighten the cooler line fitting to 22 ft. lbs. (27 Nm). For 1992–98 vehicles position the transaxle cooler line 35 degrees inward from vertical and hold while tightening.

12. Install the wiring harness plug to the fan motor electrical connector.

13. If removed, install the air cleaner/intake air duct assembly and the temperature sensor connector.

14. Connect the negative battery cable.

TESTING

▶ **See Figure 81**

When conducting tests on the electric cooling fan and circuit, use of a high impedance Digital Volt Ohm Meter (DVOM) is necessary.

Cooling Fan Inoperative Test

1. Check the 30 amp cooling fan maxifuse (labeled COOL FAN) in the underhood junction block.

2. If the fuse is OK, disconnect the cooling fan motor connector and connect a DVOM from terminal B wire (BLK/RED on 1991–94 models or LIGHT BLUE on 1995–98 models) to ground.

3. Disengage the A/C compressor clutch connector in order to prevent the air conditioning system from overpressurizing and releasing refrigerant from the pressure valve.

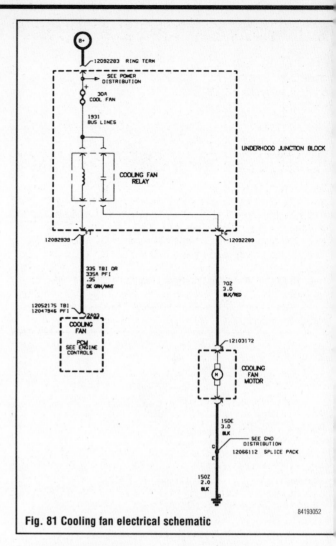

Fig. 81 Cooling fan electrical schematic

4. Start the engine and turn the air conditioning **ON**:

a. If there is voltage at the B wire, check for open at terminal A (BLK) wire to ground which would prevent circuit completion and keep an otherwise good motor from operating. If there is no open and voltage is present, the motor must be faulty. Repair the open wire or replace the faulty motor, as applicable.

b. If there is no voltage at the B wire, turn the engine **OFF** and with the DVOM still connected to terminal B, unplug the PCM connector. Install a jumper wire between the following PCM connector terminal and ground:
- 1991–94 models—PCM terminal J2A03 (DRK GRN/WHT)
- 1995 models—PCM terminal J2A03 (DRK GRN)
- 1996–98 models—PCM terminal J1C01 (DRK GRN)

5. If there is no voltage at the connector, the fan relay is faulty. If there is voltage, the PCM is bad.

6. The fan motor may also be checked by jumping 12 volts to the 2 wires (A = negative, B = positive). The motor should run while voltage is applied.

Cooling Fan Constantly On Test

1. Disconnect the negative battery cable and remove the cooling fan relay from the underhood junction block. Connect the negative battery cable and inspect the motor to see if it still operates.

2. If the motor operates with the relay removed, a short to power exists in the BLK/RED wire (1991–94) or LIGHT BLUE wire (1995–98) from the underhood junction block to the fan motor.

3. If the fan does not operate with the relay removed, substitute a new relay and watch for operation. If the fan operates with a new relay, check the DRK GRN/WHT or DRK GRN wire for a short to ground and repair, if found. If a short does not exist, yet the fan operates, the PCM is likely at fault.

Water Pump

REMOVAL & INSTALLATION

▶ **See Figures 82, 83, 84, 85 and 86**

1. Allow a sufficient amount of time for the engine to cool down.
2. Disconnect the negative battery cable. Drain the engine coolant from the radiator and block drains into a suitable clean container.

✳✳ CAUTION

When draining the coolant, keep in mind that cats and dogs are attracted by ethylene glycol antifreeze, and are quite likely to drink any that is left in an uncovered container or in puddles on the ground. This will prove fatal in sufficient quantity. Always drain the coolant into a sealable container. Coolant may be reused unless it is contaminated or several years old.

3. Remove the serpentine drive belt.
4. Raise the front of the vehicle and support safely using suitable jackstands. Remove the right front tire and inner wheel well splash shield.
5. If access to the water pump is desired from underhood, remove the air conditioning compressor bolts and position the compressor aside with the refrigerant lines intact.
6. Spray the water pump hub with penetrating oil to loosen any rust or corrosion that might bind the pulley and damage it during removal.
7. Remove the water pump pulley bolts and allow the pulley to hang freely on the hub. A 1 in. (25.4mm) block of wood or a hammer handle may be wedged between the pump and crankshaft pulleys to hold the assembly while loosening the retaining bolts.
8. Move the pulley outward or remove as necessary for access and remove the 6 water pump flange bolts. Carefully pull the pump and pulley assembly away from the engine and remove the assembly from the vehicle. If necessary, a gasket scraper may be inserted under the flange, but be careful not to damage the machined aluminum block sealing surface.

To install:

9. Thoroughly clean the gasket mating surfaces of all old gasket material. Apply a small amount of gasket sealant at the outer edges of the bolt holes to hold the gasket in place, then install the gasket onto the water pump assembly.
10. Install the pump assembly with the small bump located next to one of the attaching bolts in the 11 o'clock position. Install and tighten the bolts in a crisscross sequence as shown to 22 ft. lbs. (30 Nm).

Fig. 83 Remove the water pump mounting fasteners . . .

Fig. 84 . . . and remove the water pump from the vehicle

Fig. 82 After removing the water pump pulley mounting bolts, remove the pulley

Fig. 85 Remove and discard the water pump gasket

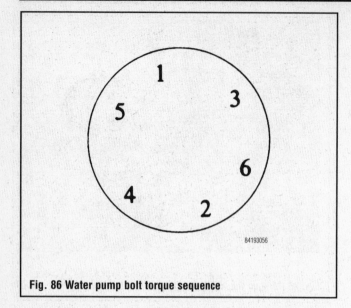

84193056

Fig. 86 Water pump bolt torque sequence

11. Install or reposition the pump pulley, as applicable and tighten the bolts to 19 ft. lbs. (25 Nm). If the pump hub exposed through the pulley is rusty, clean it with a wire brush and apply a thin coat of primer to prevent the pulley from rusting onto the hub.

12. Install the serpentine drive belt, the right splash shield and right tire assembly.

13. If repositioned, install the air conditioning compressor.

14. Close the radiator drain plug and install the cylinder block drain plug. Tighten the block plug to 26 ft. lbs. (35 Nm).

15. Connect the negative battery cable and properly fill the engine cooling system.

16. Operate the engine and check for coolant leaks.

Cylinder Head

REMOVAL & INSTALLATION

♦ See Figures 87 thru 94

❊❊ WARNING

Only remove the cylinder head when the engine is cold. Warpage may result if the cylinder head is removed while the engine is hot.

1. Disconnect the negative battery cable and remove the coolant bottle cap. Drain the engine coolant from the radiator and block drains into a suitable clean container.

❊❊ CAUTION

When draining the coolant, keep in mind that cats and dogs are attracted by ethylene glycol antifreeze, and are quite likely to drink any that is left in an uncovered container or in puddles on the ground. This will prove fatal in sufficient quantity. Always drain the coolant into a sealable container. Coolant may be reused unless it is contaminated or several years old.

2. Remove the air cleaner/inlet duct assembly. Disconnect the PCV valve and fresh air hose from the camshaft cover.

3. Disconnect the accelerator cable from the throttle body and the bracket from the intake manifold.

4. Properly relieve the fuel system pressure.

5. Label and disconnect the following electrical connectors from the cylinder head assembly and components, as applicable. Long nose pliers are necessary to disconnect the coolant temperature connectors. When disconnected, position the electrical harness over the underhood junction block and battery cover.

a. Coolant temperature and PCM connectors. These connectors are located on the rear side of the cylinder head for DOHC engines.

b. The single injector connector (TBI Engine) or the 4 injector connectors (MFI Engine).

c. Idle Air Control (IAC) valve.

d. Manifold Air Pressure (MAP) sensor.

e. Throttle Position Sensor (TPS).

f. Exhaust Gas Recirculation (EGR) solenoid.

g. Label and disconnect the spark plug wires from the plugs.

h. Oxygen sensor.

i. Air conditioning compressor.

6. Label and disconnect the following vacuum hoses, as applicable, from the area around the cylinder head assembly.

a. Canister purge valve.

b. EGR valve.

c. MAP sensor, for the 1991–94 SOHC engine only.

d. Brake booster vacuum hose at the intake manifold or the brake booster.

e. Throttle Body Injection (TBI) unit assembly on 1991–94 SOHC Engines or the throttle body for all MFI engines.

f. Fuel regulator, for 1991–97 DOHC or 1995–97 SOHC engines.

7. Disconnect the upper radiator hose at the cylinder head outlet, the de-aeration hose at the intake manifold, or the heater hose at the intake manifold or front of dash.

8. Remove the bolt which retains the fuel lines to the intake manifold assembly. Disconnect the fuel feed and return lines from the fuel rail or throttle body, as applicable.

9. For 1992–98 vehicles with a torque axis mount system, unclip the lower splash shield for access. Place a 1 in. **x** 1 in. **x** 2 in. long block of wood between the torque strut and cradle to ease removal and installation of the torque engine mount. Remove the 3 right side upper engine torque axis to front cover nuts and the 2 mount to midrail bracket nuts, allowing the powertrain to rest on the block of wood.

➡**Placing a block of wood under the torque axis mount prior to removing the upper mount will allow the engine to rest on the wood, thus preventing the engine from shifting. This will allow you to install the mount during assembly without jacking or raising the engine.**

10. Remove the serpentine drive belt and belt tensioner. It is not necessary to remove the water pump pulley, however, for 1992–98 vehicles it will be necessary to remove the idler pulley to access the engine front cover.

11. For SOHC engines, disconnect the de-aeration line at the cylinder head water outlet and from the support bracket.

12. Remove the fasteners and the camshaft cover, then inspect the cover's silicone insulators for cracks or deterioration and replace as necessary. Be sure to cover the valve train area to prevent dirt and debris from entering the engine.

13. Remove the power steering pump bracket attaching bolts (3 for the SOHC engine or 5 for the DOHC engine) and position the assembly next to the right side front of the dash panel away from the intake manifold and cylinder head. It is not necessary to remove the water pump pulley.

14. If equipped, remove the 3 air conditioning compressor front bracket bolts attached to the cylinder head and block, then remove the rear bracket bolts from the compressor or engine. Do not discharge the system or disconnect the refrigerant lines. Support the compressor out of the way, from the vehicle front support bar.

15. Raise and support the front of the vehicle safely using jackstands.

16. Drain the engine oil into a suitable drainpan.

17. Remove the right side tire and splash shield.

18. For DOHC engines, remove the intake manifold support brace bolt ached to the intake manifold next to the alternator.

19. Remove the crankshaft damper/pulley assembly. Use a strap wrench or a block of wood wedged between the pulley spoke and the rear lower side of the front cover to hold the assembly while removing the bolt. Then use a 3-jaw puller on the jaw slots cast into the pulley to remove the assembly.

20. Disconnect the exhaust pipe from the manifold, then remove and discard the gasket.

21. Install crankshaft sprocket retainer tool SA9104E or equivalent, with the flat side toward the sprocket. Properly remove the engine front/timing chain cover; refer to the Timing Chain Cover procedure in this section.

✳✳ WARNING

Be sure to properly install the crankshaft sprocket retainer tool. Failure to hold the crankshaft timing sprocket in place will cause timing chain damage.

22. Rotate the crankshaft clockwise to the position 90 degrees from Top Dead Center (TDC) so the timing mark and keyway align with the main bearing cap split line. This will make sure pistons will not contact the valves upon assembly.

23. Remove the timing chain, tensioner, guides, camshaft sprocket(s) and chain. Use a ⅞ in. (21mm) wrench to hold the camshaft when removing the sprocket bolts.

24. For the SOHC TBI engine, remove the throttle body assembly and cover the intake manifold opening. Be sure to remove and discard the old throttle body gasket.

25. Use a 6 point socket to remove the 10 cylinder head bolts in several passes of the proper sequence. Failure to follow the proper sequence or removal of the head when hot could result in head warpage or cracking. Also, the use of a 12 point socket on the cylinder head bolts may round the bolt heads.

26. Lift the cylinder head from the dowels, if necessary use a small pry-bar for leverage between the cylinder head and block bosses. Be careful not to damage the sealing surfaces if prying is necessary to remove the head from the block.

27. If necessary, remove the intake manifold or the exhaust manifold by loosening the mounting nuts in the proper sequence. If any cylinder head studs back out, the threads should be cleaned, the studs carefully installed and then tightened to 106 inch lbs. (12 Nm).

To install:

➡Before installing the cylinder head, it should be cleaned and inspected for excessive wear or damage. Information on cleaning and inspecting the cylinder head can be found under the ENGINE RECONDITIONING heading, later in this section.

28. If removed, install the intake manifold and/or the exhaust manifold and new gasket(s). Tighten to specification in the proper sequence.

29. Clean the gasket mating surfaces. Be careful not to damage the aluminum components. Make sure the block bolt holes are clean of any residual sealer, oil or foreign matter.

30. Using a dial gauge at 4 points around each cylinder, check that the cylinder liners are flush or do not deviate more than 0.0005 in. (0.013mm).

31. Make sure the crankshaft is still 90 degrees past TDC and that the camshaft(s) are properly positioned with the dowel pin(s) at the 12 o'clock position to prevent valve damage. Install the cylinder head gasket and carefully guide the head into place over the dowels.

32. If the head bolts and/or the block were replaced, install the bolts and tighten in sequence to 48 ft. lbs. (65 Nm) to insure proper clamp load, then remove the bolts.

33. Coat the cylinder head bolts with clean engine oil and thread the bolts by hand until finger-tight. Tighten the bolts in sequence to 22 ft. lbs. (30 Nm).

34. Tighten the cylinder head bolts again, in sequence to 33 ft. lbs. (45 Nm) for SOHC engines or to 37 ft. lbs. (50 Nm) for DOHC engines. Install Snap-on® torque angle gauge tool 360, or equivalent, and calibrate the tool to zero. In sequence, tighten each cylinder head bolt an additional 90 degrees.

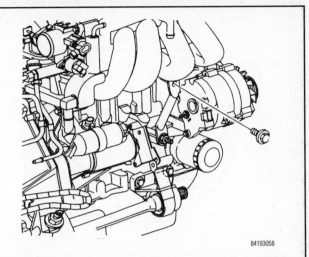

Fig. 87 Remove the intake manifold support brace bolt, located next to the alternator—DOHC engine

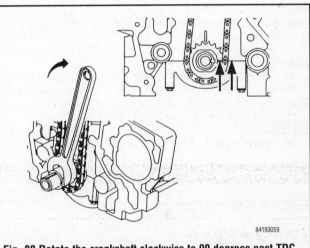

Fig. 88 Rotate the crankshaft clockwise to 90 degrees past TDC (the crankshaft sprocket timing mark will be at 3 o'clock) to prevent valve damage during assembly

Fig. 89 Use a 6-point socket to remove the 10 cylinder head bolts—make several passes in the proper sequence

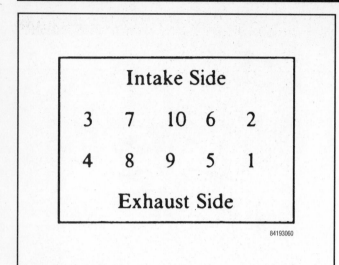

```
        Intake Side

  3    7    10   6    2

  4    8    9    5    1

        Exhaust Side
```
84193060

Fig. 90 Cylinder head bolt removal sequence

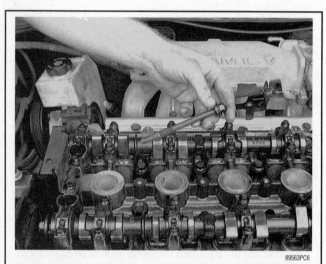
89563PC6

Fig. 91 Remove the cylinder head bolts. Examine each bolt for signs of stretching or damage and replace, if necessary

89563PC7

Fig. 92 With an assistant, remove the cylinder head carefully from the top of the engine block

35. Install the timing chain, sprockets, guides and tensioner. Then install the front cover assembly. Refer to the appropriate procedures in this section.

36. Position a new gasket, then connect the exhaust pipe to the manifold. Install and tighten the fasteners to 23 ft. lbs. (31 Nm).

37. If not already done, remove the crankshaft sprocket retainer tool. Apply a thin film of RTV sealant to the damper/pulley assembly flange and washer only. Install the crankshaft damper/pulley assembly and tighten the bolt to 158 ft. lbs. (214 Nm) while holding the pulley with a strap wrench or block of wood.

38. For DOHC vehicles, install the intake manifold support brace bolts next to the alternator, then tighten the block bolt to 33 ft. lbs. (45 Nm) and tighten the manifold bolt to 22 ft. lbs. (30 Nm).

39. Apply a small drop of RTV across the cylinder head and front cover T-joints. Inspect the old camshaft cover gasket and replace if damaged. Install the gasket and the camshaft cover. Tighten the fasteners uniformly to 22 ft. lbs. (30 Nm) for SOHC vehicles or in proper sequence to 89 inch lbs. (10 Nm) for DOHC vehicles.

40. Install the drive belt tensioner and tighten the bolt to 22 ft. lbs. (30 Nm). For 1992–98 vehicles, install the idler pulley and tighten the fasteners to 20 ft. lbs. (27 Nm).

41. If not done during removal, drain the engine oil and change the filter, then install the drain plug and tighten to 26 ft. lbs. (35 Nm).

42. If removed, verify the gaps on all spark plugs and install. Tighten to 20 ft. lbs. (27 Nm).

43. For SOHC TBI engines, install a new gasket and the throttle body assembly. Tighten the assembly retainers to 24 ft. lbs. (33 Nm).

44. Install the power steering pump assembly to the bracket, then tighten the bolts to 22 ft. lbs. (30 Nm).

45. If equipped, install the air conditioning compressor and bolts. Tighten the rear bracket bolts to 19 ft. lbs. (25 Nm), then tighten the front bracket bolts to 35 ft. lbs. (47 Nm).

46. Install the accessory drive belt making sure the belt is properly aligned on the pulley.

47. For 1992–98 vehicles with a torque axis mounting, install the 2 mount to midrail bracket nuts and tighten to 37 ft. lbs. (50 Nm). Install the 3 upper mount to engine front cover nuts and tighten them uniformly to 37 ft. lbs. (50 Nm). Remove the support block of wood after the assembly is installed.

48. Install the splash shield, then install the tire and wheel assembly. Tighten the lug nuts, in a crisscross pattern, to 103 ft. lbs. (140 Nm).

49. Connect the following applicable vacuum hoses disconnected during removal.
   a. Canister purge valve.
   b. EGR valve.
   c. MAP sensor, for the SOHC TBI engine only.
   d. Brake booster vacuum hose at the intake manifold or the brake booster.
   e. TBI unit assembly on 1991–94 SOHC Engines or the throttle body for all MFI engines.
   f. Fuel regulator, for 1991–97 DOHC or 1995–97 SOHC engines.

50. Position the wiring harness and install the following applicable wire connectors removed during disassembly:
   a. Coolant temperature and PCM connectors. These connectors are located on the rear side of the cylinder head for DOHC engines.
   b. The single injector connector (TBI Engine) or the 4 injector connectors (MFI Engine).
   c. IAC valve.
   d. MAP sensor.
   e. TPS.
   f. EGR solenoid.
   g. Spark plug wires from the plugs.
   h. Oxygen sensor.
   i. Air conditioning compressor.

51. Install the throttle cable bracket and tighten the fastener to 19 ft. lbs. (25 Nm). Connect the cable, then verify that it is properly routed and not binding.

52. Connect the upper radiator hose at the cylinder head outlet, the de-aeration hose at the intake manifold, or the heater hose to the intake manifold or front of dash.

Fig. 93 Remove and discard the cylinder head gasket

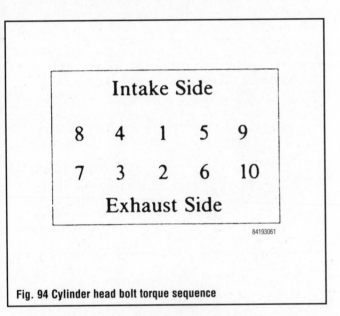

Fig. 94 Cylinder head bolt torque sequence

Intake Side

| 8 | 4 | 1 | 5 | 9 |
| 7 | 3 | 2 | 6 | 10 |

Exhaust Side

Oil Pan

REMOVAL & INSTALLATION

♦ **See Figures 95 and 96**

1. Raise the front of the vehicle and support it safely using jackstands. Remove the plug and drain the engine oil from the pan and crankcase.

2. Disconnect the fasteners from the exhaust manifold flange and the pipe rear flange, then remove the front exhaust pipe and gaskets from the vehicle.

3. Remove the engine stiffening bracket and the flywheel cover.

4. Remove the right wheel and splash shield, then loosen the 4 front motor mount bolts. Back the bolts out about ½ in. (12mm).

5. Remove all the oil pan bolts. For vehicles with a manual transaxle, an 8mm flex socket may be used to access the rear oil pan bolts located next to the flywheel.

6. Using SA9123E, or an equivalent RTV cutter tool, separate the oil pan from the engine. Drive the tool around the pan to shear the RTV seam, then tap the pan sideways with a rubber mallet to loosen.

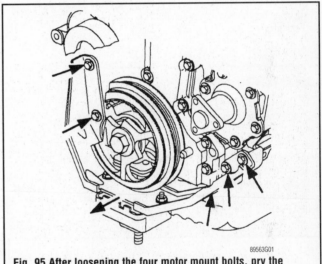

Fig. 95 After loosening the four motor mount bolts, pry the mount away from the engine block for better clearance

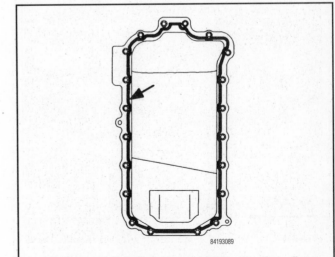

Fig. 96 Apply a 0.16 in (4mm) bead of RTV to the oil pan flange to the inner side of the bolt holes

53. Apply a few drops of clean engine oil to the male fuel line fittings. Connect any fuel line fittings and install the feed/return lines.

54. On TBI engines, tighten the throttle body fittings to 19 ft. lbs. (25 Nm). On MFI engines, tighten the fuel rail and pressure regulator fittings to 133 inch lbs. (15 Nm), as applicable. Install and tighten the fuel bracket retaining bolt.

55. Install the air cleaner and intake duct assembly.

56. Add engine oil and properly fill the engine cooling system.

57. Thoroughly inspect the engine compartment area, especially the cylinder head, to be sure that all wires, hoses and lines have been connected. Also inspect to make sure that all components and fasteners have been properly installed.

58. Connect the negative battery cable.

59. Prime the fuel system by cycling the ignition a few times without cranking the engine, then start the engine and check for leaks.

60. Operate the engine at idle for 3–5 minutes and listen or unusual noises. If the lifters are noisy or the cylinders are misfiring, warm the engine to normal operating temperature running at less than 2000 rpm. Once the engine is warm and the thermostat has opened, cycle the engine between idle and 3000 rpm for 10 minutes or drive the vehicle at least 5 miles to purge air from the lifters. If air cannot be purged, the faulty lifters must be replaced.

61. Verify proper coolant level. Add coolant, if necessary, after the engine has cooled.

7. Pry the engine mount away from the engine as necessary and remove the oil pan. Be careful not to damage or score component surfaces when prying.

To install:

8. Carefully clean the gasket mating surfaces with a scraper and solvent.

9. Apply a 0.16 in. (4mm) bead of RTV sealer to the pan flange. Make sure the RTV is applied to the inner side of the flange from the bolt holes as shown.

10. Install the oil pan within 3 minutes of RTV application and tighten the bolts to 80 inch lbs. (9 Nm).

11. Tighten the front mount bolts to 37 ft. lbs. (50 Nm).

12. Install the right splash shield and wheel.

13. Install the engine stiffening bracket and the flywheel cover.

14. Install the exhaust pipe. Tighten the pipe to manifold nuts in a crosswise pattern to 23 ft. lbs. (31 Nm) and the pipe to converter bolts to 33 ft. lbs. (45 Nm).

15. Remove the jackstands and carefully lower the vehicle, then fill the engine crankcase with clean engine oil immediately in order to prevent an attempt to start the engine without oil.

16. Start the engine and check for leaks.

Oil Pump

The oil pump is located in a housing built into the lower portion of the timing chain/engine front cover. The pump may be serviced after the cover has been removed from the vehicle.

REMOVAL

▶ See Figures 97 and 98

1. Disconnect the negative battery cable and drain the engine oil.

2. Remove the timing chain front cover. See the procedure in this section.

3. Remove the oil pump cover Torx® bolts using a suitable impact driver. Because the pump cover screws are coated with a sealant to prevent oil leakage, they must be replaced when removed.

4. Remove the driven and drive rotors.

5. If necessary, remove the relief valve using regulator valve removal tool SA9103E or an equivalent puller, to withdraw the valve from the bore.

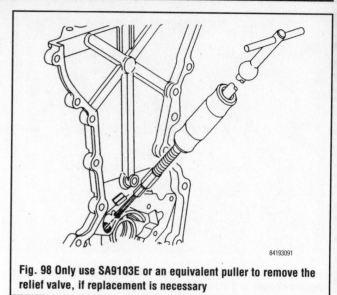

Fig. 98 Only use SA9103E or an equivalent puller to remove the relief valve, if replacement is necessary

Whenever the valve is removed, it must be replaced because the puller jaws will damage the valve sealing seat.

INSPECTION

▶ See Figures 99, 100 and 101

1. With the timing chain front cover and the oil pump body cover removed, use a feeler gauge to measure the clearance between the driven rotor and pump body. Clearance should not exceed 0.011 in. (0.277mm).

2. Use a feeler gauge to measure the clearance between the both rotor tips. Clearance should not exceed 0.006 in. (0.150mm).

3. Using Plastigage® or a feeler gauge, temporarily install the pump cover and measure the rotor-to-cover clearance. Clearance should not exceed 0.005 in. (0.128mm).

4. If necessary, replace the pump components and/or the front cover assembly.

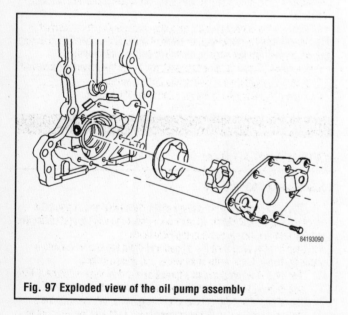

Fig. 97 Exploded view of the oil pump assembly

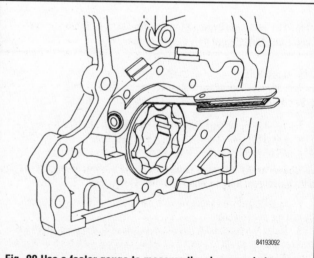

Fig. 99 Use a feeler gauge to measure the clearance between the driven rotor and pump body

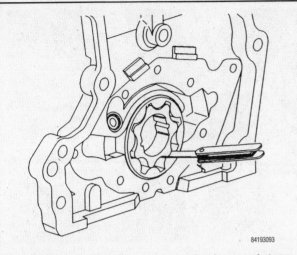

Fig. 100 Use a feeler gauge to measure the clearance between both rotor tips

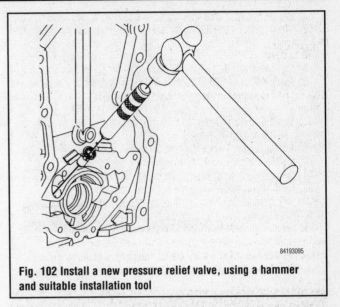

Fig. 102 Install a new pressure relief valve, using a hammer and suitable installation tool

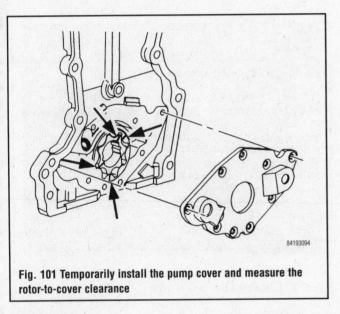

Fig. 101 Temporarily install the pump cover and measure the rotor-to-cover clearance

INSTALLATION

◆ See Figure 102

1. If removed, install a new relief valve into the cover bore. Coat the valve with clean engine oil and tap it into the bore using a hammer and SA9103E or an equivalent installer tool.

➡ Whenever the oil pump is installed, the assembly must be packed with petroleum jelly in order to prime the pump.

2. Install the driven and drive rotors into the pump with the chamfer toward the front oil seal.

3. Install the pump body cover and secure with new bolts which are covered with a sealant to prevent oil leakage. Tighten the bolts to 97 inch lbs. (11 Nm).

4. Install the timing chain front cover.

5. Properly fill the engine crankcase, start the engine and check for leaks.

Crankshaft Damper/Pulley

REMOVAL & INSTALLATION

1. Raise the front of the vehicle and support it safely using jackstands, then remove the plug and drain the engine oil from the pan and crankcase.

2. Remove the wheel and splash shield from the right front of the vehicle.

3. Remove the serpentine drive belt and, if necessary, remove the drive belt tensioner.

4. Using a strap wrench or a piece of wood wedged between the damper spoke and the lower side of the engine front cover, hold the damper and remove the bolt. With a suitable 3-jaw puller and the slots cast into the damper, pull the crankshaft damper/pulley assembly from the crankshaft.

To install:

5. Apply a thin film of RTV between the damper/pulley assembly flange and washer only, the washer and bolt head flange are designed to prevent oil leakage.

6. Position the crankshaft damper/pulley assembly to the crankshaft, then secure using a strap wrench or piece of wood (as accomplished during removal) and tighten the retaining bolt to 159 ft. lbs. (215 Nm).

7. If removed, install the drive belt tensioner. Install the serpentine drive belt.

8. Install the right front splash shield and the wheel.

9. Carefully remove the jackstands and lower the vehicle.

Timing Chain Cover

REMOVAL & INSTALLATION

◆ See Figures 103 thru 123

1. Disconnect the negative battery cable, then raise and support the front of the vehicle sufficiently to work both under the vehicle and under the hood. Be sure to properly position the jackstands.

2. Remove the plug from the oil pan and drain the engine oil into a suitable container. Remove the right wheel and splash shield.

3. For 1992–98 vehicles with a torque axis mount system, place a 1 in. x 1 in. x 2 in. long block of wood between the torque strut and cradle to ease removal and installation of the torque engine mount. Remove the 3 right side upper engine torque axis to front cover nuts and the 2 mount to midrail bracket nuts, allowing the powertrain to rest on the block of wood.

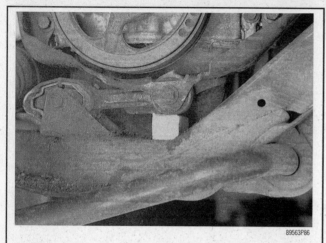

Fig. 103 Place a 1 in. x 1 in. x 2 in. long piece of wood between the torque strut and cradle before removal of the torque engine mount

Fig. 104 Loosen the torque engine mount-to-cylinder head retaining bolts . . .

Fig. 105 . . . then loosen the torque engine mount-to-body retaining bolts

Fig. 106 Remove the torque engine mount from the engine

Fig. 107 Loosen the belt tension idler pulley bolt . . .

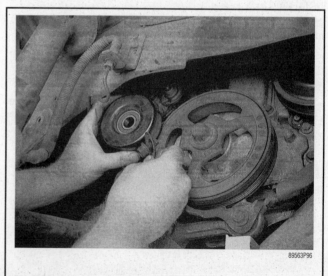

Fig. 108 . . . and remove the pulley

➡Placing a block of wood under the torque axis mount prior to removing the upper mount will allow the engine to rest on the wood, thus preventing the engine from shifting. This will allow you to install the mount during assembly without jacking or raising the engine.

4. Remove the serpentine drive belt and belt tensioner. It is not necessary to remove the water pump pulley; however, for 1992–98 vehicles the idler pulley must be removed to access the engine front cover.

5. Remove the power steering pump attaching bolts and support the assembly aside with the lines attached. If equipped, and if necessary for access to the front cover bolts, separate the A/C compressor from the bracket and support the assembly aside. Again, keep the compressor lines attached and do not discharge the system.

6. Remove the camshaft cover. Cover the valve train assemblies to protect them from foreign debris or dirt.

7. Using a strap wrench or a piece of wood wedged between the damper spoke and the lower side of the engine front cover, hold the damper and remove the bolt. With a suitable 3-jaw puller and the slots cast into the damper, pull the crankshaft damper/pulley assembly from the crankshaft.

8. Install the special oil seal replacement tool SA9104E or equivalent, to make sure the front crankshaft timing sprocket is held firmly in place and prevent guide damage. Install with the flat side towards the crankshaft sprocket.

9. Remove the front 4 oil pan bolts, then using a suitable RTV cutting tool, cut the front seal away from the front cover.

10. Spray the 2 dowel pin holes with penetrating oil to facilitate front cover removal from the dowel pins.

11. Remove the front cover bolts. For 1992–98 vehicles, one bolt is located above the serpentine drive belt pulley, under the torque axis mount flange.

12. Using a small suitable tool, carefully pry the cover away from the cylinder block at the pry location tabs which are provided. Remove the cover from under the hood or through the wheel well. Be sure to cover the front of the engine to prevent debris or dirt from entering the oil galley openings and oil pan. If necessary, pry the front cover oil seal from the cover for replacement.

To install:

13. Make sure the oil galleys are clear. Carefully clean the gasket mating surfaces with a scraper or wire brush and carburetor solvent, brake clean or alcohol. Use a ³⁄₁₆ in. drill bit and tap handle to clean the front cover holes. If removed, seat a new front cover oil seal using the installation tool with a suitable press, or wait until the cover is installed, then use a threaded installation tool and the crankshaft's threads to pull the seal into position. For additional details, refer to the Timing Chain Cover Oil Seal replacement procedure, later in this section.

➡If the engine front cover casting or assembly is replaced on 1992–98 vehicles, the 3 torque axis mount studs should also be replaced. Tighten the new studs to 19 ft. lbs. (25 Nm).

14. Apply a 0.08 in. (2mm) bead of RTV sealer along the vertical sealing surfaces of the front cover to the inside of the bolt holes and to the front of the oil pan. Extra sealer is necessary at the oil pan and cylinder head joints. For DOHC engines, apply a thin bead around the one center cover bolt hole on 1991 vehicles or the 2 inner cover bolt holes on 1992–98 vehicles. Be sure to assemble the front cover to the engine within 3 minutes of RTV application.

15. If removed, install the crankshaft sprocket retaining tool to align the oil pump and crankshaft during cover installation. Position the front cover to the engine and install the bolts. Tighten the perimeter bolts starting at the center and working outwards on both sides to 19 ft. lbs. (25 Nm) for SOHC engines or to 22 ft. lbs. (30 Nm) for DOHC engines.

16. Install and tighten the front cover center or inner bolts to 89 inch lbs. (10 Nm) except for the upper inside bolt on 1992–98 DOHC engines which should be tightened to 22 ft. lbs. (30 Nm). Install the 4 oil pan front bolts and tighten to 80 inch lbs. (9 Nm).

17. After front cover installation, spray 6–12 squirts of oil through the front oil seal drain back hole to verify that it is not plugged.

Fig. 109 Loosen the serpentine belt tensioner mounting bolts . . .

Fig. 110 . . . and remove the belt tensioner from the vehicle

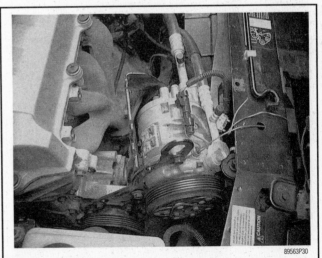

Fig. 111 With the refrigerant lines still attached, unfasten the A/C compressor and support the assembly aside

Fig. 112 Remove the camshaft cover. Place a cover over the valve train assembly to protect it from dirt and debris

Fig. 115 Remove the damper pulley from the crankshaft stub

Fig. 113 After wedging a piece of wood between the damper spoke and lower side of the engine front cover, loosen . . .

Fig. 116 Install a special tool to hold the crankshaft timing sprocket firmly in place

Fig. 114 . . . then remove the damper pulley retaining bolt

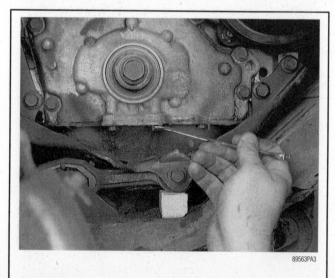

Fig. 117 Remove the 4 oil pan-to-timing cover bolts . . .

Fig. 118 . . . then, with a suitable RTV cutting tool, cut away the front seal

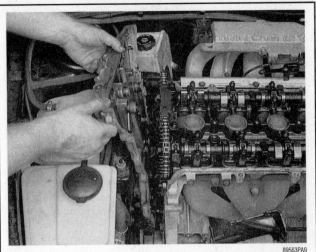

Fig. 121 Lift the timing chain front cover up and out of the vehicle

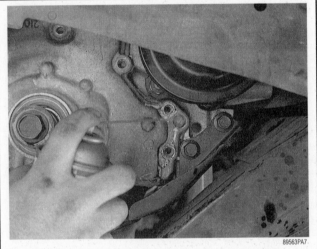

Fig. 119 Spray the 2 dowel pin holes with penetrating oil to facilitate front cover removal

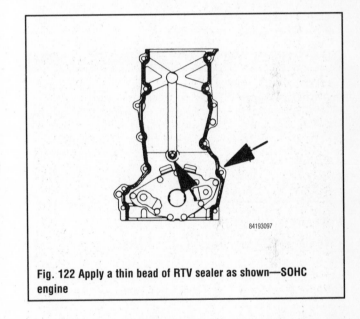

Fig. 122 Apply a thin bead of RTV sealer as shown—SOHC engine

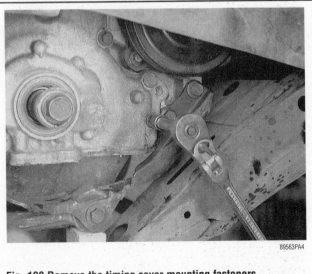

Fig. 120 Remove the timing cover mounting fasteners

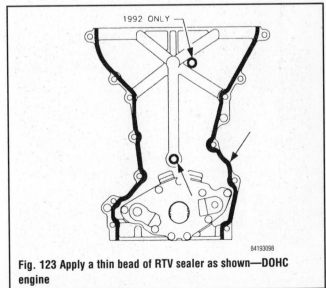

Fig. 123 Apply a thin bead of RTV sealer as shown—DOHC engine

18. Apply a thin film of RTV between the damper/pulley assembly flange and washer only; the washer and bolt head flange are designed to prevent oil leakage.

19. Remove the crankshaft retaining tool and position the crankshaft damper/pulley assembly, then secure using the wood or strap wrench (as accomplished during removal) and tighten the bolt to 159 ft. lbs. (215 Nm).

20. Apply a small drop of RTV across the cylinder head and front cover T-joints. Inspect the old camshaft cover gasket and replace if damaged. Install the gasket and the camshaft cover. Tighten the fasteners uniformly to 22 ft. lbs. (30 Nm) for SOHC vehicles or in proper sequence to 89 inch lbs. (10 Nm) for DOHC vehicles.

21. Position and install the A/C compressor assembly and/or the power steering pump assembly, as applicable.

22. Install the idler pulley if removed, then install the belt tensioner and the serpentine drive belt.

23. For 1992–98 vehicles equipped with a torque axis mount system, install the 2 engine mounts to midrail bracket nuts and tighten to 37 ft. lbs. (50 Nm). Next install the 3 mount to front cover nuts, tighten them uniformly to 37 ft. lbs. (50 Nm) in order to prevent front cover damage. Then, remove the block of wood from under the torque strut.

24. Install the splash shield and the wheel assembly, then remove the jackstands and carefully lower the vehicle.

25. Immediately fill the engine crankcase with clean engine oil and connect the negative battery cable.

26. Start the engine and check for leaks.

Timing Chain Cover Oil Seal

The oil seal can be replaced either with the front cover installed or with the cover off of the engine. The seal should never be installed by tapping with a hammer; in both cases, an appropriate installation tool should be used. If the cover is removed, a press can be used with the tool installed to position the seal. The safest method to prevent cover damage is probably to install the seal with the front cover on the engine, using the tool and crankshaft threads to draw the seal into position.

REPLACEMENT

With Cover Removed

▶ See Figures 124 and 125

1. Note the depth to which the factory seal was installed into the front cover, then use a suitable prytool to carefully remove the oil seal from the front cover. Be careful not to score or damage the front cover or crankshaft surfaces.

2. Clean the seal bore and oil drain back passage.

3. Place the engine front cover on the base of a suitable arbor press.

4. Position the seal to the front cover and place oil seal and thread service seal installer tool SA9104E, or equivalent, over the seal.

5. Press the seal into the engine front cover approximately 0.04 in. (1mm) further into the engine front cover than the factory seal removed earlier.

6. Install the timing chain front cover to the engine.

With Cover Installed

▶ See Figure 125

1. Disconnect the negative battery cable, then raise and support the front of the vehicle sufficiently to work both under the vehicle and in the wheel well. Be sure to properly position the jackstands.

2. Remove the plug from the oil pan and drain the engine oil into a suitable container. If the right side of the vehicle is positioned higher than the left, it may not be necessary to drain the oil from the crankcase. If this

Fig. 124 Remove the oil seal using a prying tool

is desired, a drain pan should be positioned below the front cover to catch any oil that does leak. Remove the right wheel and splash shield.

3. For 1992–98 vehicles with a torque axis mount system, it may be necessary to remove the engine mount for access to the components. If this is determined necessary, place a 1 in. x 1 in. x 2 in. long block of wood between the torque strut and cradle to ease removal and installation of the torque engine mount. Remove the 3 right side upper engine torque axis to front cover nuts and the 2 mount to midrail bracket nuts, allowing the powertrain to rest on the block of wood.

➡Placing a block of wood under the torque axis mount prior to removing the upper mount will allow the engine to rest on the wood, thus preventing the engine from shifting. This will allow you to install the mount during assembly without jacking or raising the engine.

4. Using a strap wrench or a piece of wood wedged between the damper spoke and the lower side of the engine front cover, hold the damper and remove the bolt. With a suitable 3-jaw puller and the slots cast into the damper, pull the crankshaft damper/pulley assembly from the crankshaft.

5. Use a suitable prytool to carefully pry the front oil seal from the front cover. Be careful not to damage the front cover or crankshaft.

6. Clean the seal bore and oil drain back passage.

To install:

7. Make sure the oil drain back is free of contamination. Position the oil seal and thread service seal installer tool SA9104E or equivalent onto the end of the crankshaft. Use the tool to draw the seal into position. Never tap on the seal or the seal installer with a hammer.

8. Apply a thin film of clean engine oil to the new seal lip.

9. Position the crankshaft damper/pulley assembly, then secure using the wood or strap wrench (as accomplished during removal) and tighten the bolt to 159 ft. lbs. (215 Nm).

10. If removed, install the 2 engine mounts to midrail bracket nuts and tighten to 37 ft. lbs. (50 Nm). Next install the 3 mount to front cover nuts, tighten them uniformly to 37 ft. lbs. (50 Nm) in order to prevent front cover damage. Then remove the block of wood from under the torque strut.

11. Install the splash shield and the wheel assembly, then remove the jackstands and carefully lower the vehicle.

12. Immediately fill the engine crankcase with clean engine oil and connect the negative battery cable.

13. Start the engine and check for leaks.

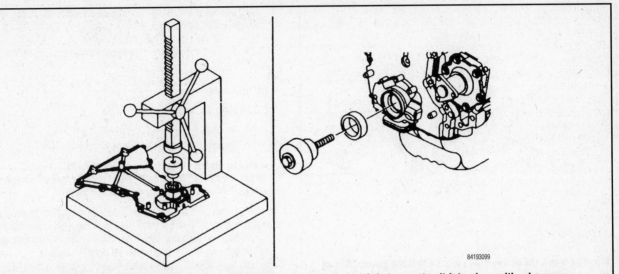

Fig. 125 Install the new front cover oil seal using a press or the threaded installation tool, but never tap it into place with a hammer

Timing Chain and Sprockets

REMOVAL & INSTALLATION

SOHC Engine

▶ See Figures 126, 127, 128 and 129

1. Disconnect the negative battery cable.
2. Remove the timing chain front cover.

➡ During timing chain and sprocket removal, position the crankshaft 90 degrees past Top Dead Center (TDC), to make sure the pistons will not contact the valves upon assembly.

3. Carefully rotate the crankshaft clockwise so the timing mark on the crankshaft sprocket and keyway align with the main bearing cap split line (90 degrees past TDC).
4. Remove bolts, then remove the timing guides and tensioner.
5. Remove the camshaft sprocket bolt, using a 7/8 in. (21mm) wrench to hold the camshaft. Then remove the timing chain and camshaft sprocket. Remove the crankshaft sprocket, if necessary.

To install:

6. Inspect the chain for wear and damage. Check the inside diameter of the chain, it should be no more than 16.77 in. (426mm). Inspect the chain guides for wear or cracks and the timing sprockets for teeth or key wear. Replace components as necessary.
7. Verify that the crankshaft is positioned 90 degrees clockwise past TDC from the keyway (keyway at 3 o'clock).
8. Bring the camshaft up to No. 1 TDC by loosely installing the sprocket and rotating the sprocket until the timing pin can be inserted. The camshaft contains wrench flats to assist in turning the shaft. The dowel pin should be at 12 o'clock when the camshaft is at TDC and a timing pin (3/16 in. drill bit) should then install at about the 8 o'clock position.
9. If removed, install the crankshaft sprocket, then rotate the crankshaft counterclockwise 90 degrees up to No. 1 TDC (keyway at 12 o'clock).
10. Position the chain under the crankshaft sprocket and over the camshaft sprocket. If necessary remove the camshaft sprocket, then slide the camshaft sprocket into position with the chain already engaged. The timing chain should be positioned so that one silver link plate aligns with the reference mark on the camshaft sprocket and the other aligns with the downward tooth (at the 6 o'clock position) on the crankshaft sprocket. The letters FRT on the camshaft sprocket must face forward, away from the

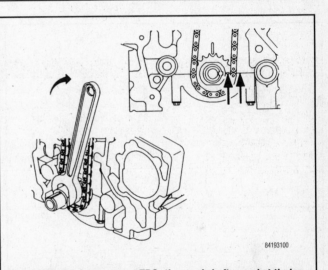

Fig. 126 At 90 degrees past TDC, the crankshaft sprocket timing mark and keyway will align with the main bearing cap split line

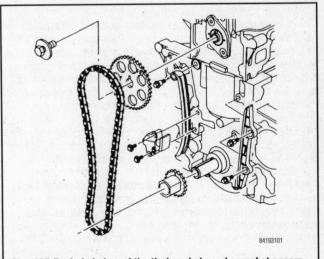

Fig. 127 Exploded view of the timing chain and sprocket assembly—SOHC engine

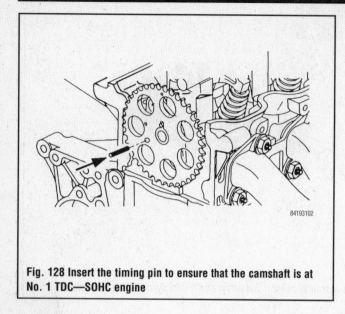

Fig. 128 Insert the timing pin to ensure that the camshaft is at No. 1 TDC—SOHC engine

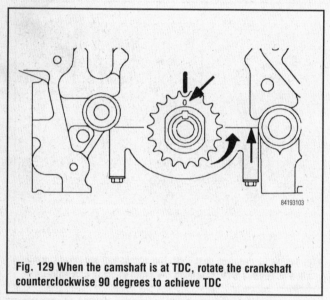

Fig. 129 When the camshaft is at TDC, rotate the crankshaft counterclockwise 90 degrees to achieve TDC

cylinder head and excess chain slack should be located on the tensioner side of the block.

11. Temporarily install the timing pin to verify proper alignment of the camshaft and sprocket, then install and tighten the sprocket bolt to 75 ft. lbs. (102 Nm). Again, use a wrench on the camshaft flats to hold the shaft in position while tightening the bolt. Do not allow the camshaft retaining bolt to torque against the timing pin or cylinder head damage will result.

12. Install the chain guides with the words FRONT facing out. Install the fixed guide first and verify the chain is snug against the guide, then install the pivot guide. Tighten the bolts to 19 ft. lbs. (26 Nm) and verify that the pivot guide moves freely.

13. Retract the tensioner plunger and pin the ratchet lever using a 1/8 in. No. 31 drill bit inserted in the alignment hole at the bottom front of the component. Install the tensioner and tighten the bolts to 14 ft. lbs. (19 Nm), then remove the drill bit.

14. Make one final check to verify all components are properly timed, then remove all timing pins.
15. Install the timing chain front cover.
16. Connect the negative battery cable, start the engine and check for leaks.

DOHC Engine

▶ **See Figures 130 thru 139**

1. Disconnect the negative battery cable.
2. Remove the timing chain front cover.

➡**During timing chain and sprocket removal, position the crankshaft 90 degrees past Top Dead Center (TDC) to make sure the pistons will not contact the valves upon assembly.**

3. Carefully rotate the crankshaft clockwise so the timing mark on the crankshaft sprocket and keyway align with the main bearing cap split line.
4. Remove the bolts, then remove the timing guides and tensioner.
5. Remove the camshaft sprocket bolts, using a 7/8 in. (21mm) wrench to hold the camshaft. Then remove the timing chain and camshaft sprocket. Remove the crankshaft sprocket, if necessary.

Fig. 130 Rotate the crankshaft clockwise until the timing mark on the crankshaft sprocket and keyway align with the main bearing cap split line

To install:

6. Inspect the chain for wear and damage. Check the inside diameter of the chain, it should be no more than 23.15 in. (588mm). Inspect the chain guides for wear or cracks and the timing sprockets for teeth or key wear. Replace components as necessary.

7. Verify that the crankshaft is positioned 90 degrees clockwise past TDC. The crankshaft keyway should be at 3 o'clock aligned with the main bearing cap split line to prevent piston and valve damage.

8. Install the camshaft sprockets, retaining bolts and washers. Make sure the letters FRT on the sprockets face forward, away from the cylinder block. Use the wrench flats provided on the camshafts to hold the shaft and tighten the bolts to 75 ft. lbs. (102 Nm).

9. Bring the camshafts up to No. 1 TDC by rotating the camshafts and sprocket until the dowel pins are at 12 o'clock. Install a 1/8 in. drill bit into the hole in the sprocket about 9 o'clock.

10. If removed, install the crankshaft sprocket, then rotate the crankshaft counterclockwise 90 degree up to No. 1 TDC (keyway and sprocket timing mark at 12 o'clock, in alignment with the block timing mark).

11. Position the timing chain under the crankshaft sprocket and over the camshaft sprockets so 2 silver link plates align with the reference marks on the camshaft sprockets and another 2 plates align with the downward tooth (at 6 o'clock position) on the crankshaft sprocket. Excess chain slack should be located on the tensioner side of the cylinder block.

12. Verify that the crankshaft reference mark aligns with the cylinder block mark at 12 o'clock and that the timing pins are installed in the holes at about the 9 o'clock position. Remove the timing pins from the camshaft sprockets.

13. Install the timing chain fixed guide to the right of the block face toward the water pump. Tighten the bolts to 21 ft. lbs. (28 Nm) and verify the chain is snug against the guide.

14. Install the pivoting chain guide and check for clearance between the block and head. Tighten the bolt to 19 ft. lbs. (26 Nm) and verify the guide pivots freely.

15. Install the 2 forward camshaft bearing caps and the upper timing chain guide, then tighten the retaining bolts to 124 inch lbs. (14 Nm).

16. Retract the tensioner plunger and pin the ratchet lever using a ⅛ in. (3.18mm) No. 31 drill bit inserted in the alignment hole at the lower front of the component. Install the tensioner and tighten the bolts to 14 ft. lbs. (19 Nm), then remove the drill bit.

Fig. 133 Retract the tensioner plunger and pin the ratchet lever with a ⅛ in. (3.18mm) drill bit inserted in the alignment hole

Fig. 131 Loosen the timing chain tensioner mounting bolt . . .

Fig. 134 Remove the mounting fasteners from both timing guides . . .

Fig. 132 . . . and remove the tensioner from the engine block

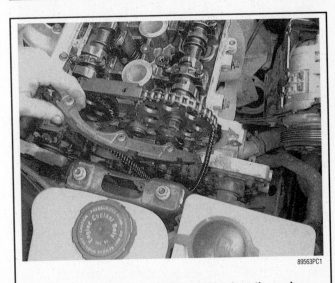

Fig. 135 . . . and remove the timing guides from the engine

Fig. 136 Remove the camshaft sprocket bolts, using a 7/8 in. (21mm) wrench to hold the camshaft

Fig. 137 Remove the timing chain and camshaft sprockets

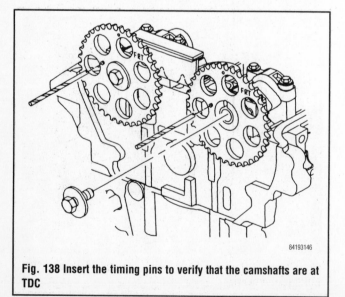

Fig. 138 Insert the timing pins to verify that the camshafts are at TDC

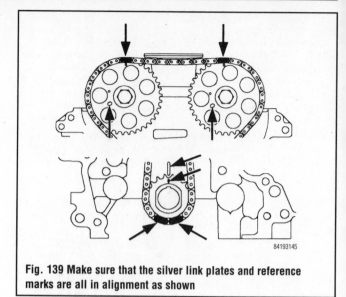

Fig. 139 Make sure that the silver link plates and reference marks are all in alignment as shown

17. Make one final check to verify all components are properly timed, then remove all timing pins.

18. Install the timing chain front cover.

19. Connect the negative battery cable, start the engine and check for leaks.

Camshaft

REMOVAL & INSTALLATION

SOHC Engine

▶ See Figure 140

1. Disconnect the negative, then the positive battery cable. Remove the battery cover and battery from the vehicle.

2. Remove the timing chain front cover.

3. Remove the timing chain and camshaft sprocket.

4. Remove the rocker arm/shaft assemblies.

5. Remove the lifters, and label or position them for assembly in their original locations.

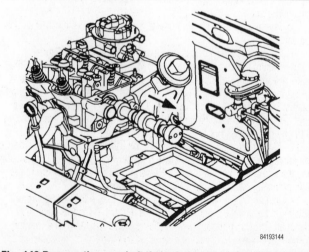

Fig. 140 Remove the camshaft through the oversized hole at the rear of the cylinder head—SOHC engine

6. Drive the camshaft plug inward, then remove it from the cylinder head with a magnet.

7. Carefully pull the camshaft from the rear of the cylinder head through the oversized camshaft plug hole. Turn the camshaft back and forth slowly while withdrawing to help prevent journal or bearing damage.

To install:

8. Clean and inspect all parts prior to installation. Lubricate the camshaft and carefully insert it through the hole at the rear of the cylinder head.

9. Coat a new rear cylinder head plug with Loctite® 242 or equivalent and install it using a standard bushing driver.

10. Install the valve lifters into their original bores, or if the camshaft has been replaced, install new lifters.

11. Install the rocker arm/shaft assemblies.

12. Install the timing chain and camshaft sprocket.

13. Install the timing chain front cover.

14. Install the battery and tighten the battery hold-down nut and screw to 80 inch lbs. (9 Nm). Connect the positive battery cable only, at this time.

15. Connect the positive, then the negative battery cable, start the engine and check for leaks.

DOHC Engine

♦ See Figures 141, 142, 143 and 144

➡Be very careful when working around the camshaft sprockets and timing chain cover during this procedure. If a bolt or washer is accidentally dropped between the front cover and engine assembly, the cover will have to be removed for retrieval.

1. Disconnect the negative battery cable and remove the serpentine drive belt.

2. Disconnect the spark plug wires from the plugs, remove the EGR valve solenoid attachment screw and remove the PCV fresh air hose.

3. Remove the camshaft cover, then inspect the cover's silicone insulators for cracks or deterioration, and replace as necessary.

4. Turn the crankshaft clockwise until the mark on the crankshaft pulley is in alignment with the pointer on the front cover and the No. 1 cylinder is at Top Dead Center (TDC) of the compression stroke. Both camshaft dowel pins will be at the 12 o'clock position and the timing pin holes will be aligned when the No. 1 cylinder is at TDC. If necessary, the right wheel and splash shield can be removed to help observe the timing marks.

5. Carefully remove each camshaft sprocket's retaining bolt. Use a ⅞ in. (21mm) open end wrench to hold the camshaft from turning while removing the bolts.

6. For 1992–98 vehicles, position a front angled support fixture in front of the camshaft sprockets.

7. Attach the camshaft sprocket adapters to the end of each camshaft using the pilot bolts, but do not tighten the bolts. For 1992–98 vehicles, the front angled support should come between the sprocket adapters and camshaft sprockets.

8. Remove the upper timing chain guide and both front camshaft bearing caps.

9. For 1991 vehicles, position the front angled support fixture.

10. Secure the support fixture using ⅞ in. bolts/blocks and align the 2 holes in each camshaft sprocket, adapter and the front support fixture. Install the 4 nuts, but do not tighten. The steel blocks should be installed against the rearward side of the camshaft sprocket. Tighten the sprocket pilot bolts to 19 ft. lbs. (25 Nm) while holding the camshafts from turning with an open end wrench.

11. Move each camshaft sprocket off the end of the camshaft by rocking the sprocket forward or by carefully prying between the end of the camshaft and the sprocket. Then tighten the 4 nuts and bolts with blocks from the side of the support fixture to 19 ft. lbs. (25 Nm).

12. Install the 2 bolts retaining the support fixture to the engine front cover and tighten the bolts to 89 inch lbs. (10 Nm). Then remove each camshaft sprocket pilot bolt while holding the camshafts with a wrench.

13. Carefully pry between the sprocket and the end of the camshaft to move the camshaft rearward. Pry only enough to remove its end from inside the sprocket pilot otherwise camshaft or lifter damage may occur.

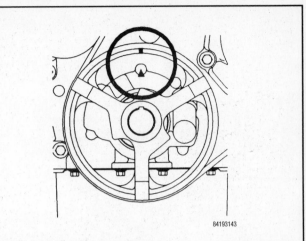

Fig. 141 To ensure proper engine timing during assembly, turn the crankshaft clockwise until the mark on the damper/pulley is in alignment with the cover pointer

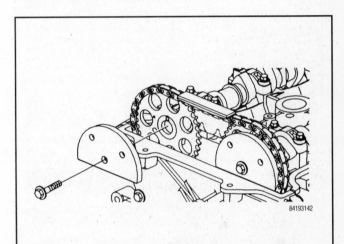

Fig. 142 For 1991 vehicles without the torque axis mount system, position the adapters directly to the sprockets—DOHC engine

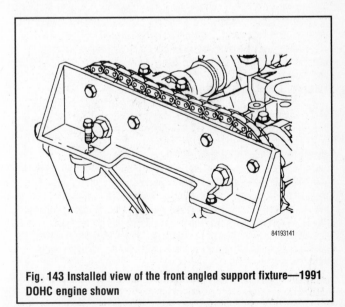

Fig. 143 Installed view of the front angled support fixture—1991 DOHC engine shown

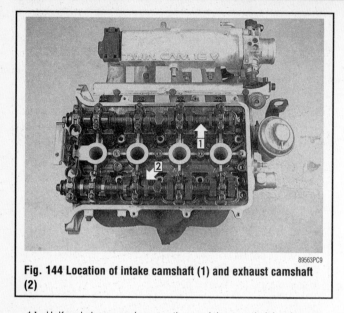

Fig. 144 Location of intake camshaft (1) and exhaust camshaft (2)

14. Uniformly loosen and remove the remaining camshaft bearing cap bolts. To prevent bolt/cap damage, do not use power tools and make several passes. Then remove each camshaft. Position the bearing caps for installation in their original locations.

To install:

15. Clean and inspect all parts prior to installation. Oil the camshaft and install with the **IN** camshaft on the intake side and **EX** camshaft on the exhaust side.

➡ The dowel pin in each camshaft must be located at the 12 o'clock position during installation to prevent valve and piston damage.

16. Install all bearing caps, except for the forward pair, in their original positions, making sure the arrows on the caps are pointing forward toward the camshaft sprockets. Lightly oil each of the cap bolts, then install and uniformly tighten the bolts to 124 inch lbs. (14 Nm).

17. Install one camshaft sprocket pilot bolt in each camshaft and tighten to 124 inch lbs. (14 Nm) in order to pull the camshaft fully forward and align the sprocket support for installation of the sprocket onto the camshaft.

18. Remove the 4 sprocket support bolt/blocks and nuts. Then, for 1991 vehicles only, remove the front angled support fixture. The torque axis mount system of the 1992–98 vehicles requires the fixture to remain in place longer.

19. Verify that the camshafts are fully positioned forward and install the 2 forward bearing caps and the upper chain guide. The caps are marked **E1** or **I1** for exhaust or intake and must be positioned with their arrows pointing towards the sprockets. Tighten the cap bolts to 124 inch lbs. (14 Nm).

20. Make sure the camshaft dowel pin aligns with the slot in each camshaft sprocket. If necessary, rotate the camshaft slightly (1–2 degrees) and move each sprocket from the adapter onto the end of the camshaft. Fully seat each sprocket on the end of each camshaft.

21. Remove the 2 sprocket pilot bolts and adapters while using a wrench on the camshaft flats to assure the camshaft cannot move.

22. For 1992–98 vehicles, remove the support angled fixture.

23. Install the camshaft sprocket retaining bolts and washers. Hold the camshafts and tighten the bolts to 76 ft. lbs. (103 Nm).

24. Verify all visible timing marks and holes are in alignment. Turn the crankshaft clockwise until the mark on the crankshaft pulley aligns with the mark on the front cover. Check timing by inserting ³⁄₁₆ in. drill bits through the camshaft sprocket alignment holes, into the cylinder head. If the alignment pins cannot be inserted, turn the crankshaft 360 degrees clockwise and repeat. If the pins cannot be inserted within 1–2 degrees of either TDC position, the camshafts are not properly timed. Do not start the engine until the camshafts are timed.

25. Apply a small drop of RTV across the cylinder head and front cover T-joints. Inspect the old camshaft cover gasket and replace if damaged. Install the gasket and the camshaft cover. Tighten the fasteners in proper sequence to 89 inch lbs. (10 Nm).

26. Install the right splash shield and wheel, if removed to observe the timing marks.

27. Install the PCV and fresh air hoses, the EGR valve solenoid attaching screw and the spark plug wires.

28. Install the serpentine drive belt and connect the negative battery cable.

29. Start the engine and check for leaks.

INSPECTION

▶ **See Figures 145, 146, 147 and 148**

1. Clean the camshaft in solvent and allow it to dry.

2. Inspect the camshaft for obvious signs of wear: scores, nicks or pits on the journals or lobes. Light scuffs or nicks can be removed with an oil stone.

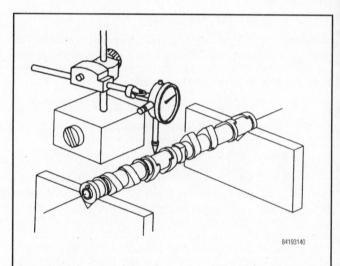

Fig. 145 Use a dial indicator to measure camshaft run-out at the center journal

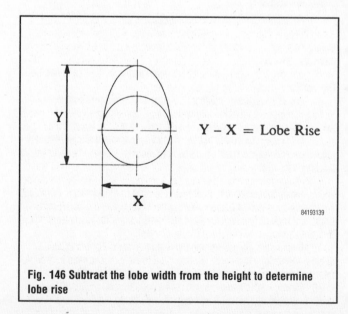

$$Y - X = \text{Lobe Rise}$$

Fig. 146 Subtract the lobe width from the height to determine lobe rise

7. Subtract the journal diameter measurements from their respective bore diameter measurements to calculate oil clearance. Replace the camshaft and/or cylinder head if clearance is more than 0.0054 in. (0.138mm) for SOHC engines or 0.005 in. (0.125mm) for DOHC engines.

8. Position a precision straightedge across the bottom of the cylinder head camshaft contact surfaces. Use a feeler gauge to inspect for warpage; replace a cylinder head with warpage of more than 0.0025 in. (0.064mm) for SOHC engines or 0.003 in. (0.075mm) for DOHC engines.

Rear Main Seal

REMOVAL & INSTALLATION

▶ **See Figures 149 and 150**

Both engines use a one-piece round seal mounted in a seal carrier. The seal may be replaced with the carrier installed or removed from the engine. To replace the seal with the carrier installed:
1. Disconnect the negative battery cable.
2. Remove the transaxle assembly from the vehicle. Refer to Section 7 of this manual.

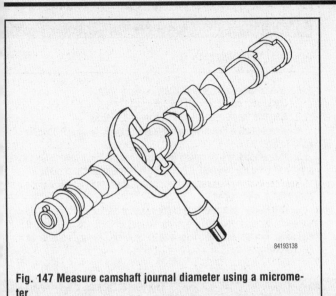

Fig. 147 Measure camshaft journal diameter using a micrometer

Fig. 148 Measuring camshaft contact surface warpage—DOHC engine shown

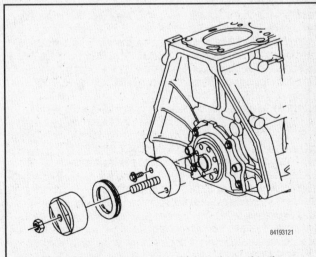

Fig. 149 Use the pry tangs provided in the carrier to remove the rear seal

Fig. 150 The seal installation tool will draw the seal to the proper depth in the carrier

3. Position the camshaft in V-blocks with the front and rear journals riding on the blocks. Check if the camshaft is bent using a dial indicator on the center bearing journal. The run-out limit at the center journal is 0.004 in. (0.1mm) for DOHC engines or 0.0028 in. (0.07mm) for SOHC engines. Replace the camshaft if run-out is excessive.

4. Using a micrometer, measure the camshaft lobes across their maximum and minimum lobe height dimensions. Subtract the lobe width from the lobe height to arrive at lobe rise. Replace a SOHC camshaft if any lobes have a rise of less than 0.252 in. (6.4mm). Replace a DOHC camshaft if any intake rise is less than 0.351 in. (8.91mm), or any exhaust rise is less than 0.339 in. (8.61mm).

5. Using a micrometer, measure the diameter of the journals and replace any camshaft containing a journal that is less than the minimum. For SOHC engines, journal diameter should be greater than 1.747 in. (44.375mm). DOHC engines should have a minimum camshaft journal diameter of 1.139 in. (28.925mm).

6. Measure the diameter of the camshaft bearings. For the DOHC engine, temporarily install the bearing caps to the cylinder head in order to take the measurement. Bore diameter must be less than 1.753 in. (44.513mm) for SOHC engines or 1.144 in. (29.05mm) for DOHC engines, or the cylinder head must be replaced.

3. As applicable, remove the clutch and flywheel assembly or the flex-plate.

4. Use the prying tangs provided in the carrier to remove the seal with a small suitable prybar and hammer. Be careful not to damage the crankshaft oil seal lip contact surface.

To install:

5. Clean the carrier and crankshaft with solvent and a rag to prevent seal lip damage during installation. Check for scores or damage to the sealing surfaces.

6. Apply a light coat of clean engine oil to the seal lip and the carrier inner diameter. Install using SA9121E or an equivalent seal installer. The tool is designed to prevent seal lip from rolling during installation and will seat the seal 0.04 in. (1mm) lower than the factory seal. Never tap on the seal or seal installer with a hammer.

7. Install the flywheel or flexplate assembly.

8. Install the transaxle assembly into the vehicle.

9. Connect the negative battery cable, start the engine and check for leaks.

EXHAUST SYSTEM

Inspection

▶ See Figures 151 thru 159

➡Safety glasses should be worn at all times when working on or near the exhaust system. Older exhaust systems will almost always be covered with loose rust particles which will shower you when disturbed. These particles are more than a nuisance and could injure your eyes.

Flywheel/Flexplate

REMOVAL & INSTALLATION

1. Disconnect the negative battery cable.

2. Raise and safely support the vehicle.

3. Remove the transaxle assembly. Refer to Section 7.

4. If equipped with a manual transaxle, remove the clutch assembly. Refer to Section 7.

5. If equipped with a manual transaxle, remove the flywheel-to-crankshaft bolts and the flywheel. If equipped with an automatic transaxle, remove the flexplate-to-crankshaft bolts and the flexplate. Remove the flexplate shims, if equipped.

6. Installation is the reverse of the removal procedure. If equipped with a manual transaxle, tighten the flywheel bolts to 59 ft. lbs. (80 Nm). If equipped with an automatic transaxle, tighten the flexplate bolts to 44 ft. lbs. (60 Nm).

✳✳ CAUTION

Do NOT perform exhaust repairs or inspection with the engine or exhaust hot. Allow the system to cool completely before attempting any work. Exhaust systems are noted for sharp edges, flaking metal and rusted bolts. Gloves and eye protection are required. A healthy supply of penetrating oil and rags is highly recommended.

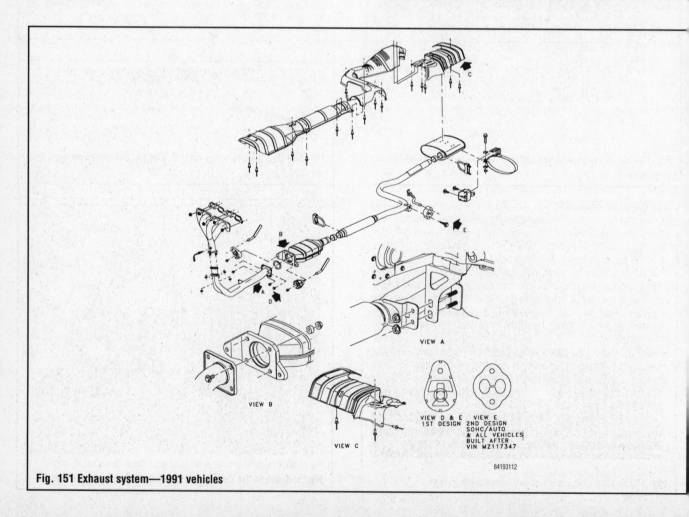

Fig. 151 Exhaust system—1991 vehicles

84193112

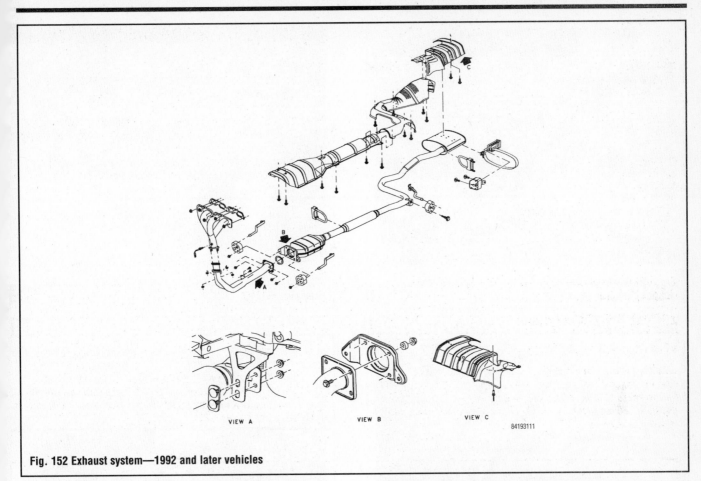

VIEW A   VIEW B   VIEW C

84193111

Fig. 152 Exhaust system—1992 and later vehicles

Your vehicle must be raised and supported safely to inspect the exhaust system properly. By placing 4 safety stands under the vehicle for support should provide enough room for you to slide under the vehicle and inspect the system completely. Start the inspection at the exhaust manifold where the header pipe is attached and work your way to the back of the vehicle. Check the complete exhaust system for open seams, holes loose connections, or other deterioration which could permit exhaust fumes to seep into the passenger compartment. Inspect all mounting brackets and hangers for deterioration, some models may have rubber O-rings that can be overstretched and non-supportive. These components will need to be replaced if

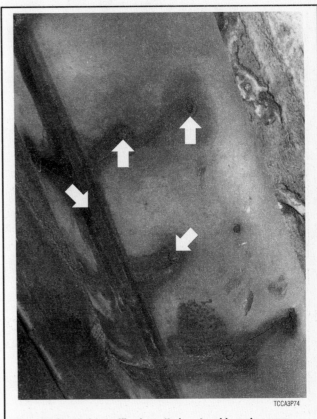

TCCA3P74

Fig. 154 Check the muffler for rotted spot welds and seams

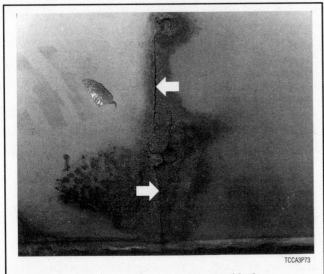

TCCA3P73

Fig. 153 Cracks in the muffler are a guaranteed leak

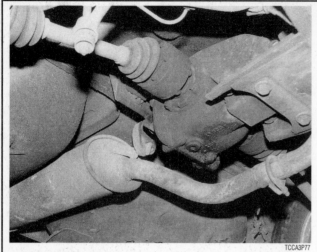

Fig. 155 Make sure the exhaust components are not contacting the body or suspension

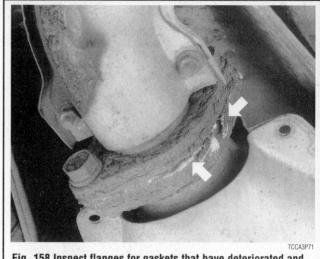

Fig. 158 Inspect flanges for gaskets that have deteriorated and need replacement

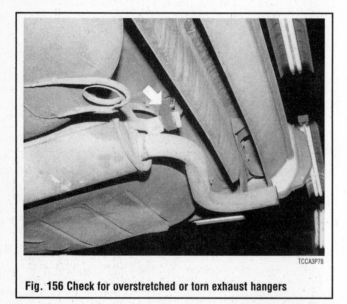

Fig. 156 Check for overstretched or torn exhaust hangers

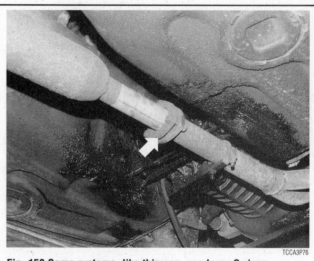

Fig. 159 Some systems, like this one, use large O-rings (donuts) in between the flanges

Fig. 157 Example of a badly deteriorated exhaust pipe

found. It has always been a practice to use a pointed tool to poke up into the exhaust system where the deterioration spots are to see whether or not they crumble. Some models may have heat shield covering certain parts of the exhaust system , it will be necessary to remove these shields to have the exhaust visible for inspection also.

REPLACEMENT

▶ **See Figure 160**

There are basically two types of exhaust systems. One is the flange type where the component ends are attached with bolts and a gasket in-between. The other exhaust system is the slip joint type. These components slip into one another using clamps to retain them together.

✳✳ CAUTION

Allow the exhaust system to cool sufficiently before spraying a solvent exhaust fasteners. Some solvents are highly flammable and could ignite when sprayed on hot exhaust components.

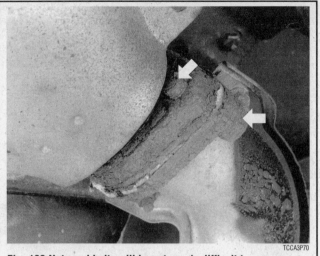

Fig. 160 Nuts and bolts will be extremely difficult to remove when deteriorated with rust

Fig. 161 Example of a flange type exhaust system joint

Before removing any component of the exhaust system, ALWAYS squirt a liquid rust dissolving agent onto the fasteners for ease of removal. A lot of knuckle skin will be saved by following this rule. It may even be wise to spray the fasteners and allow them to sit overnight.

Flange Type

▶ See Figure 161

✳✳ CAUTION

Do NOT perform exhaust repairs or inspection with the engine or exhaust hot. Allow the system to cool completely before attempting any work. Exhaust systems are noted for sharp edges, flaking metal and rusted bolts. Gloves and eye protection are required. A healthy supply of penetrating oil and rags is highly recommended. Never spray liquid rust dissolving agent onto a hot exhaust component.

Before removing any component on a flange type system, ALWAYS squirt a liquid rust dissolving agent onto the fasteners for ease of removal. Start by unbolting the exhaust piece at both ends (if required). When unbolting the headpipe from the manifold, make sure that the bolts are free before trying to remove them. if you snap a stud in the exhaust manifold, the stud will have to be removed with a bolt extractor, which often means removal of the manifold itself. Next, disconnect the component from the mounting; slight twisting and turning may be required to remove the component completely from the vehicle. You may need to tap on the component with a rubber mallet to loosen the component. If all else fails, use a hacksaw to separate the parts. An oxy-acetylene cutting torch may be faster but the sparks are DANGEROUS near the fuel tank, and at the very least, accidents could happen, resulting in damage to the under-car parts, not to mention yourself.

Slip Joint Type

▶ See Figure 162

Before removing any component on the slip joint type exhaust system, ALWAYS squirt a liquid rust dissolving agent onto the fasteners for ease of removal. Start by unbolting the exhaust piece at both ends (if required). When unbolting the headpipe from the manifold, make sure that the bolts are free before trying to remove them. if you snap a stud in the exhaust manifold, the stud will have to be removed with a bolt extractor, which often means removal of the manifold itself. Next, remove the mounting U-bolts from around the exhaust pipe you are extracting from the vehicle. Don't be surprised if the U-bolts break while removing the nuts. Loosen the exhaust pipe from any mounting brackets retaining it to the floor pan and separate the components.

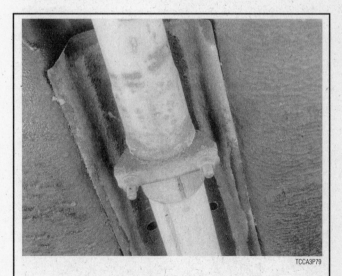

Fig. 162 Example of a common slip joint type system

ENGINE RECONDITIONING

Determining Engine Condition

Anything that generates heat and/or friction will eventually burn or wear out (ie. a light bulb generates heat, therefore its life span is limited). With this in mind, a running engine generates tremendous amounts of both; friction is encountered by the moving and rotating parts inside the engine and heat is created by friction and combustion of the fuel. However, the engine has systems designed to help reduce the effects of heat and friction and provide added longevity. The oiling system reduces the amount of friction encountered by the moving parts inside the engine, while the cooling system reduces heat created by friction and combustion. If either system is not maintained, a break-down will be inevitable. Therefore, you can see how regular maintenance can affect the service life of your vehicle. If you do not drain, flush and refill your cooling system at the proper intervals, deposits will begin to accumulate in the radiator, thereby reducing the amount of heat it can extract from the coolant. The same applies to your oil and filter; if it is not changed often enough it becomes laden with contaminates and is unable to properly lubricate the engine. This increases friction and wear.

There are a number of methods for evaluating the condition of your engine. A compression test can reveal the condition of your pistons, piston rings, cylinder bores, head gasket(s), valves and valve seats. An oil pressure test can warn you of possible engine bearing, or oil pump failures. Excessive oil consumption, evidence of oil in the engine air intake area and/or bluish smoke from the tail pipe may indicate worn piston rings, worn valve guides and/or valve seals. As a general rule, an engine that uses no more than one quart of oil every 1000 miles is in good condition. Engines that use one quart of oil or more in less than 1000 miles should first be checked for oil leaks. If any oil leaks are present, have them fixed before determining how much oil is consumed by the engine, especially if blue smoke is not visible at the tail pipe.

COMPRESSION TEST

▶ See Figure 163

A noticeable lack of engine power, excessive oil consumption and/or poor fuel mileage measured over an extended period are all indicators of internal engine wear. Worn piston rings, scored or worn cylinder bores, blown head gaskets, sticking or burnt valves, and worn valve seats are all possible culprits. A check of each cylinder's compression will help locate the problem.

➡**A screw-in type compression gauge is more accurate than the type you simply hold against the spark plug hole. Although it takes slightly longer to use, it's worth the effort to obtain a more accurate reading.**

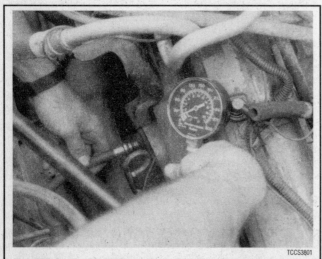

TCCS3801

Fig. 163 A screw-in type compression gauge is more accurate and easier to use without an assistant

1. Make sure that the proper amount and viscosity of engine oil is in the crankcase, then ensure the battery is fully charged.
2. Warm-up the engine to normal operating temperature, then shut the engine **OFF**.
3. Disable the ignition system.
4. Label and disconnect all of the spark plug wires from the plugs.
5. Thoroughly clean the cylinder head area around the spark plug ports, then remove the spark plugs.
6. Set the throttle plate to the fully open (wide-open throttle) position. You can block the accelerator linkage open for this, or you can have an assistant fully depress the accelerator pedal.
7. Install a screw-in type compression gauge into the No. 1 spark plug hole until the fitting is snug.

✳✳ WARNING

Be careful not to crossthread the spark plug hole.

8. According to the tool manufacturer's instructions, connect a remote starting switch to the starting circuit.
9. With the ignition switch in the **OFF** position, use the remote starting switch to crank the engine through at least five compression strokes (approximately 5 seconds of cranking) and record the highest reading on the gauge.
10. Repeat the test on each cylinder, cranking the engine approximately the same number of compression strokes and/or time as the first.
11. Compare the highest readings from each cylinder to that of the others. The indicated compression pressures are considered within specifications if the lowest reading cylinder is within 75 percent of the pressure recorded for the highest reading cylinder. For example, if your highest reading cylinder pressure was 150 psi (1034 kPa), then 75 percent of that would be 113 psi (779 kPa). So the lowest reading cylinder should be no less than 113 psi (779 kPa).
12. If a cylinder exhibits an unusually low compression reading, pour a tablespoon of clean engine oil into the cylinder through the spark plug hole and repeat the compression test. If the compression rises after adding oil, it means that the cylinder's piston rings and/or cylinder bore are damaged or worn. If the pressure remains low, the valves may not be seating properly (a valve job is needed), or the head gasket may be blown near that cylinder. If compression in any two adjacent cylinders is low, and if the addition of oil doesn't help raise compression, there is leakage past the head gasket. Oil and coolant in the combustion chamber, combined with blue or constant white smoke from the tail pipe, are symptoms of this problem. However, don't be alarmed by the normal white smoke emitted from the tail pipe during engine warm-up or from cold weather driving. There may be evidence of water droplets on the engine dipstick and/or oil droplets in the cooling system if a head gasket is blown.

OIL PRESSURE TEST

Check for proper oil pressure at the sending unit passage with an externally mounted mechanical oil pressure gauge (as opposed to relying on a factory installed dash-mounted gauge). A tachometer may also be needed, as some specifications may require running the engine at a specific rpm.

1. With the engine cold, locate and remove the oil pressure sending unit.
2. Following the manufacturer's instructions, connect a mechanical oil pressure gauge and, if necessary, a tachometer to the engine.
3. Start the engine and allow it to idle.
4. Check the oil pressure reading when cold and record the number. You may need to run the engine at a specified rpm, so check the specifications chart located earlier in this section.
5. Run the engine until normal operating temperature is reached (upper radiator hose will feel warm).
6. Check the oil pressure reading again with the engine hot and record the number. Turn the engine **OFF**.
7. Compare your hot oil pressure reading to that given in the chart. If the reading is low, check the cold pressure reading against the chart. If the

cold pressure is well above the specification, and the hot reading was lower than the specification, you may have the wrong viscosity oil in the engine. Change the oil, making sure to use the proper grade and quantity, then repeat the test.

Low oil pressure readings could be attributed to internal component wear, pump related problems, a low oil level, or oil viscosity that is too low. High oil pressure readings could be caused by an overfilled crankcase, too high of an oil viscosity or a faulty pressure relief valve.

Buy or Rebuild?

Now that you have determined that your engine is worn out, you must make some decisions. The question of whether or not an engine is worth rebuilding is largely a subjective matter and one of personal worth. Is the engine a popular one, or is it an obsolete model? Are parts available? Will it get acceptable gas mileage once it is rebuilt? Is the car it's being put into worth keeping? Would it be less expensive to buy a new engine, have your engine rebuilt by a pro, rebuild it yourself or buy a used engine from a salvage yard? Or would it be simpler and less expensive to buy another car? If you have considered all these matters and more, and have still decided to rebuild the engine, then it is time to decide how you will rebuild it.

➡**The editors at Chilton feel that most engine machining should be performed by a professional machine shop. Don't think of it as wasting money, rather, as an assurance that the job has been done right the first time. There are many expensive and specialized tools required to perform such tasks as boring and honing an engine block or having a valve job done on a cylinder head. Even inspecting the parts requires expensive micrometers and gauges to properly measure wear and clearances. Also, a machine shop can deliver to you clean, and ready to assemble parts, saving you time and aggravation. Your maximum savings will come from performing the removal, disassembly, assembly and installation of the engine and purchasing or renting only the tools required to perform the above tasks. Depending on the particular circumstances, you may save 40 to 60 percent of the cost doing these yourself.**

A complete rebuild or overhaul of an engine involves replacing all of the moving parts (pistons, rods, crankshaft, camshaft, etc.) with new ones and machining the non-moving wearing surfaces of the block and heads. Unfortunately, this may not be cost effective. For instance, your crankshaft may have been damaged or worn, but it can be machined undersize for a minimal fee.

So, as you can see, you can replace everything inside the engine, but, it is wiser to replace only those parts which are really needed, and, if possible, repair the more expensive ones. Later in this section, we will break the engine down into its two main components: the cylinder head and the engine block. We will discuss each component, and the recommended parts to replace during a rebuild on each.

Engine Overhaul Tips

Most engine overhaul procedures are fairly standard. In addition to specific parts replacement procedures and specifications for your individual engine, this section is also a guide to acceptable rebuilding procedures. Examples of standard rebuilding practice are given and should be used along with specific details concerning your particular engine.

Competent and accurate machine shop services will ensure maximum performance, reliability and engine life. In most instances it is more profitable for the do-it-yourself mechanic to remove, clean and inspect the component, buy the necessary parts and deliver these to a shop for actual machine work.

Much of the assembly work (crankshaft, bearings, piston rods, and other components) is well within the scope of the do-it-yourself mechanic's tools and abilities. You will have to decide for yourself the depth of involvement you desire in an engine repair or rebuild.

TOOLS

The tools required for an engine overhaul or parts replacement will depend on the depth of your involvement. With a few exceptions, they will be the tools found in a mechanic's tool kit (see Section 1 of this manual). More in-depth work will require some or all of the following:

- A dial indicator (reading in thousandths) mounted on a universal base
- Micrometers and telescope gauges
- Jaw and screw-type pullers
- Scraper
- Valve spring compressor
- Ring groove cleaner
- Piston ring expander and compressor
- Ridge reamer
- Cylinder hone or glaze breaker
- Plastigage®
- Engine stand

The use of most of these tools is illustrated in this section. Many can be rented for a one-time use from a local parts jobber or tool supply house specializing in automotive work.

Occasionally, the use of special tools is called for. See the information on Special Tools and the Safety Notice in the front of this book before substituting another tool.

OVERHAUL TIPS

Aluminum has become extremely popular for use in engines, due to its low weight. Observe the following precautions when handling aluminum parts:
- Never hot tank aluminum parts (the caustic hot tank solution will eat the aluminum.
- Remove all aluminum parts (identification tag, etc.) from engine parts prior to the tanking.
- Always coat threads lightly with engine oil or anti-seize compounds before installation, to prevent seizure.
- Never overtighten bolts or spark plugs especially in aluminum threads.

When assembling the engine, any parts that will be exposed to frictional contact must be prelubed to provide lubrication at initial start-up. Any product specifically formulated for this purpose can be used, but engine oil is not recommended as a prelube in most cases.

When semi-permanent (locked, but removable) installation of bolts or nuts is desired, threads should be cleaned and coated with Loctite® or another similar, commercial non-hardening sealant.

CLEANING

▶ **See Figures 164, 165, 166 and 167**

Before the engine and its components are inspected, they must be thoroughly cleaned. You will need to remove any engine varnish, oil sludge and/or carbon deposits from all of the components to insure an accurate inspection. A crack in the engine block or cylinder head can easily become overlooked if hidden by a layer of sludge or carbon.

Most of the cleaning process can be carried out with common hand tools and readily available solvents or solutions. Carbon deposits can be chipped away using a hammer and a hard wooden chisel. Old gasket material and varnish or sludge can usually be removed using a scraper and/or cleaning solvent. Extremely stubborn deposits may require the use of a power drill with a wire brush. If using a wire brush, use extreme care around any critical machined surfaces (such as the gasket surfaces, bearing saddles, cylinder bores, etc.). USE OF A WIRE BRUSH IS NOT RECOMMENDED ON ANY ALUMINUM COMPONENTS. Always follow any safety recommendations given by the manufacturer of the tool and/or solvent. You should always wear eye protection during any cleaning process involving scraping, chipping or spraying of solvents.

An alternative to the mess and hassle of cleaning the parts yourself is to drop them off at a local garage or machine shop. They will, more than likely, have the necessary equipment to properly clean all of the parts for a nominal fee.

✳✳ CAUTION

Always wear eye protection during any cleaning process involving scraping, chipping or spraying of solvents.

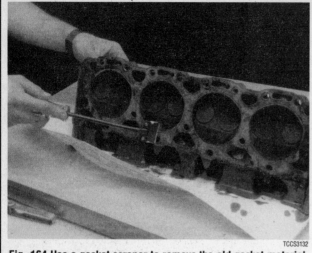

Fig. 164 Use a gasket scraper to remove the old gasket material from the mating surfaces

Fig. 165 Use a ring expander tool to remove the piston rings

Fig. 167 . . . use a piece of an old ring to clean the grooves. Be careful, the ring can be quite sharp

Remove any oil galley plugs, freeze plugs and/or pressed-in bearings and carefully wash and degrease all of the engine components including the fasteners and bolts. Small parts such as the valves, springs, etc., should be placed in a metal basket and allowed to soak. Use pipe cleaner type brushes, and clean all passageways in the components. Use a ring expander and remove the rings from the pistons. Clean the piston ring grooves with a special tool or a piece of broken ring. Scrape the carbon off of the top of the piston. You should never use a wire brush on the pistons. After preparing all of the piston assemblies in this manner, wash and degrease them again.

✳✳ WARNING

Use extreme care when cleaning around the cylinder head valve seats. A mistake or slip may cost you a new seat.

When cleaning the cylinder head, remove carbon from the combustion chamber with the valves installed. This will avoid damaging the valve seats.

REPAIRING DAMAGED THREADS

▸ See Figures 168, 169, 170, 171 and 172

Several methods of repairing damaged threads are available. Heli-Coil® (shown here), Keenserts® and Microdot® are among the most widely used. All involve basically the same principle—drilling out stripped threads, tapping the hole and installing a prewound insert—making welding, plugging and oversize fasteners unnecessary.

Two types of thread repair inserts are usually supplied: a standard type for most inch coarse, inch fine, metric course and metric fine thread sizes and a spark lug type to fit most spark plug port sizes. Consult the individual tool manufacturer's catalog to determine exact applications. Typical thread repair kits will contain a selection of prewound threaded inserts, a tap (corresponding to the outside diameter threads of the insert) and an installation tool. Spark plug inserts usually differ because they require a tap equipped with pilot threads and a combined reamer/tap section. Most manufacturers also supply blister-packed thread repair inserts separately in addition to a master kit containing a variety of taps and inserts plus installation tools.

Before attempting to repair a threaded hole, remove any snapped, broken or damaged bolts or studs. Penetrating oil can be used to free frozen threads. The offending item can usually be removed with locking pliers or using a screw/stud extractor. After the hole is clear, the thread can be repaired, as shown in the series of accompanying illustrations and in the kit manufacturer's instructions.

Fig. 166 Clean the piston ring grooves using a ring groove cleaner tool, or . . .

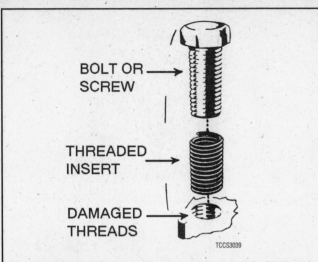

Fig. 168 Damaged bolt hole threads can be replaced with thread repair inserts

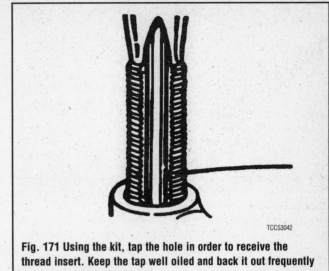

Fig. 171 Using the kit, tap the hole in order to receive the thread insert. Keep the tap well oiled and back it out frequently to avoid clogging the threads

Fig. 169 Standard thread repair insert (left), and spark plug thread insert

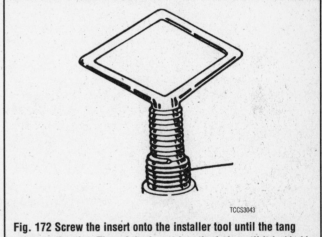

Fig. 172 Screw the insert onto the installer tool until the tang engages the slot. Thread the insert into the hole until it is ¼–½ turn below the top surface, then remove the tool and break off the tang using a punch

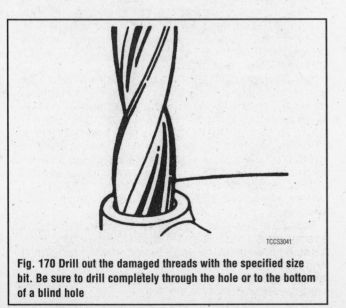

Fig. 170 Drill out the damaged threads with the specified size bit. Be sure to drill completely through the hole or to the bottom of a blind hole

Engine Preparation

To properly rebuild an engine, you must first remove it from the vehicle, then disassemble and diagnose it. Ideally you should place your engine on an engine stand. This affords you the best access to the engine components. Follow the manufacturer's directions for using the stand with your particular engine. Remove the flywheel or flexplate before installing the engine to the stand.

Now that you have the engine on a stand, and assuming that you have drained the oil and coolant from the engine, it's time to strip it of all but the necessary components. Before you start disassembling the engine, you may want to take a moment to draw some pictures, or fabricate some labels or containers to mark the locations of various components and the bolts and/or studs which fasten them. Modern day engines use a lot of little brackets and clips which hold wiring harnesses and such, and these holders are often mounted on studs and/or bolts that can be easily mixed up. The manufacturer spent a lot of time and money designing your vehicle, and they wouldn't have wasted any of it by haphazardly placing brackets, clips or fasteners on the vehicle. If it's present when you disassemble it, put it back when you assemble, you will regret not remembering that little bracket which holds a wire harness out of the path of a rotating part.

You should begin by unbolting any accessories still attached to the engine, such as the water pump, power steering pump, alternator, etc. Then, unfasten any manifolds (intake or exhaust) which were not removed during the engine removal procedure. Finally, remove any covers remaining on the engine such as the rocker arm, front or timing cover and oil pan. Some front covers may require the vibration damper and/or crank pulley to be removed beforehand. The idea is to reduce the engine to the bare necessities (cylinder head(s), valve train, engine block, crankshaft, pistons and connecting rods), plus any other 'in block' components such as oil pumps, balance shafts and auxiliary shafts.

Finally, remove the cylinder head(s) from the engine block and carefully place on a bench. Disassembly instructions for each component follow later in this section.

Cylinder Head

There are two basic types of cylinder heads used on today's automobiles: the Overhead Valve (OHV) and the Overhead Camshaft (OHC). The latter can also be broken down into two subgroups: the Single Overhead Camshaft (SOHC) and the Dual Overhead Camshaft (DOHC). Generally, if there is only a single camshaft on a head, it is just referred to as an OHC head. Also, an engine with a OHV cylinder head is also known as a pushrod engine.

Most cylinder heads these days are made of an aluminum alloy due to its light weight, durability and heat transfer qualities. However, cast iron was the material of choice in the past, and is still used on many vehicles today. Whether made from aluminum or iron, all cylinder heads have valves and seats. Some use two valves per cylinder, while the more hi-tech engines will utilize a multi-valve configuration using 3, 4 and even 5 valves per cylinder. When the valve contacts the seat, it does so on precision machined surfaces, which seals the combustion chamber. All cylinder heads have a valve guide for each valve. The guide centers the valve to the seat and allows it to move up and down within it. The clearance between the valve and guide can be critical. Too much clearance and the engine may consume oil, lose vacuum and/or damage the seat. Too little, and the valve can stick in the guide causing the engine to run poorly if at all, and possibly causing severe damage. The last component all cylinder heads have are valve springs. The spring holds the valve against its seat. It also returns the valve to this position when the valve has been opened by the valve train or camshaft. The spring is fastened to the valve by a retainer and valve locks (sometimes called keepers). Aluminum heads will also have a valve spring shim to keep the spring from wearing away the aluminum.

An ideal method of rebuilding the cylinder head would involve replacing all of the valves, guides, seats, springs, etc. with new ones. However, depending on how the engine was maintained, often this is not necessary. A major cause of valve, guide and seat wear is an improperly tuned engine. An engine that is running too rich, will often wash the lubricating oil out of the guide with gasoline, causing it to wear rapidly. Conversely, an engine which is running too lean will place higher combustion temperatures on the valves and seats allowing them to wear or even burn. Springs fall victim to the driving habits of the individual. A driver who often runs the engine rpm to the redline will wear out or break the springs faster then one that stays well below it. Unfortunately, mileage takes it toll on all of the parts. Generally, the valves, guides, springs and seats in a cylinder head can be machined and re-used, saving you money. However, if a valve is burnt, it may be wise to replace all of the valves, since they were all operating in the same environment. The same goes for any other component on the cylinder head. Think of it as an insurance policy against future problems related to that component.

Unfortunately, the only way to find out which components need replacing, is to disassemble and carefully check each piece. After the cylinder head(s) are disassembled, thoroughly clean all of the components.

DISASSEMBLY

OHC Heads

▶ See Figures 173 and 174

Whether it is a single or dual overhead camshaft cylinder head, the disassembly procedure is relatively unchanged. One aspect to pay attention to

Fig. 173 Exploded view of a valve, seal, spring, retainer and locks from an OHC cylinder head

TCCA3P54

Fig. 174 Example of a multi-valve cylinder head. Note how it has 2 intake and 2 exhaust valve ports

TCCA3P62

is careful labeling of the parts on the dual camshaft cylinder head. There will be an intake camshaft and followers as well as an exhaust camshaft and followers and they must be labeled as such. In some cases, the components are identical and could easily be installed incorrectly. DO NOT MIX THEM UP! Determining which is which is very simple; the intake camshaft and components are on the same side of the head as was the intake manifold. Conversely, the exhaust camshaft and components are on the same side of the head as was the exhaust manifold.

CUP TYPE CAMSHAFT FOLLOWERS

◗ **See Figures 175, 176 and 177**

Most cylinder heads with cup type camshaft followers will have the valve spring, retainer and locks recessed within the follower's bore. You will need a C-clamp style valve spring compressor tool, an OHC spring removal tool (or equivalent) and a small magnet to disassemble the head.

1. If not already removed, remove the camshaft(s) and/or followers. Mark their positions for assembly.
2. Position the cylinder head to allow use of a C-clamp style valve spring compressor tool.

➡It is preferred to position the cylinder head gasket surface facing you with the valve springs facing the opposite direction and the head laying horizontal.

Fig. 177 Position the OHC spring tool in the follower bore, then compress the spring with a C-clamp type tool

Fig. 175 C-clamp type spring compressor and an OHC spring removal tool (center) for cup type followers

Fig. 176 Most cup type follower cylinder heads retain the camshaft using bolt-on bearing caps

3. With the OHC spring removal adapter tool positioned inside of the follower bore, compress the valve spring using the C-clamp style valve spring compressor.
4. Remove the valve locks. A small magnetic tool or screwdriver will aid in removal.
5. Release the compressor tool and remove the spring assembly.
6. Withdraw the valve from the cylinder head.
7. If equipped, remove the valve seal.

➡Special valve seal removal tools are available. Regular or needle nose type pliers, if used with care, will work just as well. If using ordinary pliers, be sure not to damage the follower bore. The follower and its bore are machined to close tolerances and any damage to the bore will effect this relationship.

8. If equipped, remove the valve spring shim. A small magnetic tool or screwdriver will aid in removal.
9. Repeat Steps 3 through 8 until all of the valves have been removed.

ROCKER ARM TYPE CAMSHAFT FOLLOWERS

◗ **See Figures 178 thru 186**

Most cylinder heads with rocker arm-type camshaft followers are easily disassembled using a standard valve spring compressor. However, certain models may not have enough open space around the spring for the standard tool and may require you to use a C-clamp style compressor tool instead.

1. If not already removed, remove the rocker arms and/or shafts and the camshaft. If applicable, also remove the hydraulic lash adjusters. Mark their positions for assembly.

Fig. 178 Example of the shaft mounted rocker arms on some OHC heads

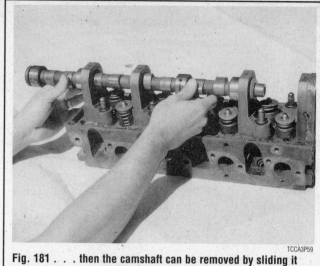

Fig. 181 . . . then the camshaft can be removed by sliding it out (shown), or unbolting a bearing cap (not shown)

Fig. 179 Another example of the rocker arm type OHC head. This model uses a follower under the camshaft

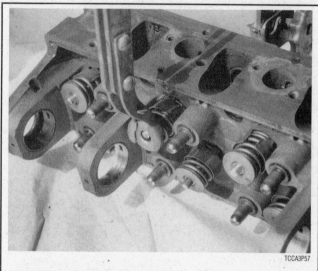

Fig. 182 Compress the valve spring . . .

Fig. 180 Before the camshaft can be removed, all of the followers must first be removed . . .

Fig. 183 . . . then remove the valve locks from the valve stem and spring retainer

Fig. 184 Remove the valve spring and retainer from the cylinder head

Fig. 185 Remove the valve seal from the guide. Some gentle prying or pliers may help to remove stubborn ones

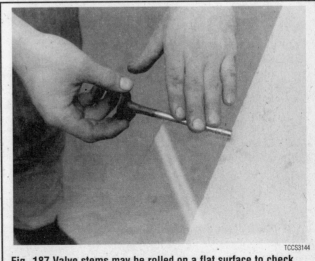

Fig. 186 All aluminum and some cast iron heads will have these valve spring shims. Remove all of them as well

2. Position the cylinder head to allow access to the valve spring.

3. Use a valve spring compressor tool to relieve the spring tension from the retainer.

➡ **Due to engine varnish, the retainer may stick to the valve locks. A gentle tap with a hammer may help to break it loose.**

4. Remove the valve locks from the valve tip and/or retainer. A small magnet may help in removing the small locks.

5. Lift the valve spring, tool and all, off of the valve stem.

6. If equipped, remove the valve seal. If the seal is difficult to remove with the valve in place, try removing the valve first, then the seal. Follow the steps below for valve removal.

7. Position the head to allow access for withdrawing the valve.

➡ **Cylinder heads that have seen a lot of miles and/or abuse may have mushroomed the valve lock grove and/or tip, causing difficulty in removal of the valve. If this has happened, use a metal file to carefully remove the high spots around the lock grooves and/or tip. Only file it enough to allow removal.**

8. Remove the valve from the cylinder head.

9. If equipped, remove the valve spring shim. A small magnetic tool or screwdriver will aid in removal.

10. Repeat Steps 3 though 9 until all of the valves have been removed.

INSPECTION

Now that all of the cylinder head components are clean, it's time to inspect them for wear and/or damage. To accurately inspect them, you will need some specialized tools:
- A 0–1 in. micrometer for the valves
- A dial indicator or inside diameter gauge for the valve guides
- A spring pressure test gauge

If you do not have access to the proper tools, you may want to bring the components to a shop that does.

Valves

◆ **See Figures 187 and 188**

The first thing to inspect are the valve heads. Look closely at the head, margin and face for any cracks, excessive wear or burning. The margin is the best place to look for burning. It should have a squared edge with an even width all around the diameter. When a valve burns, the margin will look melted and the edges rounded. Also inspect the valve head for any signs of tulipping. This will show as a lifting of the edges or dishing in the center of the head and will usually not occur to all of the valves. All of the heads should look the same, any

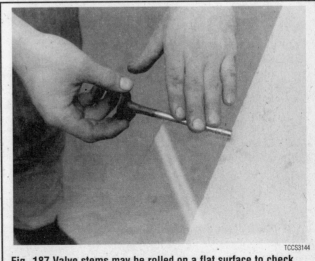

Fig. 187 Valve stems may be rolled on a flat surface to check for bends

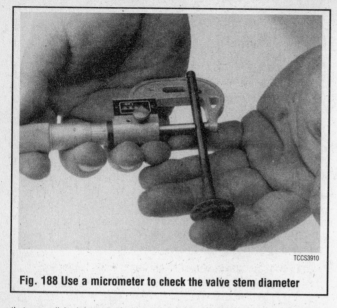

Fig. 188 Use a micrometer to check the valve stem diameter

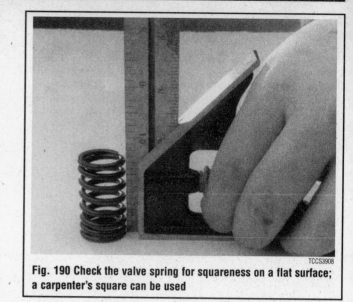

Fig. 190 Check the valve spring for squareness on a flat surface; a carpenter's square can be used

that seem dished more than others are probably bad. Next, inspect the valve lock grooves and valve tips. Check for any burrs around the lock grooves, especially if you had to file them to remove the valve. Valve tips should appear flat, although slight rounding with high mileage engines is normal. Slightly worn valve tips will need to be machined flat. Last, measure the valve stem diameter with the micrometer. Measure the area that rides within the guide, especially towards the tip where most of the wear occurs. Take several measurements along its length and compare them to each other. Wear should be even along the length with little to no taper. If no minimum diameter is given in the specifications, then the stem should not read more than 0.001 in. (0.025mm) below the specification. Any valves that fail these inspections should be replaced.

Springs, Retainers and Valve Locks

▶ See Figures 189 and 190

The first thing to check is the most obvious, broken springs. Next check the free length and squareness of each spring. If applicable, insure to distinguish between intake and exhaust springs. Use a ruler and/or carpenters square to measure the length. A carpenters square should be used to check the springs for squareness. If a spring pressure test gauge is available, check each springs rating and compare to the specifications chart. Check the readings against the specifications given. Any springs that fail these inspections should be replaced.

The spring retainers rarely need replacing, however they should still be checked as a precaution. Inspect the spring mating surface and the valve lock retention area for any signs of excessive wear. Also check for any signs of cracking. Replace any retainers that are questionable.

Valve locks should be inspected for excessive wear on the outside contact area as well as on the inner notched surface. Any locks which appear worn or broken and its respective valve should be replaced.

Cylinder Head

There are several things to check on the cylinder head: valve guides, seats, cylinder head surface flatness, cracks and physical damage.

VALVE GUIDES

▶ See Figure 191

Now that you know the valves are good, you can use them to check the guides, although a new valve, if available, is preferred. Before you measure anything, look at the guides carefully and inspect them for any cracks, chips or breakage. Also if the guide is a removable style (as in most aluminum heads), check them for any looseness or evidence of movement. All of the guides should appear to be at the same height from the spring seat. If any seem lower (or higher) from another, the guide has moved. Mount a dial indicator onto the

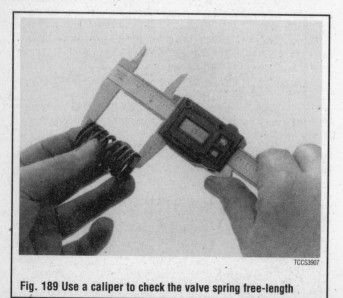

Fig. 189 Use a caliper to check the valve spring free-length

Fig. 191 A dial gauge may be used to check valve stem-to-guide clearance; read the gauge while moving the valve stem

spring side of the cylinder head. Lightly oil the valve stem and insert it into the cylinder head. Position the dial indicator against the valve stem near the tip and zero the gauge. Grasp the valve stem and wiggle towards and away from the dial indicator and observe the readings. Mount the dial indicator 90 degrees from the initial point and zero the gauge and again take a reading. Compare the two readings for a out of round condition. Check the readings against the specifications given. An Inside Diameter (I.D.) gauge designed for valve guides will give you an accurate valve guide bore measurement. If the I.D. gauge is used, compare the readings with the specifications given. Any guides that fail these inspections should be replaced or machined.

VALVE SEATS

A visual inspection of the valve seats should show a slightly worn and pitted surface where the valve face contacts the seat. Inspect the seat carefully for severe pitting or cracks. Also, a seat that is badly worn will be recessed into the cylinder head. A severely worn or recessed seat may need to be replaced. All cracked seats must be replaced. A seat concentricity gauge, if available, should be used to check the seat run-out. If run-out exceeds specifications the seat must be machined (if no specification is given use 0.002 in. or 0.051mm).

CYLINDER HEAD SURFACE FLATNESS

▶ See Figures 192 and 193

After you have cleaned the gasket surface of the cylinder head of any old gasket material, check the head for flatness.

Place a straightedge across the gasket surface. Using feeler gauges, determine the clearance at the center of the straightedge and across the cylinder head at several points. Check along the centerline and diagonally on the head surface. If the warpage exceeds 0.003 in. (0.076mm) within a 6.0 in. (15.2cm) span, or 0.006 in. (0.152mm) over the total length of the head, the cylinder head must be resurfaced. After resurfacing the heads of a V-type engine, the intake manifold flange surface should be checked, and if necessary, milled proportionally to allow for the change in its mounting position.

CRACKS AND PHYSICAL DAMAGE

Generally, cracks are limited to the combustion chamber, however, it is not uncommon for the head to crack in a spark plug hole, port, outside of the head or in the valve spring/rocker arm area. The first area to inspect is always the hottest: the exhaust seat/port area.

A visual inspection should be performed, but just because you don't see a crack does not mean it is not there. Some more reliable methods for inspecting for cracks include Magnaflux®, a magnetic process or Zyglo®, a dye penetrant. Magnaflux® is used only on ferrous metal (cast iron) heads. Zyglo® uses a spray on fluorescent mixture along with a black light to reveal the cracks. It is strongly recommended to have your cylinder head

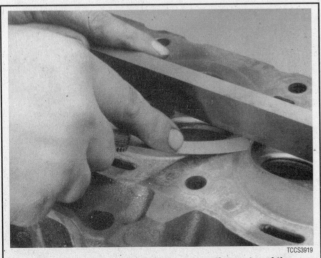

Fig. 192 Check the head for flatness across the center of the head surface using a straightedge and feeler gauge

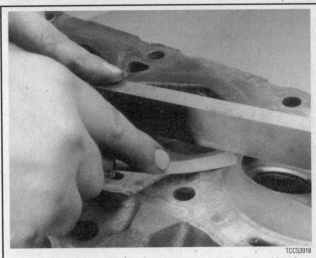

Fig. 193 Checks should also be made along both diagonals of the head surface

checked professionally for cracks, especially if the engine was known to have overheated and/or leaked or consumed coolant. Contact a local shop for availability and pricing of these services.

Physical damage is usually very evident. For example, a broken mounting ear from dropping the head or a bent or broken stud and/or bolt. All of these defects should be fixed or, if unrepairable, the head should be replaced.

Camshaft and Followers

Inspect the camshaft(s) and followers as described earlier in this section.

REFINISHING & REPAIRING

Many of the procedures given for refinishing and repairing the cylinder head components must be performed by a machine shop. Certain steps, if the inspected part is not worn, can be performed yourself inexpensively. However, you spent a lot of time and effort so far, why risk trying to save a couple bucks if you might have to do it all over again?

Valves

Any valves that were not replaced should be refaced and the tips ground flat. Unless you have access to a valve grinding machine, this should be done by a machine shop. If the valves are in extremely good condition, as well as the valve seats and guides, they may be lapped in without performing machine work.

It is a recommended practice to lap the valves even after machine work has been performed and/or new valves have been purchased. This insures a positive seal between the valve and seat.

LAPPING THE VALVES

➡Before lapping the valves to the seats, read the rest of the cylinder head section to insure that any related parts are in acceptable enough condition to continue.

➡Before any valve seat machining and/or lapping can be performed, the guides must be within factory recommended specifications.

1. Invert the cylinder head.
2. Lightly lubricate the valve stems and insert them into the cylinder head in their numbered order.
3. Raise the valve from the seat and apply a small amount of fine lapping compound to the seat.
4. Moisten the suction head of a hand-lapping tool and attach it to the head of the valve.
5. Rotate the tool between the palms of both hands, changing the position of the valve on the valve seat and lifting the tool often to prevent grooving.

6. Lap the valve until a smooth, polished circle is evident on the valve and seat.

7. Remove the tool and the valve. Wipe away all traces of the grinding compound and store the valve to maintain its lapped location.

✳✳ WARNING

Do not get the valves out of order after they have been lapped. They must be put back with the same valve seat they were lapped with.

Springs, Retainers and Valve Locks

There is no repair or refinishing possible with the springs, retainers and valve locks. If they are found to be worn or defective, they must be replaced with new (or known good) parts.

Cylinder Head

Most refinishing procedures dealing with the cylinder head must be performed by a machine shop. Read the sections below and review your inspection data to determine whether or not machining is necessary.

VALVE GUIDE

➡**If any machining or replacements are made to the valve guides, the seats must be machined.**

Unless the valve guides need machining or replacing, the only service to perform is to thoroughly clean them of any dirt or oil residue.

There are only two types of valve guides used on automobile engines: the replaceable-type (all aluminum heads) and the cast-in integral-type (most cast iron heads). There are four recommended methods for repairing worn guides.

- Knurling
- Inserts
- Reaming oversize
- Replacing

Knurling is a process in which metal is displaced and raised, thereby reducing clearance, giving a true center, and providing oil control. It is the least expensive way of repairing the valve guides. However, it is not necessarily the best, and in some cases, a knurled valve guide will not stand up for more than a short time. It requires a special knurlizer and precision reaming tools to obtain proper clearances. It would not be cost effective to purchase these tools, unless you plan on rebuilding several of the same cylinder head.

Installing a guide insert involves machining the guide to accept a bronze insert. One style is the coil-type which is installed into a threaded guide. Another is the thin-walled insert where the guide is reamed oversize to accept a split-sleeve insert. After the insert is installed, a special tool is then run through the guide to expand the insert, locking it to the guide. The insert is then reamed to the standard size for proper valve clearance.

Reaming for oversize valves restores normal clearances and provides a true valve seat. Most cast-in type guides can be reamed to accept an valve with an oversize stem. The cost factor for this can become quite high as you will need to purchase the reamer and new, oversize stem valves for all guides which were reamed. Oversizes are generally 0.003 to 0.030 in. (0.076 to 0.762mm), with 0.015 in. (0.381mm) being the most common.

To replace cast-in type valve guides, they must be drilled out, then reamed to accept replacement guides. This must be done on a fixture which will allow centering and leveling off of the original valve seat or guide, otherwise a serious guide-to-seat misalignment may occur making it impossible to properly machine the seat.

Replaceable-type guides are pressed into the cylinder head. A hammer and a stepped drift or punch may be used to install and remove the guides. Before removing the guides, measure the protrusion on the spring side of the head and record it for installation. Use the stepped drift to hammer out the old guide from the combustion chamber side of the head. When installing, determine whether or not the guide also seals a water jacket in the head, and if it does, use the recommended sealing agent. If there is no water jacket, grease the valve guide and its bore. Use the stepped drift, and hammer the new guide into the cylinder head from the spring side of the

cylinder head. A stack of washers the same thickness as the measured protrusion may help the installation process.

VALVE SEATS

➡**Before any valve seat machining can be performed, the guides must be within factory recommended specifications.**

➡**If any machining or replacements were made to the valve guides, the seats must be machined.**

If the seats are in good condition, the valves can be lapped to the seats, and the cylinder head assembled. See the valves section for instructions on lapping.

If the valve seats are worn, cracked or damaged, they must be serviced by a machine shop. The valve seat must be perfectly centered to the valve guide, which requires very accurate machining.

CYLINDER HEAD SURFACE

If the cylinder head is warped, it must be machined flat. If the warpage is extremely severe, the head may need to be replaced. In some instances, it may be possible to straighten a warped head enough to allow machining. In either case, contact a professional machine shop for service.

➡**Any OHC cylinder head that shows excessive warpage should have the camshaft bearing journals align bored after the cylinder head has been resurfaced.**

✳✳ WARNING

Failure to align bore the camshaft bearing journals could result in severe engine damage including but not limited to: valve and piston damage, connecting rod damage, camshaft and/or crankshaft breakage.

CRACKS AND PHYSICAL DAMAGE

Certain cracks can be repaired in both cast iron and aluminum heads. For cast iron, a tapered threaded insert is installed along the length of the crack. Aluminum can also use the tapered inserts, however welding is the preferred method. Some physical damage can be repaired through brazing or welding. Contact a machine shop to get expert advice for your particular dilemma.

ASSEMBLY

▶ **See Figure 194**

The first step for any assembly job is to have a clean area in which to work. Next, thoroughly clean all of the parts and components that are to be

Fig. 194 Once assembled, check the valve clearance and correct as needed

assembled. Finally, place all of the components onto a suitable work space and, if necessary, arrange the parts to their respective positions.

Cup Type Camshaft Followers

To install the springs, retainers and valve locks on heads which have these components recessed into the camshaft follower's bore, you will need a small screwdriver-type tool, some clean white grease and a lot of patience. You will also need the C-clamp style spring compressor and the OHC tool used to disassemble the head.

1. Lightly lubricate the valve stems and insert all of the valves into the cylinder head. If possible, maintain their original locations.
2. If equipped, install any valve spring shims which were removed.
3. If equipped, install the new valve seals, keeping the following in mind:
• If the valve seal presses over the guide, lightly lubricate the outer guide surfaces.
• If the seal is an O-ring type, it is installed just after compressing the spring but before the valve locks.
4. Place the valve spring and retainer over the stem.
5. Position the spring compressor and the OHC tool, then compress the spring.
6. Using a small screwdriver as a spatula, fill the valve stem side of the lock with white grease. Use the excess grease on the screwdriver to fasten the lock to the driver.
7. Carefully install the valve lock, which is stuck to the end of the screwdriver, to the valve stem then press on it with the screwdriver until the grease squeezes out. The valve lock should now be stuck to the stem.
8. Repeat Steps 6 and 7 for the remaining valve lock.
9. Relieve the spring pressure slowly and insure that neither valve lock becomes dislodged by the retainer.
10. Remove the spring compressor tool.
11. Repeat Steps 2 through 10 until all of the springs have been installed.
12. Install the followers, camshaft(s) and any other components that were removed for disassembly.

Rocker Arm Type Camshaft Followers

1. Lightly lubricate the valve stems and insert all of the valves into the cylinder head. If possible, maintain their original locations.
2. If equipped, install any valve spring shims which were removed.
3. If equipped, install the new valve seals, keeping the following in mind:
• If the valve seal presses over the guide, lightly lubricate the outer guide surfaces.
• If the seal is an O-ring type, it is installed just after compressing the spring but before the valve locks.
4. Place the valve spring and retainer over the stem.
5. Position the spring compressor tool and compress the spring.
6. Assemble the valve locks to the stem.
7. Relieve the spring pressure slowly and insure that neither valve lock becomes dislodged by the retainer.
8. Remove the spring compressor tool.
9. Repeat Steps 2 through 8 until all of the springs have been installed.
10. Install the camshaft(s), rockers, shafts and any other components that were removed for disassembly.

Engine Block

GENERAL INFORMATION

A thorough overhaul or rebuild of an engine block would include replacing the pistons, rings, bearings, timing belt/chain assembly and oil pump. For OHV engines also include a new camshaft and lifters. The block would then have the cylinders bored and honed oversize (or if using removable cylinder sleeves, new sleeves installed) and the crankshaft would be cut undersize to provide new wearing surfaces and perfect clearances. However, your particular engine may not have everything worn out. What if only the piston rings have worn out and the clearances on everything else are still within factory specifications? Well, you could just replace the rings and put

it back together, but this would be a very rare example. Chances are, if one component in your engine is worn, other components are sure to follow, and soon. At the very least, you should always replace the rings, bearings and oil pump. This is what is commonly called a "freshen up".

Cylinder Ridge Removal

Because the top piston ring does not travel to the very top of the cylinder, a ridge is built up between the end of the travel and the top of the cylinder bore.

Pushing the piston and connecting rod assembly past the ridge can be difficult, and damage to the piston ring lands could occur. If the ridge is not removed before installing a new piston or not removed at all, piston ring breakage and piston damage may occur.

➡It is always recommended that you remove any cylinder ridges before removing the piston and connecting rod assemblies. If you know that new pistons are going to be installed and the engine block will be bored oversize, you may be able to forego this step. However, some ridges may actually prevent the assemblies from being removed, necessitating its removal.

There are several different types of ridge reamers on the market, none of which are inexpensive. Unless a great deal of engine rebuilding is anticipated, borrow or rent a reamer.

1. Turn the crankshaft until the piston is at the bottom of its travel.
2. Cover the head of the piston with a rag.
3. Follow the tool manufacturer's instructions and cut away the ridge, exercising extreme care to avoid cutting too deeply.
4. Remove the ridge reamer, the rag and as many of the cuttings as possible. Continue until all of the cylinder ridges have been removed.

DISASSEMBLY

▶ See Figures 195 and 196

The engine disassembly instructions following assume that you have the engine mounted on an engine stand. If not, it is easiest to disassemble the engine on a bench or the floor with it resting on the bell housing or transaxle mounting surface. You must be able to access the connecting rod fasteners and turn the crankshaft during disassembly. Also, all engine covers (timing, front, side, oil pan, whatever) should have already been removed. Engines which are seized or locked up may not be able to be completely disassembled, and a core (salvage yard) engine should be purchased.

If not done during the cylinder head removal, remove the timing chain and/or sprocket assembly. Remove the oil pick-up and pump assembly and, if necessary, the pump drive. If equipped, remove any balance or auxiliary

Fig. 195 Place rubber hose over the connecting rod studs to protect the crankshaft and cylinder bores from damage

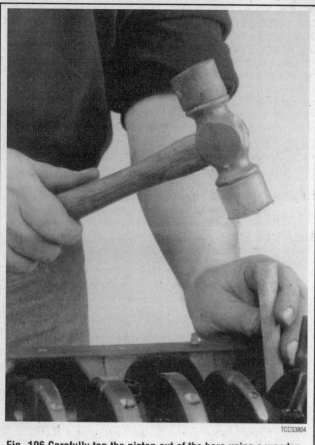

Fig. 196 Carefully tap the piston out of the bore using a wooden dowel

shafts. If necessary, remove the cylinder ridge from the top of the bore. See the cylinder ridge removal procedure earlier in this section.

Rotate the engine over so that the crankshaft is exposed. Use a number punch or scribe and mark each connecting rod with its respective cylinder number. The cylinder closest to the front of the engine is always number 1. However, depending on the engine placement, the front of the engine could either be the flywheel or damper/pulley end. Generally the front of the engine faces the front of the vehicle. Use a number punch or scribe and also mark the main bearing caps from front to rear with the front most cap being number 1 (if there are five caps, mark them 1 through 5, front to rear).

✳✳ WARNING

Take special care when pushing the connecting rod up from the crankshaft because the sharp threads of the rod bolts/studs will score the crankshaft journal. Insure that special plastic caps are installed over them, or cut two pieces of rubber hose to do the same.

Again, rotate the engine, this time to position the number one cylinder bore (head surface) up. Turn the crankshaft until the number one piston is at the bottom of its travel, this should allow the maximum access to its connecting rod. Remove the number one connecting rods fasteners and cap and place two lengths of rubber hose over the rod bolts/studs to protect the crankshaft from damage. Using a sturdy wooden dowel and a hammer, push the connecting rod up about 1 in. (25mm) from the crankshaft and remove the upper bearing insert. Continue pushing or tapping the connecting rod up until the piston rings are out of the cylinder bore. Remove the piston and rod by hand, put the upper half of the bearing insert back into the rod, install the cap with its bearing insert installed, and hand-tighten the cap fasteners. If the parts are kept in order in this manner, they will not get lost and you will be able to tell which bearings came form what cylinder if any

problems are discovered and diagnosis is necessary. Remove all the other piston assemblies in the same manner. On V-style engines, remove all of the pistons from one bank, then reposition the engine with the other cylinder bank head surface up, and remove that banks piston assemblies.

The only remaining component in the engine block should now be the crankshaft. Loosen the main bearing caps evenly until the fasteners can be turned by hand, then remove them and the caps. Remove the crankshaft from the engine block. Thoroughly clean all of the components.

INSPECTION

Now that the engine block and all of its components are clean, it's time to inspect them for wear and/or damage. To accurately inspect them, you will need some specialized tools:

• Two or three separate micrometers to measure the pistons and crankshaft journals
• A dial indicator
• Telescoping gauges for the cylinder bores
• A rod alignment fixture to check for bent connecting rods

If you do not have access to the proper tools, you may want to bring the components to a shop that does.

Generally, you shouldn't expect cracks in the engine block or its components unless it was known to leak, consume or mix engine fluids, it was severely overheated, or there was evidence of bad bearings and/or crankshaft damage. A visual inspection should be performed on all of the components, but just because you don't see a crack does not mean it is not there. Some more reliable methods for inspecting for cracks include Magnaflux®, a magnetic process or Zyglo®, a dye penetrant. Magnaflux® is used only on ferrous metal (cast iron). Zyglo® uses a spray on fluorescent mixture along with a black light to reveal the cracks. It is strongly recommended to have your engine block checked professionally for cracks, especially if the engine was known to have overheated and/or leaked or consumed coolant. Contact a local shop for availability and pricing of these services.

Engine Block

ENGINE BLOCK BEARING ALIGNMENT

Remove the main bearing caps and, if still installed, the main bearing inserts. Inspect all of the main bearing saddles and caps for damage, burrs or high spots. If damage is found, and it is caused from a spun main bearing, the block will need to be align-bored or, if severe enough, replacement. Any burrs or high spots should be carefully removed with a metal file.

Place a straightedge on the bearing saddles, in the engine block, along the centerline of the crankshaft. If any clearance exists between the straightedge and the saddles, the block must be align-bored.

Align-boring consists of machining the main bearing saddles and caps by means of a flycutter that runs through the bearing saddles.

DECK FLATNESS

The top of the engine block where the cylinder head mounts is called the deck. Insure that the deck surface is clean of dirt, carbon deposits and old gasket material. Place a straightedge across the surface of the deck along its centerline and, using feeler gauges, check the clearance along several points. Repeat the checking procedure with the straightedge placed along both diagonals of the deck surface. If the reading exceeds 0.003 in. (0.076mm) within a 6.0 in. (15.2cm) span, or 0.006 in. (0.152mm) over the total length of the deck, it must be machined.

CYLINDER BORES

▶ See Figure 197

The cylinder bores house the pistons and are slightly larger than the pistons themselves. A common piston-to-bore clearance is 0.0015–0.0025 in. (0.0381mm–0.0635mm). Inspect and measure the cylinder bores. The bore should be checked for out-of-roundness, taper and size. The results of this inspection will determine whether the cylinder can be used in its existing size and condition, or a rebore to the next oversize is required (or in the case of removable sleeves, have replacements installed).

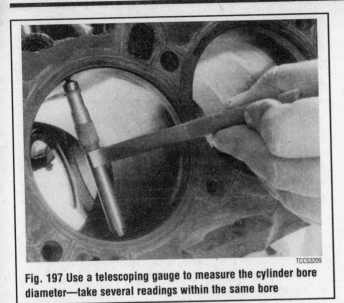

Fig. 197 Use a telescoping gauge to measure the cylinder bore diameter—take several readings within the same bore

The amount of cylinder wall wear is always greater at the top of the cylinder than at the bottom. This wear is known as taper. Any cylinder that has a taper of 0.0012 in. (0.305mm) or more, must be rebored. Measurements are taken at a number of positions in each cylinder: at the top, middle and bottom and at two points at each position; that is, at a point 90 degrees from the crankshaft centerline, as well as a point parallel to the crankshaft centerline. The measurements are made with either a special dial indicator or a telescopic gauge and micrometer. If the necessary precision tools to check the bore are not available, take the block to a machine shop and have them mike it. Also if you don't have the tools to check the cylinder bores, chances are you will not have the necessary devices to check the pistons, connecting rods and crankshaft. Take these components with you and save yourself an extra trip.

For our procedures, we will use a telescopic gauge and a micrometer. You will need one of each, with a measuring range which covers your cylinder bore size.

1. Position the telescopic gauge in the cylinder bore, loosen the gauges lock and allow it to expand.

➡ Your first two readings will be at the top of the cylinder bore, then proceed to the middle and finally the bottom, making a total of six measurements.

2. Hold the gauge square in the bore, 90 degrees from the crankshaft centerline, and gently tighten the lock. Tilt the gauge back to remove it from the bore.

3. Measure the gauge with the micrometer and record the reading.

4. Again, hold the gauge square in the bore, this time parallel to the crankshaft centerline, and gently tighten the lock. Again, you will tilt the gauge back to remove it from the bore.

5. Measure the gauge with the micrometer and record this reading. The difference between these two readings is the out-of-round measurement of the cylinder.

6. Repeat Steps 1 through 5, each time going to the next lower position, until you reach the bottom of the cylinder. Then go to the next cylinder, and continue until all of the cylinders have been measured.

The difference between these measurements will tell you all about the wear in your cylinders. The measurements which were taken 90 degrees from the crankshaft centerline will always reflect the most wear. That is because at this position is where the engine power presses the piston against the cylinder bore the hardest. This is known as thrust wear. Take your top, 90 degree measurement and compare it to your bottom, 90 degree measurement. The difference between them is the taper. When you measure your pistons, you will compare these readings to your piston sizes and determine piston-to-wall clearance.

Crankshaft

Inspect the crankshaft for visible signs of wear or damage. All of the journals should be perfectly round and smooth. Slight scores are normal

for a used crankshaft, but you should hardly feel them with your fingernail. When measuring the crankshaft with a micrometer, you will take readings at the front and rear of each journal, then turn the micrometer 90 degrees and take two more readings, front and rear. The difference between the front-to-rear readings is the journal taper and the first-to-90 degree reading is the out-of-round measurement. Generally, there should be no taper or out-of-roundness found, however, up to 0.0005 in. (0.0127mm) for either can be overlooked. Also, the readings should fall within the factory specifications for journal diameters.

If the crankshaft journals fall within specifications, it is recommended that it be polished before being returned to service. Polishing the crankshaft insures that any minor burrs or high spots are smoothed, thereby reducing the chance of scoring the new bearings.

Pistons and Connecting Rods

PISTONS

◆ See Figure 198

The piston should be visually inspected for any signs of cracking or burning (caused by hot spots or detonation), and scuffing or excessive wear on the skirts. The wristpin attaches the piston to the connecting rod. The piston should move freely on the wrist pin, both sliding and pivoting. Grasp the connecting rod securely, or mount it in a vise, and try to rock the piston back and forth along the centerline of the wristpin. There should not be any excessive play evident between the piston and the pin. If there are C-clips retaining the pin in the piston then you have wrist pin bushings in the rods. There should not be any excessive play between the wrist pin and the rod bushing. Normal clearance for the wrist pin is approx. 0.001–0.002 in. (0.025mm–0.051mm).

Use a micrometer and measure the diameter of the piston, perpendicular to the wrist pin, on the skirt. Compare the reading to its original cylinder measurement obtained earlier. The difference between the two readings is the piston-to-wall clearance. If the clearance is within specifications, the piston may be used as is. If the piston is out of specification, but the bore is not, you will need a new piston. If both are out of specification, you will need the cylinder rebored and oversize pistons installed. Generally if two or more pistons/bores are out of specification, it is best to rebore the entire block and purchase a complete set of oversize pistons.

Fig. 198 Measure the piston's outer diameter, perpendicular to the wrist pin, with a micrometer

CONNECTING ROD

You should have the connecting rod checked for straightness at a machine shop. If the connecting rod is bent, it will unevenly wear the bearing and piston, as well as place greater stress on these components. Any bent or twisted connecting rods must be replaced. If the rods are straight and the wrist pin clearance is within specifications, then only the bearing end of the rod need be checked. Place the connecting rod into a vice, with the bearing inserts in place, install the cap to the rod and tighten the fasteners to specifications. Use a telescoping gauge and carefully measure the inside diameter of the bearings. Compare this reading to the rods original crankshaft journal diameter measurement. The difference is the oil clearance. If the oil clearance is not within specifications, install new bearings in the rod and take another measurement. If the clearance is still out of specifications, and the crankshaft is not, the rod will need to be reconditioned by a machine shop.

➡You can also use Plastigage® to check the bearing clearances. The assembling section has complete instructions on its use.

Camshaft

Inspect the camshaft and lifters/followers, as described earlier in this section.

Bearings

All of the engine bearings should be visually inspected for wear and/or damage. The bearing should look evenly worn all around with no deep scores or pits. If the bearing is severely worn, scored, pitted or heat blued, then the bearing, and the components that use it, should be brought to a machine shop for inspection. Full-circle bearings (used on most camshafts, auxiliary shafts, balance shafts, etc.) require specialized tools for removal and installation, and should be brought to a machine shop for service.

Oil Pump

➡The oil pump is responsible for providing constant lubrication to the whole engine and so it is recommended that a new oil pump be installed when rebuilding the engine.

Completely disassemble the oil pump and thoroughly clean all of the components. Inspect the oil pump gears and housing for wear and/or damage. Insure that the pressure relief valve operates properly and there is no binding or sticking due to varnish or debris. If all of the parts are in proper working condition, lubricate the gears and relief valve, and assemble the pump.

REFINISHING

♦ See Figure 199

Almost all engine block refinishing must be performed by a machine shop. If the cylinders are not to be rebored, then the cylinder glaze can be removed with a ball hone. When removing cylinder glaze with a ball hone, use a light or penetrating type oil to lubricate the hone. Do not allow the hone to run dry as

Fig. 199 Use a ball type cylinder hone to remove any glaze and provide a new surface for seating the piston rings

this may cause excessive scoring of the cylinder bores and wear on the hone. If new pistons are required, they will need to be installed to the connecting rods. This should be performed by a machine shop as the pistons must be installed in the correct relationship to the rod or engine damage can occur.

Pistons and Connecting Rods

♦ See Figure 200

Only pistons with the wrist pin retained by C-clips are serviceable by the home-mechanic. Press fit pistons require special presses and/or heaters to remove/install the connecting rod and should only be performed by a machine shop.

All pistons will have a mark indicating the direction to the front of the engine and the must be installed into the engine in that manner. Usually it is a notch or arrow on the top of the piston, or it may be the letter F cast or stamped into the piston.

Fig. 200 Most pistons are marked to indicate positioning in the engine (usually a mark means the side facing the front)

C-CLIP TYPE PISTONS

1. Note the location of the forward mark on the piston and mark the connecting rod in relation.
2. Remove the C-clips from the piston and withdraw the wrist pin.

➡Varnish build-up or C-clip groove burrs may increase the difficulty of removing the wrist pin. If necessary, use a punch or drift to carefully tap the wrist pin out.

3. Insure that the wrist pin bushing in the connecting rod is usable, and lubricate it with assembly lube.
4. Remove the wrist pin from the new piston and lubricate the pin bores on the piston.
5. Align the forward marks on the piston and the connecting rod and install the wrist pin.
6. The new C-clips will have a flat and a rounded side to them. Install both C-clips with the flat side facing out.
7. Repeat all of the steps for each piston being replaced.

ASSEMBLY

Before you begin assembling the engine, first give yourself a clean, dirt free work area. Next, clean every engine component again. The key to a good assembly is cleanliness.

Mount the engine block into the engine stand and wash it one last time using water and detergent (dishwashing detergent works well). While washing it, scrub the cylinder bores with a soft bristle brush and thoroughly clean all of the oil passages. Completely dry the engine and spray the entire assembly down with an anti-rust solution such as WD-40® or similar product. Take a clean lint-free rag and wipe up any excess anti-rust solution from the bores, bearing saddles, etc. Repeat the final cleaning process on the crankshaft. Replace any freeze or oil galley plugs which were removed during disassembly.

Crankshaft

▶ See Figures 201, 202, 203 and 204

1. Remove the main bearing inserts from the block and bearing caps.
2. If the crankshaft main bearing journals have been refinished to a definite undersize, install the correct undersize bearing. Be sure that the bearing inserts and bearing bores are clean. Foreign material under inserts will distort bearing and cause failure.
3. Place the upper main bearing inserts in bores with tang in slot.

➡The oil holes in the bearing inserts must be aligned with the oil holes in the cylinder block.

4. Install the lower main bearing inserts in bearing caps.
5. Clean the mating surfaces of block and rear main bearing cap.
6. Carefully lower the crankshaft into place. Be careful not to damage bearing surfaces.
7. Check the clearance of each main bearing by using the following procedure:

   a. Place a piece of Plastigage® or its equivalent, on bearing surface across full width of bearing cap and about ¼ in. off center.
   b. Install cap and tighten bolts to specifications. Do not turn crankshaft while Plastigage® is in place.
   c. Remove the cap. Using the supplied Plastigage® scale, check width of Plastigage® at widest point to get maximum clearance. Difference between readings is taper of journal.

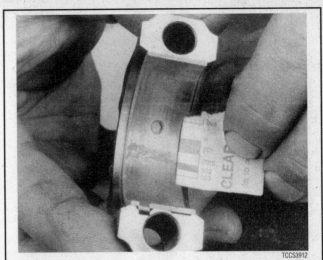

Fig. 202 After the cap is removed again, use the scale supplied with the gauging material to check the clearance

Fig. 203 A dial gauge may be used to check crankshaft end-play

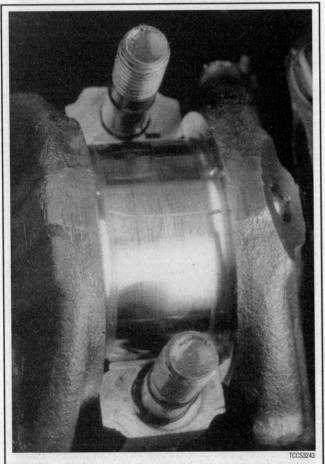

Fig. 201 Apply a strip of gauging material to the bearing journal, then install and tighten the cap

Fig. 204 Carefully pry the crankshaft back and forth while reading the dial gauge for end-play

d. If clearance exceeds specified limits, try a 0.001 in. or 0.002 in. undersize bearing in combination with the standard bearing. Bearing clearance must be within specified limits. If standard and 0.002 in. undersize bearing does not bring clearance within desired limits, refinish crankshaft journal, then install undersize bearings.

8. After the bearings have been fitted, apply a light coat of engine oil to the journals and bearings. Install the rear main bearing cap. Install all bearing caps except the thrust bearing cap. Be sure that main bearing caps are installed in original locations. Tighten the bearing cap bolts to specifications.

9. Install the thrust bearing cap with bolts finger-tight.

10. Pry the crankshaft forward against the thrust surface of upper half of bearing.

11. Hold the crankshaft forward and pry the thrust bearing cap to the rear. This aligns the thrust surfaces of both halves of the bearing.

12. Retain the forward pressure on the crankshaft. Tighten the cap bolts to specifications.

13. Install the rear main seal.

14. Measure the crankshaft end-play as follows:

a. Mount a dial gauge to the engine block and position the tip of the gauge to read from the crankshaft end.

b. Carefully pry the crankshaft toward the rear of the engine and hold it there while you zero the gauge.

c. Carefully pry the crankshaft toward the front of the engine and read the gauge.

d. Confirm that the reading is within specifications. If not, install a new thrust bearing and repeat the procedure. If the reading is still out of specifications with a new bearing, have a machine shop inspect the thrust surfaces of the crankshaft, and if possible, repair it.

15. Rotate the crankshaft so as to position the first rod journal to the bottom of its stroke.

Pistons and Connecting Rods

▶ See Figures 205, 206, 207 and 208

1. Before installing the piston/connecting rod assembly, oil the pistons, piston rings and the cylinder walls with light engine oil. Install connecting rod bolt protectors or rubber hose onto the connecting rod bolts/studs. Also perform the following:

a. Select the proper ring set for the size cylinder bore.

b. Position the ring in the bore in which it is going to be used.

c. Push the ring down into the bore area where normal ring wear is not encountered.

d. Use the head of the piston to position the ring in the bore so that the ring is square with the cylinder wall. Use caution to avoid damage to the ring or cylinder bore.

e. Measure the gap between the ends of the ring with a feeler gauge. Ring gap in a worn cylinder is normally greater than specification. If the ring gap is greater than the specified limits, try an oversize ring set.

f. Check the ring side clearance of the compression rings with a feeler gauge inserted between the ring and its lower land according to specification. The gauge should slide freely around the entire ring circumference without binding. Any wear that occurs will form a step at the inner portion of the lower land. If the lower lands have high steps, the piston should be replaced.

2. Unless new pistons are installed, be sure to install the pistons in the cylinders from which they were removed. The numbers on the connecting rod and bearing cap must be on the same side when installed in the cylinder bore. If a connecting rod is ever transposed from one engine or cylinder to another, new bearings should be fitted and the connecting rod should be numbered to correspond with the new cylinder number. The notch on the piston head goes toward the front of the engine.

3. Install all of the rod bearing inserts into the rods and caps.

4. Install the rings to the pistons. Install the oil control ring first, then the second compression ring and finally the top compression ring. Use a piston ring expander tool to aid in installation and to help reduce the chance of breakage.

5. Make sure the ring gaps are properly spaced around the circumference of the piston. Fit a piston ring compressor around the piston and slide the piston and connecting rod assembly down into the cylinder bore, pushing it in with the wooden hammer handle. Push the piston down until it is only slightly below the top of the cylinder bore. Guide the connecting rod onto the crankshaft bearing journal carefully, to avoid damaging the crankshaft.

TCCS3923

Fig. 205 Checking the piston ring-to-ring groove side clearance using the ring and a feeler gauge

TCCS3917

Fig. 206 The notch on the side of the bearing cap matches the tang on the bearing insert

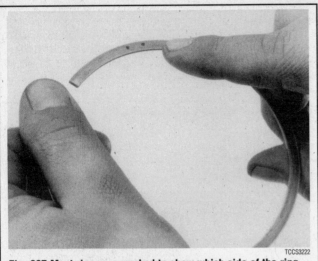

Fig. 207 Most rings are marked to show which side of the ring should face up when installed to the piston

Fig. 208 Install the piston and rod assembly into the block using a ring compressor and the handle of a hammer

6. Check the bearing clearance of all the rod bearings, fitting them to the crankshaft bearing journals. Follow the procedure in the crankshaft installation above.

7. After the bearings have been fitted, apply a light coating of assembly oil to the journals and bearings.

8. Turn the crankshaft until the appropriate bearing journal is at the bottom of its stroke, then push the piston assembly all the way down until the connecting rod bearing seats on the crankshaft journal. Be careful not to allow the bearing cap screws to strike the crankshaft bearing journals and damage them.

9. After the piston and connecting rod assemblies have been installed, check the connecting rod side clearance on each crankshaft journal.

10. Prime and install the oil pump and the oil pump intake tube.

11. Install the cylinder head(s) using new gaskets.
12. Install the timing sprockets and chain assembly.

Engine Covers and Components

Install the timing cover(s) and oil pan. Refer to your notes and drawings made prior to disassembly and install all of the components that were removed. Install the engine into the vehicle.

Engine Start-up and Break-in

STARTING THE ENGINE

Now that the engine is installed and every wire and hose is properly connected, go back and double check that all coolant and vacuum hoses are connected. Check that you oil drain plug is installed and properly tightened. If not already done, install a new oil filter onto the engine. Fill the crankcase with the proper amount and grade of engine oil. Fill the cooling system with a 50/50 mixture of coolant/water.

1. Connect the vehicle battery.
2. Start the engine. Keep your eye on your oil pressure indicator; if it does not indicate oil pressure within 10 seconds of starting, turn the vehicle off.

✳✳ WARNING

Damage to the engine can result if it is allowed to run with no oil pressure. Check the engine oil level to make sure that it is full. Check for any leaks and if found, repair the leaks before continuing. If there is still no indication of oil pressure, you may need to prime the system.

3. Confirm that there are no fluid leaks (oil or other).
4. Allow the engine to reach normal operating temperature (the upper radiator hose will be hot to the touch).
5. If necessary, set the ignition timing.
6. Install any remaining components such as the air cleaner (if removed for ignition timing) or body panels which were removed.

BREAKING IT IN

Make the first miles on the new engine, easy ones. Vary the speed but do not accelerate hard. Most importantly, do not lug the engine, and avoid sustained high speeds until at least 100 miles. Check the engine oil and coolant levels frequently. Expect the engine to use a little oil until the rings seat. Change the oil and filter at 500 miles, 1500 miles, then every 3000 miles past that.

KEEP IT MAINTAINED

Now that you have just gone through all of that hard work, keep yourself from doing it all over again by thoroughly maintaining it. Not that you may not have maintained it before, heck you could have had one to two hundred thousand miles on it before doing this. However, you may have bought the vehicle used, and the previous owner did not keep up on maintenance. Which is why you just went through all of that hard work. See?

TORQUE SPECIFICATIONS

Component	Ft. lbs.	Nm
Valve Cover		
SOHC Engine		
Valve cover fasteners	22	30
DOHC Engine		
Camshaft cover fasteners	89 inch lbs.	10
Rocker Arms/Shafts		
Rocker arm bolts	18	25
Thermostat		
Housing bolts	22	30
Cylinder block drain plug	26	35
Intake Manifold		
SOHC Engine		
Intake manifold nuts	22	30
Power steering pump fasteners	27	38
Manifold support bracket bolt	22	30
DOHC Engine		
Intake manifold nuts	22	30
Power steering pump fasteners	28	38
Manifold support brackets bolts		
Right block bolt	41	55
Left block bolt	22	30
Engine block drain plug	26	35
Exhaust Manifold		
SOHC Engine		
Manifold nuts	16	22
DOHC Engine		
Manifold nuts	23	31
Oxygen sensor	18	25
A/C compressor		
Except the front bracket-to-compressor fasteners	25	18
Front bracket-to-compressor fasteners	40	54
Pipe and manifold fasteners	23	31
Electric Cooling Fan		
Motor and fan blade		
Left-hand threaded fan nut	27-44 inch lbs.	3-5
Water Pump		
Pump assembly bolts	22	30
Pump pulley bolts	19	25
Cylinder Head		
Cylinder head bolts		
First pass	22	30
Second pass		
SOHC Engines	33	45
DOHC Engines	37	50
Third Pass	All bolts an additional 90 degrees	
Drive belt tensioner	22	30
Idler Pulley fasteners		
1992-93 vehicles	33	45
Oil pan drain plug	26	35
Spark plugs	20	27
TBI assembly retainers	24	33

TORQUE SPECIFICATIONS

Component	Ft. lbs.	Nm
Cylinder Head		
Models with torque axis mounting		
Mount-to-midrail bracket nuts		
1992-93 vehicles	52	70
Upper mount-to-engine front cover nuts	52	70
Wheel lug nuts	103	140
Accelerator cable bracket fastener	19	25
Oil Pan		
Oil pan bolts	80 inch lbs.	9
Front mount bolts	40	54
Exhaust pipe-to-manifold nuts	23	31
Exhaust pipe-to-converter bolts	33	45
Oil Pump		
Pump body cover bolts	97 inch lbs.	11
Crankshaft Damper/Pulley		
Crankshaft damper/pulley bolt	159	215
Timing Chain Front Cover		
Front cover-to-engine bolts		
SOHC Engine	19	25
DOHC Engine	22	30
① Front cover center or inner bolts	89 inch lbs.	10
Timing Chain and Sprockets		
SOHC Engine		
Sprocket bolt	75	102
Tensioner bolts	14	19
DOHC Engine		
Camshaft gears bolts	75	102
Timing chain fixed guide bolts	21	28
Pivoting chain guide bolt	19	26
Tensioner bolts	14	19
Camshaft		
DOHC Engine		
Bearing cap bolts	124 inch lbs.	14
Camshaft sprocket pilot bolts	124 inch lbs.	14
Camshaft sprocket bolts	76	103
Flywheel/Flexplate		
Manual transaxle		
Flywheel bolts	59	80
Automatic transaxle		
Flexplate bolts	44	60

① Except upper inside bolt on 1992-93 DOHC models: 22 ft. lbs. (30 Nm)

89563C06

USING A VACUUM GAUGE

White needle = steady needle   *Dark needle = drifting needle*

The vacuum gauge is one of the most useful and easy-to-use diagnostic tools. It is inexpensive, easy to hook up, and provides valuable information about the condition of your engine.

Indication: Normal engine in good condition

Gauge reading: Steady, from 17–22 in./Hg.

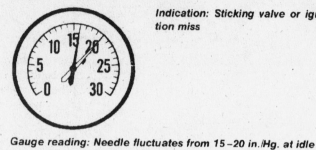

Indication: Sticking valve or ignition miss

Gauge reading: Needle fluctuates from 15–20 in./Hg. at idle

Indication: Late ignition or valve timing, low compression, stuck throttle valve, leaking carburetor or manifold gasket.

Gauge reading: Low (15–20 in./Hg.) but steady

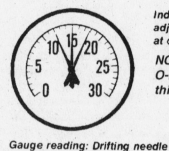

Indication: Improper carburetor adjustment, or minor intake leak at carburetor or manifold

NOTE: Bad fuel injector O-rings may also cause this reading.

Gauge reading: Drifting needle

Indication: Weak valve springs, worn valve stem guides, or leaky cylinder head gasket (vibrating excessively at all speeds).

NOTE: A plugged catalytic converter may also cause this reading.

Gauge reading: Needle fluctuates as engine speed increases

Indication: Burnt valve or improper valve clearance. The needle will drop when the defective valve operates.

Gauge reading: Steady needle, but drops regularly

Indication: Choked muffler or obstruction in system. Speed up the engine. Choked muffler will exhibit a slow drop of vacuum to zero.

Gauge reading: Gradual drop in reading at idle

Indication: Worn valve guides

Gauge reading: Needle vibrates excessively at idle, but steadies as engine speed increases

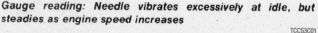

TCCS3C01

Troubleshooting Engine Mechanical Problems

Problem	Cause	Solution
External oil leaks	• Cylinder head cover RTV sealant broken or improperly seated	• Replace sealant; inspect cylinder head cover sealant flange and cylinder head sealant surface for distortion and cracks
	• Oil filler cap leaking or missing	• Replace cap
	• Oil filter gasket broken or improperly seated	• Replace oil filter
	• Oil pan side gasket broken, improperly seated or opening in RTV sealant	• Replace gasket or repair opening in sealant; inspect oil pan gasket flange for distortion
	• Oil pan front oil seal broken or improperly seated	• Replace seal; inspect timing case cover and oil pan seal flange for distortion
	• Oil pan rear oil seal broken or improperly seated	• Replace seal; inspect oil pan rear oil seal flange; inspect rear main bearing cap for cracks, plugged oil return channels, or distortion in seal groove
	• Timing case cover oil seal broken or improperly seated	• Replace seal
	• Excess oil pressure because of restricted PCV valve	• Replace PCV valve
	• Oil pan drain plug loose or has stripped threads	• Repair as necessary and tighten
	• Rear oil gallery plug loose	• Use appropriate sealant on gallery plug and tighten
	• Rear camshaft plug loose or improperly seated	• Seat camshaft plug or replace and seal, as necessary
Excessive oil consumption	• Oil level too high	• Drain oil to specified level
	• Oil with wrong viscosity being used	• Replace with specified oil
	• PCV valve stuck closed	• Replace PCV valve
	• Valve stem oil deflectors (or seals) are damaged, missing, or incorrect type	• Replace valve stem oil deflectors
	• Valve stems or valve guides worn	• Measure stem-to-guide clearance and repair as necessary
	• Poorly fitted or missing valve cover baffles	• Replace valve cover
	• Piston rings broken or missing	• Replace broken or missing rings
	• Scuffed piston	• Replace piston
	• Incorrect piston ring gap	• Measure ring gap, repair as necessary
	• Piston rings sticking or excessively loose in grooves	• Measure ring side clearance, repair as necessary
	• Compression rings installed upside down	• Repair as necessary
	• Cylinder walls worn, scored, or glazed	• Repair as necessary

TCCS3C02

Troubleshooting Engine Mechanical Problems

Problem	Cause	Solution
Excessive oil consumption (cont.)	• Piston ring gaps not properly staggered	• Repair as necessary
	• Excessive main or connecting rod bearing clearance	• Measure bearing clearance, repair as necessary
No oil pressure	• Low oil level	• Add oil to correct level
	• Oil pressure gauge, warning lamp or sending unit inaccurate	• Replace oil pressure gauge or warning lamp
	• Oil pump malfunction	• Replace oil pump
	• Oil pressure relief valve sticking	• Remove and inspect oil pressure relief valve assembly
	• Oil passages on pressure side of pump obstructed	• Inspect oil passages for obstruction
	• Oil pickup screen or tube obstructed	• Inspect oil pickup for obstruction
	• Loose oil inlet tube	• Tighten or seal inlet tube
Low oil pressure	• Low oil level	• Add oil to correct level
	• Inaccurate gauge, warning lamp or sending unit	• Replace oil pressure gauge or warning lamp
	• Oil excessively thin because of dilution, poor quality, or improper grade	• Drain and refill crankcase with recommended oil
	• Excessive oil temperature	• Correct cause of overheating engine
	• Oil pressure relief spring weak or sticking	• Remove and inspect oil pressure relief valve assembly
	• Oil inlet tube and screen assembly has restriction or air leak	• Remove and inspect oil inlet tube and screen assembly. (Fill inlet tube with lacquer thinner to locate leaks.)
	• Excessive oil pump clearance	• Measure clearances
	• Excessive main, rod, or camshaft bearing clearance	• Measure bearing clearances, repair as necessary
High oil pressure	• Improper oil viscosity	• Drain and refill crankcase with correct viscosity oil
	• Oil pressure gauge or sending unit inaccurate	• Replace oil pressure gauge
	• Oil pressure relief valve sticking closed	• Remove and inspect oil pressure relief valve assembly
Main bearing noise	• Insufficient oil supply	• Inspect for low oil level and low oil pressure
	• Main bearing clearance excessive	• Measure main bearing clearance, repair as necessary
	• Bearing insert missing	• Replace missing insert
	• Crankshaft end-play excessive	• Measure end-play, repair as necessary
	• Improperly tightened main bearing cap bolts	• Tighten bolts with specified torque
	• Loose flywheel or drive plate	• Tighten flywheel or drive plate attaching bolts
	• Loose or damaged vibration damper	• Repair as necessary

TCCS3C03

Troubleshooting Engine Mechanical Problems

Problem	Cause	Solution
Connecting rod bearing noise	• Insufficient oil supply	• Inspect for low oil level and low oil pressure
	• Carbon build-up on piston	• Remove carbon from piston crown
	• Bearing clearance excessive or bearing missing	• Measure clearance, repair as necessary
	• Crankshaft connecting rod journal out-of-round	• Measure journal dimensions, repair or replace as necessary
	• Misaligned connecting rod or cap	• Repair as necessary
	• Connecting rod bolts tightened improperly	• Tighten bolts with specified torque
Piston noise	• Piston-to-cylinder wall clearance excessive (scuffed piston)	• Measure clearance and examine piston
	• Cylinder walls excessively tapered or out-of-round	• Measure cylinder wall dimensions, rebore cylinder
	• Piston ring broken	• Replace all rings on piston
	• Loose or seized piston pin	• Measure piston-to-pin clearance, repair as necessary
	• Connecting rods misaligned	• Measure rod alignment, straighten or replace
	• Piston ring side clearance excessively loose or tight	• Measure ring side clearance, repair as necessary
	• Carbon build-up on piston is excessive	• Remove carbon from piston
Valve actuating component noise	• Insufficient oil supply	• Check for: (a) Low oil level (b) Low oil pressure (c) Wrong hydraulic tappets (d) Restricted oil gallery (e) Excessive tappet to bore clearance
	• Rocker arms or pivots worn	• Replace worn rocker arms or pivots
	• Foreign objects or chips in hydraulic tappets	• Clean tappets
	• Excessive tappet leak-down	• Replace valve tappet
	• Tappet face worn	• Replace tappet; inspect corresponding cam lobe for wear
	• Broken or cocked valve springs	• Properly seat cocked springs; replace broken springs
	• Stem-to-guide clearance excessive	• Measure stem-to-guide clearance, repair as required
	• Valve bent	• Replace valve
	• Loose rocker arms	• Check and repair as necessary
	• Valve seat runout excessive	• Regrind valve seat/valves
	• Missing valve lock	• Install valve lock
	• Excessive engine oil	• Correct oil level

TCCS3C04

Troubleshooting Engine Performance

Problem	Cause	Solution
Hard starting (engine cranks normally)	• Faulty engine control system component	• Repair or replace as necessary
	• Faulty fuel pump	• Replace fuel pump
	• Faulty fuel system component	• Repair or replace as necessary
	• Faulty ignition coil	• Test and replace as necessary
	• Improper spark plug gap	• Adjust gap
	• Incorrect ignition timing	• Adjust timing
	• Incorrect valve timing	• Check valve timing; repair as necessary
Rough idle or stalling	• Incorrect curb or fast idle speed	• Adjust curb or fast idle speed (If possible)
	• Incorrect ignition timing	• Adjust timing to specification
	• Improper feedback system operation	• Refer to Chapter 4
	• Faulty EGR valve operation	• Test EGR system and replace as necessary
	• Faulty PCV valve air flow	• Test PCV valve and replace as necessary
	• Faulty TAC vacuum motor or valve	• Repair as necessary
	• Air leak into manifold vacuum	• Inspect manifold vacuum connections and repair as necessary
	• Faulty distributor rotor or cap	• Replace rotor or cap (Distributor systems only)
	• Improperly seated valves	• Test cylinder compression, repair as necessary
	• Incorrect ignition wiring	• Inspect wiring and correct as necessary
	• Faulty ignition coil	• Test coil and replace as necessary
	• Restricted air vent or idle passages	• Clean passages
	• Restricted air cleaner	• Clean or replace air cleaner filter element
Faulty low-speed operation	• Restricted idle air vents and passages	• Clean air vents and passages
	• Restricted air cleaner	• Clean or replace air cleaner filter element
	• Faulty spark plugs	• Clean or replace spark plugs
	• Dirty, corroded, or loose ignition secondary circuit wire connections	• Clean or tighten secondary circuit wire connections
	• Improper feedback system operation	• Refer to Chapter 4
	• Faulty ignition coil high voltage wire	• Replace ignition coil high voltage wire (Distributor systems only)
	• Faulty distributor cap	• Replace cap (Distributor systems only)
Faulty acceleration	• Incorrect ignition timing	• Adjust timing
	• Faulty fuel system component	• Repair or replace as necessary
	• Faulty spark plug(s)	• Clean or replace spark plug(s)
	• Improperly seated valves	• Test cylinder compression, repair as necessary
	• Faulty ignition coil	• Test coil and replace as necessary

Troubleshooting Engine Performance

Problem	Cause	Solution
Faulty acceleration (cont.)	• Improper feedback system operation	• Refer to Chapter 4
Faulty high speed operation	• Incorrect ignition timing • Faulty advance mechanism	• Adjust timing (if possible) • Check advance mechanism and repair as necessary (Distributor systems only)
	• Low fuel pump volume • Wrong spark plug air gap or wrong plug • Partially restricted exhaust manifold, exhaust pipe, catalytic converter, muffler, or tailpipe • Restricted vacuum passages • Restricted air cleaner	• Replace fuel pump • Adjust air gap or install correct plug • Eliminate restriction • Clean passages • Cleaner or replace filter element as necessary
	• Faulty distributor rotor or cap	• Replace rotor or cap (Distributor systems only)
	• Faulty ignition coil • Improperly seated valve(s)	• Test coil and replace as necessary • Test cylinder compression, repair as necessary
	• Faulty valve spring(s)	• Inspect and test valve spring tension, replace as necessary
	• Incorrect valve timing	• Check valve timing and repair as necessary
	• Intake manifold restricted	• Remove restriction or replace manifold
	• Worn distributor shaft	• Replace shaft (Distributor systems only)
	• Improper feedback system operation	• Refer to Chapter 4
Misfire at all speeds	• Faulty spark plug(s) • Faulty spark plug wire(s) • Faulty distributor cap or rotor	• Clean or relace spark plug(s) • Replace as necessary • Replace cap or rotor (Distributor systems only)
	• Faulty ignition coil • Primary ignition circuit shorted or open intermittently • Improperly seated valve(s)	• Test coil and replace as necessary • Troubleshoot primary circuit and repair as necessary • Test cylinder compression, repair as necessary
	• Faulty hydraulic tappet(s) • Improper feedback system operation • Faulty valve spring(s)	• Clean or replace tappet(s) • Refer to Chapter 4 • Inspect and test valve spring tension, repair as necessary
	• Worn camshaft lobes • Air leak into manifold	• Replace camshaft • Check manifold vacuum and repair as necessary
	• Fuel pump volume or pressure low • Blown cylinder head gasket • Intake or exhaust manifold passage(s) restricted	• Replace fuel pump • Replace gasket • Pass chain through passage(s) and repair as necessary
Power not up to normal	• Incorrect ignition timing • Faulty distributor rotor	• Adjust timing • Replace rotor (Distributor systems only)

TCCS3C06

Troubleshooting Engine Performance

Problem	Cause	Solution
Power not up to normal (cont.)	• Incorrect spark plug gap	• Adjust gap
	• Faulty fuel pump	• Replace fuel pump
	• Faulty fuel pump	• Replace fuel pump
	• Incorrect valve timing	• Check valve timing and repair as necessary
	• Faulty ignition coil	• Test coil and replace as necessary
	• Faulty ignition wires	• Test wires and replace as necessary
	• Improperly seated valves	• Test cylinder compression and repair as necessary
	• Blown cylinder head gasket	• Replace gasket
	• Leaking piston rings	• Test compression and repair as necessary
	• Improper feedback system operation	• Refer to Chapter 4
Intake backfire	• Improper ignition timing	• Adjust timing
	• Defective EGR component	• Repair as necessary
	• Defective TAC vacuum motor or valve	• Repair as necessary
Exhaust backfire	• Air leak into manifold vacuum	• Check manifold vacuum and repair as necessary
	• Faulty air injection diverter valve	• Test diverter valve and replace as necessary
	• Exhaust leak	• Locate and eliminate leak
Ping or spark knock	• Incorrect ignition timing	• Adjust timing
	• Distributor advance malfunction	• Inspect advance mechanism and repair as necessary (Distributor systems only)
	• Excessive combustion chamber deposits	• Remove with combustion chamber cleaner
	• Air leak into manifold vacuum	• Check manifold vacuum and repair as necessary
	• Excessively high compression	• Test compression and repair as necessary
	• Fuel octane rating excessively low	• Try alternate fuel source
	• Sharp edges in combustion chamber	• Grind smooth
	• EGR valve not functioning properly	• Test EGR system and replace as necessary
Surging (at cruising to top speeds)	• Low fuel pump pressure or volume	• Replace fuel pump
	• Improper PCV valve air flow	• Test PCV valve and replace as necessary
	• Air leak into manifold vacuum	• Check manifold vacuum and repair as necessary
	• Incorrect spark advance	• Test and replace as necessary
	• Restricted fuel filter	• Replace fuel filter
	• Restricted air cleaner	• Clean or replace air cleaner filter element
	• EGR valve not functioning properly	• Test EGR system and replace as necessary
	• Improper feedback system operation	• Refer to Chapter 4

TCCS3C07

Troubleshooting the Serpentine Drive Belt

Problem	Cause	Solution
Tension sheeting fabric failure (woven fabric on outside circumference of belt has cracked or separated from body of belt)	• Grooved or backside idler pulley diameters are less than minimum recommended • Tension sheeting contacting (rubbing) stationary object • Excessive heat causing woven fabric to age • Tension sheeting splice has fractured	• Replace pulley(s) not conforming to specification • Correct rubbing condition • Replace belt • Replace belt
Noise (objectional squeal, squeak, or rumble is heard or felt while drive belt is in operation)	• Belt slippage • Bearing noise • Belt misalignment • Belt-to-pulley mismatch • Driven component inducing vibration • System resonant frequency inducing vibration	• Adjust belt • Locate and repair • Align belt/pulley(s) • Install correct belt • Locate defective driven component and repair • Vary belt tension within specifications. Replace belt.
Rib chunking (one or more ribs has separated from belt body)	• Foreign objects imbedded in pulley grooves • Installation damage • Drive loads in excess of design specifications • Insufficient internal belt adhesion	• Remove foreign objects from pulley grooves • Replace belt • Adjust belt tension • Replace belt
Rib or belt wear (belt ribs contact bottom of pulley grooves)	• Pulley(s) misaligned • Mismatch of belt and pulley groove widths • Abrasive environment • Rusted pulley(s) • Sharp or jagged pulley groove tips • Rubber deteriorated	• Align pulley(s) • Replace belt • Replace belt • Clean rust from pulley(s) • Replace pulley • Replace belt
Longitudinal belt cracking (cracks between two ribs)	• Belt has mistracked from pulley groove • Pulley groove tip has worn away rubber-to-tensile member	• Replace belt • Replace belt
Belt slips	• Belt slipping because of insufficient tension • Belt or pulley subjected to substance (belt dressing, oil, ethylene glycol) that has reduced friction • Driven component bearing failure • Belt glazed and hardened from heat and excessive slippage	• Adjust tension • Replace belt and clean pulleys • Replace faulty component bearing • Replace belt
"Groove jumping" (belt does not maintain correct position on pulley, or turns over and/or runs off pulleys)	• Insufficient belt tension • Pulley(s) not within design tolerance • Foreign object(s) in grooves	• Adjust belt tension • Replace pulley(s) • Remove foreign objects from grooves

TCCS3C09

Troubleshooting the Serpentine Drive Belt

Problem	Cause	Solution
"Groove jumping" (belt does not maintain correct position on pulley, or turns over and/or runs off pulleys)	• Excessive belt speed • Pulley misalignment • Belt-to-pulley profile mismatched • Belt cordline is distorted	• Avoid excessive engine acceleration • Align pulley(s) • Install correct belt • Replace belt
Belt broken (Note: identify and correct problem before replacement belt is installed)	• Excessive tension • Tensile members damaged during belt installation • Belt turnover • Severe pulley misalignment • Bracket, pulley, or bearing failure	• Replace belt and adjust tension to specification • Replace belt • Replace belt • Align pulley(s) • Replace defective component and belt
Cord edge failure (tensile member exposed at edges of belt or separated from belt body)	• Excessive tension • Drive pulley misalignment • Belt contacting stationary object • Pulley irregularities • Improper pulley construction • Insufficient adhesion between tensile member and rubber matrix	• Adjust belt tension • Align pulley • Correct as necessary • Replace pulley • Replace pulley • Replace belt and adjust tension to specifications
Sporadic rib cracking (multiple cracks in belt ribs at random intervals)	• Ribbed pulley(s) diameter less than minimum specification • Backside bend flat pulley(s) diameter less than minimum • Excessive heat condition causing rubber to harden • Excessive belt thickness • Belt overcured • Excessive tension	• Replace pulley(s) • Replace pulley(s) • Correct heat condition as necessary • Replace belt • Replace belt • Adjust belt tension

TCCS3C10

Troubleshooting the Cooling System

Problem	Cause	Solution
High temperature gauge indication—overheating	• Coolant level low • Improper fan operation • Radiator hose(s) collapsed • Radiator airflow blocked • Faulty pressure cap • Ignition timing incorrect • Air trapped in cooling system • Heavy traffic driving • Incorrect cooling system component(s) installed • Faulty thermostat • Water pump shaft broken or impeller loose • Radiator tubes clogged • Cooling system clogged • Casting flash in cooling passages • Brakes dragging • Excessive engine friction • Antifreeze concentration over 68% • Missing air seals • Faulty gauge or sending unit • Loss of coolant flow caused by leakage or foaming • Viscous fan drive failed	• Replenish coolant • Repair or replace as necessary • Replace hose(s) • Remove restriction (bug screen, fog lamps, etc.) • Replace pressure cap • Adjust ignition timing • Purge air • Operate at fast idle in neutral intermittently to cool engine • Install proper component(s) • Replace thermostat • Replace water pump • Flush radiator • Flush system • Repair or replace as necessary. Flash may be visible by removing cooling system components or removing core plugs. • Repair brakes • Repair engine • Lower antifreeze concentration percentage • Replace air seals • Repair or replace faulty component • Repair or replace leaking component, replace coolant • Replace unit
Low temperature indication—undercooling	• Thermostat stuck open • Faulty gauge or sending unit	• Replace thermostat • Repair or replace faulty component
Coolant loss—boilover	• Overfilled cooling system • Quick shutdown after hard (hot) run • Air in system resulting in occasional "burping" of coolant • Insufficient antifreeze allowing coolant boiling point to be too low • Antifreeze deteriorated because of age or contamination • Leaks due to loose hose clamps, loose nuts, bolts, drain plugs, faulty hoses, or defective radiator	• Reduce coolant level to proper specification • Allow engine to run at fast idle prior to shutdown • Purge system • Add antifreeze to raise boiling point • Replace coolant • Pressure test system to locate source of leak(s) then repair as necessary

TCCS3C11

Troubleshooting the Cooling System (cont.)

Problem	Cause	Solution
Coolant loss—boilover	• Faulty head gasket • Cracked head, manifold, or block • Faulty radiator cap	• Replace head gasket • Replace as necessary • Replace cap
Coolant entry into crankcase or cylinder(s)	• Faulty head gasket • Crack in head, manifold or block	• Replace head gasket • Replace as necessary
Coolant recovery system inoperative	• Coolant level low • Leak in system • Pressure cap not tight or seal missing, or leaking • Pressure cap defective • Overflow tube clogged or leaking • Recovery bottle vent restricted	• Replenish coolant to FULL mark • Pressure test to isolate leak and repair as necessary • Repair as necessary • Replace cap • Repair as necessary • Remove restriction
Noise	• Fan contacting shroud • Loose water pump impeller • Glazed fan belt • Loose fan belt • Rough surface on drive pulley • Water pump bearing worn • Belt alignment	• Reposition shroud and inspect engine mounts (on electric fans inspect assembly) • Replace pump • Apply silicone or replace belt • Adjust fan belt tension • Replace pulley • Remove belt to isolate. Replace pump. • Check pulley alignment. Repair as necessary.
No coolant flow through heater core	• Restricted return inlet in water pump • Heater hose collapsed or restricted • Restricted heater core • Restricted outlet in thermostat housing • Intake manifold bypass hole in cylinder head restricted • Faulty heater control valve • Intake manifold coolant passage restricted	• Remove restriction • Remove restriction or replace hose • Remove restriction or replace core • Remove flash or restriction • Remove restriction • Replace valve • Remove restriction or replace intake manifold

NOTE: *Immediately after shutdown, the engine enters a condition known as heat soak. This is caused by the cooling system being inoperative while engine temperature is still high. If coolant temperature rises above boiling point, expansion and pressure may push some coolant out of the radiator overflow tube. If this does not occur frequently it is considered normal.*

TCCS3C12

4

DRIVEABILTIY AND EMISSION CONTROLS

AIR POLLUTION

The earth's atmosphere, at or near sea level, consists approximately of 78 percent nitrogen, 21 percent oxygen and 1 percent other gases. If it were possible to remain in this state, 100 percent clean air would result. However, many varied sources allow other gases and particulates to mix with the clean air, causing our atmosphere to become unclean or polluted.

Some of these pollutants are visible while others are invisible, with each having the capability of causing distress to the eyes, ears, throat, skin and respiratory system. Should these pollutants become concentrated in a specific area and under certain conditions, death could result due to the displacement or chemical change of the oxygen content in the air. These pollutants can also cause great damage to the environment and to the many man made objects that are exposed to the elements.

To better understand the causes of air pollution, the pollutants can be categorized into 3 separate types, natural, industrial and automotive.

Natural Pollutants

Natural pollution has been present on earth since before man appeared and continues to be a factor when discussing air pollution, although it causes only a small percentage of the overall pollution problem. It is the direct result of decaying organic matter, wind born smoke and particulates from such natural events as plain and forest fires (ignited by heat or lightning), volcanic ash, sand and dust which can spread over a large area of the countryside.

Such a phenomenon of natural pollution has been seen in the form of volcanic eruptions, with the resulting plume of smoke, steam and volcanic ash blotting out the sun's rays as it spreads and rises higher into the atmosphere. As it travels into the atmosphere the upper air currents catch and carry the smoke and ash, while condensing the steam back into water vapor. As the water vapor, smoke and ash travel on their journey, the smoke dissipates into the atmosphere while the ash and moisture settle back to earth in a trail hundreds of miles long. In some cases, lives are lost and millions of dollars of property damage result.

Industrial Pollutants

Industrial pollution is caused primarily by industrial processes, the burning of coal, oil and natural gas, which in turn produce smoke and fumes. Because the burning fuels contain large amounts of sulfur, the principal ingredients of smoke and fumes are sulfur dioxide and particulate matter. This type of pollutant occurs most severely during still, damp and cool weather, such as at night. Even in its less severe form, this pollutant is not confined to just cities. Because of air movements, the pollutants move for miles over the surrounding countryside, leaving in its path a barren and unhealthy environment for all living things.

Working with Federal, State and Local mandated regulations and by carefully monitoring emissions, big business has greatly reduced the amount of pollutant introduced from its industrial sources, striving to obtain an acceptable level. Because of the mandated industrial emission clean up, many land areas and streams in and around the cities that were formerly barren of vegetation and life, have now begun to move back in the direction of nature's intended balance.

Automotive Pollutants

The third major source of air pollution is automotive emissions. The emissions from the internal combustion engines were not an appreciable problem years ago because of the small number of registered vehicles and the nation's small highway system. However, during the early 1950's, the trend of the American people was to move from the cities to the surrounding suburbs. This caused an immediate problem in transportation because the majority of suburbs were not afforded mass transit conveniences. This lack of transportation created an attractive market for the automobile manufacturers, which resulted in a dramatic increase in the number of vehicles produced and sold, along with a marked increase in highway construction between cities and the suburbs. Multi-vehicle families emerged with a growing emphasis placed on an individual vehicle per family member. As the increase in vehicle ownership and usage occurred, so did pollutant levels in and around the cities, as suburbanites drove daily to their businesses and employment, returning at the end of the day to their homes in the suburbs.

It was noted that a smoke and fog type haze was being formed and at times, remained in suspension over the cities, taking time to dissipate. At first this "smog," derived from the words "smoke" and "fog," was thought to result from industrial pollution but it was determined that automobile emissions shared the blame. It was discovered that when normal automobile emissions were exposed to sunlight for a period of time, complex chemical reactions would take place.

It is now known that smog is a photo chemical layer which develops when certain oxides of nitrogen (NOx) and unburned hydrocarbons (HC) from automobile emissions are exposed to sunlight. Pollution was more severe when smog would become stagnant over an area in which a warm layer of air settled over the top of the cooler air mass, trapping and holding the cooler mass at ground level. The trapped cooler air would keep the emissions from being dispersed and diluted through normal air flows. This type of air stagnation was given the name "Temperature Inversion."

TEMPERATURE INVERSION

In normal weather situations, surface air is warmed by heat radiating from the earth's surface and the sun's rays. This causes it to rise upward, into the atmosphere. Upon rising it will cool through a convection type heat exchange with the cooler upper air. As warm air rises, the surface pollutants are carried upward and dissipated into the atmosphere.

When a temperature inversion occurs, we find the higher air is no longer cooler, but is warmer than the surface air, causing the cooler surface air to become trapped. This warm air blanket can extend from above ground level to a few hundred or even a few thousand feet into the air. As the surface air is trapped, so are the pollutants, causing a severe smog condition. Should this stagnant air mass extend to a few thousand feet high, enough air movement with the inversion takes place to allow the smog layer to rise above ground level but the pollutants still cannot dissipate. This inversion can remain for days over an area, with the smog level only rising or lowering from ground level to a few hundred feet high. Meanwhile, the pollutant levels increase, causing eye irritation, respiratory problems, reduced visibility, plant damage and in some cases, even disease.

This inversion phenomenon was first noted in the Los Angeles, California area. The city lies in terrain resembling a basin and with certain weather conditions, a cold air mass is held in the basin while a warmer air mass covers it like a lid.

Because this type of condition was first documented as prevalent in the Los Angeles area, this type of trapped pollution was named Los Angeles Smog, although it occurs in other areas where a large concentration of automobiles are used and the air remains stagnant for any length of time.

HEAT TRANSFER

Consider the internal combustion engine as a machine in which raw materials must be placed so a finished product comes out. As in any machine operation, a certain amount of wasted material is formed. When we relate this to the internal combustion engine, we find that through the input of air and fuel, we obtain power during the combustion process to drive the vehicle. The by-product or waste of this power is, in part, heat and exhaust gases with which we must dispose.

The heat from the combustion process can rise to over 4000°F (2204°C). The dissipation of this heat is controlled by a ram air effect, the use of cooling fans to cause air flow and a liquid coolant solution surrounding the combustion area to transfer the heat of combustion through the cylinder walls and into the coolant. The coolant is then directed to a thin-finned, multi-tubed radiator, from which the excess heat is transferred to the atmosphere by 1 of the 3 heat transfer methods, conduction, convection or radiation.

The cooling of the combustion area is an important part in the control of exhaust emissions. To understand the behavior of the combustion and transfer of its heat, consider the air/fuel charge. It is ignited and the flame front burns progressively across the combustion chamber until the burning charge reaches the cylinder walls. Some of the fuel in contact with the walls is not hot enough to burn, thereby snuffing out or quenching the combustion process. This leaves unburned fuel in the combustion chamber. This unburned fuel is then forced out of the cylinder and into the exhaust system, along with the exhaust gases.

Many attempts have been made to minimize the amount of unburned fuel in the combustion chambers due to quenching, by increasing the coolant temperature and lessening the contact area of the coolant around the combustion area. However, design limitations within the combustion chambers prevent the complete burning of the air/fuel charge, so a certain amount of the unburned fuel is still expelled into the exhaust system, regardless of modifications to the engine.

AUTOMOTIVE EMISSIONS

Before emission controls were mandated on internal combustion engines, other sources of engine pollutants were discovered along with the exhaust emissions. It was determined that engine combustion exhaust produced approximately 60 percent of the total emission pollutants, fuel evaporation from the fuel tank and carburetor vents produced 20 percent, with the final 20 percent being produced through the crankcase as a by-product of the combustion process.

Exhaust Gases

The exhaust gases emitted into the atmosphere are a combination of burned and unburned fuel. To understand the exhaust emission and its composition, we must review some basic chemistry.

When the air/fuel mixture is introduced into the engine, we are mixing air, composed of nitrogen (78 percent), oxygen (21 percent) and other gases (1 percent) with the fuel, which is 100 percent hydrocarbons (HC), in a semi-controlled ratio. As the combustion process is accomplished, power is produced to move the vehicle while the heat of combustion is transferred to the cooling system. The exhaust gases are then composed of nitrogen, a diatomic gas (N_2), the same as was introduced in the engine, carbon dioxide (CO_2), the same gas that is used in beverage carbonation, and water vapor (H_2O). The nitrogen (N_2), for the most part, passes through the engine unchanged, while the oxygen (O_2) reacts (burns) with the hydrocarbons (HC) and produces the carbon dioxide (CO_2) and the water vapors (H_2O). If this chemical process would be the only process to take place, the exhaust emissions would be harmless. However, during the combustion process, other compounds are formed which are considered dangerous. These pollutants are hydrocarbons (HC), carbon monoxide (CO), oxides of nitrogen (NOx) oxides of sulfur (SOx) and engine particulates.

HYDROCARBONS

Hydrocarbons (HC) are essentially fuel which was not burned during the combustion process or which has escaped into the atmosphere through fuel evaporation. The main sources of incomplete combustion are rich air/fuel mixtures, low engine temperatures and improper spark timing. The main sources of hydrocarbon emission through fuel evaporation on most vehicles used to be the vehicle's fuel tank and carburetor float bowl.

To reduce combustion hydrocarbon emission, engine modifications were made to minimize dead space and surface area in the combustion chamber. In addition, the air/fuel mixture was made more lean through the improved control which feedback carburetion and fuel injection offers and by the addition of external controls to aid in further combustion of the hydrocarbons outside the engine. Two such methods were the addition of air injection systems, to inject fresh air into the exhaust manifolds and the installation of catalytic converters, units that are able to burn traces of hydrocarbons without affecting the internal combustion process or fuel economy.

To control hydrocarbon emissions through fuel evaporation, modifications were made to the fuel tank to allow storage of the fuel vapors during periods of engine shut-down. Modifications were also made to the air intake system so that at specific times during engine operation, these vapors may be purged and burned by blending them with the air/fuel mixture.

CARBON MONOXIDE

Carbon monoxide is formed when not enough oxygen is present during the combustion process to convert carbon (C) to carbon dioxide (CO_2). An increase in the carbon monoxide (CO) emission is normally accompanied by an increase in the hydrocarbon (HC) emission because of the lack of oxygen to completely burn all of the fuel mixture.

Carbon monoxide (CO) also increases the rate at which the photo chemical smog is formed by speeding up the conversion of nitric oxide (NO) to nitrogen dioxide (NO_2). To accomplish this, carbon monoxide (CO) combines with oxygen (O_2) and nitric oxide (NO) to produce carbon dioxide (CO_2) and nitrogen dioxide (NO_2). ($CO + O_2 + NO = CO_2 + NO_2$).

The dangers of carbon monoxide, which is an odorless and colorless toxic gas are many. When carbon monoxide is inhaled into the lungs and passed into the blood stream, oxygen is replaced by the carbon monoxide in the red blood cells, causing a reduction in the amount of oxygen supplied to the many parts of the body. This lack of oxygen causes headaches, lack of coordination, reduced mental alertness and, should the carbon monoxide concentration be high enough, death could result.

NITROGEN

Normally, nitrogen is an inert gas. When heated to approximately 2500°F (1371°C) through the combustion process, this gas becomes active and causes an increase in the nitric oxide (NO) emission.

Oxides of nitrogen (NOx) are composed of approximately 97–98 percent nitric oxide (NO). Nitric oxide is a colorless gas but when it is passed into the atmosphere, it combines with oxygen and forms nitrogen dioxide (NO_2). The nitrogen dioxide then combines with chemically active hydrocarbons (HC) and when in the presence of sunlight, causes the formation of photochemical smog.

Ozone

To further complicate matters, some of the nitrogen dioxide (NO_2) is broken apart by the sunlight to form nitric oxide and oxygen. ($NO_2 + $ sunlight $= NO + O$). This single atom of oxygen then combines with diatomic (meaning 2 atoms) oxygen (O_2) to form ozone (O_3). Ozone is one of the smells associated with smog. It has a pungent and offensive odor, irritates the eyes and lung tissues, affects the growth of plant life and causes rapid deterioration of rubber products. Ozone can be formed by sunlight as well as electrical discharge into the air.

The most common discharge area on the automobile engine is the secondary ignition electrical system, especially when inferior quality spark plug cables are used. As the surge of high voltage is routed through the secondary cable, the circuit builds up an electrical field around the wire, which acts upon the oxygen in the surrounding air to form the ozone. The faint glow along the cable with the engine running that may be visible on a dark night, is called the "corona discharge." It is the result of the electrical field passing from a high along the cable, to a low in the surrounding air, which forms the ozone gas. The combination of corona and ozone has been a major cause of cable deterioration. Recently, different and better quality insulating materials have lengthened the life of the electrical cables.

Although ozone at ground level can be harmful, ozone is beneficial to the earth's inhabitants. By having a concentrated ozone layer called the "ozonosphere," between 10 and 20 miles (16–32 km) up in the atmosphere, much of the ultra violet radiation from the sun's rays are absorbed and screened. If this ozone layer were not present, much of the earth's surface would be burned, dried and unfit for human life.

OXIDES OF SULFUR

Oxides of sulfur (SOx) were initially ignored in the exhaust system emissions, since the sulfur content of gasoline as a fuel is less than $\frac{1}{10}$ of 1 percent. Because of this small amount, it was felt that it contributed very little to the overall pollution problem. However, because of the difficulty in solving the sulfur emissions in industrial pollutions and the introduction of catalytic converter to the automobile exhaust systems, a change was mandated. The automobile exhaust system, when equipped with a catalytic converter, changes the sulfur dioxide (SO_2) into sulfur trioxide (SO_3).

When this combines with water vapors (H_2O), a sulfuric acid mist (H_2SO_4) is formed and is a very difficult pollutant to handle since it is extremely corrosive. This sulfuric acid mist that is formed, is the same mist that rises from the vents of an automobile battery when an active chemical reaction takes place within the battery cells.

When a large concentration of vehicles equipped with catalytic converters are operating in an area, this acid mist may rise and be distributed over a large ground area causing land, plant, crop, paint and building damage.

PARTICULATE MATTER

A certain amount of particulate matter is present in the burning of any fuel, with carbon constituting the largest percentage of the particulates. In gasoline, the remaining particulates are the burned remains of the various other compounds used in its manufacture. When a gasoline engine is in good internal condition, the particulate emissions are low but as the engine wears internally, the particulate emissions increase. By visually inspecting the tail pipe emissions, a determination can be made as to where an engine defect may exist. An engine with light gray or blue smoke emitting from the tail pipe normally indicates an increase in the oil consumption through burning due to internal engine wear. Black smoke would indicate a defective fuel delivery system, causing the engine to operate in a rich mode. Regardless of the color of the smoke, the internal part of the engine or the fuel delivery system should be repaired to prevent excess particulate emissions.

Diesel and turbine engines emit a darkened plume of smoke from the exhaust system because of the type of fuel used. Emission control regulations are mandated for this type of emission and more stringent measures are being used to prevent excess emission of the particulate matter. Electronic components are being introduced to control the injection of the fuel at precisely the proper time of piston travel, to achieve the optimum in fuel ignition and fuel usage. Other particulate after-burning components are being tested to achieve a cleaner emission.

Good grades of engine lubricating oils should be used, which meet the manufacturers specification. Cut-rate oils can contribute to the particulate emission problem because of their low flash or ignition temperature point. Such oils burn prematurely during the combustion process causing emission of particulate matter.

The cooling system is an important factor in the reduction of particulate matter. The optimum combustion will occur, with the cooling system operating at a temperature specified by the manufacturer. The cooling system must be maintained in the same manner as the engine oiling system, as each system is required to perform properly in order for the engine to operate efficiently for a long time.

Crankcase Emissions

Crankcase emissions are made up of water, acids, unburned fuel, oil fumes and particulates. These emissions are classified as hydrocarbons (HC) and are formed by the small amount of unburned, compressed air/fuel mixture entering the crankcase from the combustion area (between the cylinder walls and piston rings) during the compression and power strokes. The head of the compression and combustion help to form the remaining crankcase emissions.

Since the first engines, crankcase emissions were allowed into the atmosphere through a road draft tube, mounted on the lower side of the engine block. Fresh air came in through an open oil filler cap or breather. The air passed through the crankcase mixing with blow-by gases. The motion of the vehicle and the air blowing past the open end of the road draft tube caused a low pressure area (vacuum) at the end of the tube. Crankcase emissions were simply drawn out of the road draft tube into the air.

To control the crankcase emission, the road draft tube was deleted. A hose and/or tubing was routed from the crankcase to the intake manifold so the blow-by emission could be burned with the air/fuel mixture. However, it was found that intake manifold vacuum, used to draw the crankcase emissions into the manifold, would vary in strength at the wrong time and not allow the proper emission flow. A regulating valve was needed to control the flow of air through the crankcase.

Testing, showed the removal of the blow-by gases from the crankcase as quickly as possible, was most important to the longevity of the engine. Should large accumulations of blow-by gases remain and condense, dilution of the engine oil would occur to form water, soots, resins, acids and lead salts, resulting in the formation of sludge and varnishes. This condensation of the blow-by gases occurs more frequently on vehicles used in numerous starting and stopping conditions, excessive idling and when the engine is not allowed to attain normal operating temperature through short runs.

Evaporative Emissions

Gasoline fuel is a major source of pollution, before and after it is burned in the automobile engine. From the time the fuel is refined, stored, pumped and transported, again stored until it is pumped into the fuel tank of the vehicle, the gasoline gives off unburned hydrocarbons (HC) into the atmosphere. Through the redesign of storage areas and venting systems, the pollution factor was diminished, but not eliminated, from the refinery standpoint. However, the automobile still remained the primary source of vaporized, unburned hydrocarbon (HC) emissions.

Fuel pumped from an underground storage tank is cool but when exposed to a warmer ambient temperature, will expand. Before controls were mandated, an owner might fill the fuel tank with fuel from an underground storage tank and park the vehicle for some time in warm area, such as a parking lot. As the fuel would warm, it would expand and should no provisions or area be provided for the expansion, the fuel would spill out of the filler neck and onto the ground, causing hydrocarbon (HC) pollution and creating a severe fire hazard. To correct this condition, the vehicle manufacturers added overflow plumbing and/or gasoline tanks with built in expansion areas or domes.

However, this did not control the fuel vapor emission from the fuel tank. It was determined that most of the fuel evaporation occurred when the vehicle was stationary and the engine not operating. Most vehicles carry 5–25 gallons (19–95 liters) of gasoline. Should a large concentration of vehicles be parked in one area, such as a large parking lot, excessive fuel vapor emissions would take place, increasing as the temperature increases.

To prevent the vapor emission from escaping into the atmosphere, the fuel systems were designed to trap the vapors while the vehicle is stationary, by sealing the system from the atmosphere. A storage system is used to collect and hold the fuel vapors from the carburetor (if equipped) and the fuel tank when the engine is not operating. When the engine is started, the storage system is then purged of the fuel vapors, which are drawn into the engine and burned with the air/fuel mixture.

EMISSION CONTROLS

The emission control system begins at the air intake and ends at the tailpipe. The emission control system includes various sub-systems such as the positive crankcase ventilation system, evaporative emission control system, the exhaust gas recirculation system and exhaust catalyst, as well as the electronic controls that govern the fuel and ignition system. These components are combined to control engine operation for maximum engine efficiency and minimal exhaust emissions.

Positive Crankcase Ventilation System

OPERATION

▶ **See Figures 1 and 2**

The system consists of a tube from the air filter housing to the rocker/camsahft cover and a second tube from the rocker/camshaft cover to the intake manifold. Under normal operating conditions, clean air flows from the air filter into the rocker/camshaft cover where it mixes with crankcase oil vapors. These vapors are drawn through the PCV valve and

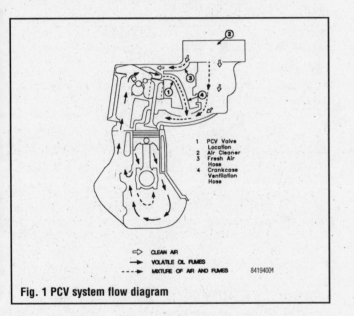

Fig. 1 PCV system flow diagram

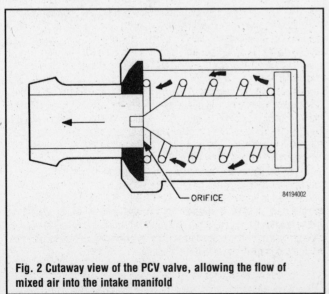

Fig. 2 Cutaway view of the PCV valve, allowing the flow of mixed air into the intake manifold

into the intake manifold to be burned with the air/fuel mixture. The flow to the intake manifold is metered by the PCV valve. When manifold vacuum is high, the valve is pulled closed and flow is restricted to maintain a smooth idle. If crankcase pressure is very high, vapors can flow directly into the air filter housing.

A plugged PCV system will cause oil leaks or a build up of sludge in the engine. An air filter coated with engine oil indicates excessive crankcase pressure. A leaking valve or hose might cause rough or high idle, engine stalling and/or Powertrain Control Module (PCM) trouble codes.

COMPONENT TESTING

1. Visually inspect the PCV valve hose, the fresh air supply hose and their attaching nipples or grommets for splits, cuts, damage, clogging, or restrictions. Repair or replace, as necessary.
2. If the hoses pass inspection, start the engine and allow it to warm until normal operating temperature is reached.
3. Remove the PCV valve from the rocker arm cover, but leave it connected to the hose. With the engine at idle, feel the end of the valve for manifold vacuum. If there is no vacuum, check for a plugged or leaking hose, PCV valve or manifold port. Replace a plugged or damaged hose.
4. Stop the engine and remove the PCV valve. It should rattle when shaken. If the valve does not rattle or it is plugged, replace the valve.

REMOVAL & INSTALLATION

▶ **See Figure 3**

1. Remove the PCV valve from the mounting grommet in the rocker/camshaft cover.
2. Disconnect the valve from the PCV hose and remove the valve from the vehicle.
3. Installation is the reverse of the removal procedure.

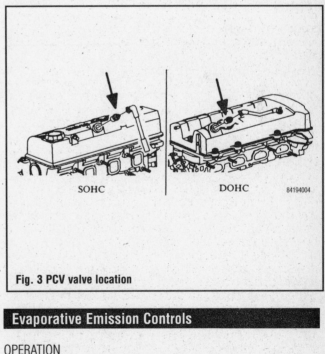

Fig. 3 PCV valve location

Evaporative Emission Controls

OPERATION

▶ **See Figure 4**

The Evaporative Emission Control System (EECS) limits the amount of fuel vapors allowed to escape into the air. When the engine is not running, fuel vapors from the sealed fuel tank flow through the single vapor line to

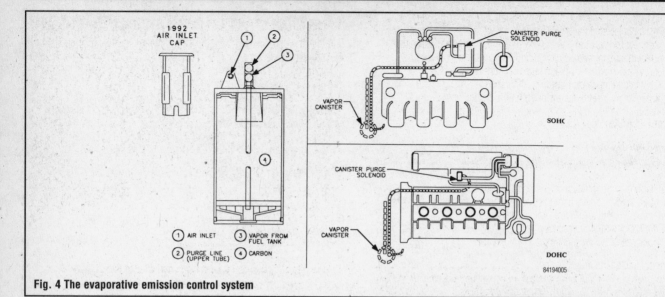

Fig. 4 The evaporative emission control system

the charcoal canister mounted against the inner wheel well, in the right front of the engine compartment on 1991–97 models, or on top of the fuel tank on 1998 models. The charcoal absorbs the fuel vapors and retains them until they are purged with fresh air.

Any time the ignition switch is **ON**, the canister purge solenoid valve is supplied with 12 volts. Under the appropriate engine operating conditions, the PCM completes the circuit to ground, opening the valve. With the valve open, intake manifold vacuum is applied to the canister drawing in fresh air and purging the vapors. The purged vapors are then drawn into the engine and burned efficiently during the normal combustion process.

The purge valve is opened when the coolant temperature is above 158°F (70°C), vehicle speed is greater than 1 mph and throttle position is more than 4 percent open.

COMPONENT TESTING

Carefully check for cracks or leaks in the vacuum lines or in the canister itself. The lines and fittings can be reached without removing the canister. Cracks or leaks in the system may cause poor idle, stalling, poor driveability, fuel loss or a fuel vapor odor.

Vapor odor and fuel loss may also be caused by: fuel leaking from the lines, tank or injectors, loose, disconnected or kinked lines, or an improperly seated air cleaner and gasket.

If the system passes the visual inspection and a problem is still suspected, proceed as follows:

1. With the ignition switch in the **OFF** position, unplug the electrical connector from the canister purge solenoid. Use an ohmmeter to check the solenoid coil resistance; it should be 19–31 ohms.

2. Turn the ignition switch **ON** then, disconnect the vacuum line from the intake manifold and connect a hand vacuum pump to the valve.

3. Apply a vacuum of about 10 in. Hg (34 kPa) to the hose and solenoid, then observe if the valve holds the vacuum. If the valve does not hold, apply vacuum directly to the solenoid to determine if the solenoid or hose is at fault, then replace the component which allowed vacuum to release.

4. If the vacuum holds in the initial test, jumper the test terminals on the ALDL (terminals A and B are the test terminals and are the two uppermost right terminals of the ALDL connector), and observe if the valve activates, releasing the vacuum. If the valve does not activate and release the vacuum, replace the solenoid.

REMOVAL & INSTALLATION

Canister Purge Solenoid Valve

▶ See Figure 5

The purge solenoid valve is mounted on the engine, towards the rear and below the intake manifold.

1. As an added safety precaution, you may wish to disable the air bag(s), if equipped. Refer to Section 6 for the air bag disabling procedure.

2. Disconnect the negative battery cable.

3. Remove the air cleaner/intake tube assembly.

➡**For DOHC engines, the solenoid can be reached through the access hole located next to the intake manifold support bracket.**

4. Unplug the solenoid vacuum hoses and electrical connector.

5. Remove the attaching bolt, then remove the canister purge solenoid.

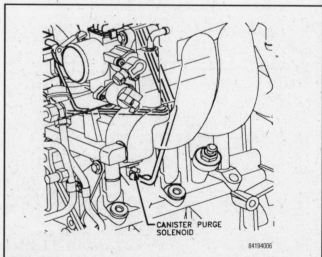

Fig. 5 The canister purge solenoid is mounted below the intake manifold

To install:

6. Install the solenoid and tighten the retaining bolt to 19 ft. lbs. (25 Nm).

7. Connect the vacuum lines to the solenoid, then install and push in the electrical connector until it clicks firmly into place.

8. Install the air cleaner/intake tube assembly.

9. Connect the negative battery cable.

10. If necessary, enable the air bag(s). Refer to Section 6.

Charcoal Canister

1991–97 MODELS

▶ See Figures 6, 7, 8 and 9

1. Raise the front of vehicle sufficiently for underhood and undervehicle service, then support safely using jackstands.

2. Remove the right front tire from the vehicle.

3. Remove the right inner fender well. The fasteners at the front corner should be carefully pried free using a small prybar.

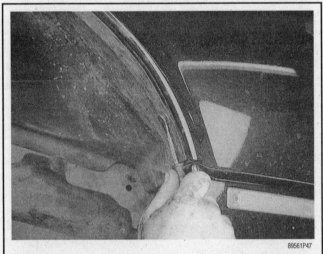

Fig. 6 Lift out center of plastic fastener with a small bladed screwdriver

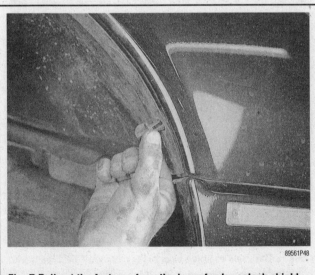

Fig. 7 Pull out the fastener from the inner fender splash shield

Fig. 8 Location of the evaporative canister at the right front of the engine compartment—1991–97 models

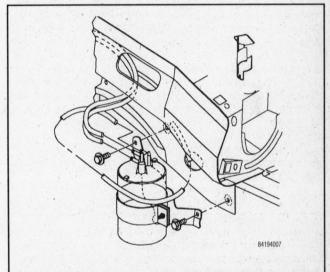

Fig. 9 Installation and vapor hose routing for the charcoal canister

4. Disconnect the vacuum hoses from the canister.

5. Remove the bracket assembly fasteners or the band clamp fasteners, as necessary, then remove the canister from the vehicle.

To install:

6. Install the canister into the bracket assembly with the ports correctly positioned, but do not tighten the bracket fasteners until the vapor lines are installed.

7. Connect the vacuum hoses to the top of the canister, making sure that no lines are kinked or damaged. Tighten the canister/bracket assembly fasteners to 22 ft. lbs. (30 Nm). The canister air inlet is routed into the vehicle frame rail on 1991 vehicles while 1992–97 vehicles are equipped with a special air inlet cap.

➡Be careful not to overtighten the assembly fasteners or the charcoal canister may be damaged.

8. Install the inner fender well and the right front tire.

9. Remove the jackstands and carefully lower the vehicle.

1998 MODELS

Saturn models for 1998 have the evaporative charcoal canister mounted on top of the fuel tank assembly. The fuel tank must be removed to access the charcoal canister. For fuel tank removal procedures, refer to Section 5.

Exhaust Gas Recirculation System

OPERATION

The Exhaust Gas Recirculation (EGR) valve is used to allow a controlled amount of exhaust gas to be recirculated into the intake system. This limits peak flame temperature in the combustion chamber so the engine produces less NOx (oxides of nitrogen).

A negative backpressure EGR valve is used to control the amount of exhaust gas which is recirculated. Intake manifold vacuum is supplied directly to the top of the diaphragm to pull open the normally closed valve. Exhaust backpressure pushing against the valve keeps the diaphragm pushed against the bleed hole. When the rpm is high but the throttle is closed, exhaust backpressure becomes negative and the diaphragm is pulled down just enough to uncover the bleed hole. The vacuum on top of the diaphragm leaks off and the valve slowly closes.

The EGR vacuum is controlled by the PCM through a solenoid valve. The PCM energizes or de-energizes the solenoid by providing or withholding ground at the appropriate times. When the solenoid is energized, it prevents vacuum from reaching the EGR valve by venting it to the atmosphere. Once the proper conditions have been met, the PCM removes the ground, thus de-energizing the solenoid and allowing vacuum to open the EGR valve. The valve is open only when the throttle is open more than 4 percent and coolant temperature is above 104°F (40°C) for the DOHC engine with automatic transaxle, or 122°F (50°C) for all others.

TESTING

Vacuum Operated EGR Valve

▶ See Figures 10 and 11

If fault Code 32 is present along with Code 26, troubleshoot Code 26 first and then clear the code memory. Drive the vehicle to see if Code 32 sets again, then proceed to troubleshoot the EGR system. A complete listing of trouble codes can be found later in this section.

If code 32 resets, or a problem is still suspected with the EGR system, test the EGR valve and solenoid as follows:

1. Make sure the ignition is **OFF** so that the solenoid is not energized.
2. Disconnect the EGR valve vacuum hose from the solenoid and attach a hand vacuum pump. Apply vacuum and observe the EGR valve for movement. The valve should easily open and vacuum should be held for a minimum of 20 seconds. If the valve does not open, check for damaged, plugged or kinked lines, or for restrictions in the valve vacuum port which would prevent vacuum from opening the valve. Clean restrictions and/or replace the lines or valve, as necessary, to correct the condition.
3. If the valve opened and vacuum held, disconnect the vacuum hose from the EGR valve port and attach the hand vacuum pump directly to the EGR valve. Apply vacuum and observe the valve opening again. While you watch the valve and the vacuum gauge, have an assistant start the engine. The valve should close and vacuum should drop immediately as the engine is started. If this does not occur, remove the valve and either clean the plugged passages or replace the faulty valve.
4. If the valve is working, but system trouble is still suspected, test the solenoid. Turn the ignition **OFF** and reconnect the vacuum lines to the valve and solenoid.
5. Disconnect the solenoid vacuum line from the throttle body and attach the vacuum pump to the line. Apply vacuum to the lines and observe

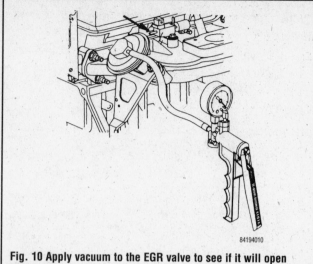

Fig. 10 Apply vacuum to the EGR valve to see if it will open properly

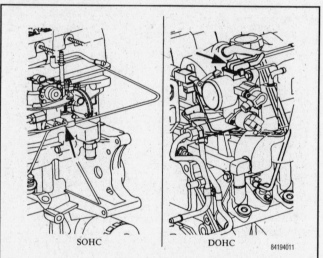

SOHC   DOHC   84194011

Fig. 11 Apply vacuum to the throttle body vacuum line to see if the solenoid is causing a problem with the EGR system

to see if it holds for at least 20 seconds. If the vacuum does not hold, replace the leaking lines or the bad solenoid, whichever was at fault. Remember, the non-energized solenoid should allow the EGR valve to open.

6. A suspected bad solenoid can also be checked using an ohmmeter. Resistance across the solenoid's terminals should be 22–42 ohms.

Linear EGR Valve

1. Connect a scan tool to the DLC connector, below the left side of the instrument panel.
2. With the engine running, open the linear EGR valve to the full open position.
3. If the engine runs rough when the valve is open, check the malfunction diagnostic history. Also, check the EGR pintle for excessive carbon build-up.
4. If the engine does not run rough with the valve fully open, check to see if the pintle position increases when the valve is open. If this happens, remove and check the valve for foreign material build-up. Clean or replace the linear EGR valve if necessary.

REMOVAL & INSTALLATION

EGR Solenoid

▶ See Figure 12

1. As an added safety precaution, you may wish to disable the air bag(s), if equipped. Refer to Section 6 for disabling procedure.
2. Disconnect the negative battery cable.
3. Remove the air cleaner assembly for SOHC engines or the air intake tube and resonator for DOHC engines.
4. Unplug the solenoid electrical connector and vacuum line.
5. Unfasten the solenoid attaching bolt, then remove the solenoid from the engine.

To install:

6. Install the solenoid and tighten the bolt to 89 inch lbs. (10 Nm) for DOHC engines, or to 18 ft. lbs. (25 Nm) for SOHC engines.
7. Connect the vacuum lines, then install the electrical connector and push until it clicks firmly into place.
8. Install the air cleaner assembly or the intake tube and resonator, as applicable.
9. Connect the negative battery cable.
10. If necessary, enable the air bag(s). Refer to Section 6.

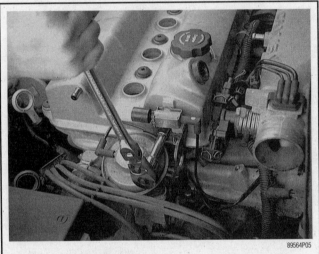

Fig. 12 Loosen the EGR solenoid attaching bolt, then remove the solenoid from the engine

Vacuum Operated EGR Valve

▶ See Figures 13, 14 and 15

1. Disconnect the negative battery cable.
2. Remove the air cleaner assembly for SOHC engines or the air intake tube and resonator for DOHC engines.
3. Disconnect the vacuum line from the valve.
4. For SOHC engines, remove the intake manifold brace and fuel line clips, in order to access the EGR valve fasteners.
5. Remove the valve fasteners.
6. Remove the valve from the vehicle, then remove and discard the gasket from the mating surfaces.

To install:

7. Inspect the EGR passages on the engine for excessive deposits and remove using a screwdriver or length of odometer cable. Use a wire brush or wheel to remove deposits from the EGR valve and engine mounting surfaces. Make sure the surfaces are free of scores or cracks.
8. Install the valve using a new gasket and tighten the fasteners to 19 ft. lbs. (25 Nm). For the SOHC engine, install the intake manifold brace and the fuel line clips.
9. Connect the vacuum hose to the valve.

Fig. 13 Disconnect the vacuum line from the EGR valve

Fig. 14 Remove the EGR valve mounting fasteners

Fig. 15 Remove the valve from the vehicle, then remove and discard the mounting gasket

10. Install the air cleaner assembly or the air intake tube and resonator, as applicable.

11. Connect the negative battery cable.

Linear EGR Valve

1. Disconnect the negative battery cable.
2. Remove the air cleaner/intake tube assembly.
3. Disconnect the EGR valve wiring harness.
4. Remove the valve fasteners.
5. Remove the valve from the vehicle, then remove and discard the gasket from the mating surfaces.

ELECTRONIC ENGINE CONTROLS

General Information

The fuel injection system, described in detail in Section 5, operates in conjunction with the ignition system to obtain optimum performance and fuel economy, along with a minimum of exhaust emissions. The various sensors described below are used by the Powertrain Control Module (PCM) for feedback to determine and regulate proper engine operating conditions.

➡**Although many of the following tests may be conducted using a Digital Volt/Ohm Meter (DVOM), certain steps or procedures may require the use of a specialized Saturn tester known as a Portable Diagnostic Tool (PDT), or an equivalent scan/testing tool. If these testers are not available, the vehicle should be taken to a reputable service station which has the appropriate equipment.**

Powertrain Control Module (PCM)

OPERATION

The PCM is located in the upper dashboard area, in back of the instrument cluster assembly. The PCM utilizes signals from a variety of sensors to optimize control of the engine, transaxle and various accessories. The PCM is a microprocessor-based computer that is the main component of the powertrain control system. The PCM contains either one or two separate controllers located in one unit. The manual transaxle version contains only an engine controller, while the automatic transaxle version contains both an engine and transaxle controller.

The PCM contains an Electronicaly Erasable Programmable Read Only Memory (EEPROM), which stores information necessary for proper vehicle operation. Whenever the PCM is replaced, this memory must be reprogrammed in order for the vehicle to operate. The vehicle may not even start if this is not done. Only a dealer/serivce station equipped with the Saturn Service Stall System, or an equivalent EEPROM reprogramming computer, will be able to properly calibrate the PCM.

The PCM module contains a built-in learning ability that allows it to maximize driveability by correcting for minor variations in the fuel system. Any time the PCM is replaced or the power is removed from the PCM, the module must go through a relearning process. Maximum performance may not return until after this is accomplished. To enable the PCM to relearn the fuel system:

1. Start the vehicle and warm it to normal operating temperature.
2. Drive the vehicle at part throttle, with moderate acceleration and idle conditions, until normal performance returns.
3. Park the vehicle and engage the parking brake.
4. For automatic transaxles, shift into **D**. For manual transaxles, shift into **N**.
5. Allow the engine to idle at normal operating temperature for about 2 minutes, until the engine stabilizes.

To install:

6. Inspect the EGR passages on the engine for excessive deposits and remove using a screwdriver or length of odometer cable. Use a wire brush or wheel to remove deposits from the EGR valve and engine mounting surfaces. Make sure the surfaces are free of scores or cracks.

7. Install the valve using a new gasket and tighten the fasteners to 19 ft. lbs. (25 Nm).

8. Connect the EGR valve wiring harness to the valve.
9. Install the air cleaner/intake tube assembly.
10. Connect the negative battery cable.

REMOVAL & INSTALLATION

1991–94 Models

▶ See Figure 16

1. If equipped, properly disable the SIR system as follows:
    a. Align the steering wheel so the tires are in the straight-ahead position, then turn the ignition **OFF**.
    b. Remove the SIR fuse from the Instrument Panel Junction Block (IPJB).
    c. Remove the Connector Position Assurance (CPA), then detach the yellow 2-way SIR connector at the base of the steering column.
2. Disconnect the negative battery cable.
3. Remove the driver's side kick panel and/or the knee bolster for access to the PCM.
4. Rotate the retaining screw ¼ turn counterclockwise and remove the PCM from the carrier.
5. Unplug the electrical connectors from the PCM and remove the unit from the vehicle.

To install:

6. Install the wiring harness connectors to the PCM, then position the unit into the carrier.
7. Push the unit upward in the carrier until 2 clicks are heard or felt.
8. Connect the negative battery cable and properly reprogram the EEPROM.
9. If equipped, enable the SIR system as follows:
    a. Verify that the ignition switch is **OFF**, then attach the SIR electrical connector at the base of the steering column. Install the CPA device to the connector.

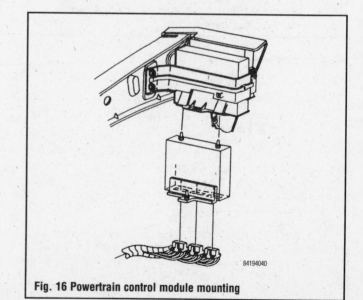

Fig. 16 Powertrain control module mounting

b. Install the SIR fuse to the IPJB and install the fuse box cover.

c. Turn the ignition **ON** and verify that the AIR BAG indicator lamp flashes 7–9 times, then goes out. If the light does not flash as indicated, inspect the system for malfunction.

1995–98 Models

1. Disable the driver side SIR system as follows:

a. Align the steering wheel so the tires are in the straight-ahead position, then turn the ignition to the **OFF** or **LOCK** position and remove the key.

b. Remove the SIR fuse from the Instrument Panel Junction Block (IPJB).

c. Remove the Connector Position Assurance (CPA) device, then detach the yellow 2-way SIR connector at the base of the steering column.

2. Disable the passenger side SIR system as follows:

a. Using a small, flat bladed screwdriver, pry off the upper instrument panel pad screw caps. Remove the trim panel retaining screws.

b. Disengage the rear trim panel clips by lifting up. Then, disengage the front trim panel clips by lifting up at each outer corner and pulling rearward. Be careful not to damage the trim panel during its removal from the vehicle.

c. Remove the upper trim panel insulator.

d. Remove the CPA device, then detach the yellow 2-way SIR connector on the pigtail from the passenger side air bag module.

3. Disconnect the negative battery cable.

4. Unplug the PCM electrical connector from below the instrument panel.

5. Remove the PCM mounting screws at the cross-car beam and lift upward, towards the windshield, to disengage it from the mounting bracket.

To install:

6. Lower the PCM unit into the mounting bracket. Be sure to guide the attaching stud into the slot in the cross-car beam.

7. Install and tighten the mounting nut to 89 inch lbs. (10 Nm).

8. Install the PCM connectors.

9. Connect the negative battery cable.

10. Enable the driver side SIR system:

a. Verify that the ignition switch is in the **OFF** or **LOCK** position, then attach the SIR electrical connector at the base of the steering column. Install the CPA device to the connector.

11. Enable the passenger side SIR system:

a. Install the insulator and upper trim panel assembly. Be sure that all flaps are tucked and clips engaged. Install and tighten the retaining screws to 20 inch lbs. (2.3 Nm). Install the trim panel screw caps.

b. Connect the yellow 2-way inflator module pigtail and install the CPA device to the connector.

12. Install the SIR fuse to the IPJB and install the junction block cover.

13. Turn the ignition **ON** and verify that the AIR BAG indicator lamp flashes 7–9 times, then extinguishes. If the light does not flash as indicated, inspect the system for malfunction.

Oxygen Sensor

OPERATION

This sensor is used to report the concentration of oxygen in the exhaust. It consists of an arrangement of platinum and zirconia plates, all protected with a slotted outer shield. At about 600°F (318°C), the presence of oxygen in the exhaust will cause voltage to be generated across the dissimilar metals, up to a maximum of just under one volt. The PCM reads this signal and adjusts the fuel injector pulse width to maintain the right amount of oxygen in the exhaust for the catalytic converter to work properly. To quickly reach and stay at operating temperature, the sensor is threaded directly into the exhaust manifold. The threads are coated with an anti-seize compound. When replacing the sensor, be careful not to get anti-seize compound in the slots of the outer shield.

TESTING

▶ See Figure 17

The only way to properly test just the oxygen sensor and circuit is with a scan tool connected to the ALDL. If an exhaust gas analyzer is available, a faulty sensor can be found if the rest of the fuel injection system works properly, and the sensor sends the PCM a message of too lean or too rich. Information about the sensor and engine operation can also be gained by running the engine in the Field Service Mode, described later in this manual.

1. Using an ohmmeter, check the resistance of the oxygen sensor circuit to ground. Detach the connector for the purple wire from the oxygen sensor and read the wire resistance to ground; it should be at least 200 ohms.

2. With a scan tool connected, run the engine to normal operating temperature and follow the tool's menu to read oxygen sensor voltage. Sensor output should vary from approximately 350–550 millivolts.

3. Stop the engine and disconnect the sensor wire. With the wire from

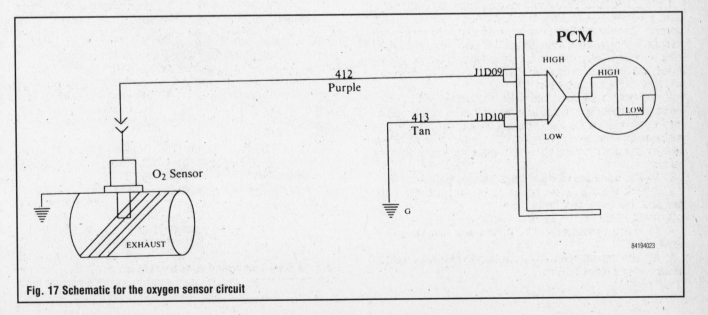

84194023

Fig. 17 Schematic for the oxygen sensor circuit

the PCM grounded and the ignition switch **ON**, the scan tool should show less than 100 millivolts. If not, the wiring or the PCM is faulty.

4. To test with an exhaust gas analyzer:

a. No other faults can exist.

b. Run the engine at normal operating temperature for more than one minute.

c. The TPS must read more than 6.5 percent throttle opening.

5. When the above conditions are met and all other engine systems work properly, the exhaust analyzer will change when the sensor wire is disconnected to simulate Open Loop Mode.

REMOVAL & INSTALLATION

▶ **See Figures 18, 19 and 20**

➡The oxygen sensor uses a permanently attached pigtail and connector, removal of which may affect the proper operation of the sensor.

1. As an added safety precaution, you may wish to disable the air bag(s), if equipped. Refer to Section 6 for disabling procedures.

2. Disconnect the negative battery cable.

3. Unplug the sensor electrical connector.

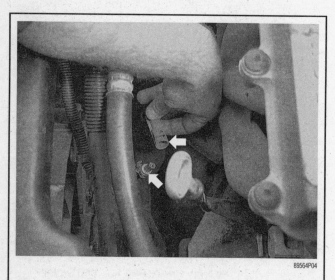

Fig. 18 Unplug the oxygen sensor connector

Fig. 19 Using a 19mm, 6-point crow's foot wrench, loosen the oxygen sensor

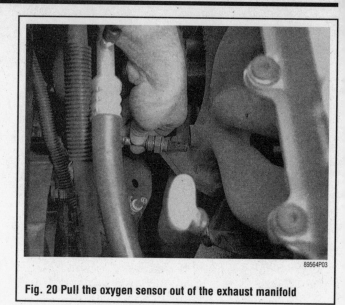

Fig. 20 Pull the oxygen sensor out of the exhaust manifold

4. Using a 19mm, 6-point crow's foot wrench, loosen and remove the oxygen sensor from the exhaust manifold.

To install:

➡Be careful not to drop or damage the oxygen sensor when handling. The pigtail, connector and louvered end should be kept free of dirt or contaminants, but avoid using a solvent to clean the sensor.

5. Coat the oxygen sensor threads with a nickel based anti-seize compound that does not contain silicone.

6. Install the oxygen sensor into the exhaust manifold and tighten to 18 ft. lbs. (25 Nm).

7. Attach the sensor electrical connector and install the Connector Position Assurance (CPA).

8. Connect the negative battery cable.

9. If necessary, enable the air bag(s). Refer to Section 6.

Idle Air Control Valve

OPERATION

▶ **See Figure 21**

The IAC valve is a 2-coil electric stepper motor that opens and closes a valve in the throttle body idle air passage. Engine idle speed is a function of total air flow into the engine based on the IAC valve pintle position, plus the throttle valve opening and calibrated vacuum loss through accessories. The PCM will operate the valve pintle in order to maintain the correct idle speed under all engine loads and conditions. The pintle is moved inward or outward, based on a number of steps or counts sent from the PCM in the form of voltage pulses.

Whenever the ignition is turned **OFF**, the PCM will command the IAC valve pintle to the fully inward position or 0 count position. Then, the PCM will move the pintle to the 121 count position where it is parked while the ignition remains **OFF**. Should the valve be disconnected with the engine running, the resulting IAC counts may no longer correspond to the valve pintle position and a starting or idle problem may result. In order to give the PCM time to park the pintle, always allow a minimum of 10 seconds after the engine is stopped before unplugging the IAC valve connector.

There is a base idle speed adjustment to give the PCM a starting position from which to work, but that adjustment should not be made unless the throttle body has been replaced. There is a sealed throttle stop screw on the throttle body that determines the base idle speed setting. This screw is set on a flow bench at the factory.

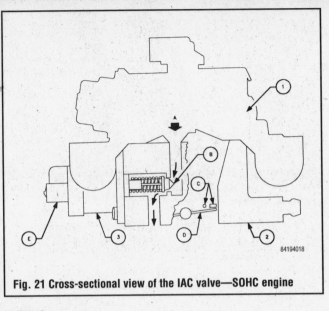

Fig. 21 Cross-sectional view of the IAC valve—SOHC engine

Fig. 23 Checking the resistance of the IAC valve terminals

TESTING

▶ See Figures 22 and 23

1. To check minimum idle speed adjustment, start and warm the engine to its normal operating temperature. Use a suitable scan tool to bottom the IAC pintle, then locate and plug the IAC valve inlet in the throttle body. To plug the inlet, use special tool SA9196E for SOHC systems, SA9106E for DOHC systems, or use an equivalent air plug.

2. With the engine at operating temperature, the transaxle in **N** or **P** and all accessories and fans **OFF**, base idle speed should be 450–650 rpm. If not, check for a vacuum leak, bad electrical connection, clogged air or PCV passage, or misadjustment of the cruise control or throttle cables.

3. To check power to the IAC valve, turn the ignition switch **OFF** and wait at least 10 seconds for the PCM to park the pintle, then unplug the IAC valve connector. With the ignition switch **ON** but the engine not running, connect a voltmeter to terminals **A** and **B**. Voltage should fluctuate from 0–10.75 volts. Connect the voltmeter to terminals **C** and **D** and look for the same results.

4. If there is no voltage to the IAC valve, turn the ignition **OFF** and use the PCM pin-out chart in this section to check the wiring between the PCM and the valve for short or open circuits.

5. If power is reaching the IAC valve, check the resistance across all possible valve terminal combinations. There should be at least 200 ohms between the terminals.

REMOVAL & INSTALLATION

▶ See Figures 24, 25 and 26

1. As an added safety precaution, you may wish to disable the air bag(s). Refer to Section 6 for the disabling procedures.

2. Disconnect the negative battery cable, then remove the air cleaner assembly for SOHC engines or the air intake tube and resonator for DOHC engines.

3. Detach the IAC valve electrical connector.

4. Remove the mounting screws, then remove the IAC valve from the throttle body. Remove and discard the valve O-ring.

To install:

5. Carefully clean the IAC valve mounting surface to ensure a proper O-ring seal.

➡When replacing the IAC valve, make sure to only use an identical part. The pintle shape and diameter are essential to proper valve operation and may vary from application to application.

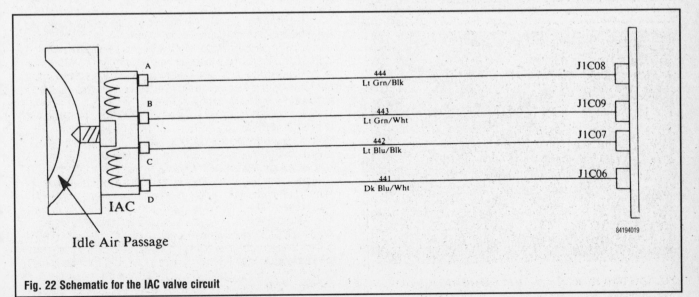

Fig. 22 Schematic for the IAC valve circuit

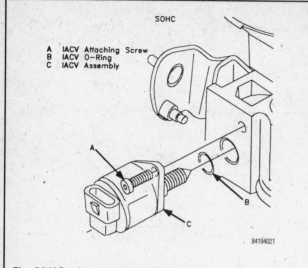

Fig. 24 IAC valve mounting—SOHC engine

A   IACV Attaching Screw
B   IACV O-Ring
C   IACV Assembly

SOHC

84194021

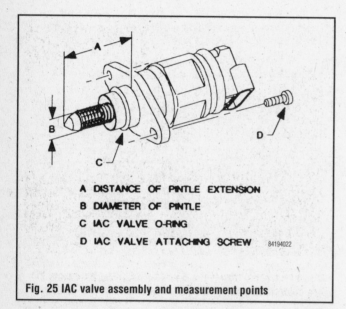

A   DISTANCE OF PINTLE EXTENSION
B   DIAMETER OF PINTLE
C   IAC VALVE O-RING
D   IAC VALVE ATTACHING SCREW

84194022

Fig. 25 IAC valve assembly and measurement points

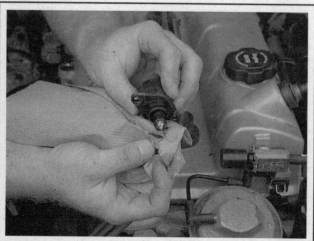

89564P16

Fig. 26 Carefully clean the IAC valve pintle valve and mounting surface to ensure a proper O-ring seal

6. Retract the pintle of a new IAC valve using a suitable scan tool, then lubricate the new O-ring with clean engine oil and install on the IAC valve.

7. Install the valve to the throttle body. Coat the mounting screws with Loctite® 242 or an equivalent threadlock, then install the screws and tighten to 27 inch lbs. (3 Nm).

8. Attach the IAC valve electrical connector.

9. Install the air cleaner assembly or the air intake tube and resonator, as applicable.

10. Connect the negative battery cable.

11. If necessary, enable the air bag(s). Refer to Section 6.

Coolant Temperature Sensor (CTS)

OPERATION

This sensor is a thermistor type, meaning as the temperature increases, resistance decreases. The PCM sends a 5 volt reference signal to the sensor (pin B) and a sensor reference ground to the sensor (pin A). The PCM calculates coolant temperature depending on the return signal.

TESTING

▶ See Figures 27, 28 and 29

If Code 14 (1991–95 models), P0117/P1114 (1996–97 models) or P0118 (1998 models) is set, the PCM will turn the coolant fan **ON** at all times. The thermostat opens at 190°F (88°C) and coolant temperature should stabilize there at idle.

If Code 17 (1991–95 models) or P1628 (1996–98 models) has been set, the fault is in the PCM and cannot be repaired. Code 14, P0117/P1114 or P0118 indicates that the CTS signal is above sensor range. Code 15 (1991–95 models), P0118/P1115 (1996–97 models) or P0117 (1998 models) indicates that the signal is below sensor range. Test the sensor first.

➡ **The sensor is threaded into the coolant passage in the back of the cylinder head.**

1. Unplug the connector from the sensor and use the chart to check the sensor resistance with an ohmmeter. This can also work if the sensor is removed and immersed in a container of water.

2. With the ignition switch **ON**, there should be some voltage across the connector terminals. If not, check the wiring between the sensor and the PCM.

3. If using a scan tool, the reading should be the same as the actual coolant temperature with the ignition switch **ON**.

   a. If the reading is less than –32°F (–35°C), unplug the sensor connector and jumper the PCM circuit terminals together. If the reading goes to above 266°F (130°C), the problem is in the PCM connector, ground or in the sensor. If the reading does not change or go high enough, the problem is in the wiring or the PCM.

   b. If the reading is more than 284°F (140°C), unplug the sensor connector. If the reading goes below –32°F (–35°C), the problem is in the sensor. If it does not change or does not go low enough, the problem is in the wiring or the PCM.

4. If a scan tool is available, allow the engine too cool to the ambient temperature overnight, then compare the CTS and the ATS readings. Temperatures should be within 4 degrees of each other with the ignition **ON** and the engine not running.

REMOVAL & INSTALLATION

▶ See Figures 30 and 31

The sensor is threaded into the coolant passage in the back of the cylinder head.

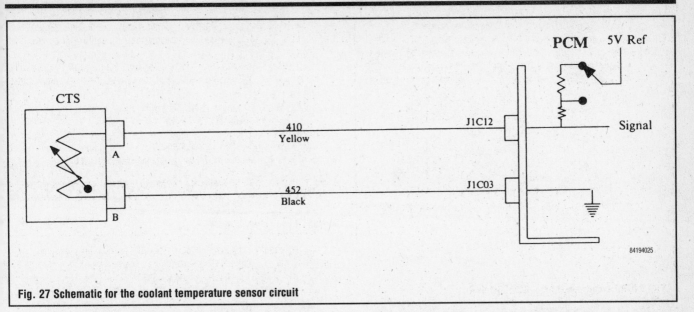

Fig. 27 Schematic for the coolant temperature sensor circuit

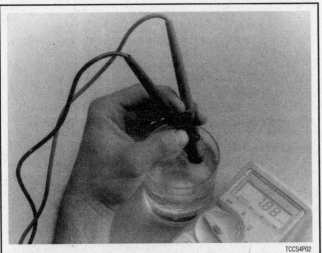

Fig. 28 Test the coolant temperature sensor, and compare results to the sensor resistance chart

Fig. 30 A small prying tool can be used to gently disengage the coolant temperature sensor connector

DEGREES (°C)	DEGREES (°F)	SENSOR RESISTANCE (OHMS)
–40	–40	77k – 109k
–29	–20	39k – 53k
–18	0	21k – 27k
–7	20	11k – 15k
4	40	6.6k – 8.4k
16	60	3.9k – 4.5k
27	80	2.4k – 2.7k
38	100	1.5k – 1.7k
49	120	.98k – 1.1k
60	140	650 – 730
72	160	430 – 480
83	180	302 – 334
94	200	215 – 235
105	220	159 – 172
120	248	104 – 113
140	284	63 – 68

Fig. 29 Sensor resistance chart for the CTS and ATS

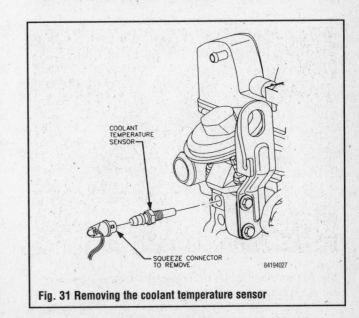

COOLANT TEMPERATURE SENSOR

SQUEEZE CONNECTOR TO REMOVE.

Fig. 31 Removing the coolant temperature sensor

➡**On 1991–95 Saturn models, two coolant temperature "sensors" are utilized. Be careful not to confuse the single-wired coolant temperature *sender* used by the instrument cluster temperature gauge (which is covered in Section 2) with the 2-wired sensor used by the PCM. On 1996–98 Saturn models, only one 2-wired sensor is used to perform both functions.**

1. Turn the ignition to the **OFF** position.
2. Disconnect the negative battery cable.
3. Position a clean container under the engine or radiator plug and drain the engine coolant to a level below the sensor.
4. Using your hand, a pair of pliers, or with the aid of a small prying tool, gently squeeze the sides of the sensor electrical connector, then remove the connector from the sensor. Do not pull on the wires.
5. Using an appropriate sized deep well socket, remove the sensor from the engine.

To install:

6. Apply a coat of Loctite® 242 or an equivalent threadlock to the coolant temperature sensor threads, then install the sensor and tighten to 71 inch lbs. (8 Nm).
7. Install the sensor electrical connector. Push in until a click is heard and slightly pull back to confirm a positive engagement.
8. Connect the negative battery cable.
9. Refill the engine cooling system.

Air Temperature Sensor (ATS)

OPERATION

This sensor is a thermistor type, meaning as the temperature increases, the sensor resistance decreases. It is the same type of sensor as the CTS. The PCM sends a 5 volt reference signal to the sensor and interprets the voltage drop across the sensor to calculate engine intake air temperature.

TESTING

▶ **See Figures 29 and 32**

1. If a code has been set and the vehicle has not been exposed to temperatures lower than −22°F (−30°), unplug the connector from the sensor and use the chart to check the sensor resistance with an ohmmeter.
2. With the ignition switch **ON**, check for 5 volts to the sensor at connector terminal **A**, the tan wire.
3. If using a scan tool connected to the ALDL, the reading should be the same as the actual air temperature with the ignition switch **ON**.

a. If the reading is less than −22°F (−30°C), unplug the sensor connector and jumper the PCM terminals together. If the reading goes to above 266°F (130°C), the problem is in the sensor. If the reading does not change or go high enough, the problem is in the wiring or the PCM.

b. If the reading is more than 284°F (140°C), unplug the sensor connector. If the reading goes below -22°F (-30°C), the problem is in the sensor. If it does not change or does not go low enough, the problem is in the wiring or the PCM.

4. If a scan tool is available, allow the engine to cool to the ambient temperature overnight, then compare the CTS and ATS readings. Temperatures should be within 4 degrees of each other with the ignition **ON** and the engine not running.

REMOVAL & INSTALLATION

▶ **See Figures 33 and 34**

1. As an added safety precaution, you may wish to disable the air bag(s). Refer to Section 6 for the disabling procedures.
2. Disconnect the negative battery cable.
3. Remove the air inlet tube fasteners, then rotate the tube to access the sensor.

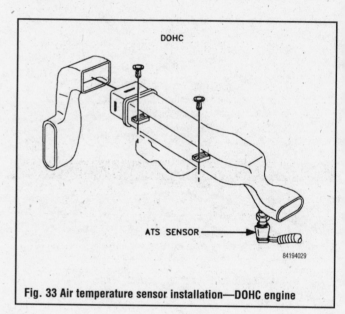

Fig. 33 Air temperature sensor installation—DOHC engine

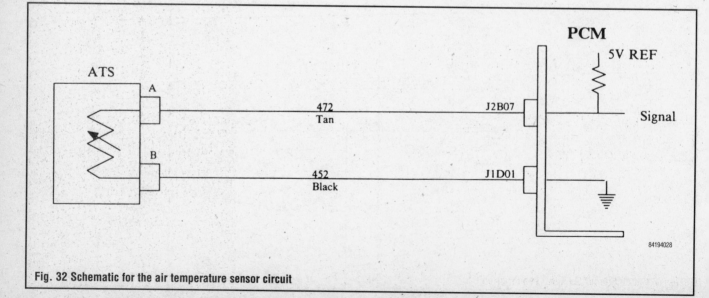

Fig. 32 Schematic for the air temperature sensor circuit

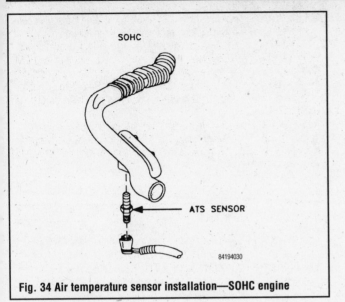

Fig. 34 Air temperature sensor installation—SOHC engine

4. Using your hand or a pair of pliers, gently squeeze the sides of the sensor electrical connector, then remove the connector from the sensor.

5. Using a 13mm deep well socket, loosen and remove the sensor from the inlet duct.

To install:

6. Install the sensor and tighten to 44 inch lbs. (5 Nm).
7. Attach the sensor electrical connector.
8. Install the air inlet tube using the fasteners.
9. Connect the negative battery cable.
10. If necessary, enable the air bag(s). Refer to Section 6.

Manifold Air Pressure (MAP) Sensor

OPERATION

This sensor is used to read the air pressure in the intake manifold, which is always positive (but less than atmospheric pressure) when the engine is running. The Powertrain Control Module (PCM) sends a reference voltage to the sensor, then a pressure sensitive resistor in the sensor reduces the voltage returning to the PCM. The portion of the voltage returned to the PCM is interpreted as engine load. The return signal is low (low load) when engine vac-

uum is high (throttle closed). As the throttle is opened and engine vacuum decreases (increased manifold air pressure), the return signal increases. While the engine is running, this type of load sensing automatically accounts for changes in altitude, so no separate altitude sensor is needed. A high pressure (14–15 psi) will allow 4–5 volts to return to the PCM while a low pressure (5–7 psi) will allow 0.5–0.9 volts. On SOHC engines with TBI, the MAP sensor is mounted on the camshaft/valve cover. On all MFI engines, the sensor is mounted directly to the end of the intake manifold.

TESTING

▶ **See Figures 35, 36 and 37**

1. Unplug the MAP sensor connector and turn the ignition switch **ON**. Check for about 4.5–5 volts between connector terminal **C** (gray wire) and ground.
2. Check for the same voltage between terminals **C** and **A** (black wire). This is a ground through the PCM.
3. Reconnect the wiring, then install a Saturn diagnostic probe, or equivalent, on the green wire and connect either a voltmeter probe or a scan tool to the ALDL. Connect a vacuum gauge to the intake manifold.
4. With the ignition switch **ON** and the engine not running, the return

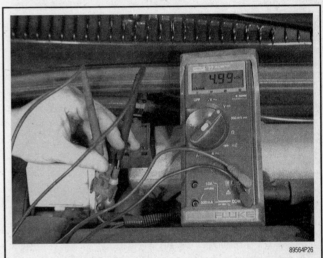

Fig. 35 With the ignition turned ON, measure the MAP sensor circuit for a reference voltage

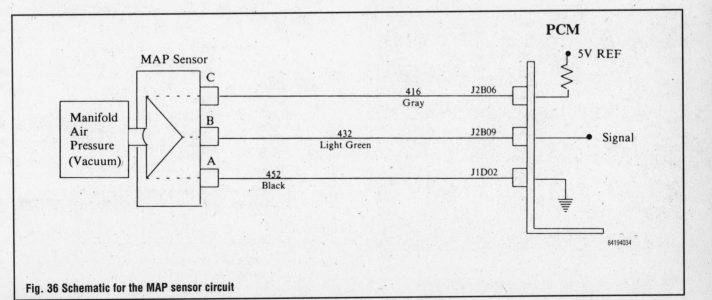

Fig. 36 Schematic for the MAP sensor circuit

ALTITUDE		VOLTAGE RANGE
Meters	Feet	
Below 305	Below 1,000	3.8 – 5.5V
305 – 610	1,000 – 2,000	3.6 – 5.3V
610 – 914	2,000 – 3,000	3.5 – 5.1V
914 – 1219	3,000 – 4,000	3.3 – 5.0V
1219 – 1524	4,000 – 5,000	3.2 – 4.8V
1524 – 1829	5,000 – 6,000	3.0 – 4.6V
1829 – 2133	6,000 – 7,000	2.9 – 4.5V
2133 – 2438	7,000 – 8,000	2.8 – 4.3V
2438 – 2743	8,000 – 9,000	2.6 – 4.2V
2743 – 3948	9,000 – 10,000	2.5 – 4.0V

LOW ALTITUDE = HIGH PRESSURE = HIGH VOLTAGE

84194035

Fig. 37 MAP sensor return voltage altitude compensation chart (with the ignition ON and the engine not running)

89564P10

Fig. 39 Pull the sensor from the vehicle after disengaging the connector

signal voltage should be close to the supply voltage in Step 1, if below a 1000 ft. (305 meters) altitude. If above this altitude, see the chart for the correct voltage.

5. With the engine at idle, there should be at least 16 in. Hg (54 kPa) of manifold vacuum and 1–1.5 volts on the signal return wire. When the throttle is opened suddenly, the signal voltage should increase. It may only change momentarily.

6. If the voltage at idle is the same as in Step 1, or if the voltage does not change when the throttle is moved suddenly, the sensor is faulty and must be replaced.

REMOVAL & INSTALLATION

▶ **See Figures 38, 39 and 40**

1. As an added safety precaution, you may wish to disable the air bag(s). Refer to Section 6 for the disabling procedures.

2. Disconnect the negative battery cable.

3. For SOHC TBI engines, remove the sensor attaching nuts and pull the sensor slightly from the engine, then disconnect the sensor vacuum hose.

4. For all MFI engines, remove the attaching bolts and pull the sensor slightly from the engine. If the port seal remained in the intake manifold, remove the seal.

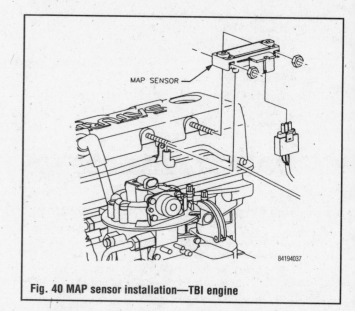

MAP SENSOR

84194037

Fig. 40 MAP sensor installation—TBI engine

5. Unplug the sensor electrical connector and remove the sensor from the vehicle.

To install:

6. On TBI engines, connect the vacuum hose to the sensor.

7. On MFI engines, install the port seal into the intake manifold. To ensure proper seating, the seal must be installed directly into the manifold, rather than onto the sensor.

8. Attach the electrical connector to the MAP sensor.

9. Position the sensor to the engine and install the fasteners. Tighten the retaining nuts to 53 inch lbs. (6 Nm) or the retaining bolts to 44 inch lbs. (5 Nm).

10. Connect the negative battery cable.

11. If necessary, enable the air bag(s). Refer to Section 6.

Throttle Position Sensor

OPERATION

The Throttle Position Sensor (TPS) is mounted to the throttle body and connected to the throttle plate shaft. The TPS is a potentiometer with a 5 volt reference input and a signal ground provided by the PCM. At a closed

89564P09

Fig. 38 Loosen the MAP sensor attaching bolts

throttle position, sensor is low (approximately 0.5 volt). As the throttle plate opens, the output signal increases so that at wide open throttle, the output voltage will be high (approximately 5 volts).

TESTING

▶ **See Figures 41 and 42**

The TPS is a potentiometer attached to the throttle plate shaft. The PCM sends a 5 volt supply signal to the sensor and the percentage of the signal returned corresponds to the percentage of throttle opening, so that at Wide Open Throttle (WOT), the return signal should be approximately 4.9 volts.

1. The sensor and signal voltages can be checked quickly using a voltmeter. Remove the air cleaner ducting as required to locate the sensor on the throttle body. First check the connectors and the wiring between the sensor and the PCM.

2. Unplug the sensor wiring harness and connect an ohmmeter to terminal A and C on the TPS. Resistance should be above 200 ohms and should change smoothly as the throttle is moved.

3. Connect a voltmeter between connector terminal A and ground. With the ignition switch **ON**, but the engine not running, there should be 5 volts, which is the supply signal directly from the PCM.

4. Turn the ignition switch **OFF** and check for continuity to ground at terminal B on the connector.

5. Reconnect the wiring and install a diagnostic connector, Saturn diagnostic probe or a voltmeter on the blue wire leading to terminal C and to ground. With the ignition switch **ON** but the engine not running, the voltage should change smoothly from about 0.4–4.7 volts as the throttle is moved from idle to full throttle.

➡**A suitable scan tool such as the Saturn PDT can be connected to the ALDL and used to monitor TPS output, instead of attaching a probe to the wire.**

REMOVAL & INSTALLATION

▶ **See Figure 43**

1. As an added safety precaution, you may wish to disable the air bag(s). Refer to Section 6 for the disabling procedure.

2. Disconnect the negative battery cable and remove the air cleaner assembly.

3. Unplug the electrical connector from the TPS.

4. Remove the TPS retaining bolts, then remove the sensor from the throttle body.

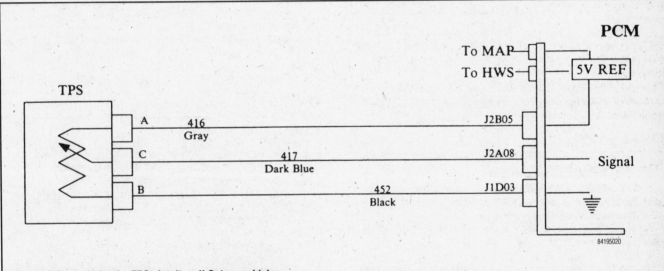

Fig. 41 Schematic for the TPS circuit—all Saturn vehicles

Fig. 42 Checking the resistance of the TPS circuit

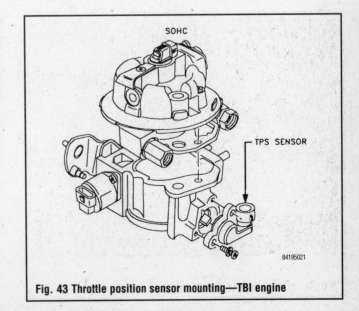

Fig. 43 Throttle position sensor mounting—TBI engine

To install:

5. Make sure the throttle valve is closed, then install the TPS to the throttle shaft.

6. Rotate the sensor counterclockwise to align the mounting holes, then install the retaining bolts and tighten to 18 inch lbs. (2 Nm).

7. Connect the wiring harness to the sensor.

8. Install the air cleaner assembly and connect the negative battery cable.

9. If necessary, enable the air bag(s). Refer to Section 6.

Crankshaft Position Sensor (CPS)

OPERATION

The crankshaft position sensor is mounted in the engine block, below the intake manifold. It is a simple pulse generator that signals the DIS module when specific crankshaft position markers move past it. The crankshaft position sensor defines the engine position. A crank pulse occurs at each occurrence of Top Dead Center (TDC), at 60 degrees After Top Dead Center (ATDC), and at 120 degrees ATDC. A seventh pulse, referred to as the sync pulse, occurs at 70 degrees ATDC at the No. 1 cylinder.

TESTING

If there is no rpm signal or no ignition output, unplug the sensor and connect an ohmmeter across the terminals. Resistance should be 700–900 ohms. The sensor and its circuit have a polarity, meaning if the sensor wiring is reversed, the PCM sees an open circuit. Reversing the CPS wires will result in a no-start condition.

➡**For more information regarding the crankshaft position sensor and the ignition system, refer to Section 2 of this manual.**

REMOVAL & INSTALLATION

1. Disconnect the negative battery cable.
2. Raise and support the vehicle safely.
3. Detach the electrical connector from the sensor, located at the lower rear of the engine block.
4. Unfasten the sensor retaining bolt, then remove the sensor from the engine.

To install:

5. Lubricate the sensor O-ring with clean engine oil and install the sensor into the engine block.

6. Install the sensor retaining bolt and tighten to 80 inch lbs. (9 Nm).
7. Attach the CPS electrical connection.
8. Lower the vehicle and connect the negative battery cable.

Vehicle Speed Sensor (VSS)

OPERATION

The VSS is a permanent magnet variable reluctance sensor, which is mounted in the differential housing of the transaxle. It detects the rotation of the differential assembly and, when vehicle speed reaches 3 mph (5 kph), transmits 16 pulses per revolution to the Powertrain Control Module (PCM). The PCM uses the vehicle speed signal from the VSS to calculate air/fuel delivery and ignition functions. Through a different circuit in the PCM, the VSS is also used to drive the speedometer.

TESTING

▶ **See Figure 44**

A trouble code will be set if engine speed is above idle, the vehicle is in gear, the MAP sensor output is less than 0.5 volts, and the VSS signal is less than 1 mph (1.6 kph).

1. If the speedometer is faulty, but there is no diagnostic code set, the problem is most likely in the instrument cluster.

2. If a code is set, unplug the VSS connector and connect an ohmmeter to the sensor terminals. The sensor resistance should be 700–900 ohms.

3. Check the continuity of the wiring between the sensor and the PCM.

REMOVAL & INSTALLATION

▶ **See Figure 45**

1. As an added safety precaution, you may wish to disable the air bag(s), if equipped. Refer to Section 6 for the disabling procedure.
2. Disconnect the negative battery cable.
3. Raise the front of the vehicle and support it safely using jackstands.
4. Unplug the electrical connector from the sensor.
5. Unscrew and remove the sensor from the transaxle housing.

To install:

6. Install the VSS to the transaxle.
7. Attach the sensor electrical connector.
8. Lower the vehicle and connect the negative battery cable.
9. If necessary, enable the air bag(s). Refer to Section 6.

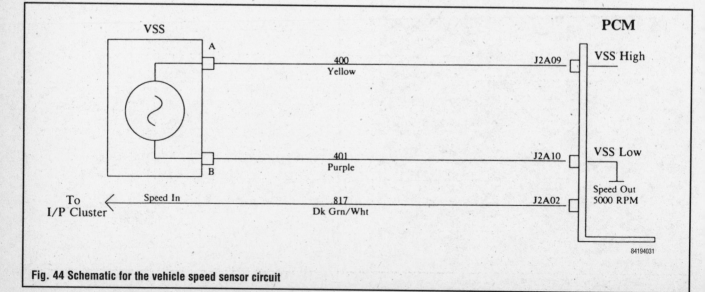

Fig. 44 Schematic for the vehicle speed sensor circuit

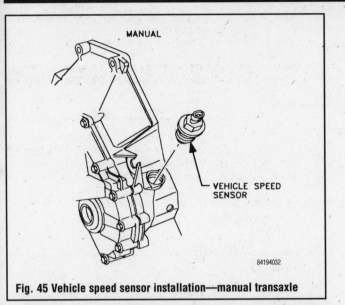

Fig. 45 Vehicle speed sensor installation—manual transaxle

Quad Driver Module (QDM)

OPERATION

The Powertrain Control Module (PCM) operates its output devices, such as solenoid valves, relays and indicator lights, by commanding a driver module. A driver module is an electronic switch that turns **ON** to complete the ground circuit and operate the 12 volt device. A Quad Driver Module operates 4 electronic switches for 4 output devices. There are 2 QDMs in the PCM. Each has a feedback circuit that detects an open or short circuit on an output device that it controls.

TESTING

◆ See Figure 46

1. All relays should have 70–90 ohms resistance. All solenoids should have 22–42 ohms resistance.
2. Turn the ignition switch **ON** and watch the lights on the instrument panel. The SERVICE ENGINE SOON and coolant temperature lights should

turn **ON**, then the coolant temperature light should go out. If not, check the fuses, bulbs and ground circuits.

3. Jumper terminals **A** and **B** on the ALDL. The cooling fan on the radiator should run.
4. Remove the jumper wire, then start the engine and operate the vehicle as required to test the remaining components.

Knock Sensor

TESTING

◆ See Figure 47

1. Disengage the wiring connector from the knock sensor and measure the resistance between the sensor and ground. It should be 3300–4500 ohms.

➡**The correct sensor installation torque is 11 ft. lbs. (15 Nm). Torque and sensor position are critical to proper operation.**

2. Reconnect the wiring. With the ignition switch **ON**, there should be 1.5–3.5 volts going to the sensor.

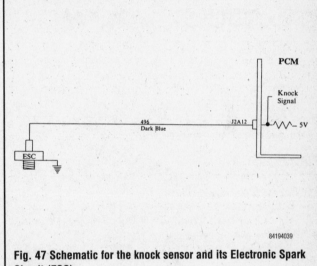

Fig. 47 Schematic for the knock sensor and its Electronic Spark Circuit (ESC)

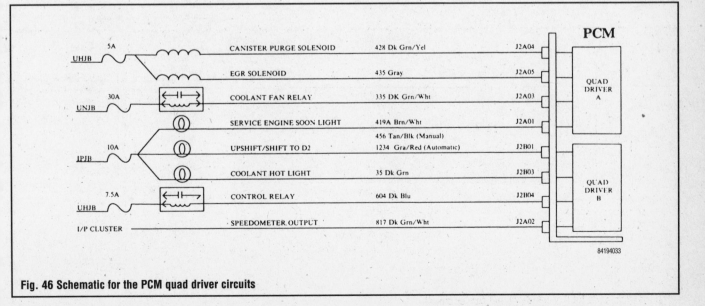

Fig. 46 Schematic for the PCM quad driver circuits

REMOVAL & INSTALLATION

▶ **See Figure 48**

The knock sensor is located in the rear center of the block, facing the firewall. The sensor is slightly above the oil pressure sender, and toward the driver's side of the vehicle.

1. If equipped, properly disable the SIR system, then disconnect the negative battery cable.
2. Raise and safely support the vehicle.
3. Squeeze the sides of the sensor's electrical connector and carefully pull it free of the sensor. Do not remove the connector by pulling on the wires, or damage may occur.
4. Loosen and remove the sensor from the engine block using a suitable box wrench or socket.

To install:

5. Install the sensor to the engine block and tighten to 133 inch lbs. (15 Nm).
6. Push the electrical connector onto the sensor until a click is heard, then gently pull back to verify a firm connection.
7. Remove the supports and carefully lower the vehicle.
8. Connect the negative battery cable and, if equipped, properly arm the SIR system.

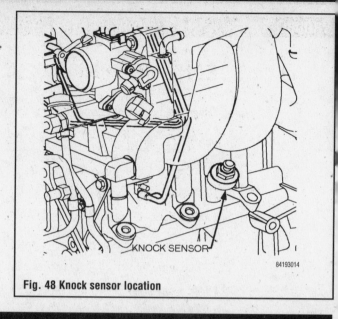

Fig. 48 Knock sensor location

COMPONENT LOCATIONS

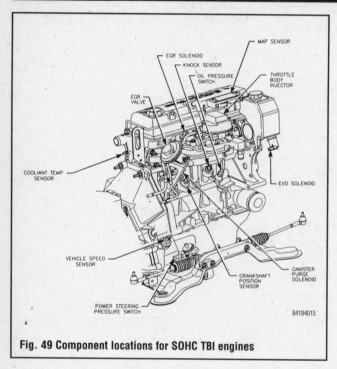

Fig. 49 Component locations for SOHC TBI engines

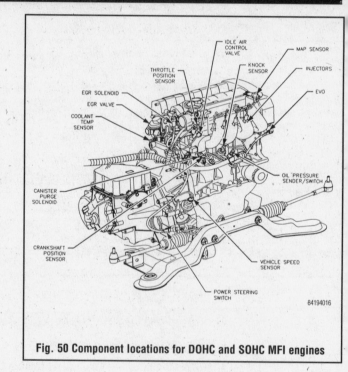

Fig. 50 Component locations for DOHC and SOHC MFI engines

ENGINE DIAGNOSTIC TROUBLE CODES

Diagnostic trouble codes cause the malfunction indicator lamp to illuminate.

I.   On-Board Diagnostic System Check
     (OBD System Check)

II. Engine Cranks, but Won't Start

III. No Malfunction Indicator Lamp

IV. No Scan Data or Won't Flash DTC 12

11   Transmission Diagnostic Trouble Codes Present

12   Diagnostic Check Only (Flash DTC)

13   Oxygen Sensor Circuit — Open/Not Ready

14   ECT Circuit — Temperature Out of Range High

15   ECT Circuit — Temperature Out of Range Low

17   PCM Fault — Pull-up Resistor

19   6X Signal Fault – No 6X Signal Between
     Reference Pulses – 1992 Vehicles

21   TP Sensor Circuit — Voltage Out of Range High

22   TP Sensor Circuit — Voltage Out of Range Low

23   IAT Circuit — Temperature Out of Range Low

24   VSS Circuit — No Signal

25   IAT Circuit — Temperature Out of Range High

26   Quad Driver Output Fault

32   EGR System Fault

33   MAP Circuit — Voltage Out of Range High

34   MAP Circuit — Voltage Out of Range Low

35   Idle Air Control (IAC) — RPM Out of Range

41   IC Control Circuit — Open or Shorted

42   Bypass Circuit — Open or Shorted

41 – 42 IC Circuit — Grounded/Bypass Open

43   Knock Sensor Circuit (KS) — Open or Shorted

44   Oxygen Sensor Indicates System Lean

45   Oxygen Sensor Indicates System Rich

46   Power Steering Pressure Circuit — Open

49   RPM — High Idle (Vacuum Leak)

51   PCM Memory Error

55   A/D Error

81   ABS Message Fault

82   PCM — Internal Communication Fault

89564G20

PCM DIAGNOSTIC TROUBLE CODES

DTC #	SYSTEM	FAULT DESCRIPTION
P0105	MAP/TP Circuit	Range/Performance
P0107	MAP Sensor Subsystem	Voltage Out of Range, Low
P0108	MAP Sensor Subsystem	Voltage Out of Range, High
P0112	IAT Sensor Subsystem	Temperature Out of Range, High
P0113	IAT Sensor Subsystem	Temperature Out of Range, Low
P0117	ECT Sensor Subsystem	Temperature Out of Range, High
P0118	ECT Sensor Subsystem	Temperature Out of Range, Low
P0122	TP Sensor Subsystem	Voltage Out of Range, Low
P0123	TP Sensor Subsystem	Voltage Out of Range, High
P0125	Engine Cooling System	Temperature Low
P0131	O2 Sensor Subsystem	Low Voltage
P0132	O2 Sensor Subsystem	High Voltage
P0133	O2 Sensor Subsystem	Slow Response
P0134	O2 Sensor Subsystem	No Activity Detected
P0137	Heated O2 Sensor (Rear)	Low Voltage
P0138	Heated O2 Sensor (Rear)	High Voltage
P0140	Heated O2 Sensor (Rear)	No Activity Detected
P0141	Heated O2 Sensor (Rear)	Heater Circuit Fault
P0171	Fuel Trim	System Lean
P0172	Fuel Trim	System Rich
P0201	Cylinder 1	Injector Circuit Fault
P0202	Cylinder 2	Injector Circuit Fault
P0203	Cylinder 3	Injector Circuit Fault
P0204	Cylinder 4	Injector Circuit Fault
P0217	Hot Light Requested	Engine Over Temperature
P0218	Hot Light Requested	Transaxle Over Temperature
P0300	Random Misfire	
P0301	Cylinder 1	Misfire Detected
P0302	Cylinder 2	Misfire Detected
P0303	Cylinder 3	Misfire Detected
P0304	Cylinder 4	Misfire Detected
P0326	Knock Sensor	Too Much Knock Present
P0327	Knock Sensor	Circuit Fault

89564G21

PCM DIAGNOSTIC TROUBLE CODES

DTC #	SYSTEM	FAULT DESCRIPTION
P0336	CKP Sensor	Signal Erratic
P0340	CAM Location Signal	Signal Missing
P0341	CAM Location Signal	Error
P0351	IC Cylinder 1 & 4 Circuit	Faulty
P0352	IC Cylinder 2 & 3 Circuit	Faulty
P0401	EGR	Insufficient Flow
P0404	EGR Circuit	Closed Valve Fault
P0405	EGR Circuit	Open/Shorted
P0420	Catalyst Converter	Efficiency Fault
P0440	EVAP Canister Purge	Large Leak Detected
P0442	EVAP Canister Purge	Small Leak Detected
P0443	EVAP Canister Purge	Circuit Fault
P0446	EVAP Canister Purge	Canister Vent Blockage
P0448	EVAP Canister Purge	Circuit Fault
P0452	EVAP Canister Purge	Voltage Low
P0453	EVAP Canister Purge	Voltage High
P0462	Fuel Gauge Circuit	Voltage Low
P0463	Fuel Gauge Circuit	Voltage High
P0500	Vehicle Speed Sensor	No Signal
P0506	Idle RPM	Too Low
P0507	Idle RPM	Too High
P0560	TCM System Voltage	Out of Range
P0561	System Voltage	Unstable
P0562	System Voltage	Out of Range, Low
P0563	System Voltage	Out of Range, High
P0565	Cruise Control Switch	Circuit Fault
P0571	Brake Switch	Circuit Fault
P0572	Brake Switch	Circuit Input Voltage, Low
P0573	Brake Switch	Circuit Input Voltage, High
P0600	PCM Serial Data Link	Fault
P0601	EPROM/Flash	Fault
P0602	DVT Checksum	Fault
P0603	TCM Novram	Fault
P0604	TCM RAM	Fault
P0605	TCM EEPROM	Fault

89564G22

PCM DIAGNOSTIC TROUBLE CODES

DTC #	SYSTEM	FAULT DESCRIPTION
P0606	TCM Processor	Fault
P0656	Fuel Gauge Circuit	Gauge Circuit Fault
P0702	Transaxle Actuator	Low Voltage
P0705	Transaxle Selector Switch	Incorrect Data
P0706	Transaxle Selector Switch	Invalid Data
P0708	Transaxle Selector Switch	No Data, Circuit Open
P0710	Transaxle Fluid Temperature Sensor	Range/Performance (Low Temp)
P0711	Transaxle Fluid Temperature Sensor	Range/Performance (High Temp)
P0712	Transaxle Fluid Temperature Sensor	Out of Range, Low Voltage
P0713	Transaxle Fluid Temperature Sensor	Out of Range, High Voltage
P0714	Transaxle Fluid Temperature Sensor	Intermittent Circuit
P0716	Turbine Speed Sensor	Signal Noisy
P0717	Turbine Speed Sensor	No Signal
P0721	Vehicle Speed Sensor	Signal Noisy
P0722	Vehicle Speed Sensor	No Signal
P0727	Engine Speed Signal	RPM Signal Fault
P0730	Transaxle	No Gears Available
P0731	Transaxle	No 1st Gear
P0732	Transaxle	No 2nd Gear
P0733	Transaxle	No 3rd Gear
P0734	Transaxle	No 4th Gear
P0740	Transaxle TCC Actuator	Circuit Open or Grounded
P0741	Transaxle Torque Conv. Clutch	Fault
P0742	Transaxle Torque Conv. Clutch	Stuck On
P0743	Transaxle TCC Actuator	Circuit Shorted to Voltage
P0744	Transaxle TCC Actuator	Circuit Intermittent Fault
P0745	Transaxle Line Pressure	Low
P0746	Transaxle Line Actuator	Circuit Open or Grounded
P0747	Transaxle Line Actuator	Circuit Shorted to Voltage
P0748	Transaxle Line Pressure	High
P0749	Transaxle Line Actuator	Circuit Intermittent Fault
P0756	Transaxle 2nd Gear Actuator	Circuit Open or Grounded
P0758	Transaxle 2nd Gear Actuator	Circuit Shorted to Voltage
P0759	Transaxle 2nd Gear Actuator	Circuit Intermittent Fault
P0761	Transaxle 3rd Gear Actuator	Circuit Open or Grounded
P0763	Transaxle 3rd Gear Actuator	Circuit Shorted to Voltage

89564G23

PCM DIAGNOSTIC TROUBLE CODES

DTC #	SYSTEM	FAULT DESCRIPTION
P0764	Transaxle 3rd Gear Actuator	Circuit Intermittent Fault
P0766	Transaxle 4th Gear Actuator	Circuit Open or Grounded
P0768	Transaxle 4th Gear Actuator	Circuit Shorted to Voltage
P0769	Transaxle 4th Gear Actuator	Circuit Intermittent Fault
P0781	Transaxle 2nd Gear	Stuck On
P0782	Transaxle 3rd Gear	Stuck On
P0783	Transaxle 4th Gear	Stuck On
P1106	MAP Sensor Subsystem	Voltage Intermittently, High
P1107	MAP Sensor Subsystem	Voltage Intermittently, Low
P1111	IAT Sensor Subsystem	Temperature Intermittently, Low
P1112	IAT Sensor Subsystem	Temperature Intermittently High
P1133	O2 Sensor Subsystem	Too Few Switches
P1134	O2 Sensor Subsystem	Rich/Lean Lean/Rich Ratio Error
P1215	Generic Field Driver (GFD)	Detected a Fault
P1336	Misfire Diagnostics	Crankshaft Adaptive Not Learned
P1351	IC Cylinder 1 & 4 Circuit	High Voltage
P1352	IC Cylinder 2 & 3 Circuit	High Voltage
P1380	ABS Controller	Faults in ABS Circuits
P1404	EGR Pintle Position	Circuit Fault
P1441	EVAP Canister Purge	Purge Valve Leak
P1508	Idle Air Control	Stuck Closed
P1509	Idle Air Control	Stuck Open
P1555	EVO	Circuit Fault
P1580	Cruise Control	Move Circuit Low Voltage
P1581	Cruise Control	Move Circuit High Voltage
P1582	Cruise Control	Direction Circuit Low Voltage
P1583	Cruise Control	Direction Circuit High Voltage
P1584	Cruise Control	Disabled
P1599	Engine	Stall Detected
P1601	PCM Serial Communications	Fault
P1602	PCM Serial Communications	Loss of ABS Serial Data
P1620	Engine	Coolant Low

89564G24

PCM DIAGNOSTIC TROUBLE CODES

DTC #	SYSTEM	FAULT DESCRIPTION
P1621	Memory Error	Memory Error
P1623	Transaxle Temp. Sensor	Pullup Resistor Fault
P1624	Customer Snap-Shot	Data Available
P1625	PCM Checksum	Transaxle Memory Error
P1627	PCM/EC A/D	Fault
P1628	PCM ECT Pullup Resistor	Fault
P1635	Sensor Five Volt Reference	Circuits Low or High
P1639	TCM A/D Signal	Fault
P1640	Quad Driver Module A	Detected a Fault
P1641	Quad Driver Module A	Quick Set Detected a Fault
P1650	Quad Driver Module B	Detected a Fault
P1651	Quad Driver Module B	Quick Set Detected a Fault
P1660	TCM GFD	Fault
P1691	Coolant Gauge	Circuit Fault
P1693	Tachometer	Circuit Fault
P1695	Remote Keyless Entry	Circuit Low
P1696	Remote Keyless Entry	Voltage High

89564G25

General Information

▶ See Figures 51, 52 and 53

The Powertrain Control Module (PCM) performs a continual self-diagnosis on many circuits of the engine control system and, if equipped, on the auto- matic transaxle control system. If a problem or irregularity is detected, the PCM will set either a code or a flag. A code indicates a suspected failure that currently exists or that has existed in an engine system. A currently existing code will illuminate the SERVICE ENGINE SOON light. A flag is a diagnostic aid that indicates an intermittent problem or irregularity, but is does not neces- sarily indicate a failure. A flag will not normally illuminate the indicator light.

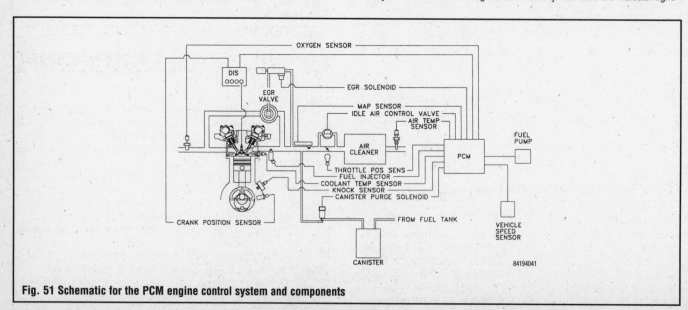

84194041

Fig. 51 Schematic for the PCM engine control system and components

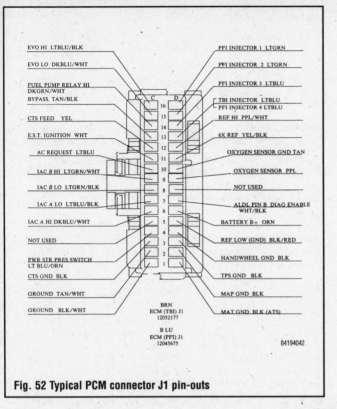

Fig. 52 Typical PCM connector J1 pin-outs

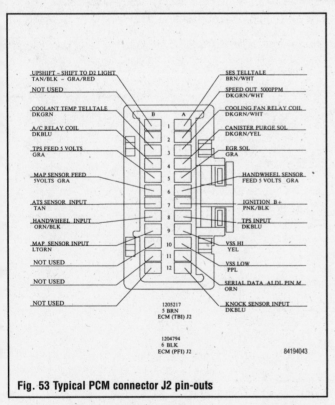

Fig. 53 Typical PCM connector J2 pin-outs

There are 2 levels of PCM memory, "General Information" and "Malfunction History." Codes are stored in both places, while flags are only stored in malfunction history. Any code stored in General Information can be read by observing the flashing SERVICE ENGINE SOON lamp while the system is in diagnostic mode. Flags and codes that are stored in Malfunction History can only be retrieved with a suitable scan tool. Most tools will be menu-driven and, although their use is not specifically covered in this section,

most described tests are useful for interpreting scan tool test results. Unlike flags for the engine control system, flags which are designated for the automatic transaxle will flash on the SHIFT TO D2 lamp (1991–1992 models) or HOT lamp (1993–98 models), along with any stored transaxle codes.

Assembly Line Diagnostic Link (ALDL) or Data Link Connector (DLC)

▶ **See Figure 54**

There are many Saturn test procedures that require the connection of a scan tool. In addition to that, 1991–95 models have the added ability to connect a jumper wire to terminals A and B of the ALDL (the main diagnostic tool connector). The ALDL (1991–95 models) or DLC (1996–98 models) is used during vehicle assembly to test the engine before it leaves the assembly plant, and after the leaving the factory to communicate with the PCM. The ALDL is located under the left side of the dashboard, near the hood release handle. Terminal A is an internal ground and is the top right terminal in the connector. Terminal B, directly adjacent and to the left of Terminal A, is the main diagnostic terminal. When these terminals are jumpered together, and the ignition switch turned **ON**, the SERVICE ENGINE SOON light will flash Code 12. This indicates that the internal diagnostic system is operating and that specific output device signals are being generated.

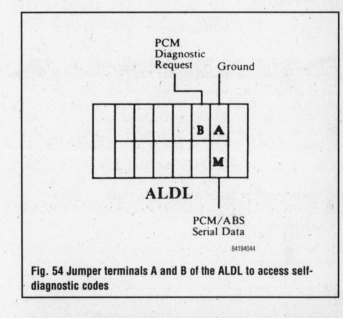

Fig. 54 Jumper terminals A and B of the ALDL to access self-diagnostic codes

Reading Codes

▶ **See Figure 55**

To enter the self-diagnostic mode, either connect a scan tool to the ALDL or DLC. On 1991–95 models, use a jumper wire to connect terminals A and B of the ALDL. Turn the ignition switch **ON**, then the PCM will enter the diagnostic program and report trouble codes on the scan tool or by flashing the SERVICE ENGINE SOON light.

All codes are 2 digits between 11 and 99. Codes are displayed on the SERVICE ENGINE SOON light by flashing the light with short and long pauses to distinguish digits of one code from digits of another. The short pause is used between digits of the same code, long pauses are between different codes. For example, the Code 12 sequence will be: flash, pause, flash-flash, long pause.

1. When the diagnostic mode is entered, Code 12 is displayed 3 times. This indicates that the internal diagnostic system is operating. If Code 12 is not displayed, the self-diagnostic program is not functioning properly. If only Code 12 is displayed, no system malfunctions have been stored.

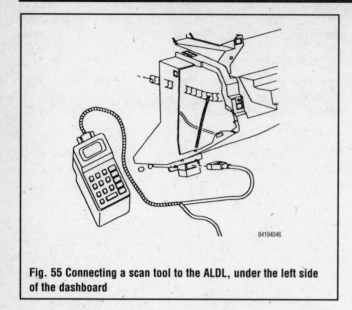

Fig. 55 Connecting a scan tool to the ALDL, under the left side of the dashboard

2. Any existing system fault codes are displayed in order from low to high, except for Code 11. Each code is displayed 3 times, followed by the next code, if any.

3. On vehicles with an automatic transaxle and on which exist stored transaxle codes, Code 11 will be displayed last, then the SHIFT TO D2 (1991–1992 models) or HOT (1993–98 models) light will begin flashing transaxle codes or flags.

4. When all engine and transaxle codes have been displayed, Code 12 will flash again. At this point, all output devices are driven, except the fuel pump, so these circuits can be checked.

5. This procedure can be repeated as required by cycling the ignition switch **OFF**, then **ON** again with the ALDL terminals jumpered together. The code display will begin again as in Step 1.

Clearing Codes

Removing power from the PCM will clear Codes from the General Information portion of the PCM memory. Flags and codes stored in Malfunction History, however, will remain in the PCM memory even if power is disconnected. Data in Malfunction History can only be read or cleared using a scan tool.

To clear General Information without cutting PCM power, therefore preserving other on-board data such as radio presets or the PCM learning ability, a scan tool may be used. If no scan tool is available, proceed as follows:

1. Turn the ignition switch **ON**, then jumper terminals A and B of the ALDL together 3 times within 5 seconds. The SERVICE ENGINE SOON light should stop flashing and any codes should be cleared.

2. Enter the self-diagnostic mode to be certain that all codes have been erased. If necessary, repeat Step 1 to ensure that all codes are removed from General Information memory.

3. If all faults have been repaired, cycling the ignition switch 50 times will automatically clear General Information of fault codes.

VACUUM DIAGRAMS

Following are vacuum diagrams for most of the engine and emissions package combinations covered by this manual. Because vacuum circuits will vary based on various engine and vehicle options, always refer first to the vehicle emission control information label, if present. Should the label be missing, or should vehicle be equipped with a different engine from the vehicle's original equipment, refer to the diagrams below for the same or similar configuration.

If you wish to obtain a replacement emissions label, most manufacturers make the labels available for purchase. The labels can usually be ordered from a local dealer.

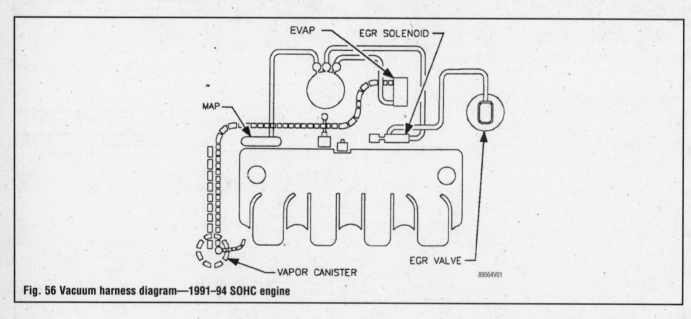

Fig. 56 Vacuum harness diagram—1991–94 SOHC engine

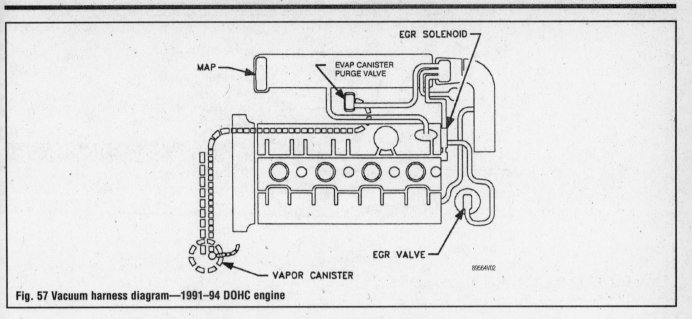

Fig. 57 Vacuum harness diagram—1991–94 DOHC engine

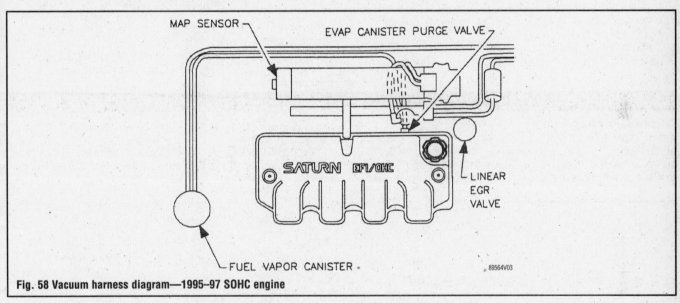

Fig. 58 Vacuum harness diagram—1995–97 SOHC engine

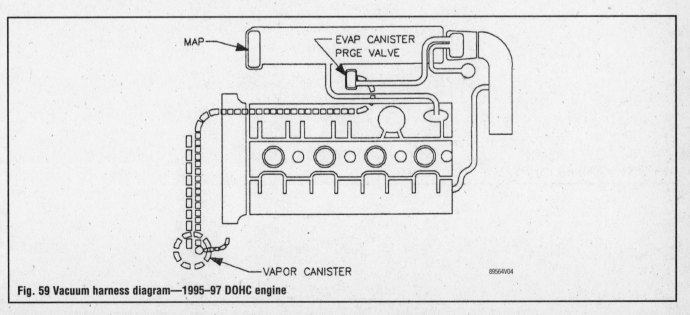

Fig. 59 Vacuum harness diagram—1995–97 DOHC engine

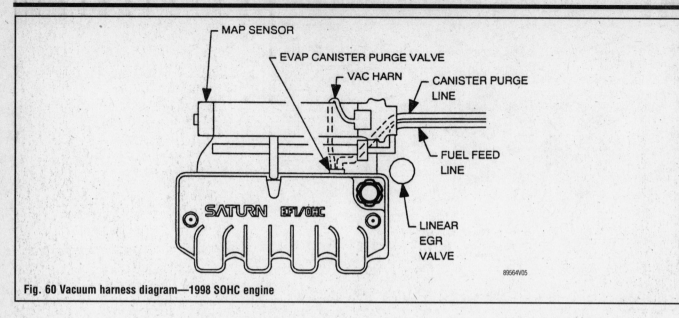

Fig. 60 Vacuum harness diagram—1998 SOHC engine

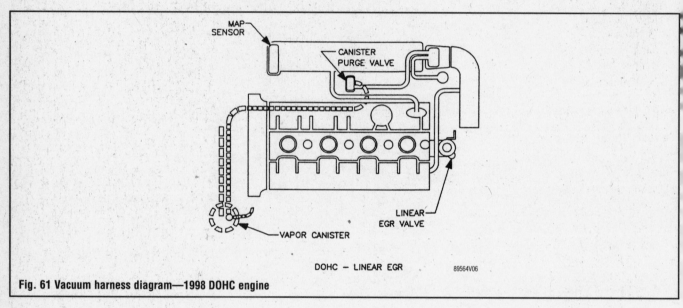

Fig. 61 Vacuum harness diagram—1998 DOHC engine

5

FUEL SYSTEM

BASIC FUEL SYSTEM DIAGNOSIS

When there is a problem starting or driving a vehicle, two of the most important checks involve the ignition and the fuel systems. The questions most mechanics attempt to answer first, "is there spark?" and "is there fuel?" will often lead to solving most basic problems. For ignition system diagnosis and testing, please refer to the information on engine electrical components and ignition systems in Section 2 of this manual. If the ignition system checks out (there is spark), then you must determine if the fuel system is operating properly (is there fuel?).

FUEL LINES AND FITTINGS

For fuel lines and fittings, all 1991–94 Saturn vehicles utilized a spring-lock coupler fitting at the engine side of the fuel filter and a quick-connect fitting for the remaining fuel line connections. However, all 1995–98 Saturn vehicles utilize quick-connect fittings throughout the entire fuel line system.

The spring-lock coupler fitting is an integral part of the fuel filter assembly that requires the use of a special tool, usually provided with the filter replacement kit.

The quick-connect fitting consists of a male and female plastic connection. These two connectors can be snapped into place (or unsnapped) without the use of any special tools.

Spring-Lock Coupler Fitting

REMOVAL & INSTALLATION

1. Using the special tool supplied with the fuel filter kit, or service tool SA9157E, disengage the spring-lock coupler fitting by installing the tool onto the male end of the fuel line at the connector.

➡**To aid in the disengagement procedure, it may be necessary to spray the female cavity end of the connector with penetrating oil to remove any debris.**

2. Using hand pressure on the tool, firmly pull the tool into the connector until it has become fully inserted. Once this has happened, pull the filter hose away from the fuel rail line.

To install:

3. Lubricate the steel male fuel line end with clean engine oil.
4. Firmly push on the female end of the spring-lock coupler until a click is heard or felt, then pull back on the fitting to confirm a positive engagement.

Quick-Connect Fitting

REMOVAL & INSTALLATION

◆ **See Figures 1 and 2**

1. Disengage the quick-connect fitting by compressing the two plastic tabs together and pulling on the fuel line. If necessary, discard the plastic retainer.

➡**To aid in the disengagement procedure, it may be necessary to spray the quick-connect fitting with penetrating oil to remove any debris.**

To install:

2. Lubricate the steel male fuel line end with clean engine oil.
3. Install the new plastic retainer and firmly push the female connection

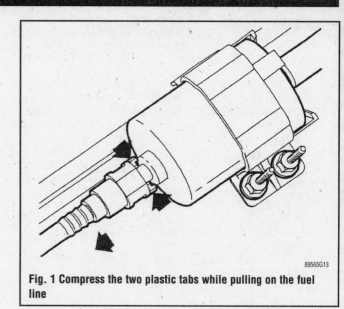

Fig. 1 Compress the two plastic tabs while pulling on the fuel line

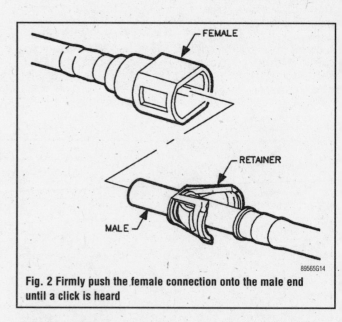

Fig. 2 Firmly push the female connection onto the male end until a click is heard

onto the male end until a click is heard. Pull back on the line to confirm a positive engagement.

THROTTLE BODY FUEL INJECTION (TBI) SYSTEM

General Information

The SOHC engine on 1991–94 models is equipped with Throttle Body Injection (TBI), which utilizes a single electric fuel injector, mounted in the throttle body unit. All fuel injection and ignition functions are controlled by the Powertrain Control Module (PCM). The PCM accepts inputs from various sensors and switches, calculates the optimum air/fuel mixture and operates the various output devices to provide peak performance within specific emissions limits. The PCM will attempt to maintain an air/fuel ratio of 14.7:1 in order to optimize catalytic converter operation. If a system failure occurs that is not serious enough to stop the engine, the PCM will illuminate the SERVICE ENGINE SOON light and operate the engine in a backup or fail-safe mode. In the backup mode, the PCM delivers fuel according to inputs from the Manifold Absolute Pressure (MAP) sensor and the Coolant Temperature Sensor (CTS). Other operating modes in the PCM program are described later.

Fuel is supplied to the engine from a pump mounted in the fuel tank. The fuel pump module includes the gauge/sending unit, which can be replaced separately. Otherwise, the module must be replaced as an assembly. The pump is operated through a relay mounted in the Instrument Panel Junction Block (IPJB), which is located under the center of the instrument panel on the passenger's side of the vehicle. A check valve in the tank unit maintains pressure in the system for a period of time after the engine is stopped in order to aid hot starting. The fuel tank must be removed to access the pump module.

Other system components include a pressure regulator, Idle Air Control (IAC) valve, Throttle Position Sensor (TPS), Air Temperature Sensor (ATS), Coolant Temperature Sensor (CTS), power steering pressure switch and an oxygen sensor. The fuel injector is a solenoid valve that the PCM pulses on and off many times per second to promote proper fuel atomization. The pulse width determines how long the injector is on each cycle and this regulates the amount of fuel supplied to the engine.

The system pressure regulator is part of the throttle body. Intake manifold pressure is supplied to the regulator diaphragm, making system pressure partly dependent on engine load. The idle air control valve is a 2-coil stepper motor that controls the amount of air allowed to bypass the throttle plate. With this valve, the PCM can closely control idle speed, even when the engine is cold or when there is a high engine load at idle.

OPERATING MODES

▶ **See Figure 3**

Starting Mode

When the ignition switch is first turned **ON**, the fuel pump relay is energized by the PCM for 2 seconds to build system pressure. When the crankshaft position signal tells the PCM that the engine is turning over or cranking, the pump will run continuously. In the start mode, the PCM checks the MAP sensor, TPS and CTS to determine the best air/fuel ratio for starting. Ratios could range from 1.5:1 at -33°F (-36°C), to 14.6:1 at 201°F (94°C) engine coolant temperature.

Clear Flood Mode

If the engine becomes flooded, it can be cleared by opening the accelerator to the full throttle position. When the throttle is open all the way and engine rpm is less than 400, the PCM will close the fuel injector while the engine is turning over in order to clear the engine of excess fuel. If throttle position is reduced below about 75 percent, the PCM will return to the start mode.

Open Loop Mode

When the engine first starts and engine speed rises above 400 rpm, the PCM operates in the Open Loop mode until specific parameters are met. Fuel requirements are calculated based on information from the MAP sensor and CTS.

Closed Loop Mode

When the correct parameters are met, the PCM will use O_2 sensor output and adjust the air/fuel mixture in order to maintain a narrow band of exhaust gas oxygen concentration. When the PCM is correcting and adjusting fuel mixture based on the oxygen sensor signal along with the other sensors, this is known as feedback air/fuel ratio control. The PCM will shift into Closed Loop mode when:
- Oxygen sensor output voltage is varied, indicating that the sensor has warmed up to operating temperature, minimum 600°F (318°C)

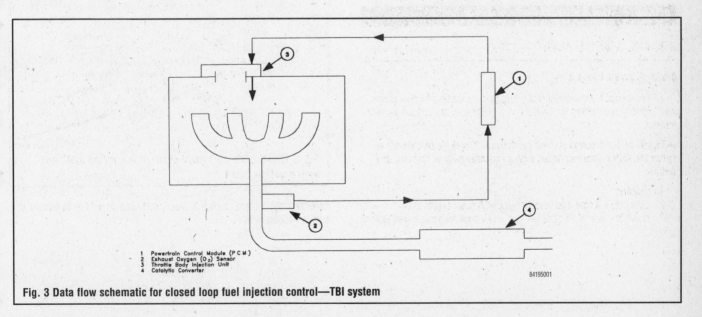

```
1  Powertrain Control Module (P C M)
2  Exhaust Oxygen (O₂) Sensor
3  Throttle Body Injection Unit
4  Catalytic Converter
```
84195001

Fig. 3 Data flow schematic for closed loop fuel injection control—TBI system

- Coolant temperature is above 68°F (20°C)
- The PCM has received an rpm signal greater than 400 for more than 1 minute
- On 1992–93 vehicles, a change in throttle position is detected

Acceleration Mode

If the throttle position is and manifold pressure is quickly increased, the PCM will provide extra fuel for smooth acceleration.

Deceleration Mode

As the throttle closes and the manifold pressure decreases, fuel flow is reduced by the PCM. If both conditions remain for a specific number of engine revolutions, the PCM decides that fuel flow is not needed and stops the flow by shutting off the injector.

Fuel Cut-Off Mode

When the PCM is receiving a Vehicle Speed Sensor (VSS) signal and engine speed goes above 6750 rpm, the injectors are shut off to prevent engine overspeed. The PCM will also shut off the injectors if the VSS signal is 0 and engine speed reaches 4000 rpm.

Battery Low Mode

If the PCM detects a low battery, it will increase injector pulse width to compensate for the low voltage and provide proper fuel delivery. It will also increase idle speed to increase alternator output.

Field Service Mode

When terminals A and B of the ALDL are jumpered with the engine running, the PCM will enter the Field Service Mode. If the engine is running in Open Loop Mode, the SERVICE ENGINE SOON light will flash quickly, about 2½ times per second. When the engine is in Closed Loop Mode, the light will flash only about once per second. If the light stays OFF most of the time in Close Loop, the engine is running lean. If the light is ON most of the time, the engine is running rich.

Relieving Fuel System Pressure

♦ See Figure 4

1. Unless battery voltage is necessary for testing, disconnect the negative battery cable. This will prevent the fuel pump from running and causing

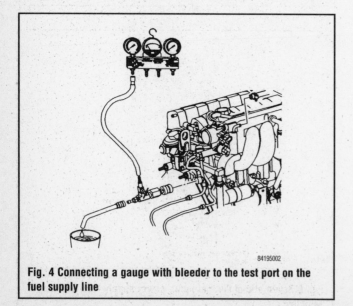

84195002

Fig. 4 Connecting a gauge with bleeder to the test port on the fuel supply line

a fuel spill through the disconnected components if the ignition key is accidentally turned on.

2. Remove the air cleaner assembly.
3. Wrap a shop rag around the fuel test port fitting, located at the lower rear of the engine, then remove the cap and connect the fuel pressure gauge tool SA9127E or equivalent.
4. Install the bleed hose from the pressure gauge into an approved container and open the valve to bleed the system pressure.
5. After the pressure is bled, remove the gauge from the test port and recap it.
6. Install the air cleaner assembly.
7. After servicing the vehicle, connect the negative battery cable and prime the fuel system as follows:
   a. Turn the ignition **ON** for 5 seconds, then **OFF** for 10 seconds.
   b. Repeat the **ON/OFF** cycle 2 more times.
   c. Crank the engine until it starts.
   d. If the engine does not readily start, repeat Steps a–c.
   e. Run the engine and check for leaks.

Fuel Pump

REMOVAL & INSTALLATION

Fuel pump replacement or service requires the removal of the fuel tank. Refer to the fuel tank procedure later in this section for proper pump replacement techniques.

TESTING

Pump Pressure Test

Trouble Code 44 or 45 could indicate pressure regulator or system pressure problems. A pump pressure test may be used to determine if the fuel pump is delivering fuel at the proper pressure or if the pump must be replaced.

✳✳ CAUTION

The following procedure will produce a small fuel spill and fumes. Make sure there is proper ventilation and be sure to take the appropriate fire safety precautions.

1. Locate the pressure test port on the fuel supply line, below and behind the EGR valve. Properly relieve any residual system pressure, then remove the fuel gauge from the test port and install the port cap.
2. Disconnect the fuel supply hose from the metal line, then connect the fuel gauge to the supply hose using a suitable adapter. Make sure the gauge is capable of reading 0–100 psi (0–690 kPa).
3. Verify that the fuel gauge shut-off valve is closed, then turn the ignition switch **ON** without starting the engine to run the fuel pump and build pressure in the system. The pump will only run for about 2 seconds without the engine running. Bleed the air out of the gauge line and cycle the fuel pump again as required to fully bleed the gauge and establish an accurate pressure reading.
4. Once maximum pressure has been achieved, allow the reading to stabilize for 30 seconds. Normal pump pressure is 58–94 psi (400–650 kPa). Allow the gauge to sit undisturbed for five minutes after the pump stops running; pressure should leak down no more than 6–8 psi (41–55 kPa) from the maximum stabilized reading. This is a fuel pump pressure test only and the results will differ from the system pressure test.
5. Bleed off the fuel system pressure and repeat the test a minimum of 2 additional times to be certain of accurate results.
6. Install the bleed hose into an approved container, then bleed the fuel system pressure and remove the gauge from the line. Lubricate the male end of the fitting with clean engine oil and reconnect the fuel line.

System Pressure Test

✳✳ CAUTION

The following procedure will produce a small fuel spill and fumes. Make sure there is proper ventilation and be sure to take the appropriate fire safety precautions.

1. Remove the air cleaner or air intake tube assembly for access, then connect fuel pressure gauge SA9127E or an equivalent gauge capable of 0–100 psi (0–690 kPa) to the fuel system test port.

2. Close the fuel gauge shut-off valve. Turn the ignition switch **ON** to run the fuel pump and build pressure in the system, then turn the ignition **OFF** again. The pump will only run for about 2 seconds without the engine running. Bleed the air out of the gauge line and cycle the fuel pump again as required to fully bleed the gauge.

3. Start and run the engine to normal operating temperature, then check the gauge for proper operating pressure. At both idle and 3000 rpm, system pressure should be 26–31 psi (179–214 kPa) for the throttle body injection system.

4. Shut the ignition **OFF** and allow the system to leak down. After about 5 minutes with the pump not running, pressure should leak down no more than 6–8 psi (41–55 kPa).

5. If pressure readings are low, do the following:
   a. Check for bent or pinched lines.
   b. Replace the fuel filter.
   c. Check for proper fuel pump pressure.
   d. Check the fuel pump for flow.
   e. Substitute a known good fuel pressure regulator.

6. If pressure readings are high, do the following:
   a. Inspect for a restricted fuel return line.
   b. Substitute a known good fuel pressure regulator.

Pump Electrical Test

▶ See Figures 5, 6, 7 and 8

1. On the Instrument Panel Junction Block (IPJB), located inside the vehicle beneath the center of the instrument panel, check the condition of fuse No. 12, the 10 amp fuel pump fuse.

2. Locate and remove the fuel pump relay located to the top left of relays in the IPJB. To test the relay, another relay from the same block may be substituted, as they are identical. Using a voltmeter or test light, check for power to the fuel pump relay. With the ignition switch **ON** or **OFF**, there should be 12 volts between terminal 30 and ground.

3. Connect a voltmeter or test light to terminal **85** and a suitable

Fig. 6 The fuel pump relay is located in the Instrument Panel Junction Block (IPJB)

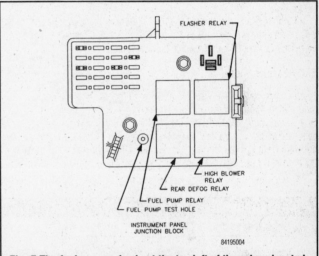

Fig. 7 The fuel pump relay is at the top left of the relays located in the IPJB—1991–94 Saturn vehicles

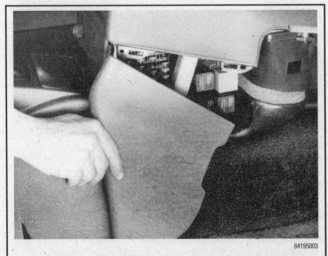

Fig. 5 Remove the access panel from the base of the dashboard center console in order to examine the fuel pump fuse and relay

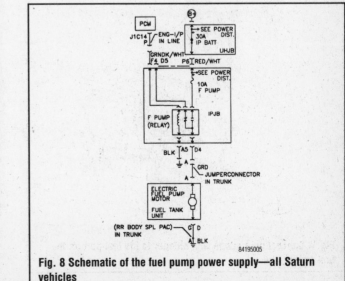

Fig. 8 Schematic of the fuel pump power supply—all Saturn vehicles

ground. When the ignition switch is first turned **ON**, there will be voltage for about 2 seconds. This is power for the relay coil from the PCM. If there is no voltage, the PCM may be faulty.

4. Locate the fuel pump wiring harness connector in the trunk. When terminals **87** and **30** are jumpered together in the relay socket, there should be 12 volts at terminal **A** on the connector. The other terminals are for the fuel gauge. The pump ground wire connects directly from the pump to chassis ground.

Throttle Body

▶ **See Figure 9**

The throttle body injection unit is made up of 2 major casting assemblies. The top piece, or fuel meter body, contains the fuel injector and pressure regulator. The bottom piece, or throttle body, contains the throttle valve, IAC valve, TPS, and the vacuum manifold or tubes which supply manifold vacuum to various engine control solenoids.

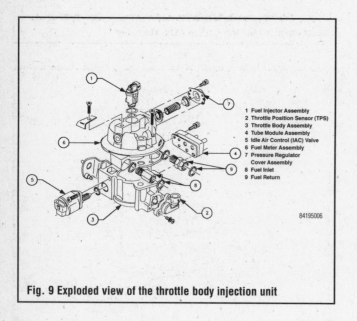

1 Fuel Injector Assembly
2 Throttle Position Sensor (TPS)
3 Throttle Body Assembly
4 Tube Module Assembly
5 Idle Air Control (IAC) Valve
6 Fuel Meter Assembly
7 Pressure Regulator
  Cover Assembly
8 Fuel Inlet
9 Fuel Return

84195006

Fig. 9 Exploded view of the throttle body injection unit

REMOVAL & INSTALLATION

▶ **See Figures 10, 11 and 12**

1. Disconnect the negative battery cable and remove the air cleaner assembly.

2. Properly relieve the fuel system pressure. Refer to the procedure in this section.

3. Unplug the electrical connectors from the following components:
   a. Idle Air Control (IAC) valve
   b. Throttle Position Sensor (TPS)
   c. Fuel injector

4. Remove the grommet (with the wires) from the throttle body.

5. Disconnect the throttle cable from the accelerator lever on the throttle body.

6. Label and disconnect the vacuum lines from the throttle body fittings.

7. Using a backup wrench to prevent stress and damage to the fittings, remove the fuel supply and return lines. Remove the O-rings from the line nuts and discard.

8. Remove the throttle body unit attaching bolts, then carefully remove the throttle body from the intake manifold.

9. Remove and discard the old gasket from the mating surface, being careful not to score the intake manifold surface. Place a clean rag or cloth over the opening to prevent debris from entering the intake manifold.

Fig. 10 Unplug the electrical connectors at these locations

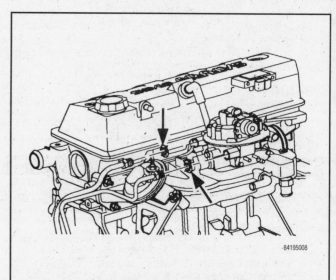

84195008

Fig. 11 Remove and discard the O-rings from fuel line nuts

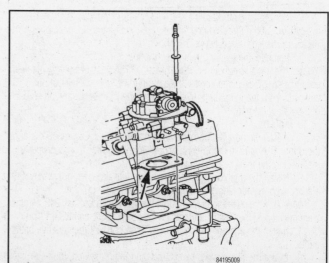

84195009

Fig. 12 Removing the throttle body assembly from the intake manifold

To install:

10. Thoroughly clean any remaining gasket material from the intake manifold and remove any old threadlock from the throttle body attaching bolt threads. Inspect the intake manifold opening for any loose parts or debris and make sure the sealing surfaces are not scored or damaged.

11. Position a new throttle body unit gasket onto the manifold flange, then install the unit. Apply a coat of Loctite® 242 or an equivalent threadlock to the attaching bolts, then install the bolts and tighten to 24 ft. lbs. (33 Nm).

12. If removed, install the fuel meter body attaching screws.

13. Even if they were not removed, torque the fuel meter body attaching screws to 35 inch lbs. (4 Nm).

14. Install new O-rings to the fuel supply and return lines, then install the lines and tighten to 19 ft. lbs. (25 Nm) using a backup wrench.

15. Connect the throttle cable to the accelerator lever on the throttle body. Make sure the cable does not hold the lever open when the accelerator pedal is in the released position.

16. Install the electrical connectors to the TPS, IAC valve and the fuel injector. Push in until a click is heard and give a slight tug on the connector to confirm positive engagement.

17. Connect the vacuum hoses from the EGR solenoid, canister purge solenoid and the MAP sensor to the TBI unit vacuum ports, as labeled during removal. Make sure the hoses are securely seated to prevent vacuum leaks.

18. Connect the negative battery cable.

19. Cycle the ignition a few times to prime the fuel system, then start the engine and check for fuel leaks.

20. Shut the engine **OFF** and install the air cleaner assembly. If damaged or deteriorated, replace the air cleaner assembly gasket.

BASE IDLE SPEED ADJUSTMENT

For information concerning the adjustment of the base idle speed, refer to Idle Speed and Mixture Adjustments in Section 1.

Fuel Injector

REMOVAL & INSTALLATION

▶ **See Figures 13, 14 and 15**

1. Disconnect the negative battery cable.
2. Remove the air cleaner assembly.
3. Properly relieve the fuel system pressure. Refer to the procedure in this section.
4. Unplug the injector electrical connector.
5. Remove the injector retaining screw and bracket.
6. Using a smooth fulcrum and a suitable prybar, carefully pry the injector out of the throttle body. Make sure the electrical connector and nozzle are protected from damage during removal.
7. Remove and discard the upper and lower injector O-rings, then inspect the injector for dirt or contamination. The injector may be cleaned using safety glasses and compressed air, but the screen may not be removed from the injector. If injector replacement is necessary, be sure to use an identical part.

To install:

8. Install the new O-rings and lubricate with clean engine oil. Be sure the upper O-ring is in the injector groove and the lower ring is properly installed in the fuel meter body cavity.

9. Install the injector, pushing it straight into the injector cavity with the electrical connector facing toward the fuel pressure regulator.

10. Install the retaining bracket. Coat the screw with Loctite® 242, or equivalent threadlock, then install and tighten the screw to 35 inch lbs. (3 Nm).

11. Install the injector wiring harness, then connect the negative battery cable.

12. Prime the fuel system by cycling the ignition switch, then start the engine and check for leaks.

13. Shut the engine **OFF** and install the air cleaner assembly.

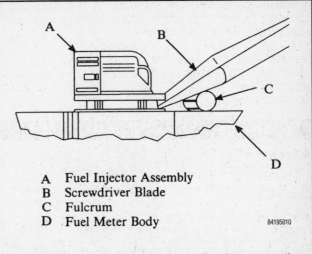

A   Fuel Injector Assembly
B   Screwdriver Blade
C   Fulcrum
D   Fuel Meter Body

84195010

Fig. 13 Using a smooth fulcrum and a small prybar, remove the fuel injector from the throttle body assembly

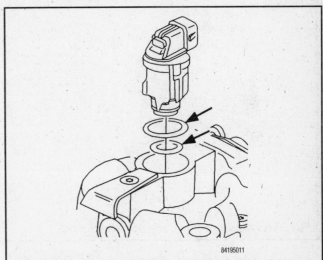

84195011

Fig. 14 Replace the upper and lower injector O-rings whenever the injector is removed

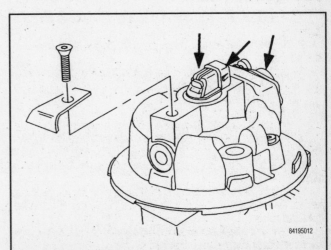

84195012

Fig. 15 Before installing the injector retaining bracket and screw, make sure the electrical connector is pointing toward the pressure regulator

TESTING

♦ See Figure 16

Electrical Test

1. Unplug the injector connector and use an ohmmeter to check the resistance of the injector coil. It should be 1–2 ohms.

2. Connect a Noid light (or an equivalent low voltage injector harness tester) to the injector connector terminal. Position the harness so the light can be seen from the driver's seat. Then, operate the starter and observe the Noid light.

3. If the Noid does not illuminate at all, check for power to the injector(s). The injectors receive power through the pink/black wire from the 7.5 Amp "INJ" fuse in the underhood junction box, any time the ignition switch is **ON**. The PCM operates the injectors by completing the ground circuit through the light blue wire.

4. If the test light stays on steadily, check for a short to ground of the wire between the injector connector and the PCM.

5. A pulsing Noid light does not necessarily indicate enough voltage to properly operate the fuel injector. If it is still questionable whether or not an injector is pulsing, connect a known good injector to the circuit and tightly grip the injector while an assistant cranks the engine. If injector pulses are felt, the circuit is good.

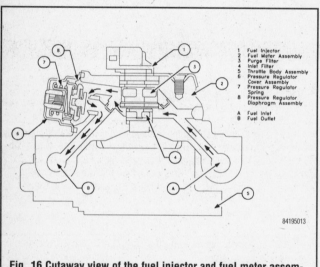

Fig. 16 Cutaway view of the fuel injector and fuel meter assemblies—1991–94 SOHC engine

Pressure Test

1. Connect the test gauge and run the fuel pump and system pressure tests as described in the Fuel Pump tests earlier in this section.

2. Disconnect the fuel return hose from the metal line and cap the line carefully, as full fuel pressure may go as high as 94 psi (650 kPa).

3. Disconnect the pressure regulator vacuum line.

4. Cycle the ignition **ON** and **OFF**, without starting the engine, to build system pressure to a minimum of 58 psi (400 kPa) on the gauge.

5. Look into the throttle body. If there is a leak, there will be drops of fuel on the throttle plate. The leak could from be the injector O-ring, the injector tip or the pressure regulator diaphragm.

6. If necessary, remove and examine the injector; a leak will be obvious.

Fuel Pressure Regulator

REMOVAL & INSTALLATION

♦ See Figures 17, 18 and 19

1. Disconnect the negative battery cable, then remove the air cleaner assembly and air inlet tubes.

2. Properly relieve the fuel system pressure. Refer to the procedure in this section.

✳✳ CAUTION

The pressure regulator contains a large spring which is under heavy pressure. Be careful to keep the regulator compressed when removing the screws, in order to prevent personal injury

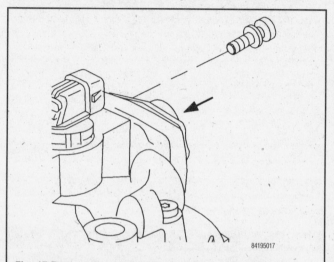

Fig. 17 Remove the regulator cover screws while holding the cover against the spring tension—1991–94 SOHC engine

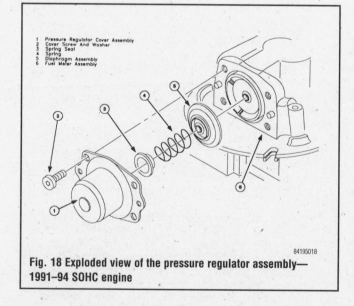

Fig. 18 Exploded view of the pressure regulator assembly—1991–94 SOHC engine

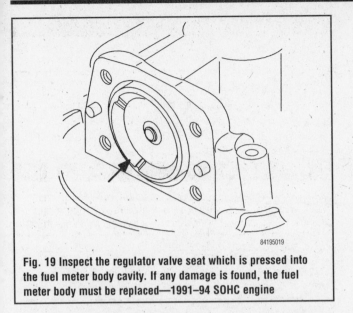

Fig. 19 Inspect the regulator valve seat which is pressed into the fuel meter body cavity. If any damage is found, the fuel meter body must be replaced—1991–94 SOHC engine

3. For assembly purposes, note the location of the cover alignment slot, then remove the 4 pressure regulator attaching screws using a No. T15 Torx® driver. Hold the regulator cover against the spring tension when removing the screws.

4. Slowly remove the cover assembly, followed by the regulator spring. Keep both parts aside to be used with the new diaphragm assembly.

5. Remove the pressure regulator diaphragm assembly, inspect for debris and then discard.

To install:

6. Thoroughly clean the pressure regulator bore and valve seat using a carburetor cleaner that does not contain methyl ethyl ketone, then allow to air dry for a few minutes before the regulator is reinstalled.

7. Inspect the regulator valve seat which is pressed into the fuel meter body cavity. If necessary, use a mirror and a magnifying glass to inspect for small particles in the throttle body bore and seat. If cracks, nicks, debris or pitting are found, the fuel meter body must be replaced.

8. Install the pressure regulator diaphragm assembly, making sure that it is properly seated in the fuel meter body groove.

9. Install the regulator spring seat into the cover, then install the cover over the spring and diaphragm assembly. Align the mounting holes and mounting pins, then align the cover as noted prior to removal.

➡**Use care to properly align the regulator cover. Fuel leaks may be caused by cover misalignment.**

10. Maintain pressure on the regulator, then install the 4 retaining screws. Tighten the screws to 22 inch lbs. (2.5 Nm).

11. Cycle the ignition a few times in order to prime the fuel system, then start the engine and check for leaks.

12. Shut the engine **OFF** and install the air cleaner assembly.

MULTI-PORT FUEL INJECTION (MFI) SYSTEM

General Information

▶ **See Figures 20 and 21**

The 1995–98 SOHC and all DOHC engines are equipped with Multi-port Fuel Injection (MFI), which utilizes one injector for each cylinder. All fuel injection and ignition functions are controlled by the Powertrain Control Module (PCM). It accepts inputs from various sensors and switches, calculates the optimum air/fuel mixture and operates the various output devices to provide peak performance within specific emissions limits. The PCM will attempt to maintain an air/fuel ratio of 14.6:1 in order to optimize catalytic converter operation. If a system failure occurs that is not serious enough to stop the engine, the PCM will illuminate the SERVICE ENGINE SOON light and operate the engine in a backup or fail-safe mode. In the backup mode, the PCM delivers fuel according to inputs from the Manifold Absolute Pressure (MAP) sensor and the Coolant Temperature Sensor (CTS). Other operating modes are in the PCM program and are described later.

Fuel is supplied to the engine from a pump mounted in the fuel tank. The fuel pump module includes the gauge/sending unit, which can be replaced separately. Otherwise, the module must be replaced as an assembly. The pump is operated through a relay mounted in the Instrument Panel Junction Block (IPJB), which is located under the center of the instrument panel on the passenger's side of the vehicle. A check valve in the tank unit maintains pressure in the system for a period of time after the engine is stopped to aid hot starting. The fuel tank must be removed to access the pump module.

Other system components include a pressure regulator, an Idle Air Control (IAC) valve, a Throttle Position Sensor (TPS), Air Temperature Sensor (ATS), Coolant Temperature Sensor (CTS), a power steering pressure switch and an oxygen sensor. The fuel injectors are solenoid valves that the PCM pulses on and off many times per second, in order to promote good fuel atomization. The pulse width determines how long the injector is on each cycle, and this regulates the amount of fuel supplied to the engine. Fuel injectors are operated simultaneously, not sequentially.

On 1991–97 models, the system pressure regulator is mounted on the end of the fuel rail that feeds the injectors. For 1998, all models have a new "Returnless" fuel system, which makes the pressure regulator an integral component of the fuel filter, which is mounted below the vehicle near the fuel tank. Intake manifold pressure is supplied to the regulator diaphragm, making system pressure partly dependent on engine load. The idle air control valve is a 2-coil stepper motor that controls the amount of air allowed to bypass the throttle plate. With this valve, the PCM can closely control idle speed even when the engine is cold or when there is a high engine load at idle.

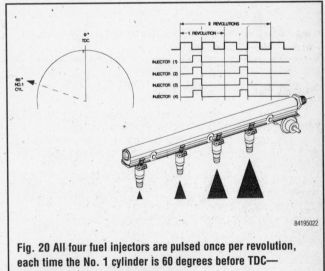

Fig. 20 All four fuel injectors are pulsed once per revolution, each time the No. 1 cylinder is 60 degrees before TDC—1995–98 SOHC and all DOHC engines

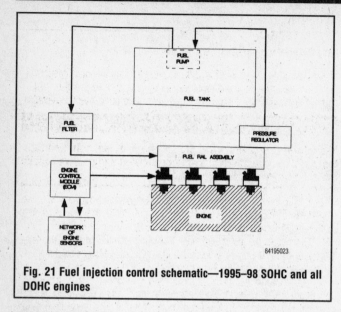

Fig. 21 Fuel injection control schematic—1995–98 SOHC and all DOHC engines

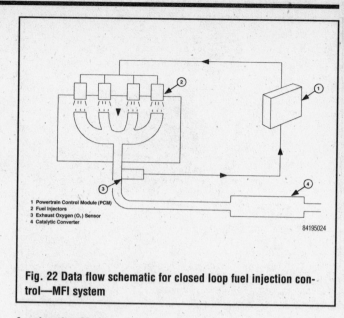

Fig. 22 Data flow schematic for closed loop fuel injection control—MFI system

OPERATING MODES

♦ See Figure 22

Starting Mode

When the ignition switch is first turned **ON**, the fuel pump relay is energized by the PCM for 2 seconds to build system pressure. When the crankshaft position signal tells the PCM that the engine is "turning over" or cranking, the pump will run continuously. In the start mode, the PCM checks the TPS and CTS to determine the best air/fuel ratio for starting. Ratios could range from 0.8:1 at -40°F (-40°C), to 14.6:1 at 220°F (104°C) engine coolant temperature.

Clear Flood Mode

If the engine becomes flooded, it can be cleared by opening the accelerator to the full throttle position. When the throttle is open all the way and engine rpm is less than 400, the PCM will close the fuel injectors while the engine is cranking, in order to clear the engine of excess fuel. If throttle position is reduced below about 75 percent, the PCM will return to the start mode.

Open Loop Mode

When the engine first starts and engine speed rises above 400 rpm, the PCM operates in the Open Loop mode until specific parameters are met. Fuel requirements are calculated based on information from the MAP sensor and CTS.

Closed Loop Mode

When the correct parameters are met, the PCM will use O_2 sensor output and adjust the air/fuel mixture, in order to maintain a narrow band of exhaust gas oxygen concentration. When the PCM is correcting and adjusting fuel mixture based on the oxygen sensor signal, along with the other sensors, this is known as "feedback air/fuel ratio control."

The PCM will shift into Closed Loop mode when:

Oxygen sensor output voltage varies, indicating that the sensor has warmed to operating temperature; that is, a minimum of 600°F (318°C)

Coolant temperature is above 68°F (20°C)

The PCM has received an rpm signal greater than 400 for more than 1 minute

➡ **To operate in closed loop mode, 1992 and later vehicles also require a change in throttle position.**

Acceleration Mode

If the throttle position changes and manifold pressure is quickly increased, the PCM will provide extra fuel for smooth acceleration.

Deceleration Mode

As the throttle closes and the manifold pressure decreases, fuel flow is reduced by the PCM. If both conditions remain for a specific number of engine revolutions, the PCM decides fuel flow is not needed and stops the flow by shutting off the injectors.

Fuel Cut-Off Mode

When the PCM is receiving a Vehicle Speed Sensor (VSS) signal and rpm goes above 6750, the injectors are shut off to prevent engine overspeed. The PCM will also shut off the injectors if the VSS signal is 0 and engine speed reaches 4000 rpm.

Battery Low Mode

If the PCM detects a low battery, it will increase injector pulse width to compensate for the low voltage and provide proper fuel delivery. It will also increase idle speed to increase alternator output.

Field Service Mode (1991–95 Models)

When terminals A and B of the ALDL are jumpered with the engine running, the PCM will enter the Field Service Mode. If the engine is running in Open Loop Mode, the SERVICE ENGINE SOON light will flash quickly, about 2½ times per second. When the engine is in Closed Loop Mode, the light will flash only about once per second. If the light stays OFF most of the time in Closed Loop, the engine is running lean. If the light is ON most of the time, the engine is running rich.

Relieving Fuel System Pressure

♦ See Figures 23 and 24

1. Unless battery voltage is necessary for testing, disconnect the negative battery cable. This will prevent the fuel pump from running and causing a fuel spill through the disconnected components, if the ignition switch is accidentally turned **ON**.

2. Remove the air intake tube and resonator.

3. Wrap a shop rag around the fuel test port fitting, located at the lower rear of the engine, then remove the cap and connect the fuel pressure gauge tool SA9127E or equivalent.

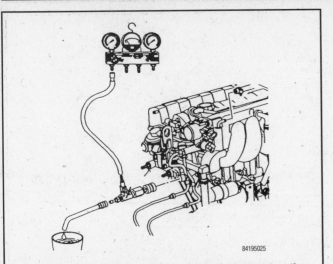

Fig. 23 Connecting a gauge with bleeder to the test port on the fuel supply line—1991–97 models

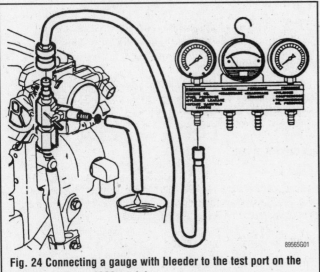

Fig. 24 Connecting a gauge with bleeder to the test port on the fuel supply line—1998 models

4. Install the bleed hose from the pressure gauge into an approved container and open the valve to bleed the system pressure.

5. After the pressure is bled, remove the gauge from the test port and recap it.

6. Install the air intake duct and resonator tube.

7. After repairs, connect the negative battery cable and prime the fuel system as follows:

    a. Turn the ignition **ON** for 5 seconds, then **OFF** for 10 seconds.

    b. Repeat the **ON/OFF** cycle 2 more times.

    c. Crank the engine until it starts. If the engine does not readily start, repeat Steps a through c.

    d. Run the engine and check for leaks.

Fuel Pump

REMOVAL & INSTALLATION

Fuel pump replacement or service requires the removal of the fuel tank. Refer to the Fuel Tank procedure later in this section for the proper pump replacement techniques.

TESTING

Fuel Pump Pressure Test

A Trouble Code 44 or 45 could indicate pressure regulator or system pressure problems. The pump pressure test may be used to determine if the fuel pump is delivering fuel at the proper pressure or if the pump must be replaced.

> ✸✸ **CAUTION**
>
> **The following procedure will produce a small fuel spill and fumes. Make sure there is proper ventilation and be sure to take the appropriate fire safety precautions.**

1. Locate the pressure test port on the fuel supply line, below and behind the EGR valve. Properly relieve any residual system pressure, then remove the fuel gauge from the test port and install the port cap.

2. Disconnect the fuel supply hose from the metal line, then connect the fuel gauge to the supply hose using a suitable adapter. Make sure the gauge is capable of reading 0–100 psi (0–690 kPa).

3. Verify that the fuel gauge shut-off valve is closed, then turn the ignition switch **ON** without starting the engine to run the fuel pump and build pressure in the system. The pump will only run for about 2 seconds without the engine running. Bleed the air out of the gauge line and cycle the fuel pump again as required to fully bleed the gauge and establish an accurate pressure reading.

4. Once maximum pressure has been achieved, allow the reading to stabilize for 30 seconds. Normal pump pressure is 58–94 psi (400–650 kPa). Allow the gauge to sit undisturbed for five minutes after the pump stops running; pressure should leak down no more than 6–8 psi (41–55 kPa) from the maximum stabilized reading. This is a fuel pump pressure test only and the results will differ from the system pressure test.

5. Bleed off the fuel system pressure and repeat the test a minimum of 2 additional times to be certain of accurate results.

6. Install the bleed hose into an approved container, then bleed the fuel system pressure and remove the gauge from the line. Lubricate the male end of the fitting with clean engine oil and reconnect the fuel line.

System Pressure Test

▶ See Figure 25

> ✸✸ **CAUTION**
>
> **The following procedure will produce a small fuel spill and fumes. Make sure there is proper ventilation and be sure to take the appropriate fire safety precautions.**

1. Remove the air cleaner or air intake tube assembly for access, then connect fuel pressure gauge SA9127E or an equivalent gauge capable of 0–100 psi (0–690 kPa) to the fuel system test port.

2. Close the fuel gauge shut-off valve. Turn the ignition switch **ON** to run the fuel pump and build pressure in the system, then turn the ignition **OFF** again. The pump will only run for about 2 seconds without the engine running. Bleed the air out of the gauge line and cycle the fuel pump again as required to fully bleed the gauge.

3. Cycle the ignition and check the gauge; pressure should be 38–44 psi (262–306 kPa) for 1991–97 models and 40–55 psi (276–379 kPa) on 1998 models. Allow the gauge to stabilize; after about 5 minutes with the pump not running, pressure should leak down no more than 6–8 psi (41–55 kPa).

4. Start and run the engine to normal operating temperature, then check the gauge for proper operating pressure. With the engine running at idle, pressure should be 31–36 psi (214–248 kPa).

5. On 1991–97 models, check the pressure regulator with the engine running at idle by disconnecting the vacuum line. With the line disconnected, the pressure should vary about 6–10 psi (40–70 kPa). If the pressure reading does not change and vacuum can be felt at the line's inlet, replace the regulator.

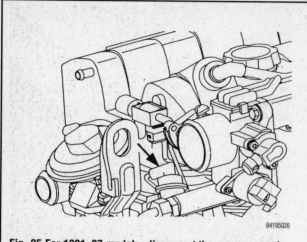

Fig. 25 For 1991–97 models, disconnect the vacuum supply from the fuel pressure regulator—if the regulator is good, fuel pressure will vary

6. If pressure readings are low, perform the following:
   a. Check for bent or pinched lines.
   b. Replace the fuel filter.
   c. Check for proper fuel pump pressure.
   d. Check the fuel pump for flow.
   e. Substitute a known good fuel pressure regulator.
7. If pressure readings are high, perform the following:
   a. Inspect for a restricted fuel return line.
   b. Substitute a known good fuel pressure regulator.

Pump Electrical Test

▶ **See Figures 5, 6, 7, 8 and 26**

1. On the Instrument Panel Junction Block (IPJB), located inside the vehicle beneath the center of the instrument panel, check the condition of the 10 amp fuel pump fuse. The inside of the access panel will provide the location of the fuel pump fuse and relay.

2. Locate and remove the fuel pump relay. On 1991–94 models, it is located to the top left of the relays in the IPJB. On 1995 and later models, it is located at the bottom right of the IPJB. To test the relay, another relay from the same block may be substituted, as they are identical. Using a voltmeter or test light, check for power to the fuel pump relay. With the ignition

switch **ON** or **OFF**, there should be 12 volts between terminal 30 and ground.

3. Connect a voltmeter or test light to terminal 85 and a suitable ground. When the ignition switch is first turned **ON**, there will be voltage for about 2 seconds. This is power for the relay coil from the PCM. If there is no voltage, the PCM may be faulty.

4. Locate the fuel pump wiring harness connector in the trunk. When terminals 87 and 30 are jumpered together in the relay socket, there should be 12 volts at terminal A on the connector. The other terminals are for the fuel gauge. The pump ground wire connects from the pump to chassis ground.

REMOVAL & INSTALLATION

▶ **See Figures 27 thru 38**

1. Disconnect the negative battery cable.
2. Remove the air intake tube and resonator.
3. Unplug the electrical connectors from the Idle Air Control (IAC) valve and the Throttle Position Sensor (TPS).
4. Disconnect the vacuum harness from the top of the throttle body.
5. Disconnect the throttle cable.
6. Remove the throttle body retaining bolts, then remove the throttle body assembly from the side of the intake manifold. Block the intake manifold entry with a clean cloth to prevent dirt or debris from entering and damaging the engine.
7. If the same unit is going to be installed on the intake manifold, examine the throttle body bore for excessive carbon build-up. If required, use carburetor cleaner, or equivalent, and a shop towel to clean the bore. Do not allow the solvent or cleaner to contact the IAC valve or TPS; remove the sensors from the throttle body, if necessary.

To install:

8. Remove the old gasket and discard, then thoroughly clean the gasket mating surfaces. Clean the old Loctite® threadlock from the throttle body mounting bolt threads.
9. Remove the cloth from the intake manifold opening, then inspect the opening for any foreign debris or pieces of gasket material.
10. Position a new flange gasket, then install the throttle body assembly to the intake manifold. Apply a coat of Loctite® 242, or an equivalent threadlock, to the mounting bolts, then install and tighten to 23 ft. lbs. (31 Nm).

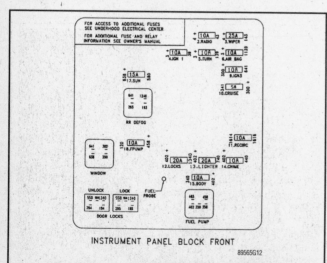

Fig. 26 The fuel pump relay is to the bottom right of the Instrument Panel Junction Block (IPJB)—1995–98 Saturn vehicles

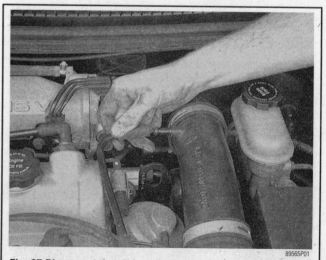

Fig. 27 Disconnect the air hose between the camshaft cover and air cleaner inlet hose

Fig. 28 Loosen the air inlet hose clamp at the throttle body

Fig. 31 Disconnect the vacuum line harness from the top of the throttle body

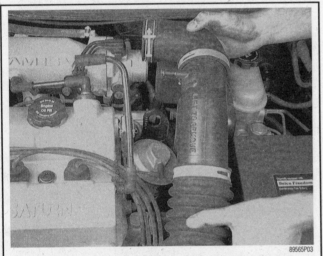

Fig. 29 Disconnect the air inlet hose at the throttle body and remove it from the vehicle

Fig. 32 Turn the throttle lever to the wide open position, then disengage the cable

Fig. 30 Unplug the IAC and TPS wiring connectors at the throttle body

Fig. 33 Loosen the throttle body mounting bolts

Fig. 34 Remove the throttle body unit from the intake manifold

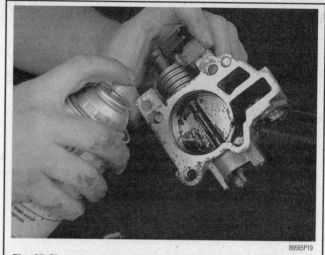

Fig. 37 Clean the throttle body unit with carburetor cleaner and a rag. Do not allow solvent to contact the IAC valve or TPS

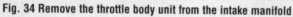

Fig. 35 Block the intake manifold opening with a clean rag to prevent any debris from entering

Fig. 38 Remove and discard the throttle body gasket. Be careful not to score the gasket sealing surface

➡️ If the throttle body is being replaced, the idle stop screw on the new unit is preset by the manufacturer and should not require adjustment. Minimum idle speed should ONLY be adjusted if all other engine systems are operating properly and idle speed still cannot be maintained.

11. Connect the throttle cable to the lever, making sure the cable does not hold the lever open when the accelerator pedal is released.
12. Connect the vacuum harness to the top of the throttle body.
13. Connect the IAC valve and TPS wiring harnesses.
14. Install the air intake tube and resonator.
15. Connect the negative battery cable.
16. Start the engine and check for vacuum leaks.

BASE IDLE SPEED ADJUSTMENT

For information concerning adjustment of the base idle speed, refer to Idle Speed and Mixture Adjustments in Section 1.

Fig. 36 Examine the throttle body bore for excessive dirt build-up

Fuel Injectors

REMOVAL & INSTALLATION

▶ See Figures 27, 28, 29, and 39 thru 54

1. Disconnect the negative battery cable.
2. Remove the air intake tube with resonator and fresh air tube.
3. Properly relieve fuel system pressure.
4. Remove the fuel line bracket bolt, then disconnect the fuel supply and return lines in the following manner:

a. On 1991–94 models, be sure to use a ¹⁵⁄₁₆ in. (24mm) backup wrench to prevent inlet port or bracket damage. If necessary, remove the fuel line bolts and rotate the rail slightly for wrench access. Remove and discard the old O-rings from the fuel lines using a suitable seal removal tool or brass pick.

b. On 1995–98 models, disengage the fuel line(s) from the rail and pressure regulator by compressing the two plastic retaining tabs of the quick-connect fitting and pulling on the line. Remove and discard the plastic retainer.

5. Disconnect the vacuum hose from the fuel pressure regulator on 1991–97 models.
6. Detach the PCV hose from the camshaft cover.
7. Remove the throttle cable bracket bolts, then disconnect the cable from the throttle lever. Lay the cable over the intake manifold and out of the way.
8. Unplug the fuel injector electrical connectors and, if not done already, remove the fuel rail bolts.
9. Remove the fuel rail assembly by carefully pulling the rail back and upward to pull the injectors from the manifold ports. Be careful not to damage the injector spray tips and the electrical connectors. Rotate the rail so the injectors point downward, then lift the rail end opposite the fuel connections to remove the rail from between the camshaft cover and intake manifold.
10. Make sure the rails and connectors are clean and free of dirt. Remove and discard the lower injector O-rings seals.
11. If injector removal is required, slide the injector retaining clip off the injector and pull the injector from the rail assembly. Remove and discard the upper injector O-rings seals.

To install:

12. If removed, lubricate and install the new injector O-rings with clean engine oil and install with the injector assembly into the fuel rail. Install the injector retaining clip.

➡There are 3 types of injector retaining clips. The 1st design, which was used on some engines built prior to January 1st, 1991, may cause injector ticking noise. If injectors are removed from the fuel rail assembly and this type is found, they must be replaced with the 2nd clip design. Do not attempt to substitute the 3rd design of clip; only 1995 and later models can use the 3rd design.

13. Lubricate the new lower injector O-rings with clean engine oil, then install.
14. For 1991–94 models, lubricate the new fuel inlet and return O-rings, then install the O-rings into the fuel outlet of the pressure regulator and inlet of the fuel rail.
15. Starting with the proper end (on 1991–97 models, the end containing the pressure regulator), and with the injectors pointing downward, guide the fuel rail assembly through the passage between the camshaft cover and the intake manifold, from the power steering pump side of the engine. Align the injectors with their respective ports, then rotate the fuel rail and carefully push the injectors into their ports.
16. Verify that the injectors are properly seated in the intake manifold.
17. Install the fuel rail retaining bolts and tighten to 22 ft. lbs. (30 Nm) on 1991–94 models or 89 inch lbs. (10 Nm) on 1995–98 models.
18. Connect the fuel injector wiring harnesses.
19. Connect the PCV valve hose to the camshaft cover, and the vacuum line to the pressure regulator, making sure they are properly seated.

Fig. 39 Loosen the supply and return lines from the pressure regulator/fuel rail assembly with the aid of a ¹⁵⁄₁₆ in. (24mm) backup wrench—1991–94 vehicles

Fig. 40 Disconnect the supply and return lines from the pressure regulator/fuel rail assembly

Fig. 41 Remove and discard the O-rings from the fuel line openings using a suitable seal removal tool or brass pick

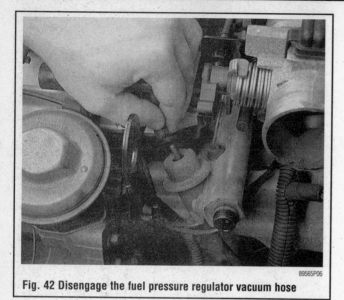

Fig. 42 Disengage the fuel pressure regulator vacuum hose

Fig. 45 Lift the cable and bracket assembly up, and lay it over the intake manifold, out of the way

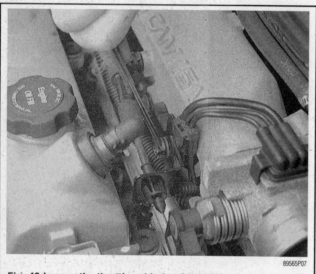

Fig. 43 Loosen the throttle cable bracket bolts . . .

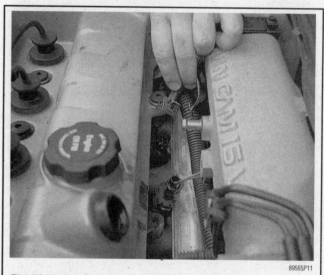

Fig. 46 Unplug the fuel injector electrical connectors . . .

Fig. 44 . . . then, while holding the throttle lever in the wide open position, disconnect the cable

Fig. 47 . . . and, if not done already, remove the fuel rail bolts

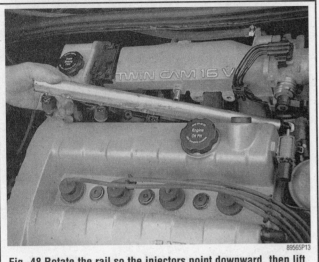

Fig. 48 Rotate the rail so the injectors point downward, then lift the rail end opposite the fuel connections to remove the rail

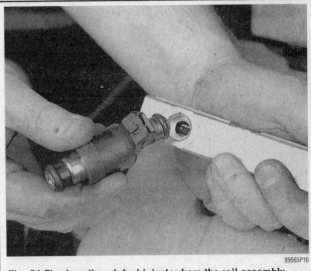

Fig. 51 Firmly pull each fuel injector from the rail assembly

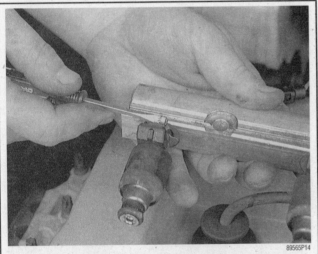

Fig. 49 Using a small bladed screwdriver, slide the injector retaining clip off the injector

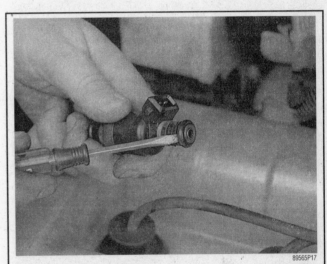

Fig. 52 Remove and discard the upper fuel injector O-ring seals. Use a small bladed screwdriver, if necessary

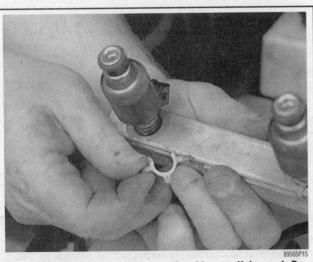

Fig. 50 The fuel injector clip is small and bouncy if dropped. Be careful not to lose it

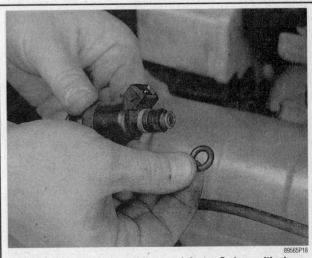

Fig. 53 Lubricate and install the new injector O-rings with clean engine oil

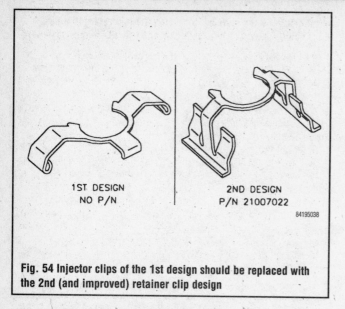

Fig. 54 Injector clips of the 1st design should be replaced with the 2nd (and improved) retainer clip design

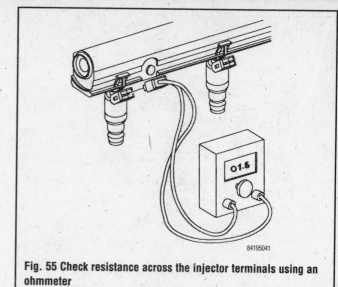

Fig. 55 Check resistance across the injector terminals using an ohmmeter

20. Install the throttle cable bracket bolts and tighten to 19–22 ft. lbs. (25–30 Nm), then connect the throttle cable.

21. Lubricate the male ends of the fuel supply and return lines with clean engine oil. Connect the fuel supply and return lines in the following manner:

a. On 1991–94 models, be sure to use a 15⁄16 in. (24mm) backup wrench to prevent inlet port and bracket damage. Tighten the fittings to 133 inch lbs. (15 Nm).

b. On 1995–98 models, install new quick-connect fitting retainers into the female cavity of the connection, then attach the fitting(s), and pull back on it (them) to confirm the connection.

22. Install the fuel line bracket bolt and tighten to 106 inch lbs. (12 Nm).

23. Connect the negative battery cable, then prime the fuel system by cycling the ignition switch **ON** and **OFF**.

24. Start the engine and check for leaks.

25. Shut the engine, then install the air intake tube and resonator assembly.

TESTING

Electrical Test

▶ See Figures 55 and 56

1. Unplug the injector connector and use an ohmmeter to check the resistance of the injector coil. It should be 11.5–12.5 ohms.

➡**A single MFI injector may short internally, causing a no-start condition, yet not blow the fuse or damage the PCM. If a multimeter is not available to test the injectors, unfasten the injector electrical connectors, one at a time, and attempt to start the engine. If the engine starts with one injector disconnected, the faulty injector has been found.**

2. Make sure all injector harnesses are unplugged, then attach a Noid light (tool SA9194E or equivalent) to the injector connector terminal so that the light can be seen from the driver's seat. Operate the starter and observe the Noid light.

a. If the Noid does not illuminate at all, check for power to the injector(s). The injectors receive power through the pink or pink/black wire from the 10 amp "INJ" fuse in the underhood junction box, any time the ignition switch is **ON**. The PCM operates the injectors by completing the ground circuit through the dark green or dark blue wires.

b. If the test light stays on steadily, check for a short to ground of the wire between the injector connector and the PCM.

c. A pulsing Noid light does not necessarily indicate enough voltage to properly operate the fuel injector. If it is still questionable whether or

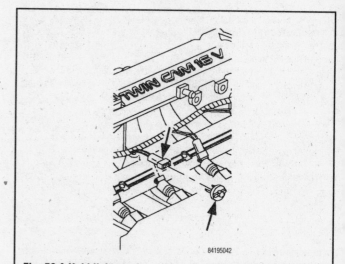

Fig. 56 A Noid light can be used to check if any electrical pulses are reaching the injector

not an injector is pulsing, connect a known good injector to the circuit and tightly grip the injector while an assistant cranks the engine. If injector pulses are felt; the circuit is good.

Injector Pressure Test

▶ See Figure 57

✳✳ CAUTION

The following procedure will produce a small fuel spill and fumes. Make sure there is proper ventilation and take the appropriate fire safety precautions.

1. Connect the test gauge and run the fuel pump and system pressure tests as described in the Fuel Pump tests.

2. For 1991–97 models, perform the following:

a. Disconnect the return line from the fuel rail. Cap the opening at the fuel rail with plug SA9410E or equivalent.

b. Disconnect the fuel pressure regulator vacuum line.

3. Disconnect the fuel line bracket retaining bolt, but leave the supply line attached to the rail assembly and connectors.

4. Remove the bolts holding the fuel rail, then pull the rail and injec-

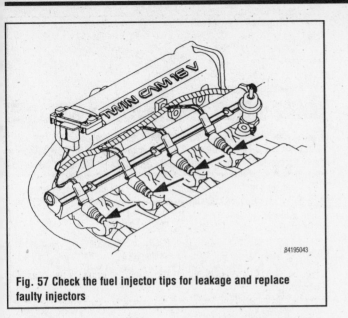

Fig. 57 Check the fuel injector tips for leakage and replace faulty injectors

tors straight out so the injector tips are visible while still in their ports. It may be necessary to rotate the throttle. Use wire to hold the assembly back against the intake manifold.

5. Wipe the injector tips free of fuel or debris and place a clean cloth under them, just contacting the injector tips to spot leaking fuel.

6. Cycle the ignition **ON** and **OFF**, without starting the engine, to build system pressure to a minimum of 46 psi (317 kPa) on the gauge for 1991–97 models or 40 psi (276 kPa) for 1998 models.

7. Allow the pressure to hold for five minutes, then check the cloth for fuel leakage. If a portion of the towel becomes wet with fuel, the faulty injector(s) must be replaced.

8. Bleed off the system pressure, cycle the ignition to rebuild fuel pressure, then allow the system to sit for five minutes. Check the cloth again for signs of leakage and, if any is found, replace the faulty injector(s).

9. Release fuel system pressure again, then remove the gauge and recap the test port.

➡Remember to lubricate the injector O-rings with clean engine oil upon installation of the fuel rail/injector assembly.

10. Install the fuel rail/injector assembly.
11. Attach the fuel line bracket retaining bolt.
12. For 1991–97 models, perform the following:
  a. Attach the fuel pressure regulator vacuum line.
  b. Remove the cap from the fuel rail opening and connect the return line to the fuel rail.
13. Start the engine and check the system for leaks.

Fuel Pressure Regulator

REMOVAL & INSTALLATION

1991–94 Models

▶ See Figures 27, 28, 29, 39, 40, 42, 58, 59, 60 and 61

1. Remove the air intake tube and resonator assembly.
2. Properly relieve the fuel system pressure. Refer to the procedure earlier in this section.
3. Remove the bolts from the fuel line clip.
4. Using a 15/16 in. (24mm) backup wrench to prevent damage to the inlet port and bracket, disconnect the fuel inlet and return lines from the pressure regulator/fuel rail assembly.

➡If difficulty is encountered accessing the fuel return line nut, loosen and/or remove the fuel supply and rail attachment fasteners.

5. Disengage the fuel pressure regulator vacuum hose.
6. Remove the regulator assembly attaching screw, then remove the regulator assembly. Plug the regulator port in the fuel rail with a clean rag to prevent system contamination.
7. Do not remove the O-ring from the retainer in the pressure regulator fuel outlet, unless it is damaged and requires replacing. Always remove and discard the O-ring from the fuel inlet.

To install:
8. Lubricate the new fuel inlet O-ring and, if necessary, the new outlet O-ring with clean engine oil, then install the O-ring(s) in the pressure regulator assembly.
9. Position the regulator assembly to the fuel rail. Coat the threads of the attaching bolt with Loctite® 242 or an equivalent threadlock, then install the bolt and tighten to 106 inch lbs. (12 Nm).
10. Connect the fuel pressure regulator vacuum hose.
11. Connect the fuel inlet and return lines, then tighten the lines to 133 inch lbs. (15 Nm) using a backup wrench to prevent port damage.

Fig. 58 Use a suitable wrench to loosen . . .

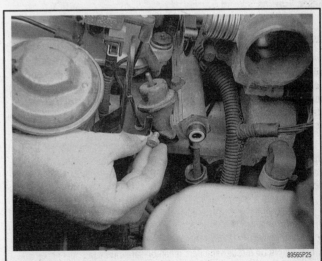

Fig. 59 . . . and remove the mounting screw from the pressure regulator/fuel rail assembly

Fig. 60 Remove the fuel pressure regulator assembly from the fuel rail

Fig. 61 Always remove and discard the O-ring from the fuel inlet

12. Cycle the ignition switch to prime the fuel system, then start the engine and check for fuel leaks.

13. Turn the ignition **OFF** and install the air intake tube and resonator assembly.

1995–97 Models

▶ **See Figures 62, 63, 64, 65 and 66**

1. Remove the air intake tube, resonator assembly and PCV hose.

2. Properly relieve the fuel system pressure. Refer to the procedure in this section.

3. Remove the throttle cable bracket mounting bolts, disconnect the cable from the throttle lever and place over the intake manifold.

4. Disengage the fuel pressure regulator vacuum hose.

5. Remove and discard the retaining clip that secures the pressure regulator to the fuel rail socket.

6. Pull upward to remove the pressure regulator from the fuel rail socket. Do not use any tools for this step, or damage to the plastic socket will occur.

7. Remove and discard the following from the fuel rail socket or pressure regulator:
- Plastic backup ring
- Upper (large) O-ring
- Filter screen
- Lower (small) O-ring

⁂ WARNING

When removing the filter screen, be careful not to drop the lower O-ring into the fuel rail. If dropped into the rail, the O-ring must be retrieved. If not retrieved, a new fuel rail must be used.

8. Inspect the fuel rail socket for foreign material and clean if necessary. Be careful not to scratch the surfaces of the fuel rail that mate to the fuel pressure regulator O-rings.

To install:

9. Be sure to lubricate the upper and lower O-rings with clean engine oil before installing them onto the pressure regulator assembly.

10. When assembling the fuel pressure regulator, install the these new components in the following order:
   a. Plastic backup ring
   b. Upper (large) O-ring
   c. Filter screen
   d. Lower (small) O-ring

11. Position the fuel pressure regulator onto the fuel rail. Using hand pressure only, install the regulator assembly into the fuel rail socket. Carefully rotating the fuel pressure regulator one quarter turn will aid in the installation process.

12. Install the new fuel pressure regulator retaining clip. Be sure to install the retaining clip correctly—the lower side of the clip surrounds the fuel rail socket (down side) and the small side of the clip (top side) surrounds the regulator. The retaining clip must be fully secured over the fuel rail socket tabs and body of the pressure regulator.

13. Position the fuel pressure regulator vacuum hose nipple at a 10° angle to the fuel rail body.

14. With the engine **OFF** and the ignition switch in the **ON** position, pressurize the fuel system and check for leaks.

15. Connect the PCV and fuel pressure regulator vacuum hoses. Be sure that they are properly seated.

16. Place the throttle cable bracket into position and install mounting fasteners. Tighten the mounting fasteners to 19 ft. lbs. (25 Nm).

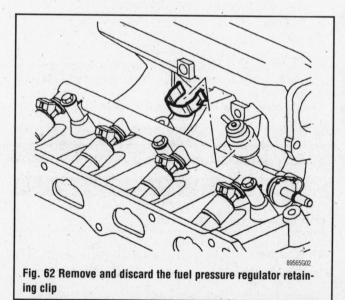

Fig. 62 Remove and discard the fuel pressure regulator retaining clip

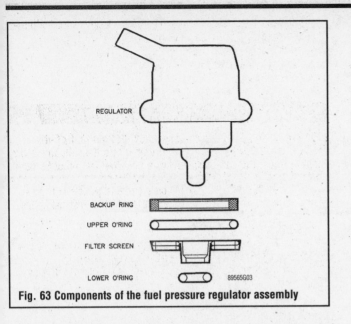

Fig. 63 Components of the fuel pressure regulator assembly

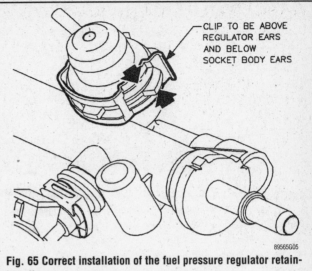

Fig. 65 Correct installation of the fuel pressure regulator retaining clip

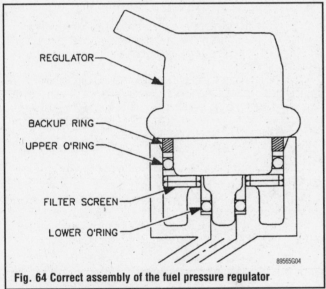

Fig. 64 Correct assembly of the fuel pressure regulator

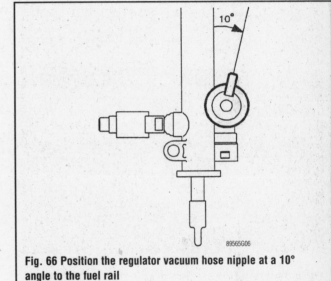

Fig. 66 Position the regulator vacuum hose nipple at a 10° angle to the fuel rail

17. Connect the cable to the throttle lever.

18. Cycle the ignition switch **ON** for 5 seconds and then **OFF** for 10 seconds to prime the fuel system. Perform this procedure 2 more times, then start the engine and check for fuel leaks.

19. Turn the ignition **OFF** and install the air intake tube and resonator assembly.

1998 Models

The fuel pressure regulator on 1998 Saturn models is an integral component of the fuel filter assembly and can only be serviced as a single component. For servicing the fuel filter/pressure regulator, refer to Section 1.

FUEL TANK/FUEL PUMP MODULE

Tank Assembly

REMOVAL & INSTALLATION

1991–97 Models

▶ See Figures 67 thru 74

To prevent excessive fuel spillage, whenever the tank is removed from the vehicle, it should be no more than ¾ full. Removal of the fuel pump module assembly requires removal of the fuel tank.

❊❊ CAUTION

The following procedure will produce a small fuel spill and fumes. Make sure there is proper ventilation and be sure to take the appropriate fire safety precautions.

1. Disconnect the negative battery cable, then properly relieve the fuel system pressure. Refer to the procedure in this section.

2. Remove the fuel filler cap, then raise and support the vehicle safely using jackstands, with the rear of the vehicle approximately 28 in. (711mm) higher than the front (to keep any fuel in the tank forward and away from the fill hose).

3. Clean the area surrounding the filler neck to avoid fuel system contamination, then position a container with a minimum 12 in. (300mm) diameter opening under the filler neck to catch any escaping fuel. Loosen the filler neck clamp at the rear of the fuel tank, wrap a shop rag around the neck tube and carefully remove the tube from the tank.

4. For 1991–92 vehicles, inside the filler on the tank is a check ball to prevent fuel from flowing out of the filler in the event of vehicle roll-over. If the tank needs to be drained on these models, the check ball must be dislodged. Use the large round end of a ½ in. drive ratchet extension (which is at least 18 in. long) to push the check ball into the fuel tank. For 1993–97 vehicles, the fuel check ball was relocated to the filler pipe, so that the pipe may simply be removed to allow siphoning. These filler pipes can be installed in 1991–92 vehicles, but the check ball must first be removed from the tank.

➡Once the check ball has been knocked into the fuel tank on 1991–92 models, the tank must be removed from the vehicle. The pump module must then be removed from the tank in order to reinstall the check ball.

5. If the check ball was dislodged (1991–92 models) or the fuel neck removed (1993–97 models) for siphoning, use a clean length of hose and/or an appropriate hand pump to siphon or pump the fuel from the tank and into an approved gasoline container.

6. Remove the filler neck bracket fastener at the left side of the rear frame rail, then loosen the fuel vent hose retaining clamp at the tank.

➡It is easier to remove hoses from the tank than to pull them from the steel vent and fill tubes.

7. Disconnect the fuel pressure and return line quick-connects by pinching the 2 plastic tangs together, then grasp both ends of one fuel line connection and twist ¼ turn in each direction while pulling them apart. Disconnect the fuel vent hose by holding the line and by pushing on the rubber connector with a small open end wrench.

➡Do not allow the fuel tank retaining straps to become bent during tank removal, otherwise strap damage and/or breakage may occur. Always utilize an assistant when removing the fuel tank to prevent damage.

8. With the aid of an assistant, remove the 2 support strap fasteners at the rear of the tank, then lower the tank and support panel approximately 8 in. (203mm). Reach upward and unplug the electrical connector from the top of the tank, then remove the tank and support panel from the vehicle.

9. If the check ball was removed or fuel pump module replacement/service is necessary:
   a. Clean the area surrounding the fuel pump module and spray the cam lockring tangs with a suitable penetrating oil to loosen the fitting.
   b. Using SA9156E, or an equivalent fuel module lockring removal tool, and a ½ in. breaker bar of approximately 18 in. (457mm) length, remove the pump unit locking ring from the tank. Attempting to use a 12 in. or shorter breaker bar may cause lockring damage.
   c. Lift and tilt the unit out at a 45 degree angle, being careful not to bend the sending unit float arm. Remove and discard the unit-to-tank O-ring.

10. The sending unit is the only portion of the module that may be serviced. The filter may be cleaned with mineral spirits, but must be replaced as an assembly with the module if damaged. If necessary, remove the sending unit from the module as follows:
   a. Unplug the 2 electrical connections using needlenose pliers or by pressing down the locking tab and pulling the connectors from the terminal.
   b. Using a small suitable tool, push in on the sender assembly attaching tang, then lift upward and remove the sender.
   c. Late 1992 and all 1993–94 vehicles use a brass float instead of plastic. Brass floats may be serviced. Use a ¼ in. flat-tipped screwdriver to carefully pry the float from the wire loop. Do not bend the float arm or deform the wire loop.

To install:
11. If removed, pinch a brass float onto the float arm wire loop and/or install the sending unit to the pump module by positioning the tang in the locator slot and snapping the unit into place. Attach the 2 sending unit electrical connectors to their terminals.

12. If removed for service or access to the check ball, install the fuel pump assembly:
   a. Before installing the pump module, remove the filler check ball and examine it for cracks or damage. Reach into the tank through the pump unit hole and carefully push the check ball into the filler tube opening from the inside. If necessary, carefully spread the plastic ball supports to ease installation.
   b. Install a new O-ring to the opening in the top of the fuel tank, then carefully insert the pump module into the tank at a 45 degree angle to prevent sending unit and float damage. The filter and flow arm must be directed toward the front of the tank.
   c. Align the pump locator tabs with the fuel tank slots, then install the cam lockring using the ring service tool.

13. With the aid of an assistant, position the tank and support panel so the wires can be connected to the module, install the module electrical connector, then secure the tank in place using the retaining straps. Tighten the strap bolts to 35 ft. lbs. (47 Nm).

14. Loosely install the filler tube, vent lines and clamps. Align the filler neck and tank so the fender will not be deflected or pushed outward by the hose. When everything is properly positioned, install the fill neck tube and bracket fastener. Vent lines should be installed into the rubber boot until the tube white marks align with the side of the boot.

15. Tighten the clamps to 18 inch lbs. (2 Nm) and tighten the fill neck bracket fastener to 53 inch lbs. (6 Nm). If the clamps must be replaced, original equipment or equivalent parts must be used, because the original parts are designed to prevent hose damage.

16. Lubricate the male ends of the fuel supply and return quick-connect fittings with a few drops of clean engine oil. Push the connectors together until the retaining tabs snap into place, then pull on opposite ends of each connection to verify that the connection is secure.

17. Remove the jackstands and carefully lower the vehicle, then install the fuel filler cap and connect the negative battery cable.

18. Prime the fuel system and check for leaks:
   a. Turn the ignition **ON** for 5 seconds, then **OFF** for 10 seconds.
   b. Repeat the **ON/OFF** cycle 2 more times.
   c. Crank the engine until it starts.
   d. If the engine does not readily start, repeat Steps a–e.
   e. Run the engine and check for leaks.

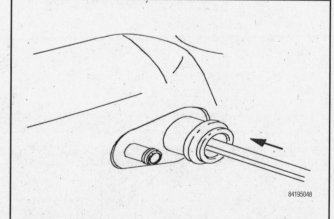

84195048

Fig. 67 For 1991–92 vehicles with the check ball installed in the fuel tank, use an 18 in. or longer ratchet extension to dislodge the fuel filler check ball, knocking it into the tank. The tank will need to be removed, in order to retrieve the ball for installation

Fig. 68 View of the fuel filler neck—1992 SC model shown

84195049

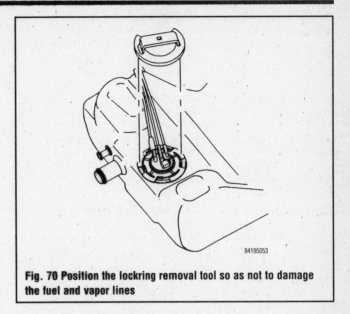

84195053

Fig. 70 Position the lockring removal tool so as not to damage the fuel and vapor lines

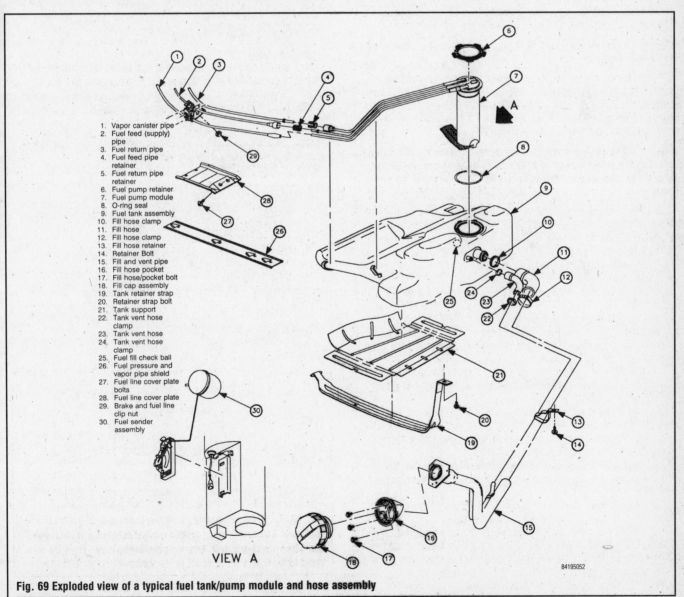

1. Vapor canister pipe
2. Fuel feed (supply) pipe
3. Fuel return pipe
4. Fuel feed pipe retainer
5. Fuel return pipe retainer
6. Fuel pump retainer
7. Fuel pump module
8. O-ring seal
9. Fuel tank assembly
10. Fill hose clamp
11. Fill hose
12. Fill hose clamp
13. Fill hose retainer
14. Retainer Bolt
15. Fill and vent pipe
16. Fill hose pocket
17. Fill hose/pocket bolt
18. Fill cap assembly
19. Tank retainer strap
20. Retainer strap bolt
21. Tank support
22. Tank vent hose clamp
23. Tank vent hose
24. Tank vent hose clamp
25. Fuel fill check ball
26. Fuel pressure and vapor pipe shield
27. Fuel line cover plate bolts
28. Fuel line cover plate
29. Brake and fuel line clip nut
30. Fuel sender assembly

VIEW A

84195052

Fig. 69 Exploded view of a typical fuel tank/pump module and hose assembly

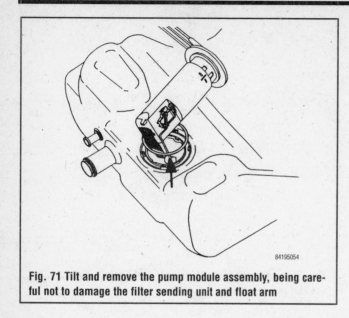

Fig. 71 Tilt and remove the pump module assembly, being careful not to damage the filter sending unit and float arm

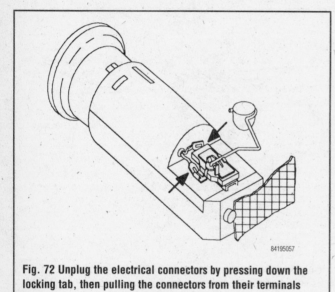

Fig. 72 Unplug the electrical connectors by pressing down the locking tab, then pulling the connectors from their terminals

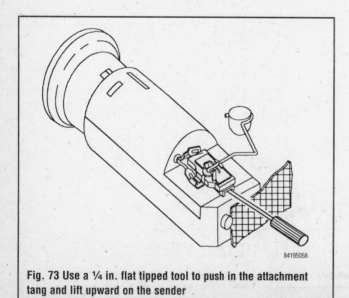

Fig. 73 Use a ¼ in. flat tipped tool to push in the attachment tang and lift upward on the sender

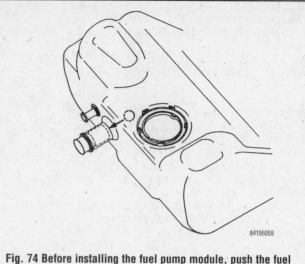

Fig. 74 Before installing the fuel pump module, push the fuel check ball into position

1998 Models

▶ See Figures 69, 70, 71, 72, and 75 thru 79

To prevent excessive fuel spillage, whenever the tank is removed from the vehicle it should be no more than ½ full. Removal of the fuel pump module assembly requires the removal of the fuel tank.

❋❋ CAUTION

The following procedure will produce a small fuel spill and fumes. Make sure there is proper ventilation and be sure to take the appropriate fire safety precautions.

1. Disconnect the negative battery cable, then properly relieve the fuel system pressure. Refer to the procedure in this section.
2. Remove the fuel filler cap and rubber closeout grommet (inside fuel filler door). Remove the mounting fastener (using a T-30 Torx® bit) at the upper end of the fuel tank filler pipe.
3. Raise and support the vehicle safely using jackstands, with the rear of the vehicle at a comfortable working height.
4. Remove the wheel house inner fender liner.
5. Disengage the wiring harness connector from the EVAP canister vent solenoid.
6. Remove the fuel tank filler pipe lower bracket mounting fastener located at the underbody left side rail.
7. Disconnect the EVAP canister vent tube at the ⅝ in. quick-connect fitting.
8. Loosen the fuel tank filler pipe hose clamp located closest to the tank.
9. Disconnect the filler pipe hose from the fuel tank, making sure that it is as straight and level as possible so as to prevent it from falling into the fuel tank. Remove the filler pipe from the vehicle. Be careful when disconnecting the filler hose as a residual amount of fuel may be left sitting in the filler pipe due to the inlet check valve on the fuel tank.
10. If there is more than 3 gallons of fuel in the tank, insert a siphon hose into the filler neck and drain into an approved gasoline container.

❋❋ CAUTION

Whenever fuel line fittings are loosened or disconnected, always wrap a shop towel around the fitting and have a suitable container available to catch any fuel spill.

11. Disengage the fuel feed line from the outlet tube of the filter/pressure regulator. If necessary, refer to Section 1 for service information on the fuel filter/pressure regulator.

12. Disengage the fuel vapor/canister purge line at the fitting adjacent to the filter/pressure regulator.

13. Remove the fuel filter/pressure regulator bracket from under the brake lines.

14. With the aid of an assistant, remove the 2 fuel tank retaining strap mounting bolts. Lower the tank just enough to disengage the 2 electrical connectors from the fuel pump and pressure sensor.

15. Carefully remove the fuel tank from the vehicle.

16. If fuel pump module replacement/service is necessary:

    a. Clean the area surrounding the fuel pump module and spray the cam lockring tangs with a suitable penetrating oil to loosen the fitting.

    b. Using SA9156E, or an equivalent fuel module lockring removal tool, and a ½ in. breaker bar of approximately 18 in. (457mm) length, remove the pump unit locking ring from the tank. Attempting to use a 12 in. or shorter breaker bar may cause lockring damage.

    c. Lift and tilt the unit out at a 45 degree angle, being careful not to bend the sending unit float arm. Remove and discard the unit to tank O-ring.

17. The sending unit is the only portion of the module that may be serviced. The filter may be cleaned with mineral spirits, but must be replaced

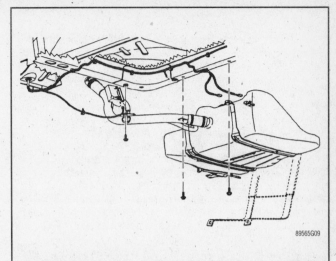

Fig. 77 With the aid of an assistant, lower the fuel tank enough to disengage the pump and pressure sensor wiring harnesses

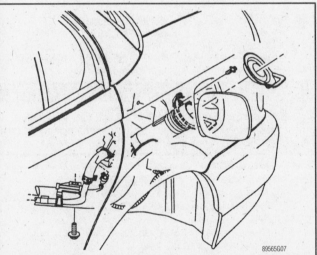

Fig. 75 Remove the rubber closeout grommet and mounting fastener at the upper end of the filler pipe

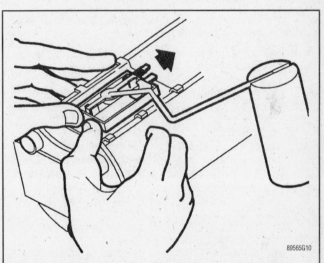

Fig. 78 Apply firm upward pressure at the bottom of the fuel sending unit with both thumbs to disengage the attachment clip

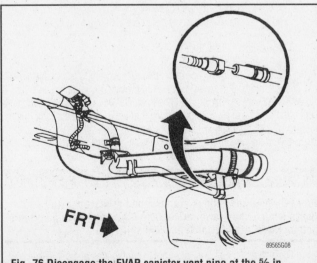

Fig. 76 Disengage the EVAP canister vent pipe at the ⅝ in. quick-connect fitting

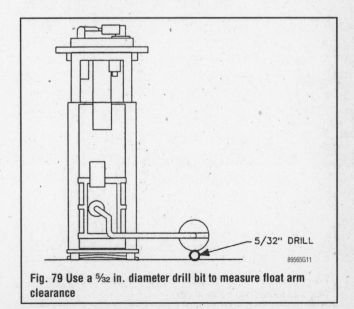

Fig. 79 Use a 5/32 in. diameter drill bit to measure float arm clearance

as an assembly with the module if damaged. If necessary, remove the sending unit from the module as follows:

   a. Unplug the electrical connector from the fuel level sending unit by disengaging the locking tab and pulling the connector from the terminal.

   b. Holding the fuel pump module in both hands, and applying firm upward pressure at the bottom of the fuel sending unit using both thumbs, push the sender toward the top of the module to disengage the bottom attachment clip.

To install:

18. Install the sending unit to the pump module by positioning the attachment clip in the locator slot and snapping the component into place. Verify that the sending unit float arm has the correct relationship to the fuel pump module by holding the component up on a flat, horizontal surface and measuring. Using a ⁵⁄₃₂ in. diameter drill bit, it must easily pass between the float and the horizontal surface with no more than ¹⁄₁₆ in. clearance. If bending the float arm is required to meet the clearance specification, it must be performed at the 90° bend near the level sender while supporting the short section of the arm.

19. Connect the wiring harness from the fuel pump module to the sending unit. Press firmly to engage.

20. Install the fuel pump assembly as follows:

   a. Before installing the pump module, inspect the fuel tank and clean the seal groove of any debris or foreign matter.

➡ **The correct fuel pump module O-ring seal is green in color. It is incorrect to use the older black seal.**

   b. Install a new O-ring seal to the opening in the top of the fuel tank, then carefully insert the pump module into the tank at a 45 degree angle to prevent sending unit and float damage.

   c. Align the pump locator tabs with the fuel tank slots by rotating the module 90° counterclockwise. The lines from the pump unit should face the 10 o'clock position.

   d. Install the cam lockring using the ring service tool.

21. Connect the fuel pump vapor line to the tank vent pipe.

22. Connect the feed and return lines to the fuel filter/pressure regulator.

23. Place the feed, return and EVAP canister purge lines into the fuel tank retaining clip and snap closed.

24. Install the loose retainer clip around the fuel feed, return and purge lines and snap shut.

25. With the aid of an assistant, position the fuel tank so the wires can be connected to the pump module and pressure sensor.

26. Properly install the fuel tank under the vehicle by making certain that the small white locator button on the left side of the tank is positioned tightly against the left side rail.

27. Install the retaining straps and shield. Tighten the strap bolts to 35 ft. lbs. (47 Nm).

28. Place the fuel filter/pressure regulator in position and install the mounting screws in the bracket. Tighten the mounting screws to 71 inch lbs. (8 Nm).

29. Install a new retainer into the female portion of the 90° quick-connect fitting of the underbody fuel line.

30. Lubricate the outlet end of the filter/pressure regulator with clean engine oil and connect to the underbody fuel feed line.

31. Be sure that the fuel filler check valve is in correct position at the end of the filler pipe. Using plain water, lightly wipe the outside of the fuel tank inlet connector.

32. Position the fuel tank filler pipe into the wheel opening with the top of the pipe protruding out of the fender/filler door opening.

33. Insert the filler pipe into the fuel tank opening and loosely install the lower bracket attachment screw. Be sure that the EVAP vent solenoid pipe is installed in the correct position on the filler pipe bracket.

34. Lower the vehicle and install the fuel tank filler pipe upper bracket mounting fastener. Tighten the mounting fastener to 31–35 inch lbs. (3.5–4.5 Nm).

35. Install the rubber closeout grommet to the filler pipe and install the fuel cap.

36. Raise the vehicle to a comfortable working height.

37. The fuel pipe connecting hose should be installed to within ¼ in. (6mm) of the stops on the fuel tank inlet connector and the hose clamp should be positioned within ³⁄₁₆ in. (4mm) of the end of the connecting hose. Tighten the hose clamp to 35 inch lbs. (4 Nm).

38. Connect the EVAP canister vent pipe to the canister vent hose.

39. Tighten the lower mounting screw on the filler pipe to the underbody. Tighten the mounting screw to 71 inch lbs. (8 Nm).

40. Engage the wiring harness connector to the EVAP vent solenoid.

41. Install the wheel house inner fender liner.

42. Remove the jackstands and carefully lower the vehicle, then install the fuel filler cap and connect the negative battery cable.

43. Prime the fuel system and check for leaks:

   a. Turn the ignition **ON** for 5 seconds, then **OFF** for 10 seconds.

   b. Repeat the **ON/OFF** cycle 2 more times.

   c. Crank the engine until it starts.

   d. If the engine does not readily start, repeat Steps a–c.

44. Run the engine and check for leaks.

6

CHASSIS ELECTRICAL

UNDERSTANDING AND TROUBLESHOOTING ELECTRICAL SYSTEMS

Basic Electrical Theory

♦ **See Figure 1**

For any 12 volt, negative ground, electrical system to operate, the electricity must travel in a complete circuit. This simply means that current (power) from the positive terminal (+) of the battery must eventually return to the negative terminal (-) of the battery. Along the way, this current will travel through wires, fuses, switches and components. If, for any reason, the flow of current through the circuit is interrupted, the component fed by that circuit will cease to function properly.

Perhaps the easiest way to visualize a circuit is to think of connecting a light bulb (with two wires attached to it) to the battery—one wire attached to the negative (-) terminal of the battery and the other wire to the positive (+) terminal. With the two wires touching the battery terminals, the circuit would be complete and the light bulb would illuminate. Electricity would follow a path from the battery to the bulb and back to the battery. It's easy to see that with longer wires on our light bulb, it could be mounted anywhere. Further, one wire could be fitted with a switch so that the light could be turned on and off.

The normal automotive circuit differs from this simple example in two ways. First, instead of having a return wire from the bulb to the battery, the current travels through the chassis of the vehicle. Since the negative (-) battery cable is attached to the chassis and the chassis is made of electrically conductive metal, the chassis of the vehicle can serve as a ground wire to complete the circuit. Secondly, most automotive circuits contain multiple components which receive power from a single circuit. This lessens the amount of wire needed to power components on the vehicle.

THE WATER ANALOGY

Electricity is the flow of electrons—hypothetical particles thought to constitute the basic "stuff" of electricity. Many people have been taught electrical theory using an analogy with water. In a comparison with water flowing through a pipe, the electrons would be the water.

The flow of electricity can be measured much like the flow of water through a pipe. The unit of measurement used is amperes, frequently abbreviated as amps (a). When connected to a circuit, an ammeter will measure the actual amount of current flowing through the circuit. When relatively few electrons flow through a circuit, the amperage is low. When many electrons flow, the amperage is high.

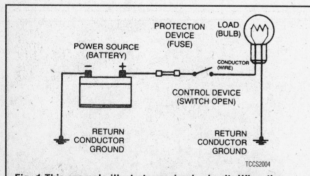

Fig. 1 This example illustrates a simple circuit. When the switch is closed, power from the positive (+) battery terminal flows through the fuse and the switch, and then to the light bulb. The light illuminates and the circuit is completed through the ground wire back to the negative (-) battery terminal. In reality, the two ground points shown in the illustration are attached to the metal chassis of the vehicle, which completes the circuit back to the battery

Just as water pressure is measured in units such as pounds per square inch (psi), electrical pressure is measured in units called volts (v). When a voltmeter is connected to a circuit, it is measuring the electrical pressure. The higher the voltage, the more current will flow through the circuit. The lower the voltage, the less current will flow.

While increasing the voltage in a circuit will increase the flow of current, the actual flow depends not only on voltage, but also on the resistance of the circuit. Resistance is the amount of force necessary to push the current through the circuit. The standard unit for measuring resistance is an ohm (W or omega). Resistance in a circuit varies depending on the amount and type of components used in the circuit. The main factors which determine resistance are:

• Material—some materials have more resistance than others. Those with high resistance are said to be insulators. Rubber is one of the best insulators available, as it allows little current to pass. Low resistance materials are said to be conductors. Copper wire is among the best conductors. Most vehicle wiring is made of copper.

• Size—the larger the wire size being used, the less resistance the wire will have. This is why components which use large amounts of electricity usually have large wires supplying current to them.

• Length—for a given thickness of wire, the longer the wire, the greater the resistance. The shorter the wire, the less the resistance. When determining the proper wire for a circuit, both size and length must be considered to design a circuit that can handle the current needs of the component.

• Temperature—with many materials, the higher the temperature, the greater the resistance. This principle is used in many of the sensors on the engine.

OHM'S LAW

The preceding definitions may lead the reader into believing that there is no relationship between current, voltage and resistance. Nothing can be further from the truth. The relationship between current, voltage and resistance can be summed up by a statement known as Ohm's law.

Voltage (E) is equal to amperage (I) times resistance (R): $E = I \times R$

Other forms of the formula are $R = E/I$ and $I = E/R$

In each of these formulas, E is the voltage in volts, I is the current in amps and R is the resistance in ohms. The basic point to remember is that as the resistance of a circuit goes up, the amount of current that flows in the circuit will go down, if voltage remains the same.

Electrical Components

POWER SOURCE

The power source for 12 volt automotive electrical systems is the battery. In most modern vehicles, the battery is a lead/acid electrochemical device consisting of six 2 volt subsections (cells) connected in series, so that the unit is capable of producing approximately 12 volts of electrical pressure. Each subsection consists of a series of positive and negative plates held a short distance apart in a solution of sulfuric acid and water.

The two types of plates are of dissimilar metals. This sets up a chemical reaction, and it is this reaction which produces current flow from the battery when its positive and negative terminals are connected to an electrical load. The power removed from the battery is replaced by the alternator, which forces electrons back through the battery, reversing the normal flow, and restoring the battery to its original chemical state.

GROUND

Two types of grounds are used in automotive electric circuits. Direct ground components are grounded through their mounting points. All other components use some sort of ground wire which is attached to the body or chassis of the vehicle. The electrical current runs through the chassis of the vehicle and returns to the battery through the ground (-) cable; if you look, you'll see that the battery ground cable connects between the battery and the body or chassis of the vehicle.

➡ **It should be noted that a good percentage of electrical problems can be traced to bad grounds.**

PROTECTIVE DEVICES

▶ See Figure 2

It is possible for large surges of current to pass through the electrical system of your vehicle. If this surge of current were to reach the load in the circuit, it could burn it out or severely damage it. To prevent this, fuses, circuit breakers and/or fusible links are connected into the supply wires of the electrical system. These items are nothing more than a built-in weak spot in the system. When an abnormal amount of current flows through the system, these protective devices work as follows to protect the circuit:

• Fuse—when an excessive electrical current passes through a fuse, the fuse "blows" (the conductor melts) and opens the circuit, preventing the passage of current.

• Circuit Breaker—a circuit breaker is basically a self-repairing fuse. It will open the circuit in the same fashion as a fuse, but when the surge subsides, the circuit breaker can be reset and does not need replacement.

• Fusible Link—a fusible link (fuse link or main link) is a short length of special, Hypalon high temperature insulated wire that acts as a fuse. When an excessive electrical current passes through a fusible link, the thin gauge wire inside the link melts, creating an intentional open to protect the circuit. To repair the circuit, the link must be replaced. Some newer type fusible links are housed in plug-in modules, which are simply replaced like a fuse, while older type fusible links must be cut and spliced if they melt. Since this link is very early in the electrical path, it's the first place to look if nothing on the vehicle works, but the battery seems to be charged and is properly connected.

✳✳ CAUTION

Always replace fuses, circuit breakers and fusible links with identically rated components. Under no circumstances should a component of higher or lower amperage rating be substituted.

SWITCHES & RELAYS

▶ See Figures 3 and 4

Switches are used in electrical circuits to control the passage of current. The most common use is to open and close circuits between the battery and the various electric devices in the system. Switches are rated according to the amount of amperage they can handle. If a sufficient amperage rated switch is not used in a circuit, the switch could overload and cause damage.

Some electrical components which require a large amount of current to operate use a special switch called a relay. Since these circuits carry a large amount of current, the thickness of the wire in the circuit is also greater. If this large wire were connected from the load to the control switch on the dashboard, the switch would have to carry the high amperage load and the dash would be twice as large to accommodate the increased size of the wiring harness. To prevent these problems, a relay is used.

Relays are composed of a coil and a switch. These two components are linked together so that when one operates, the other operates at the same time. The large wires in the circuit are connected from the battery to one side of the relay switch and from the opposite side of the relay switch to the load. Most relays are normally open, preventing current from passing through the circuit. Additional, smaller wires are connected from the relay coil to the control switch for the circuit and from the opposite side of the relay coil to ground. When the control switch is turned on, it grounds the smaller wire to the relay coil, causing the coil to operate. The coil pulls the relay switch closed, sending power to the component without routing it through the inside of the vehicle. Some common circuits which may use relays are the horn, headlights, starter, electric fuel pump and rear window defogger systems.

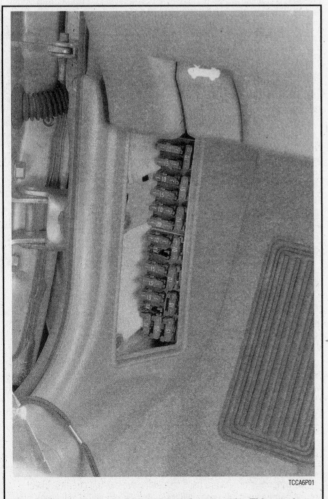

TCCA6P01

Fig. 2 Most vehicles use one or more fuse panels. This one is located in the driver's side kick panel

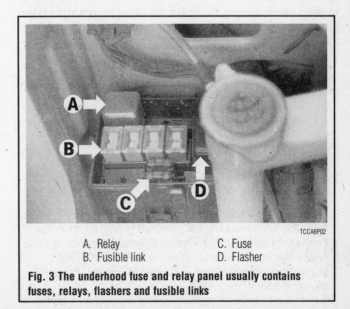

TCCA6P02

A. Relay   C. Fuse
B. Fusible link   D. Flasher

Fig. 3 The underhood fuse and relay panel usually contains fuses, relays, flashers and fusible links

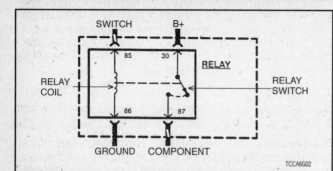

Fig. 4 Relays are composed of a coil and a switch. These two components are linked together so that when one operates, the other operates at the same time. The large wires in the circuit are connected from the battery to one side of the relay switch (B+) and from the opposite side of the relay switch to the load (component). Smaller wires are connected from the relay coil to the control switch for the circuit and from the opposite side of the relay coil to ground

LOAD

Every complete circuit must include a "load" (something to use the electricity coming from the source). Without this load, the battery would attempt to deliver its entire power supply from one pole to another. The electricity would take a short cut to ground and cause a great amount of damage to other components in the circuit by developing a tremendous amount of heat. This condition could develop sufficient heat to melt the insulation on all the surrounding wires and reduce a multiple wire cable to a lump of plastic and copper.

WIRING & HARNESSES

The average automobile contains about ½ mile of wiring, with hundreds of individual connections. To protect the many wires from damage and to keep them from becoming a confusing tangle, they are organized into bundles, enclosed in plastic or taped together and called wiring harnesses. Different harnesses serve different parts of the vehicle. Individual wires are color coded to help trace them through a harness where sections are hidden from view.

Automotive wiring or circuit conductors can be either single strand wire, multi-strand wire or printed circuitry. Single strand wire has a solid metal core and is usually used inside such components as alternators, motors, relays and other devices. Multi-strand wire has a core made of many small strands of wire twisted together into a single conductor. Most of the wiring in an automotive electrical system is made up of multi-strand wire, either as a single conductor or grouped together in a harness. All wiring is color coded on the insulator, either as a solid color or as a colored wire with an identification stripe. A printed circuit is a thin film of copper or other conductor that is printed on an insulator backing. Occasionally, a printed circuit is sandwiched between two sheets of plastic for more protection and flexibility. A complete printed circuit, consisting of conductors, insulating material and connectors for lamps or other components is called a printed circuit board. Printed circuitry is used in place of individual wires or harnesses in places where space is limited, such as behind instrument panels.

Since automotive electrical systems are very sensitive to changes in resistance, the selection of properly sized wires is critical when systems are repaired. A loose or corroded connection or a replacement wire that is too small for the circuit will add extra resistance and an additional voltage drop to the circuit.

The wire gauge number is an expression of the cross-section area of the conductor. The most common system for expressing wire size is the American Wire Gauge (AWG) system. As gauge number increases, area decreases and the wire becomes smaller. An 18 gauge wire is smaller than a 4 gauge wire. A wire with a higher gauge number will carry less current than a wire with a lower gauge number. Gauge wire size refers to the size of the strands of the conductor, not the size of the complete wire. It is possible, therefore, to have two wires of the same gauge with different diameters because one may have thicker insulation than the other.

12 volt automotive electrical systems generally use 10, 12, 14, 16 and 18 gauge wire. Main power distribution circuits and larger accessories usually use 10 and 12 gauge wire. Battery cables are usually 4 or 6 gauge, although 1 and 2 gauge wires are occasionally used.

It is essential to understand how a circuit works before trying to figure out why it doesn't. An electrical schematic shows the electrical current paths when a circuit is operating properly. Schematics break the entire electrical system down into individual circuits. In a schematic, no attempt is made to represent wiring and components as they physically appear on the vehicle; switches and other components are shown as simply as possible. Face views of harness connectors show the cavity or terminal locations in all multi-pin connectors to help locate test points.

CONNECTORS

♦ See Figures 5 and 6

Three types of connectors are commonly used in automotive applications—weatherproof, molded and hard shell.

• Weatherproof—these connectors are most commonly used in the engine compartment or where the connector is exposed to the elements. Terminals are protected against moisture and dirt by sealing rings which provide a weathertight seal. All repairs require the use of a special terminal and the tool required to service it. Unlike standard blade type terminals, these weatherproof terminals cannot be straightened once they are bent. Make certain that the connectors are properly seated and all of the sealing rings are in place when connecting leads.

• Molded—these connectors require complete replacement of the connector if found to be defective. This means splicing a new connector assembly into the harness. All splices should be soldered to insure proper contact. Use care when probing the connections or replacing terminals in them, as it is possible to create a short circuit between opposite terminals. If this happens to the wrong terminal pair, it is possible to damage certain components. Always use jumper wires between connectors for circuit checking and NEVER probe through weatherproof seals.

• Hard Shell—unlike molded connectors, the terminal contacts in hardshell connectors can be replaced. Replacement usually involves the use of a special terminal removal tool that depresses the locking tangs (barbs) on the connector terminal and allows the connector to be removed from the rear of the shell. The connector shell should be replaced if it shows any evidence of burning, melting, cracks, or breaks. Replace individual terminals that are burnt, corroded, distorted or loose.

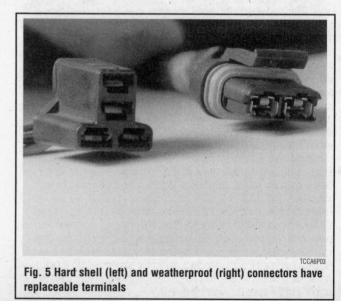

Fig. 5 Hard shell (left) and weatherproof (right) connectors have replaceable terminals

TCCA6P04

Fig. 6 Weatherproof connectors are most commonly used in the engine compartment or where the connector is exposed to the elements

Test Equipment

Pinpointing the exact cause of trouble in an electrical circuit is most times accomplished by the use of special test equipment. The following describes different types of commonly used test equipment and briefly explains how to use them in diagnosis. In addition to the information covered below, the tool manufacturer's instructions booklet (provided with the tester) should be read and clearly understood before attempting any test procedures.

JUMPER WIRES

❊❊ CAUTION

Never use jumper wires made from a thinner gauge wire than the circuit being tested. If the jumper wire is of too small a gauge, it may overheat and possibly melt. Never use jumpers to bypass high resistance loads in a circuit. Bypassing resistances, in effect, creates a short circuit. This may, in turn, cause damage and fire. Jumper wires should only be used to bypass lengths of wire.

Jumper wires are simple, yet extremely valuable, pieces of test equipment. They are basically test wires which are used to bypass sections of a circuit. Although jumper wires can be purchased, they are usually fabricated from lengths of standard automotive wire and whatever type of connector (alligator clip, spade connector or pin connector) that is required for the particular application being tested. In cramped, hard-to-reach areas, it is advisable to have insulated boots over the jumper wire terminals in order to prevent accidental grounding. It is also advisable to include a standard automotive fuse in any jumper wire. This is commonly referred to as a "fused jumper". By inserting an in-line fuse holder between a set of test leads, a fused jumper wire can be used for bypassing open circuits. Use a 5 amp fuse to provide protection against voltage spikes.

Jumper wires are used primarily to locate open electrical circuits, on either the ground (-) side of the circuit or on the power (+) side. If an electrical component fails to operate, connect the jumper wire between the component and a good ground. If the component operates only with the jumper installed, the ground circuit is open. If the ground circuit is good, but the component does not operate, the circuit between the power feed and component may be open. By moving the jumper wire successively back from the component toward the power source, you can isolate the area of the circuit where the open is located. When the component stops functioning, or the power is cut off, the open is in the segment of wire between the jumper and the point previously tested.

You can sometimes connect the jumper wire directly from the battery to the "hot" terminal of the component, but first make sure the component uses 12 volts in operation. Some electrical components, such as fuel injectors, are designed to operate on about 4 volts, and running 12 volts directly to these components will cause damage.

TEST LIGHTS

▶ See Figure 7

The test light is used to check circuits and components while electrical current is flowing through them. It is used for voltage and ground tests. To use a 12 volt test light, connect the ground clip to a good ground and probe wherever necessary with the pick. The test light will illuminate when voltage is detected. This does not necessarily mean that 12 volts (or any particular amount of voltage) is present; it only means that some voltage is present. It is advisable before using the test light to touch its ground clip and probe across the battery posts or terminals to make sure the light is operating properly.

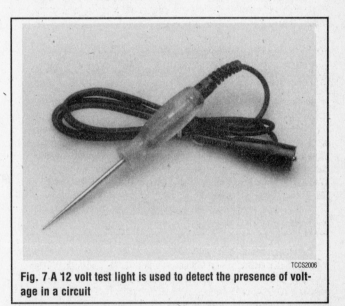

TCCS2006

Fig. 7 A 12 volt test light is used to detect the presence of voltage in a circuit

❊❊ WARNING

Do not use a test light to probe electronic ignition spark plug or coil wires. Never use a pick-type test light to probe wiring on computer controlled systems unless specifically instructed to do so. Any wire insulation that is pierced by the test light probe should be taped and sealed with silicone after testing.

Like the jumper wire, the 12 volt test light is used to isolate opens in circuits. But, whereas the jumper wire is used to bypass the open to operate the load, the 12 volt test light is used to locate the presence of voltage in a circuit. If the test light illuminates, there is power up to that point in the circuit; if the test light does not illuminate, there is an open circuit (no power). Move the test light in successive steps back toward the power source until the light in the handle illuminates. The open is between the probe and a point which was previously probed.

The self-powered test light is similar in design to the 12 volt test light, but contains a 1.5 volt penlight battery in the handle. It is most often used in place of a multimeter to check for open or short circuits when power is isolated from the circuit (continuity test).

The battery in a self-powered test light does not provide much current. A weak battery may not provide enough power to illuminate the test light even when a complete circuit is made (especially if there is high resistance in the circuit). Always make sure that the test battery is strong. To check the battery, briefly touch the ground clip to the probe; if the light glows brightly, the battery is strong enough for testing.

➡ A self-powered test light should not be used on any computer controlled system or component. The small amount of electricity transmitted by the test light is enough to damage many electronic automotive components.

MULTIMETERS

Multimeters are an extremely useful tool for troubleshooting electrical problems. They can be purchased in either analog or digital form and have a price range to suit any budget. A multimeter is a voltmeter, ammeter and ohmmeter (along with other features) combined into one instrument. It is often used when testing solid state circuits because of its high input impedance (usually 10 megaohms or more). A brief description of the multimeter main test functions follows:

• Voltmeter—the voltmeter is used to measure voltage at any point in a circuit, or to measure the voltage drop across any part of a circuit. Voltmeters usually have various scales and a selector switch to allow the reading of different voltage ranges. The voltmeter has a positive and a negative lead. To avoid damage to the meter, always connect the negative lead to the negative (-) side of the circuit (to ground or nearest the ground side of the circuit) and connect the positive lead to the positive (+) side of the circuit (to the power source or the nearest power source). Note that the negative voltmeter lead will always be black and that the positive voltmeter will always be some color other than black (usually red).

• Ohmmeter—the ohmmeter is designed to read resistance (measured in ohms) in a circuit or component. All ohmmeters will have a selector switch which permits the measurement of different ranges of resistance (usually the selector switch allows the multiplication of the meter reading by 10, 100, 1,000 and 10,000). Since the meters are powered by an internal battery, the ohmmeter can be used as a self-powered test light. When the ohmmeter is connected, current from the ohmmeter flows through the circuit or component being tested. Since the ohmmeter's internal resistance and voltage are known values, the amount of current flow through the meter depends on the resistance of the circuit or component being tested. The ohmmeter can also be used to perform a continuity test for suspected open circuits. In using the meter for making continuity checks, do not be concerned with the actual resistance readings. Zero resistance, or any ohm reading, indicates continuity in the circuit. Infinite resistance indicates an opening in the circuit. A high resistance reading where there should be none indicates a problem in the circuit. Checks for short circuits are made in the same manner as checks for open circuits, except that the circuit must be isolated from both power and normal ground. Infinite resistance indicates no continuity to ground, while zero resistance indicates a dead short to ground.

❊❊ WARNING

Never use an ohmmeter to check the resistance of a component or wire while there is voltage applied to the circuit.

• Ammeter—an ammeter measures the amount of current flowing through a circuit in units called amperes or amps. At normal operating voltage, most circuits have a characteristic amount of amperes, called "current draw" which can be measured using an ammeter. By referring to a specified current draw rating, then measuring the amperes and comparing the two values, one can determine what is happening within the circuit to aid in diagnosis. An open circuit, for example, will not allow any current to flow, so the ammeter reading will be zero. A damaged component or circuit will have an increased current draw, so the reading will be high. The ammeter is always connected in series with the circuit being tested. All of the current that normally flows through the circuit must also flow through the ammeter; if there is any other path for the current to follow, the ammeter reading will not be accurate. The ammeter itself has very little resistance to current flow and, therefore, will not affect the circuit, but it will measure current draw only when the circuit is closed and electricity is flowing. Excessive current draw can blow fuses and drain the battery, while a reduced current draw can cause motors to run slowly, lights to dim and other components to not operate properly.

Troubleshooting Electrical Systems

When diagnosing a specific problem, organized troubleshooting is a must. The complexity of a modern automotive vehicle demands that you approach any problem in a logical, organized manner. There are certain troubleshooting techniques which are standard:

• Establish when the problem occurs. Does the problem appear only under certain conditions? Were there any noises, odors or other unusual symptoms?

• Isolate the problem area. To do this, make some simple tests and observations, then eliminate the systems that are working properly. Check for obvious problems, such as broken wires and loose or dirty connections. Always check the obvious before assuming something complicated is the cause.

• Test for problems systematically to determine the cause once the problem area is isolated. Are all the components functioning properly? Is there power going to electrical switches and motors? Performing careful, systematic checks will often turn up most causes on the first inspection, without wasting time checking components that have little or no relationship to the problem.

• Test all repairs after the work is done to make sure that the problem is fixed. Some causes can be traced to more than one component, so a careful verification of repair work is important in order to pick up additional malfunctions that may cause a problem to reappear or a different problem to arise. A blown fuse, for example, is a simple problem that may require more than another fuse to repair. If you don't look for a problem that caused a fuse to blow, a shorted wire (for example) may go undetected.

Experience has shown that most problems tend to be the result of a fairly simple and obvious cause, such as loose or corroded connectors, bad grounds or damaged wire insulation which causes a short. This makes careful visual inspection of components during testing essential to quick and accurate troubleshooting.

Testing

OPEN CIRCUITS

▶ See Figure 8

1. Isolate the circuit from power and ground.
2. Connect the self-powered test light or ohmmeter ground clip to a good ground and probe sections of the circuit sequentially.
3. If the light is out or there is infinite resistance, the open is between the probe and the circuit ground.

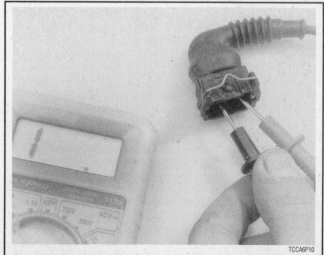

TCCA6P10

Fig. 8 The infinite reading on this multimeter (1 .) indicates that the circuit is open

4. If the light is on or the meter shows continuity, the open is between the probe and end of the circuit toward the power source.

SHORT CIRCUITS

➡**Never use a self-powered test light to perform checks for opens or shorts when power is applied to the electrical system under test. The 12 volt vehicle power will quickly burn out the light bulb in the test light.**

1. Isolate the circuit from power and ground.
2. Connect the self-powered test light or ohmmeter ground clip to a good ground and probe any easy-to-reach test point in the circuit.
3. If the light comes on or there is continuity, there is a short somewhere in the circuit.
4. To isolate the short, probe a test point at either end of the isolated circuit (the light should be on or the meter should indicate continuity).
5. Leave the test light probe engaged and sequentially open connectors or switches, remove parts, etc. until the light goes out or continuity is broken.
6. When the light goes out, the short is between the last two circuit components which were opened.

VOLTAGE

▶ **See Figures 9 and 10**

This test determines voltage available from the battery and should be the first step in any electrical troubleshooting procedure. Many electrical problems, especially on computer controlled systems, can be caused by a low state of charge in the battery. Excessive corrosion at the battery cable terminals can cause poor contact that will prevent proper charging and full battery current flow.

1. Set the voltmeter selector switch to the 20V position.
2. Connect the multimeter negative lead to the battery's negative (-) post or terminal and the positive lead to the battery's positive (+) post or terminal.
3. Turn the ignition switch **ON** to provide a load.
4. A well charged battery should register over 12 volts. If the meter reads below 11.5 volts, the battery power may be insufficient to operate the electrical system properly.

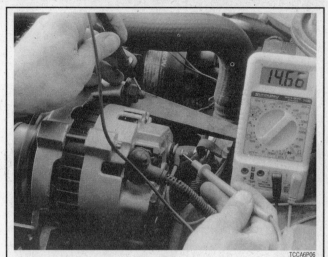
Fig. 10 Testing voltage output between the alternator's BAT terminal and ground. This voltage reading is normal

VOLTAGE DROP

▶ **See Figure 11**

When current flows through a load, the voltage beyond the load drops. This voltage drop is due to the resistance created by the load and also by small resistances created by corrosion at the connectors and damaged insulation on the wires. The maximum allowable voltage drop under load is critical, especially if there is more than one load in the circuit, since all voltage drops are cumulative.

1. Set the voltmeter selector switch to the 20 volt position.
2. Connect the multimeter negative lead to a good ground.
3. Operate the circuit and check the voltage prior to the first component (load).
4. There should be little or no voltage drop in the circuit prior to the first component. If a voltage drop exists, the wire or connectors in the circuit are suspect.
5. While operating the first component in the circuit, probe the ground side of the component with the positive meter lead and observe the voltage

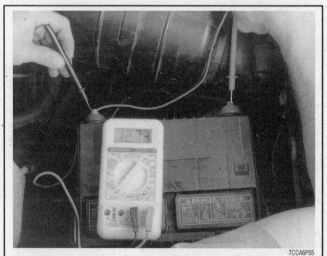
Fig. 9 Using a multimeter to check battery voltage. This battery is fully charged

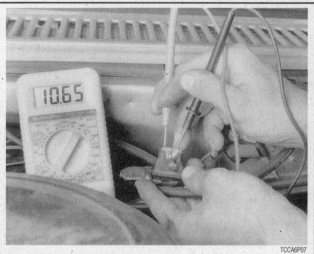
Fig. 11 This voltage drop test revealed high resistance (low voltage) in the circuit

readings. A small voltage drop should be noticed. This voltage drop is caused by the resistance of the component.

6. Repeat the test for each component (load) down the circuit.

7. If a large voltage drop is noticed, the preceding component, wire or connector is suspect.

RESISTANCE

▶ **See Figures 12 and 13**

✳✳ WARNING

Never use an ohmmeter with power applied to the circuit. The ohmmeter is designed to operate on its own power supply. The normal 12 volt automotive electrical system current could damage the meter!

Fig. 12 Checking the resistance of a coolant temperature sensor with an ohmmeter. Reading is 1.04 kilohms

1. Isolate the circuit from the vehicle's power source.

2. Ensure that the ignition key is **OFF** when disconnecting any components or the battery.

3. Where necessary, also isolate at least one side of the circuit to be checked, in order to avoid reading parallel resistances. Parallel circuit resistances will always give a lower reading than the actual resistance of either of the branches.

4. Connect the meter leads to both sides of the circuit (wire or component) and read the actual measured ohms on the meter scale. Make sure the selector switch is set to the proper ohm scale for the circuit being tested, to avoid misreading the ohmmeter test value.

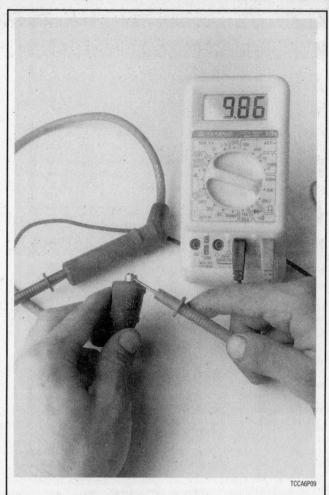
Fig. 13 Spark plug wires can be checked for excessive resistance using an ohmmeter

Wire and Connector Repair

Almost anyone can replace damaged wires, as long as the proper tools and parts are available. Automotive wire and terminals are available to fit almost any need. Even the specialized weatherproof, molded and hard shell connectors are now available from aftermarket suppliers.

Be sure the ends of all the wires are fitted with the proper terminal hardware and connectors. Wrapping a wire around a stud is never a permanent solution and will only cause trouble later. Replace wires one at a time to avoid confusion. Always route wires exactly the same as the factory.

➡**If connector repair is necessary, only attempt it if you have the proper tools. Weatherproof and hard shell connectors require special tools to release the pins inside the connector. Attempting to repair these connectors with conventional hand tools will damage them.**

BATTERY CABLES

Disconnecting the Cables

When working on any electrical component on the vehicle, it is always a good idea to disconnect the negative (-) battery cable. This will prevent potential damage to many sensitive electrical components such as the Engine Control Module (ECM), radio, alternator, etc.

➡Any time you disengage the battery cables, it is recommended that you disconnect the negative (-) battery cable first. This will prevent your accidentally grounding the positive (+) terminal to the body of the vehicle when disconnecting it, thereby preventing damage to the above mentioned components.

Before you disconnect the cable(s), first turn the ignition to the **OFF** position. This will prevent a draw on the battery which could cause arcing (electricity trying to ground itself to the body of a vehicle, just like a spark plug jumping the gap) and, of course, damaging some components such as the alternator diodes.

When the battery cable(s) are reconnected (negative cable last), be sure to check that your lights, windshield wipers and other electrically operated safety components are all working correctly. If your vehicle contains an Electronically Tuned Radio (ETR), don't forget to also reset your radio stations. Ditto for the clock.

AIR BAG (SUPPLEMENTAL INFLATABLE RESTRAINT) SYSTEM

General Information

▶ See Figure 14

The Supplemental Inflatable Restraint (SIR) system was available on Saturn vehicles produced very late in 1992 and was standard on all vehicles starting with the 1993 model year. Starting with the 1995 model year, passenger side air bags became a standard feature of the SIR system on all Saturn vehicles. The SIR or Air Bag system is designed to supplement the normal restraint of the driver's seat belt (and on 1995 and later models, the front passenger's seat belt) in the event of a frontal collision. Should an accident occur within certain specifications of force and direction, an air bag will be deployed from the center of the steering wheel (and, on 1995–98 models, above the glove box). A knee bolster is also provided to help prevent the driver or passenger from sliding downward or forward in the seat during impact.

The very name of the SIR system includes the word SUPPLEMENTAL, which indicates that the system is secondary in the protection of a driver and passenger in case of an accident. The system may not be effective at all if not used in conjunction with the lap and shoulder belts.

SERVICE PRECAUTIONS

▶ See Figures 15 and 16

✳✳ CAUTION

Whenever working on or around SIR components, always observe these general precautions to prevent the possibility of personal injury or damage to the SIR system through unwanted detonation or accidental disabling of the system.

• The inflatable restraint sensing and diagnostic module unit can maintain sufficient voltage to deploy the air bag up to 10 minutes after the ignition has been turned **OFF**, the battery disconnected, or the SIR system fuse removed. Always temporarily disable the SIR system before performing ANY work around system wiring or components.
• The SIR discriminating sensors are located in the front of the vehicle, under the hood. They are specifically calibrated and are keyed to the mounting brackets and SIR wiring harness. Never disturb the sensors or modify the keying of the sensors to the structure through the differently sized mounting holes.

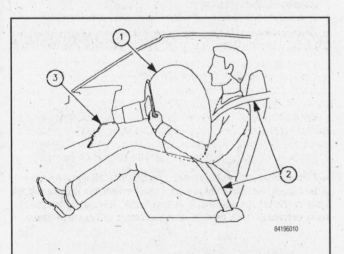

Fig. 14 The air bag (1) works along with the driver's seat belts (2) and knee bolster (3) to protect the driver from a frontal collision

Fig. 15 Vehicles equipped with an SIR system should have warning labels mounted under the hood near the latch mechanism . . .

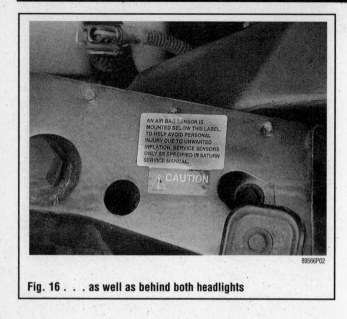

Fig. 16 . . . as well as behind both headlights

Fig. 17 Turn the ignition key to the OFF position

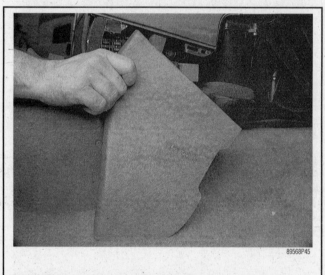

Fig. 18 Remove the right side lower console trim panel . . .

- Never strike or jar a sensor; under certain circumstances, this could cause air bag deployment or improper SIR system operation.
- In the unlikely event that the SIR module is deployed while servicing the vehicle, do not touch metal surrounding the system for at least 10 minutes, to allow the metal to cool. Consult a reputable repair shop for air bag replacement and disposal.
- In the case of deployment, it is unlikely that dangerous chemical residue will remain. Although sodium hydroxide dust (similar to lye soap) is produced during deployment, it quickly reacts with atmospheric moisture to convert to sodium carbonate and sodium bicarbonate (baking soda). Corn starch and sodium bicarbonate may rest on the surface of the bag after deployment. Always wear safety goggles and gloves as a precaution.
- Disarm the system and remove the DERM if the vehicle is to be placed in an environment where temperatures exceed approximately 176°F (80°C), such as a paint spray booth or when arc or gas welding near the control unit location in the car.

DISARMING THE SYSTEM

▶ See Figures 17, 18, 19, 20 and 21

Before working on or near any component of the SIR system, always disarm the system to prevent unwanted deployment. Replacement of inflator modules can cost in excess of $1000 plus installation and disposal of the old module.

1. Align the steering wheel so the tires are in the straight-ahead position, then turn the ignition to the **OFF** position and remove the key.
2. Remove the SIR fuse from the Instrument Panel Junction Block (IPJB).
3. Remove the Connector Position Assurance (CPA) device, then disconnect the yellow 2-way SIR connector at the base of the steering column.

If equipped with a passenger side air bag, perform the same SIR system disabling procedure with the addition of the following steps:

4. Using a small, flat bladed screwdriver, pry off the upper instrument panel pad screw caps. Remove the trim panel retaining screws.
5. Disengage the rear trim panel clips by lifting up. Then, disengage the front trim panel clips by lifting up at each outer corner and pulling rearward. Be careful not to damage the trim panel during its removal from the vehicle.
6. Remove the upper trim panel insulator.
7. Remove the CPA device, then detach the yellow 2-way SIR connector located on the pigtail from the passenger side air bag module.

Fig. 19 . . . to access the Instrument Panel Junction Block (IPJB)

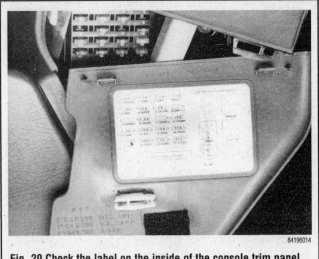

Fig. 20 Check the label on the inside of the console trim panel to locate the AIR BAG fuse in the IPJB

Fig. 21 Using a fuse puller, remove the SIR system fuse

ARMING THE SYSTEM

After completing work on or near components of the SIR system, always properly arm the system and check for proper operation by watching the AIR BAG light in the instrument cluster.

1. Verify that the ignition switch is in the **OFF** or **LOCK** position, then attach the SIR electrical connector at the base of the steering column. Install the CPA device to the connector.

2. If equipped with a passenger side air bag, connect the yellow 2-way inflator module pigtail and install the CPA device to the connector.

3. If equipped with a passenger side air bag, install the insulator and upper trim panel assembly. Be sure that all flaps are tucked and clips engaged. Install and tighten the retaining screws to 20 inch lbs. (2.3 Nm). Install the trim panel screw caps.

4. Install the SIR fuse to the IPJB and install the junction block cover.

5. Turn the ignition switch **ON** and verify that the AIR BAG indicator lamp flashes 7–9 times, then extinguishes. If the light does not flash as indicated, recheck the fuse and connector. If the light still does not function properly, consult a reputable repair shop.

HEATING & AIR CONDITIONING

Blower Motor

REMOVAL & INSTALLATION

▶ See Figures 22, 23 and 24

1. Disconnect the negative battery cable.

2. Unplug the blower motor electrical connector under the glove compartment.

3. Remove the blower motor mounting screws, then carefully lower the motor assembly from the Heating Ventilation Air Conditioning (HVAC) module.

4. Install the motor in the reverse order and check operation.

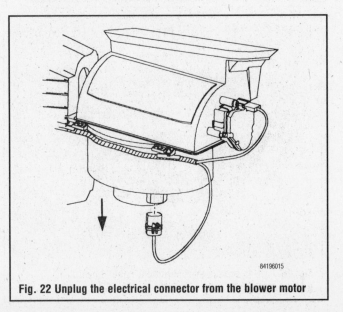

Fig. 22 Unplug the electrical connector from the blower motor

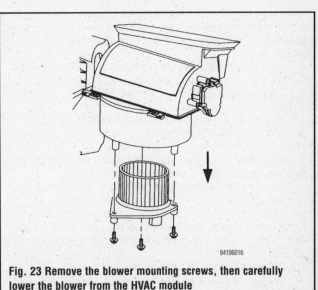

Fig. 23 Remove the blower mounting screws, then carefully lower the blower from the HVAC module

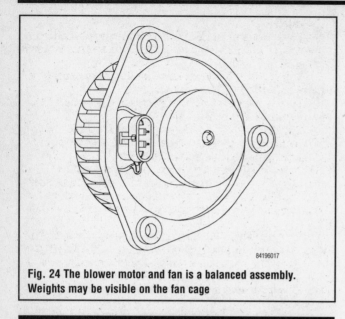

Fig. 24 The blower motor and fan is a balanced assembly. Weights may be visible on the fan cage

Heater Core

REMOVAL & INSTALLATION

▶ **See Figures 25 thru 30**

The Saturn Heating Ventilation Air Conditioning (HVAC) module is the same basic unit on all models, regardless of whether or not they are equipped with air conditioning. The module contains the blower motor, heater core, A/C evaporator (if equipped) and various valve/seal assemblies which route fresh or passenger compartment air by the A/C evaporator and/or the heater core.

1. Disconnect the negative battery cable, then drain the engine cooling system into a suitable container.

2. Raise the front of the vehicle and support it safely using jackstands.

3. While squeezing the retaining tabs, move the heater core clamps up the hoses and off the fittings, then remove the jackstands and carefully lower the vehicle.

4. For the DOHC engine, remove the air cleaner housing cover and disconnect the air induction hose at the intake manifold. For the SOHC engine, remove the air cleaner housing assembly.

5. Carefully remove the hoses from the heater core. Never pry hoses against the heater core pipes or the core may be damaged. Using com-

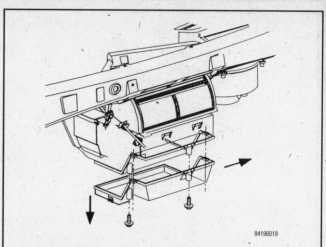

Fig. 26 After removing the lower duct fasteners, carefully drop the duct downward and slide it sideways, removing the duct from the vehicle

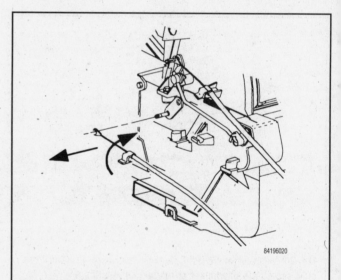

Fig. 27 Remove the temperature cable from the HVAC module

Fig. 25 Slide the retaining clamps upward on the heater core hoses until they no longer compress the hoses on the fittings

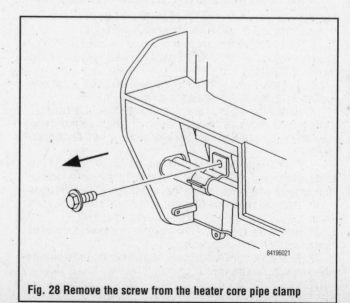

Fig. 28 Remove the screw from the heater core pipe clamp

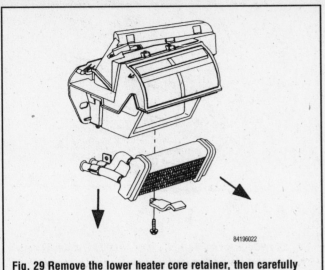

Fig. 29 Remove the lower heater core retainer, then carefully remove the heater core from the HVAC module

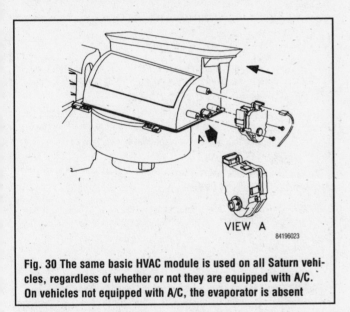

VIEW A

84196023

Fig. 30 The same basic HVAC module is used on all Saturn vehicles, regardless of whether or not they are equipped with A/C. On vehicles not equipped with A/C, the evaporator is absent

pressed air or a length of clean hose, blow the remaining coolant out of the heater core to prevent spilling it in the vehicle's interior.

6. Remove the left and right lower trim lower trim panel extensions by disconnecting the Velcro™ at the bottom of the panels and pulling them out of the upper retaining clips.

7. If equipped, remove the instrument panel lower closeout panel by pulling out at the top edge and rotating downward.

8. Remove the retaining screws and lower the heater duct straight downward. Carefully slide the duct to the side and remove it from the vehicle. Make sure the heater duct-to-rear floor duct seal is not damaged during removal.

9. Push down on the cable and lift the plastic tab to release the temperature cable hold-down clip, then disconnect the temperature cable by squeezing the valve pin and pulling the cable straight off the HVAC module.

10. Locate the heater core side cover, on the driver's side of the HVAC module, then remove the retaining screws and the side cover.

11. Remove the screws and the lower heater core cover. Remove the screw from the heater core pipe clamp.

12. Remove the screw and the lower core retainer, then carefully remove the heater core from the vehicle.

To install:

13. Install the heater core, being careful not to damage the pipe seal. Use a coating of petroleum jelly to ease installation of the pipes through the cowl.

14. Install the lower heater core retainer and the pipe clamp using the retaining screws. Install the lower and side covers.

15. Push the temperature cable over the pin and snap the cable hold-down clip over the cable holder.

16. Slide the heater duct in sideways and raise it into position, being careful not to damage the rear floor heater seal. Install the duct screws.

17. Install the left and right trim panel extensions.

18. Install the instrument panel closeout panel, if equipped.

19. Raise the front of the vehicle and support it safely using jackstands.

20. Install the heater hoses, positioning the left hose clamp tangs to the 7–8 o'clock position and the right hose clamp tangs to the 6 o'clock position.

21. If not done already, install the radiator drain plug, then install the cylinder block drain plug and tighten the block plug to 26 ft. lbs. (35 Nm).

22. Remove the jackstands and carefully lower the vehicle, then install the air cleaner housing components, as applicable.

23. Connect the negative battery cable and properly fill the engine cooling system.

24. Pressure test the cooling system or start the engine and check for leaks.

Air Conditioning Components

REMOVAL & INSTALLATION

Repair or service of air conditioning components is not covered by this manual, because of the risk of personal injury or death, and because of the legal ramifications of servicing these components without the proper EPA certification and experience. Cost, personal injury or death, environmental damage, and legal considerations (such as the fact that it is a federal crime to vent refrigerant into the atmosphere), dictate that the A/C components on your vehicle should be serviced only by a Motor Vehicle Air Conditioning (MVAC) trained, and EPA certified automotive technician.

➡**If your vehicle's A/C system uses R-12 refrigerant and is in need of recharging, the A/C system can be converted over to R-134a refrigerant (less environmentally harmful and expensive). Refer to Section 1 for additional information on R-12 to R-134a conversions, and for additional considerations dealing with your vehicle's A/C system.**

Control Cables

REMOVAL & INSTALLATION

◆ See Figures 31 thru 38

1. Disconnect the negative battery cable.

2. On 1991–94 models, carefully remove the center air outlet/trim panel by pulling outward at the clip locations. Start at the bottom and work upward to the top clips. Do not use instruments which might damage the trim panel.

3. On 1995–98 models, remove the radio/HVAC controller cover by depressing the center pins inward to release the push pin fasteners. Be careful not to push the center pins all the way through the fasteners. Remove the push pin fasteners and pull the radio/HVAC controller cover rearward.

4. If equipped, disengage the traction control/fog lamps/rear defogger electrical connector.

5. Remove the radio. Refer to the Radio removal and installation procedure, later in this section.

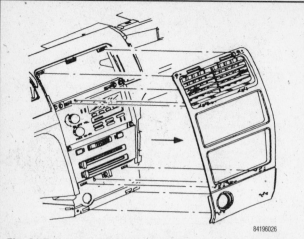

Fig. 31 Pull outward at the clip locations to remove the center console air outlet/trim panel from the dashboard—1992–93 vehicles shown

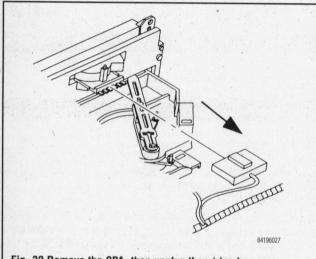

Fig. 32 Remove the CPA, then unplug the wiring harness connector from the blower motor switch

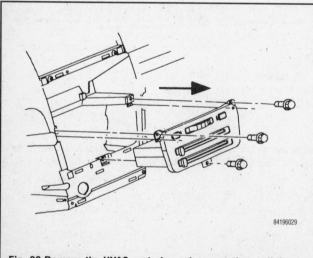

Fig. 33 Remove the HVAC control panel screws, then pull the panel forward slightly

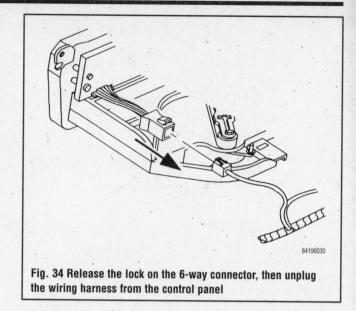

Fig. 34 Release the lock on the 6-way connector, then unplug the wiring harness from the control panel

6. Disengage the electrical connectors behind the controller face.

7. If necessary, remove the HVAC control panel screws, then carefully pull the panel forward, just enough for access.

➡ **On all Saturn models, the mode cable is black. On 1991–94 models, the temperature cable is blue, while on 1995–98 models, the temperature cable is white.**

8. Disengage the temperature and mode cables from the controller assembly by squeezing the lock tabs together while pulling the cable housing straight up. Pull the cables straight up to disengage them from the control lever pins.

➡ **The pin for the temperature cable may be easier to access if the lever is moved to the full COLD position.**

9. Reach under the instrument panel and release the cable hold-down clip by gently pushing up on the plastic tab while pushing down on the top of the cable.

10. Disconnect the mode cable by rotating the mode valve lever rearward, then turn the cable upward to clear the retention leg and slide it off the pin. Remove the mode cable.

11. Slide the temperature cable off of the temperature door hook.

To install:

12. Install the temperature cable onto the temperature door hook.

13. Install the mode cable over the valve pin and behind the retention leg.

14. Snap the cable hold-down clip(s) over the cable holder(s).

15. Install the temperature and mode cables onto the control lever pins. Install the cable housing into the channel and lock into place by pushing down.

16. If removed, install the HVAC control panel and mounting screws.

17. Adjust the cables by holding the housing and rotating the housing end. Adjustment is made as follows:

a. If the mode lever bounces back when pushed to the right side, shorten the cable housing.

b. If the mode lever bounces back when pushed to the left side, lengthen the cable housing.

c. If the temperature lever bounces back when pushed to the right side, lengthen the cable housing.

d. If the temperature lever bounces back when pushed to the left side, shorten the cable housing.

18. Reattach the electrical connectors behind the controller face.

19. Install the radio.

20. If removed, engage the traction control/fog lamps/rear defogger electrical connector.

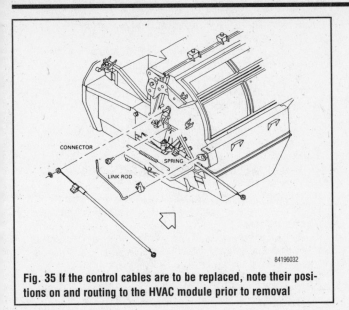

Fig. 35 If the control cables are to be replaced, note their positions on and routing to the HVAC module prior to removal

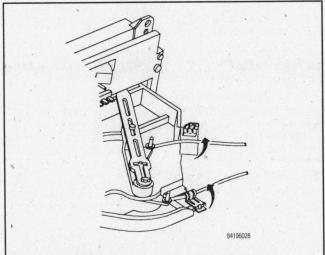

Fig. 36 Release the temperature and mode cable hold-down/adjustment clips

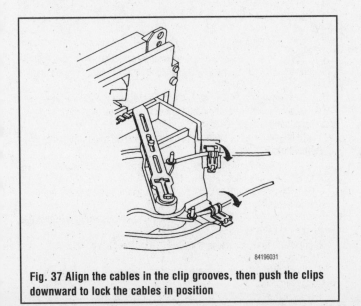

Fig. 37 Align the cables in the clip grooves, then push the clips downward to lock the cables in position

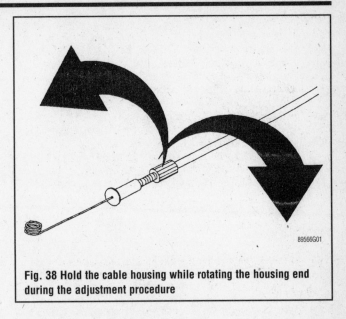

Fig. 38 Hold the cable housing while rotating the housing end during the adjustment procedure

21. On 1991–94 models, install the center air outlet/trim panel by pushing in to engage the retaining clips. Start at the top and work downward to the lower clips.

22. On 1995–98 models, install the radio/HVAC controller cover and push pin fasteners.

23. Connect the negative battery cable.

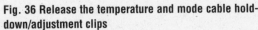
Control Panel

REMOVAL & INSTALLATION

1991–94 Models

▶ See Figures 31 thru 38

1. Disconnect the negative battery cable.

2. Carefully remove the center air outlet/trim panel by pulling outward at the clip locations. Start at the bottom and work upward to the top clips. Do not use instruments which might damage the trim panel.

3. Remove the 2 radio retaining screws located at the top of the unit, then carefully pull the radio forward and out of the dashboard. Unplug the radio electrical connector and the antenna, then remove the radio from the vehicle.

4. Remove the Connector Position Assurance (CPA) device from the blower motor switch, then unplug the wiring harness.

5. Release the temperature and mode cable hold-down/adjustment clips.

6. Remove the HVAC control panel screws, then carefully pull the panel forward, just enough for access.

7. Disconnect both cables from the control panel (black mode cable first, followed by the blue temperature cable) by squeezing the controller pin and lifting the cable straight upward and off the pin.

➡The pin for the temperature cable may be easier to access if the lever is moved to the full COLD position.

8. Slide the HVAC control panel further out from the dashboard and release the lock on the 6-way connector, then unplug the wiring harness from the panel and remove the panel from the dashboard.

To install:

9. Position the HVAC control panel to the dashboard and install the retaining screws.

10. Install the blue temperature and black mode cables over the pins.

11. Position the temperature lever to the full COLD position and the mode lever to the full VENT position (both levers to the LEFT).

12. Align the cables in the hold-down/adjustment clip grooves, then push the clips over the cables and lock them into position.

13. Check the mode and temperature levers for full and easy travel. Adjust as necessary by pushing the lever to the end that springs back, then, while holding the lever, lift up on the clip and allow the cable to adjust. When the cable is adjusted, push the clip downward to lock the cable in place.

14. Connect the wiring harnesses to the control panel and the blower motor switch. Make sure the wires do not interfere with control lever movement, then install the CPA devices to the connectors.

15. Install the radio antenna and electrical connector, then position the radio to the dashboard. Secure the radio using the 2 screws.

16. Install the center air outlet/trim panel by pushing inward at the clip locations.

17. Connect the negative battery cable.

1995–98 Models

1. Disconnect the negative battery cable.

2. Remove the radio/HVAC controller cover by depressing the center pins inward to release the push pin fasteners. Be careful not to push the center pins all the way through the fasteners.

3. Remove the push pin fasteners and pull the radio/HVAC controller cover rearward.

4. If equipped, disengage the traction control/fog lamps/rear defogger electrical connector.

5. Remove the radio. Refer to the Radio removal and installation procedure, later in this section.

6. Disengage the electrical connectors behind the controller face. Remove the wiring harness clip from the module boss.

7. Disengage the temperature and mode cables from the controller assembly by squeezing the lock tabs together while pulling the cable housing straight up. Pull the cables straight up to disengage them from the control lever pins.

8. Remove the HVAC control panel screws, then carefully pull the panel forward and remove it from the vehicle.

To install:

9. Position the HVAC control panel to the dashboard and install the retaining screws. Tighten the retaining screws to 22 inch lbs. (2.5 Nm).

10. Install the white temperature and black mode cables over the pins.

11. Install the cable housings into the clip grooves, then push down to lock them into position.

12. Reattach the electrical connectors to the back of the control panel face. Install the wiring harness clip to the module boss. Be sure that the wiring harness does not bind with control lever movement.

13. Install the radio.

14. If equipped, reattach the traction control/fog lamps/rear defogger electrical connectors.

15. Install the radio/HVAC controller cover by pushing in at the clip locations.

16. Install and lock the push pin fasteners.

17. Connect the negative battery cable.

CRUISE CONTROL

General Information

▶ **See Figure 39**

The cruise control system is utilized to maintain vehicle speed at a user selected value, with a maximum deviation of 1 mph (2 km/h) on a flat level road (0–2 percent grade).

The system is comprised of a cruise control module, which is located inside the instrument panel near the steering column, control switches, cruise brake switch and a clutch cruise switch, if equipped with a manual transaxle.

Unlike other cruise control systems, this system is completely electronic and does not require the use of engine vacuum or a servo motor to operate throttle movement. To control throttle movement, a cable from the cruise control module pulls on the accelerator linkage at the pedal.

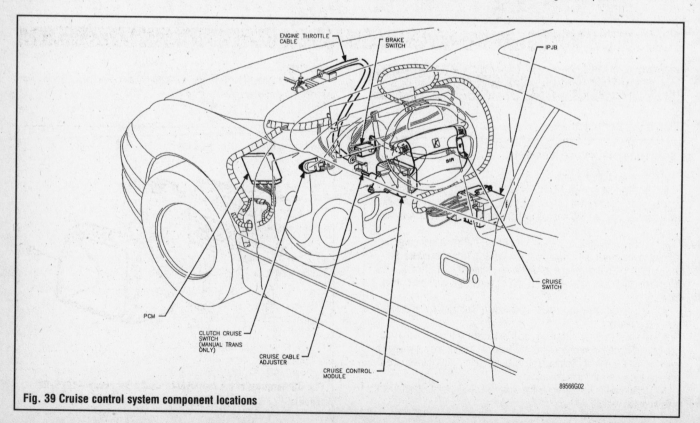

89566G02

Fig. 39 Cruise control system component locations

CRUISE CONTROL TROUBLESHOOTING

Problem	Possible Cause
Will not hold proper speed	Incorrect cable adjustment
	Binding throttle linkage
	Leaking vacuum servo diaphragm
	Leaking vacuum tank
	Faulty vacuum or vent valve
	Faulty stepper motor
	Faulty transducer
	Faulty speed sensor
	Faulty cruise control module
Cruise intermittently cuts out	Clutch or brake switch adjustment too tight
	Short or open in the cruise control circuit
	Faulty transducer
	Faulty cruise control module
Vehicle surges	Kinked speedometer cable or casing
	Binding throttle linkage
	Faulty speed sensor
	Faulty cruise control module
Cruise control inoperative	Blown fuse
	Short or open in the cruise control circuit
	Faulty brake or clutch switch
	Leaking vacuum circuit
	Faulty cruise control switch
	Faulty stepper motor
	Faulty transducer
	Faulty speed sensor
	Faulty cruise control module

Note: Use this chart as a guide. Not all systems will use the components listed.

TCCA6C01

ENTERTAINMENT SYSTEMS

Radio Receiver/Amplifier, Tape Player and Compact Disc Player

REMOVAL & INSTALLATION

▶ **See Figures 40, 41, 42, 43 and 44**

1. Disconnect the negative battery cable.
2. On 1991–94 models, carefully remove the center air outlet/trim panel by pulling outward at the clip locations. Start at the bottom and work upward to the top clips. Do not use instruments which might damage the trim panel.
3. On 1995–98 models, remove the radio/HVAC controller cover by depressing the center pins inward to release the push pin fasteners. Be careful not to push the center pins all the way through the fasteners. Remove the push pin fasteners and pull the radio/HVAC controller cover rearward.
4. If equipped, disengage the traction control/fog lamps/rear defogger electrical connector.
5. Remove the 2 radio retaining screws located at the top of the unit, then carefully pull the radio forward and from the dashboard sufficiently to reach behind the unit. If equipped, push in the spring clips through the D-holes on each side of the radio brace.
6. Unplug the radio's antenna and electrical connector, then remove the radio from the vehicle.

To install:
7. Position the radio in front of the center outlet, then attach the wiring connector and antenna plug.

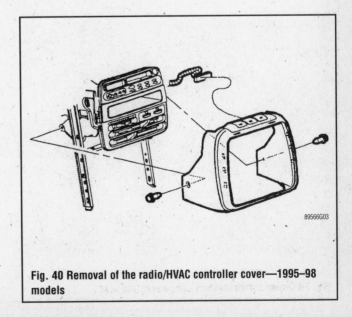

Fig. 40 Removal of the radio/HVAC controller cover—1995–98 models

89566G03

Fig. 41 Remove the center air outlet/trim panel by pulling outward at the clip locations

Fig. 42 Remove the 2 radio retaining screws located at the top of the unit

Fig. 43 Pull the radio forward sufficiently to reach and unplug the antenna connection

Fig. 44 Unplug the radio electrical connector, then remove the radio

8. Slide the radio into position and secure using the 2 retaining screws. Engage the spring clips on each side of the radio, if equipped.

9. If equipped, engage the traction control/fog lamps/rear defogger electrical connector.

10. On 1991–94 models, install the center air outlet/trim panel by pushing inward at the clip locations.

11. On 1995–98 models, install the radio/HVAC controller cover by pushing in at the clip locations. Install and lock the push pin fasteners.

12. Connect the negative battery cable.

Front Speakers

REMOVAL & INSTALLATION

▶ See Figures 45, 46 and 47

1. Disconnect the negative battery cable.
2. Remove the inner door panel. Refer to Section 10 for door panel removal procedures.
3. Remove the Torx® head screws from the speaker, then carefully pull the speaker from the mounting sufficiently to unplug the electrical connector.
4. Unplug the connector, then remove the speaker from the vehicle.

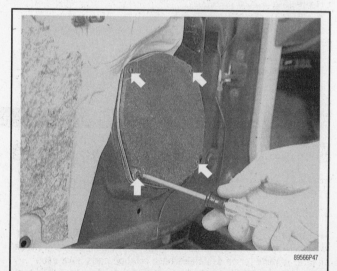

Fig. 45 Remove the Torx® head screws from the speaker

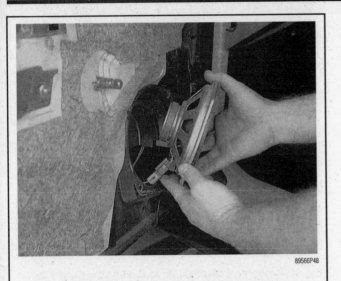

Fig. 46 Carefully pull the speaker out of the door opening

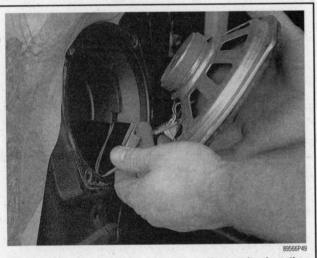

Fig. 47 Unplug the connector, then remove the speaker from the vehicle

To install:

5. Install the electrical connector to the speaker, then position the speaker into the mounting plate. Be careful not to damage the foam insulators at the top and sides of the speaker.

6. Install the speaker retaining screws.

7. Install the inner door panel.

8. Connect the negative battery cable and enjoy the tunes!

WINDSHIELD WIPERS AND WASHERS

Windshield Wiper Blade and Arm

REMOVAL & INSTALLATION

▶ **See Figures 49, 50, 51 and 52**

➡Originally, stock blades for the front wiper blade/arm assemblies were not replaceable separately. However, this may not be the case on later models (or on vehicles whose original parts have been replaced). On those models with replaceable blades, the wiper arm need not be detached.

Rear Speakers

REMOVAL & INSTALLATION

▶ **See Figure 48**

1. Disconnect the negative battery cable.

2. Access the speaker grille fasteners and electrical connector located in the trunk/cargo area. This may be accomplished from either outside the vehicle, by opening the trunk lid/liftgate, or from inside the vehicle, by folding down the rear seat.

3. Unplug the speaker electrical connector.

4. Remove the grille retainers from underneath the package shelf, then lift the grille off the speaker. Be careful not to damage the retainers, if they are to be reused.

5. Remove the Torx® head speaker retaining screws from the top of the shelf, then lift the speaker from the shelf mount.

To install:

6. Lower the speaker into the shelf mount, then install and tighten the retaining screws.

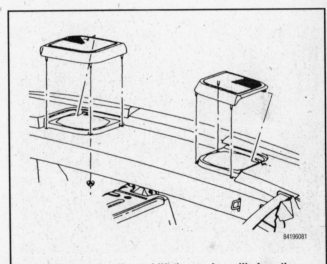

Fig. 48 Remove the clips and lift the speaker grille from the package shelf

7. Position the grille over the speaker and secure using the retainers. Be sure to replace any retainers which were damaged during removal.

8. Connect the wiring harness to the speaker.

9. Close the trunk lid/liftgate or return the rear seat back to its upright position, if applicable.

10. Connect the negative battery cable and enjoy the tunes!

1. If wiper arm reval is necessary:

a. Lift up the wiper arm finish cap, then loosen the retaining nut using a wrench.

b. Lift the wiper blade away from the windshield and remove the blade/arm assembly.

2. If a separate replacement blade is available, remove the wiper blade from the arm. Typically, this requires pivoting the wiper blade 90 degrees, then unhooking it. (Refer to the photo.) If you encounter a different design, such as a blade with a retaining clip, depress the clip to release the blade.

Fig. 49 If removing the wiper arm, open or remove the wiper arm finish cap . . .

Fig. 50 . . . then loosen the retaining nut using a wrench

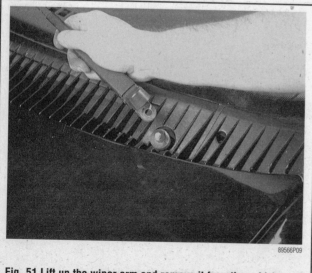

Fig. 51 Lift up the wiper arm and remove it from the vehicle

Fig. 52 If applicable, pivot the blade at a 90 degree angle and slide it off the wiper arm's hook

To install:

3. If applicable, install the replacement blade onto the wiper arm, making sure it locks into place.

4. If the wiper arm was removed, position the wiper arm/blade assembly onto the pivot shaft, then install the retaining nut and tighten to 21 ft. lbs. (28 Nm). Install or position the finish cap over the retaining nut.

Liftgate Wiper Blade and Arm

REMOVAL & INSTALLATION

▶ See Figures 53 and 54

1. On 1991–95 models, remove the rear wiper arm finish cap, then loosen the retaining nut using a wrench.

2. On 1996–98 models, use a small prying tool to disengage the wiper arm retaining latch.

3. Lift the wiper blade away from the liftgate window and remove the blade/arm assembly.

4. Pinch the wiper blade attachment clip together and slide the blade from the arm.

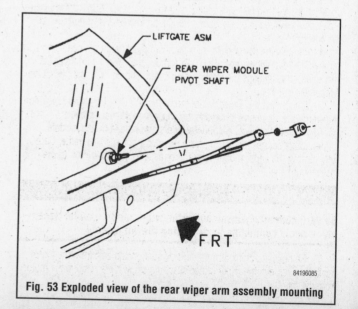

Fig. 53 Exploded view of the rear wiper arm assembly mounting

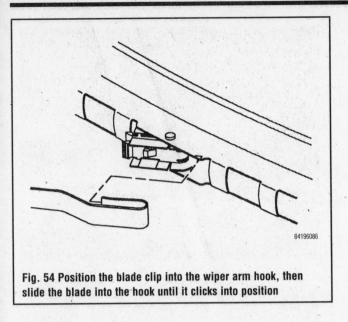

Fig. 54 Position the blade clip into the wiper arm hook, then slide the blade into the hook until it clicks into position

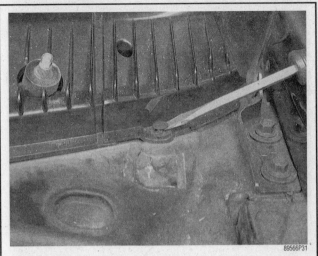

Fig. 55 Using a small prying tool, lift out the cowl trim panel's plastic fastener . . .

To install:

5. Position the blade clip into the wiper arm hook, then slide the blade into the hook until it clicks into position.

6. On 1991–95 models, position the arm onto the pivot shaft with the blade horizontal to the liftgate glass lower edge, then install the retaining nut and tighten to 13 ft. lbs. (18 Nm). Close the finish cap over the retaining nut.

7. On 1997–98 models, position the arm onto the pivot shaft with the arm just below the first defroster grid at the bottom of the liftgate glass. Push the retaining latch inward to secure the arm to the wiper motor shaft.

Windshield Wiper Motor Module

REMOVAL & INSTALLATION

▶ **See Figures 55 thru 71**

1. Make sure that the wipers are in the **PARK** position, then disconnect the negative battery cable.

2. Close the hood, then remove the wiper arm finish cap and wiper arm fastening nut. Lift the blade away from the windshield and remove the blade/arm assembly. Repeat for the other blade.

3. Remove the cowl trim fasteners at the windshield edge of the panel, then open the hood and remove the remaining fasteners. Carefully remove the cowl trim panel.

4. Remove the screw caps from the instrument panel top cover, then remove the retaining screws. Carefully remove the cover by lifting at the rear edge, to disengage the retaining clips, and sliding the panel out of the windshield clips.

5. Disconnect the defroster duct from the HVAC module and reposition it towards the glove box to expose the wiper module rear fasteners.

6. Remove the wiper module fasteners and reposition the module slightly to disconnect the wiring from the motor and module frame. Carefully remove the wiper module and motor assembly from the top of the instrument panel.

✳✳ WARNING

Be very careful when removing the wiper module/motor assembly to avoid contacting or damaging the windshield.

7. Remove the crank arm nut and disconnect the arm from the motor shaft, then remove the wiper motor attaching screws and remove the motor from the module.

Fig. 56 . . . then unfasten the trim panel retaining screw . . .

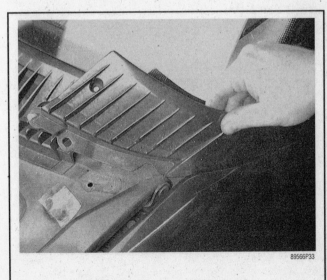

Fig. 57 . . . and remove the side cowl trim panel

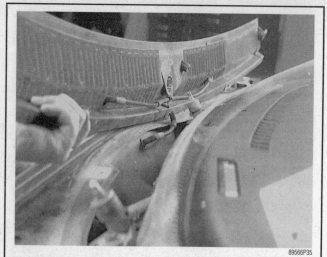

Fig. 58 Remove the center cowl trim panel's mounting fasteners and lift the panel up . . .

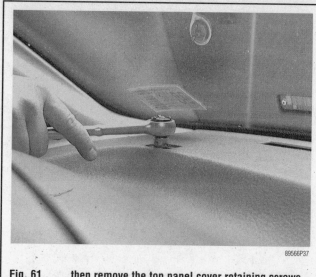

Fig. 61 . . . then remove the top panel cover retaining screws

Fig. 59 . . . then disconnect the windshield washer hose

Fig. 62 Disengage the retaining clips, then slide the instrument panel top cover out of the vehicle

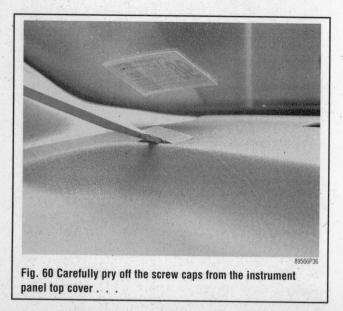

Fig. 60 Carefully pry off the screw caps from the instrument panel top cover . . .

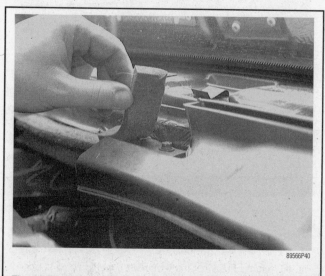

Fig. 63 Remove the instrument panel top cover insulators . . .

Fig. 64 . . . then loosen the defroster duct mounting bolts . . .

Fig. 67 . . . then remove the module mounting bolts at the cowl, outside the vehicle

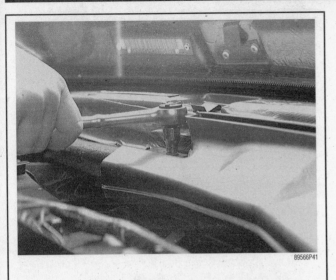

Fig. 65 . . . and remove the defroster duct from the instrument panel

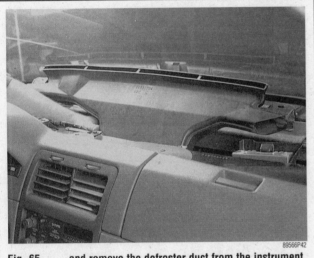

Fig. 68 Disengage the wiper motor electrical connector

Fig. 66 Remove the wiper motor module's mounting bolts from inside the vehicle . . .

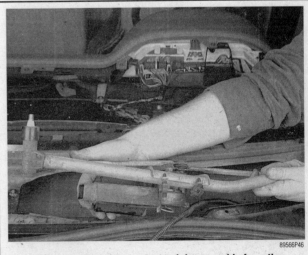

Fig. 69 Remove the wiper motor module assembly from the vehicle

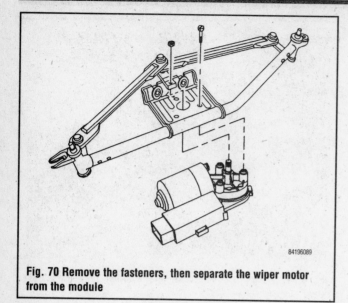

Fig. 70 Remove the fasteners, then separate the wiper motor from the module

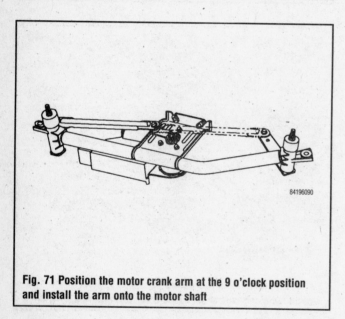

Fig. 71 Position the motor crank arm at the 9 o'clock position and install the arm onto the motor shaft

To install:

8. Verify that the motor is in the **PARK** position. If necessary, temporarily connect the motor wiring and the negative battery cable, turn the wiper control **ON** then **OFF** and the motor will move to the correct position.

9. Install the motor to the module. Position the motor crank arm at the 9 o'clock position and install the arm onto the motor shaft. Install a new retaining nut and tighten to 21 ft. lbs. (28 Nm).

10. Position the wiper module assembly into the vehicle and connect the wiring to the wiper motor and to the module frame.

➡ **Whenever possible during the remaining steps, the vehicle manufacturer recommends using new fasteners, as the torque retention of the old fasteners may be insufficient.**

11. Install the module retaining bolts and tighten to 89 inch lbs. (10 Nm).
12. Install the cowl trim panel.
13. Install the wiper arm assemblies and tighten the nuts to 21 ft. lbs. (28 Nm).
14. Position the defroster duct to the HVAC module.
15. Install the instrument panel top cover and screws, then insert the panel cover screw caps.
16. Connect the negative battery cable and verify proper system operation.

Liftgate Wiper Motor Module

REMOVAL & INSTALLATION

▶ **See Figures 72, 73 and 74**

1. Make sure that the wiper is in the **PARK** position, then disconnect the negative battery cable.
2. Remove the wiper arm/blade assembly from the liftgate.
3. Raise the liftgate, then remove the wedge block from each side of the liftgate assembly.
4. Remove the lower fasteners from the liftgate lower trim panel, by pushing in the center of each push pin approximately ⅛ in. until it clicks, then remove the fastener.
5. Insert a screwdriver or small prybar into the hole in the lower trim panel (located near the wiper pivot on the pivot hump) so that the tool sits on top of the pivot as shown. Lift up on the tool handle to disengage the trim panel upper clips, then remove the trim panel.
6. Remove the fasteners from the rear wiper module, then disconnect the washer hose from the module check valve.
7. Remove the module from the liftgate.

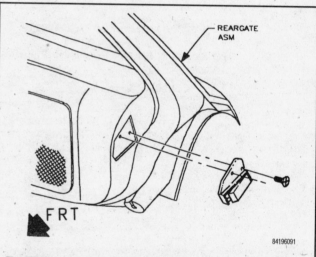

Fig. 72 Remove the wedge blocks from either end of the liftgate assembly

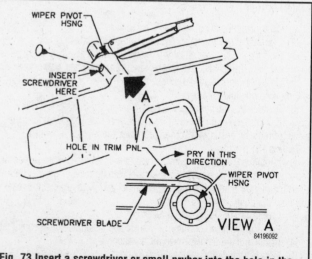

Fig. 73 Insert a screwdriver or small prybar into the hole in the lower trim panel so that the tool sits on top of the pivot

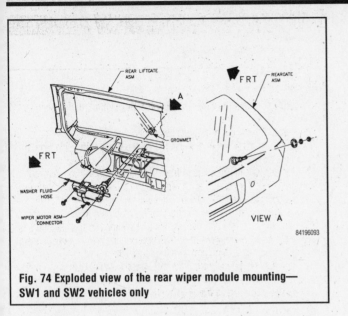

Fig. 74 Exploded view of the rear wiper module mounting—SW1 and SW2 vehicles only

Fig. 75 Typical windshield washer reservoir/pump motor assembly

To install:

8. Position the wiper module to the liftgate and connect the washer hose to the check valve.

9. Install the module fasteners and tighten to 89 inch lbs. (10 Nm).

10. Install the grommet, washer and nut on the wiper module pivot shaft, then tighten the nut to 10 ft. lbs. (14 Nm).

11. Align the upper clips on the lower trim panel to the liftgate slots, then install the panel by pushing at the clip locations.

12. Reset the trim panel push-in fasteners by spreading the center pin tabs and moving the pin so that it sits approximately ¼ in. outside the fastener. Insert the fasteners into the bottom of the trim panel and push the center pin until flush.

13. Install the liftgate wedge blocks.

14. Install the rear wiper pivot bushing.

15. Install the rear wiper arm/blade assembly.

16. Connect the negative battery cable and verify proper system operation.

Windshield Washer Reservoir/Pump Motor

REMOVAL & INSTALLATION

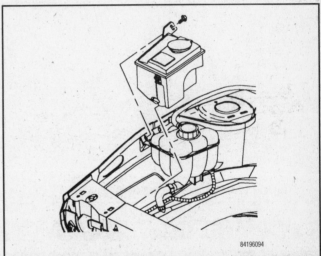

Fig. 76 The windshield washer reservoir/pump motor assembly is mounted along the fender, forward of the coolant reserve tank

▶ **See Figures 75, 76 and 77**

1. Disconnect the negative battery cable.

2. Disconnect the wiring harness from the pump motor.

3. Remove the fastener attaching the washer reservoir to the vehicle.

4. Lift the assembly and release the lower attachment tab on the reservoir.

5. Carefully remove the fluid hose from the pump. Be sure to avoid spilling fluid on painted surfaces.

6. Remove the washer reservoir/pump assembly from the vehicle.

7. If necessary, drain the fluid into a clean container and remove the pump from the reservoir. Inspect the grommet and replace it if worn.

To install:

8. If removed, install the pump motor to the reservoir. Be sure that the grommet is in place. Lubricate the grommet with light oil or grease, then insert the pump's inlet spout into the grommet and push against the reservoir until the pump is fully seated.

9. Connect the fluid hose to the pump.

10. Position the reservoir into the vehicle and fasten the engagement tab.

11. Install and tighten the reservoir fastener.

12. Connect the wiring harness to the pump.

13. If necessary, fill the reservoir with washer solvent.

14. Connect the negative battery cable and verify proper system operation.

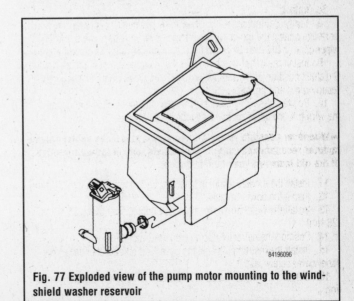

Fig. 77 Exploded view of the pump motor mounting to the windshield washer reservoir

Liftgate Washer Reservoir/Pump Motor

REMOVAL & INSTALLATION

▶ **See Figures 78 and 79**

1. Disconnect the negative battery cable.
2. Remove the passenger side, rear quarter inner trim panel for access to the assembly.
3. Remove the reservoir fasteners.
4. Pull the reservoir away from the body sufficiently to reach the wiring harness, then disconnect the harness from the pump motor.
5. Place a towel or rag underneath the pump and disconnect the hose.
6. Remove the reservoir/pump assembly from the vehicle, taking care not to spill fluid on the vehicle's interior or painted surfaces.

7. If necessary, drain the fluid from the reservoir into a clean container, then grasp the pump and pull it from the reservoir. Inspect the grommet at the reservoir's opening and replace it if worn.
   To install:
8. If removed, install the pump motor to the reservoir. Lubricate the grommet with light oil or grease, then insert the pump's inlet spout into the grommet and push against the reservoir until the pump is fully seated.
9. Connect the fluid hose and the wiring harness to the pump.
10. Position the reservoir onto the studs in the rear body and install the fasteners.
11. If necessary, fill the reservoir with washer solvent.
12. Install the rear quarter inner trim panel.
13. Connect the negative battery cable and verify proper system operation.

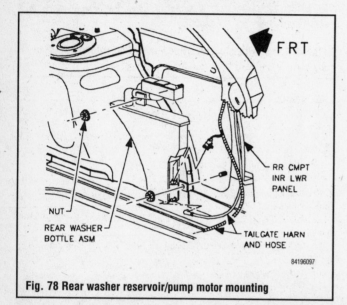

Fig. 78 Rear washer reservoir/pump motor mounting

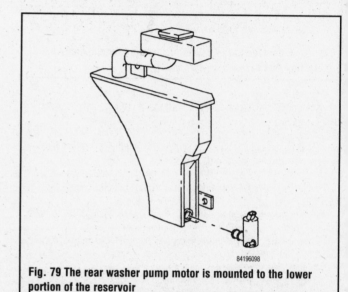

Fig. 79 The rear washer pump motor is mounted to the lower portion of the reservoir

INSTRUMENTS AND SWITCHES

Instrument Cluster

The gauges located in the instrument cluster (speedometer, tachometer, fuel gauge, voltmeter and coolant temperature gauge) are not serviceable. The entire cluster must be replaced as an assembly.

REMOVAL & INSTALLATION

1991–94 Models

▶ **See Figures 80 and 81**

1. Disconnect the negative battery cable.
2. Remove the 2 instrument panel top cover screw caps and screws. Carefully remove the cover by lifting at the rear edge to disengage the retaining clips and by sliding the panel out of the windshield clips.
3. Carefully remove the center air outlet/trim panel by pulling outward at the clip locations. Start at the bottom and move upward. Do not use tools that might damage the trim panel.
4. Open the glove box, then remove the 4 cluster trim panel attaching screws from the top of the panel.
5. Carefully pull the cluster trim panel rearward to disengage it from the retainers, then remove the CPA device and unplug the electrical connectors from the instrument panel lighting and rear window defogger switches. Remove the panel from the vehicle.

6. Remove the instrument cluster retaining screws, then pull the cluster out sufficiently to unplug the electrical connectors. Unplug the connectors by depressing the retainer legs and remove the assembly.

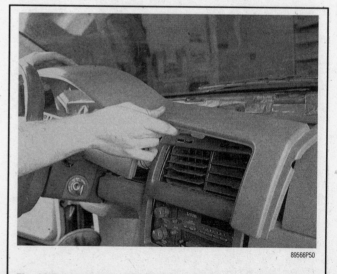

Fig. 80 Remove the instrument panel top cover from the vehicle

Fig. 81 Unplug the connectors by depressing the retainer legs, then remove the cluster assembly

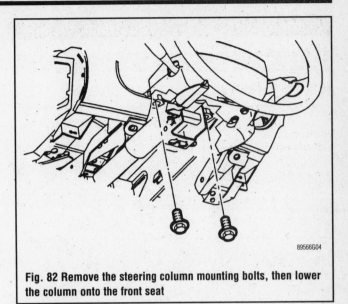

Fig. 82 Remove the steering column mounting bolts, then lower the column onto the front seat

To install:

7. Connect the wiring harnesses to the instrument cluster assembly.

8. Verify that the connectors for the instrument panel lighting and rear window defogger switches are properly positioned, then install the instrument cluster assembly and retaining screws.

9. Position the cluster trim panel and engage the electrical connectors to the panel lighting and rear defogger switches, then install the CPA devices.

10. Install the cluster trim panel into the retainers, then secure using the retaining screws.

11. Install the center air outlet/trim panel by pushing at the clip locations.

12. Install the rear of the instrument panel top cover into the windshield clips and snap the panel into position. Install the upper panel screws and screw caps.

13. Connect the negative battery cable.

1995–98 Models

▶ See Figures 82 and 83

1. Disconnect the negative battery cable.

2. Disarm the air bag system, as described earlier in this section.

3. Remove the lower steering column filler panel mounting screws.

4. Disconnect the hood release cable from the lever and remove the steering column filler panel.

5. Disengage the ignition switch electrical connector at the right steering column bolt.

6. Remove the steering column bolts and lower the column onto the front seat.

7. Remove the instrument cluster fasteners and pull back the trim bezel at the clip locations.

8. Disengage the dimmer switch connector.

9. Remove the Connector Position Assurance (CPA) devices and disengage the wiring connectors from the cluster assembly by squeezing the lock tabs on each side of the connectors.

10. Remove the mounting screws and lift the instrument cluster assembly out of the vehicle.

To install:

11. Install the instrument cluster assembly into the dashboard and fasten the mounting screws.

12. Reattach the wiring connectors behind the cluster assembly and install the CPA devices.

13. Attach the dimmer switch connector.

14. Install the instrument cluster trim bezel. Push in at the clip locations.

15. Raise the steering column into position and install the mounting bolts. Tighten the bolts to 26 ft. lbs. (35 Nm).

16. Attach the ignition switch connector. Install the hood release cable to the lever.

17. Install the lower steering column filler panel and tighten the mounting screws.

18. Arm the air bag system.

19. Connect the negative battery cable.

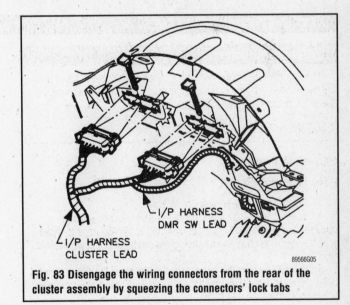

Fig. 83 Disengage the wiring connectors from the rear of the cluster assembly by squeezing the connectors' lock tabs

LIGHTING

Headlights

REMOVAL & INSTALLATION

Except 1991–96 SC2

▶ **See Figures 84 and 85**

1. Disconnect the negative battery cable and make sure the headlamp has cooled if it was operated in the past few minutes.

2. Disconnect the headlamp bulb socket from the rear of the lens by rotating the socket ¼ turn counterclockwise, then pulling the socket rearward until the bulb clears the housing. Lift the socket/harness for access.

3. Remove the bulb from the socket/harness.

4. Installation is the reverse of removal. Do not touch the glass of the bulb. Adjustment of the headlamp on these vehicles should not be necessary unless the housing assembly is removed or loosened. Nonetheless, headlight aim should always be checked and, if necessary, adjusted for safety.

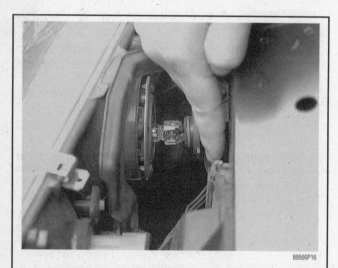

Fig. 84 Rotate the socket ¼ turn, then pull it out of the opening until the bulb clears the housing

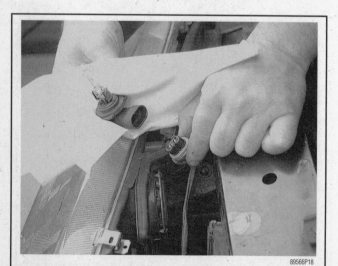

Fig. 85 When removing the bulb from the socket/harness, be careful not to touch the glass

1991–96 SC2

▶ **See Figures 86, 87, 88 and 89**

The 1991–96 SC2 model, originally known as the SC, is the only Saturn vehicle equipped with concealed headlights. The headlight doors can be opened automatically by turning the headlight switch to the headlamps **ON** position. Turning the switch back one click to the parking lights **ON** only position or 2 clicks to the lights **OFF** position will leave the doors open. To close the headlight doors, turn the switch a total of 3 clicks back to the headlight closed position.

A manual headlight control knob is located next to each headlight door and may be used to open the doors without the aid of the electric motor. To manually open the door, raise the vehicle's hood and turn the control knob until the door is fully opened. Although the knob is knurled, the top also contains a socket into which a hex key can be inserted to speed up manual headlight door operation.

1. Prior to disrupting a properly aimed headlamp, check the headlight aiming procedure for suggestions to ease adjustment. Open the headlight doors and disconnect the negative battery cable. Make sure the headlamp has cooled if it was operated in the past few minutes.

2. Remove the headlamp trim bezel screws from the sides of the lamp housing, then manually rotate the lamp sufficiently downward to provide clearance between the bezel and the lower inside corner of the lamp.

3. Manually rotate the lamp door back upward and remove the trim bezel.

4. Disconnect the spring and disengage the adjusters from the retainer, then carefully pull the lamp from the housing and disconnect the electrical harness.

5. Pull the headlamp assembly from the housing. Remove the screws, then remove the retaining ring and backing from the headlamp.

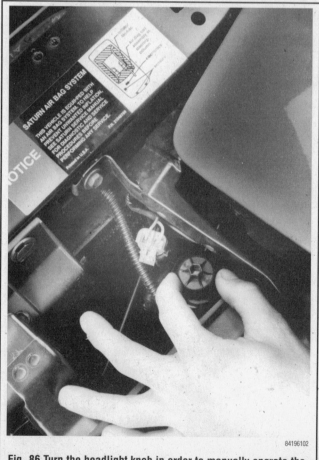

Fig. 86 Turn the headlight knob in order to manually operate the lamp door—SC or SC2

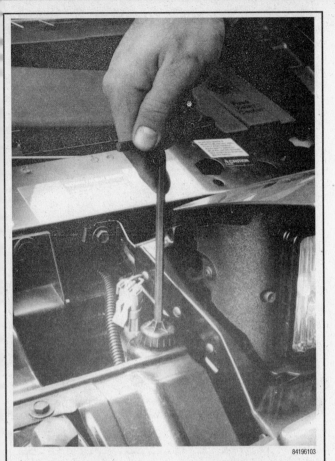

Fig. 87 The headlight knob can also be turned using a hex key—SC or SC2

To install:

6. Insert the headlamp in the backing, then position the retaining ring on the assembly and secure using the screws.

7. Connect the wiring harness to the back of the lamp assembly, then position the assembly into the housing. Engage the retaining ring adjusters and install the spring.

8. Install the trim bezel to the lamp housing and secure using the screws. Raise or lower the housing door using the manual knob, as necessary, to provide clearance for the bezel.

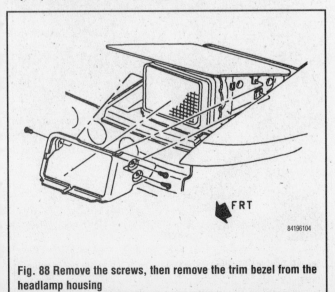

Fig. 88 Remove the screws, then remove the trim bezel from the headlamp housing

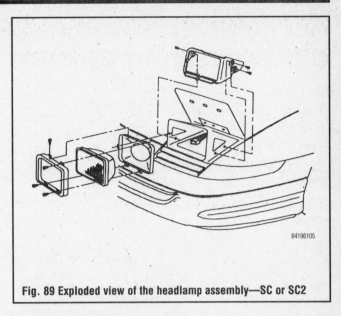

Fig. 89 Exploded view of the headlamp assembly—SC or SC2

9. Connect the negative battery cable, then check and adjust the headlight aim, as necessary, for safety.

AIMING THE HEADLIGHTS

▶ **See Figures 90, 91 and 92**

The headlights must be properly aimed to provide the best, safest road illumination. The lights should be checked for proper aim and adjusted as necessary. Certain state and local authorities have requirements for headlight aiming; these should be checked before adjustment is made.

✳✳ CAUTION

Any time that front end work is performed on your vehicle, or if the headlight aiming screws have been turned, the headlights should be accurately aimed by a reputable repair shop using the proper equipment. Headlights not properly aimed can make it virtually impossible to see and may blind other drivers on the road, possibly causing an accident. Note that the following procedure is a temporary fix, until you can take your vehicle to a repair shop for a proper adjustment.

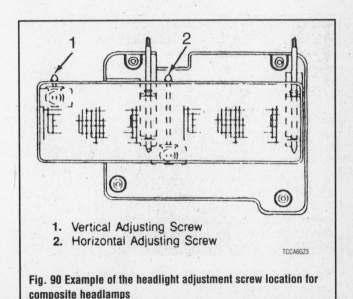

1. Vertical Adjusting Screw
2. Horizontal Adjusting Screw

Fig. 90 Example of the headlight adjustment screw location for composite headlamps

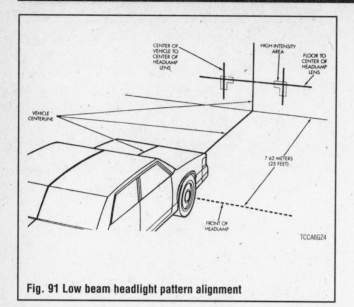

Fig. 91 Low beam headlight pattern alignment

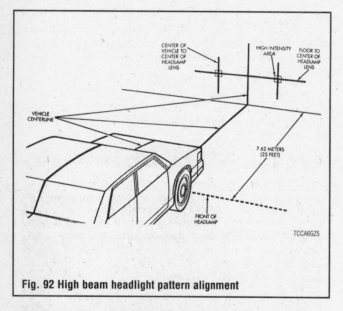

Fig. 92 High beam headlight pattern alignment

Headlight adjustment can be temporarily made using a wall, as described below, or on the rear of another vehicle. When adjusted, the lights should not glare in oncoming car or truck windshields, nor should they illuminate the passenger compartment of vehicles driving in front of you. These adjustments are rough and should always be fine-tuned by a repair shop which is equipped with headlight aiming tools. Improper adjustments may be both dangerous and illegal.

For most of the vehicles covered by this manual, horizontal and vertical aiming of each sealed beam unit is provided by two adjusting screws which move the retaining ring and adjusting plate against the tension of a coil spring. There is no adjustment for focus; this is done during headlight manufacturing.

➡Because the composite headlight assembly is bolted into position, no adjustment should be necessary or possible. Some applications, however, may be bolted to an adjuster plate or may be retained by adjusting screws. If so, follow this procedure when adjusting the lights, BUT always have the adjustment checked by a reputable shop.

Before removing the headlight bulb or disturbing the headlamp in any way, note the current settings in order to ease headlight adjustment upon reassembly. If the high or low beam setting of the old lamp still works, this can be done using the wall of a garage or a building:

1. Park the vehicle on a level surface, with the fuel tank about ½ full and with the vehicle empty of all extra cargo (unless normally carried). The vehicle should be facing a wall which is no less than 6 feet (1.8m) high and 12 feet (3.7m) wide. The front of the vehicle should be about 25 feet (7.7m) from the wall.

2. If aiming is to be performed outdoors, it is advisable to wait until dusk in order to properly see the headlight beams on the wall. If done in a garage, darken the area around the wall as much as possible by closing shades or hanging cloth over the windows.

3. Turn the headlights **ON** and mark the wall at the center of each light's low beam, then switch on the "brights" and mark the center of each light's high beam. A short length of masking tape which is visible from the front of the vehicle may be used. Although marking all four positions is advisable, marking one position from each light should be sufficient.

4. If neither beam on one side is working, and if another like-sized vehicle is available, park the second one in the exact spot where the vehicle was and mark the beams using the same-side light. Then, switch the vehicles so the one to be aimed is back in the original spot. It must be parked no closer to or farther away from the wall than the second vehicle.

5. Perform any necessary repairs, but make sure the vehicle is not moved, or is returned to the exact spot from which the lights were marked. Turn the headlights **ON** and adjust the beams to match the marks on the wall.

6. Have the headlight adjustment checked as soon as possible by a reputable repair shop.

Signal and Marker Lights

REMOVAL & INSTALLATION

Front Turn Signal and Parking Lights

SEDAN AND WAGON

▶ **See Figures 93 and 94**

1. Disconnect the turn signal bulb socket from the rear of the lens by rotating the socket ¼ turn, then pulling the socket rearward until the bulb clears the housing. Lift the socket/harness for access.
2. Remove the bulb from the socket/harness.
3. Installation is the reverse of removal.

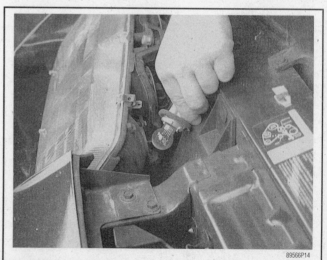

Fig. 93 Rotate the socket ¼ turn, then pull it out of the opening until the bulb clears the housing

Fig. 94 Remove the bulb from the socket/harness; this type of bulb pulls straight out

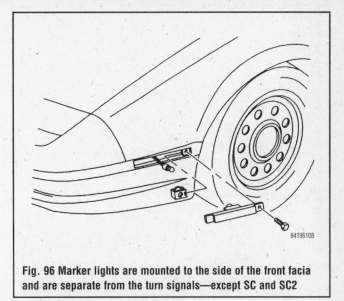

Fig. 96 Marker lights are mounted to the side of the front facia and are separate from the turn signals—except SC and SC2

COUPE

▶ See Figure 95

1. Disconnect the negative battery cable.
2. Lower the front ½ of the wheelhouse liner by removing the front retaining screws and the plastic push pin fasteners.
3. Remove the nuts and the reinforcement plate from the horizontal fender-to-facia joint and the lower rear of the facia.
4. Disconnect the lamp socket, then replace the bulb, as necessary.
5. Installation is the reverse of removal.

Side Marker Lights

SEDAN AND WAGON

▶ See Figure 96

1. Disconnect the negative battery cable.
2. Remove the retaining screws, then carefully pull the lamp out of the front facia.
3. Twist the lamp socket ¼ turn counterclockwise, then withdraw it from the lens.
4. Remove the bulb.
5. Installation is the reverse of removal.

COUPE

The side marker lights for the SC and SC2 are part of the turn signal/parking light assembly. Refer to the procedure in this section for lamp removal and bulb replacement.

Rear Turn Signal, Brake and Parking Lights

EXCEPT WAGON

▶ See Figures 97, 98, 99, 100 and 101

1. Disconnect the negative battery cable.
2. Remove the trim panel fasteners, then carefully remove the trim panel.

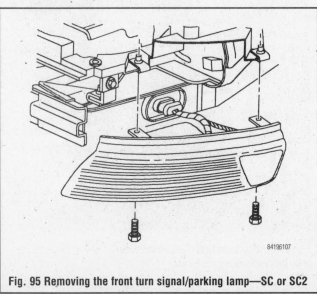

Fig. 95 Removing the front turn signal/parking lamp—SC or SC2

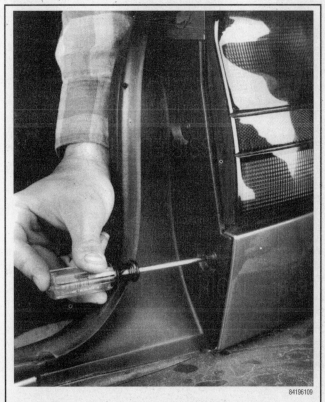

Fig. 97 Remove the trim panel fasteners—except wagon

Fig. 98 Carefully remove the trim panel—except wagon

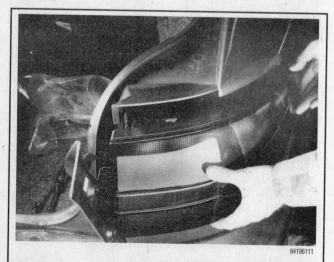

Fig. 99 Pull the lamp assembly straight back and out of the opening

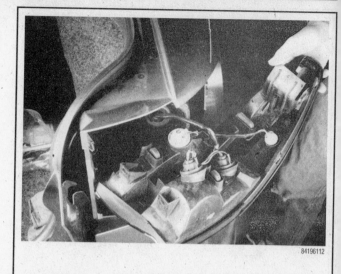

Fig. 100 Tilt the assembly downward for access to the sockets

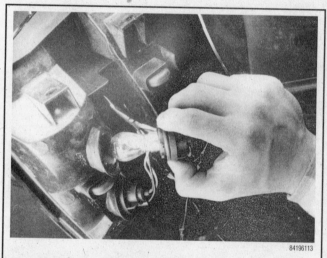

Fig. 101 Twist and remove the socket and bulb from the lamp assembly

3. Remove the 2 retaining screws from the side of the lamp assembly, then pull the lamp assembly straight back and out of the opening.

4. Tilt the assembly downward for access to the sockets, then twist and remove the socket(s) from the lamp assembly. Replace the bulb(s), as necessary.

5. Installation is the reverse of removal.

WAGON

♦ See Figure 102

1. Disconnect the negative battery cable.

2. Remove the retaining screws from the side of the lamp assembly, then pull the lamp assembly straight back and out of the opening.

3. Tilt the assembly for access to the sockets, then twist and remove the socket(s) from the lamp assembly. Replace the bulb(s), as necessary.

4. Installation is the reverse of removal.

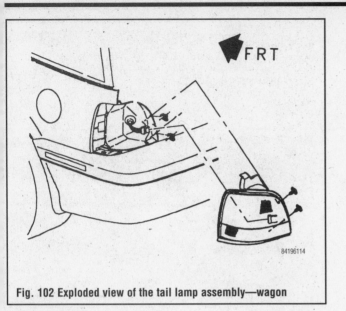

Fig. 102 Exploded view of the tail lamp assembly—wagon

High-Mount Brake Light

EXCEPT WAGON

▶ **See Figure 103**

1. Disconnect the negative battery cable.
2. Open the trunk for access to the lamps.
3. Locate the bulbs at the rear underside of the trunk lid. Remove the bulb(s) by twisting, then pulling straight out.
4. Installation is the reverse of the removal.

WAGON

▶ **See Figures 104 and 105**

1. Disconnect the negative battery cable.
2. Raise the liftgate, then remove the wedge blocks from each side of the liftgate assembly.
3. Remove the lower fasteners from the liftgate lower trim panel, by pushing in the center of each pin approximately ⅛ in. until it clicks, then remove the fastener.
4. Insert a screwdriver or small prybar into the hole in the lower trim panel (located near the wiper pivot on the pivot hump) so that the tool sits on top of the pivot. Lift up on the tool handle to disengage the trim panel upper clips, then remove the lower trim panel from the vehicle.

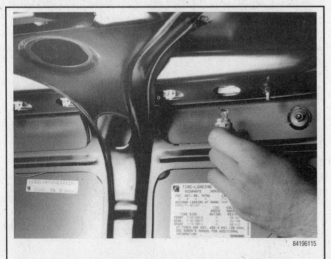

Fig. 103 Remove the high-mount brake light by twisting its socket ¼ turn counterclockwise—except wagon

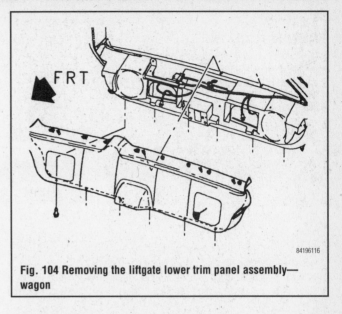

Fig. 104 Removing the liftgate lower trim panel assembly—wagon

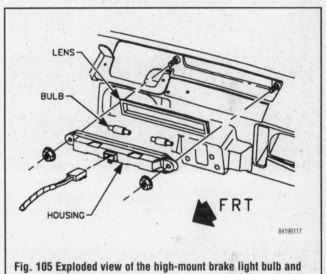

Fig. 105 Exploded view of the high-mount brake light bulb and lens assembly—wagon

5. Unplug the high-mount brake light wiring harness.
6. Remove the brake light fasteners, then remove the lamp assembly from the liftgate.
7. To remove the bulb(s), push slightly on the top edge of the lens and rotate it away from the housing. Remove the bulb(s) from the housing.

To install:

8. In order to prevent the possibility of water leaks into the liftgate, inspect the lamp gasket for damage and replace, if necessary.
9. If removed, install the bulb(s) by inserting the bottom edge of the lens into the housing and snapping it into place.
10. Position the lamp into the liftgate and tighten the fasteners to 35 inch lbs. (4 Nm).
11. Connect the wiring harness to the lamp assembly.
12. Align the upper clips on the lower trim panel to the liftgate slots, then install the panel by pushing at the clip locations.
13. Reset the trim panel push-in fasteners by spreading the center pin tabs and moving the pin so that it sits approximately ¼ in. out of the fastener. Insert the fasteners into the bottom of the trim panel and push the center pin until flush.
14. Install the liftgate wedge blocks.
15. Connect the negative battery cable and check for proper lamp operation.

Reverse Light

EXCEPT 1991–96 COUPE

For the Saturn wagons and sedans, the reverse lamp is part of the rear turn signal, brake and parking light assembly. Refer to the procedure in this section for lamp removal and bulb replacement.

1991–96 COUPE

▶ See Figure 106

1. Disconnect the negative battery cable.
2. Reach behind the bumper bar and disconnect the harness and socket from the reverse lamp. Remove the bulb from the socket.
3. If necessary, unfasten the retaining nuts and remove the lamp from the bumper bar.
4. Installation is the reverse of the removal.

Dome Light

1. Disconnect the negative battery cable.
2. Grasp the front and back of the light cover and remove by squeezing slightly, while pulling downward and tilting it toward the driver's side of the vehicle.
3. Remove the bulb from its retaining clip contacts. If the bulb has tapered ends, gently depress the spring clip/metal contact and disengage the light bulb, then pull it free of the two metal contacts.

To install:

4. Before installing the light bulb into the metal contacts, ensure that all electrical conducting surfaces are free of corrosion or dirt.
5. Position the bulb between the two metal contacts. If the contacts have small holes, be sure that the tapered ends of the bulb are situated in them.
6. To ensure that the replacement bulb functions properly, activate the applicable switch to illuminate the bulb which was just replaced. If the replacement light bulb does not illuminate, either it is faulty or there is a problem in the bulb circuit or switch. Correct as necessary.
7. Install the cover until its retaining tabs are properly engaged.
8. Connect the negative battery cable.

Trunk/Cargo Light

1. Disconnect the negative battery cable.
2. Remove the bulb from its electrical connector.
3. Installation is the reverse of removal.

License Plate Lights

▶ See Figure 107

1. Disconnect the negative battery cable.
2. Remove the license plate light lens.
3. Remove the bulb from its electrical connector.
4. Installation is the reverse of removal.

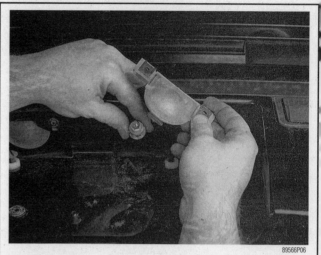

Fig. 107 After removing the license plate lamp lens, pull the bulb straight out

Fog Lights

REMOVAL & INSTALLATION

▶ See Figure 108

1. Disconnect the negative battery cable.
2. Remove the fasteners on the front of the bezel, then unplug the wiring connectors from the lamp.
3. Remove the lens from the housing.
4. Remove the retainer clip, then remove the bulb from the housing.
5. Installation is the reverse of removal.

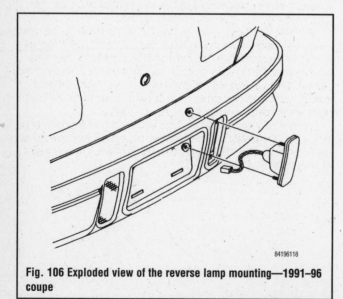

Fig. 106 Exploded view of the reverse lamp mounting—1991–96 coupe

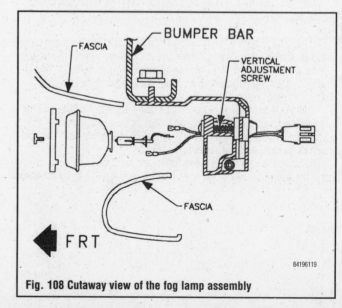

Fig. 108 Cutaway view of the fog lamp assembly

LIGHT BULBS

EXTERIOR LAMPS

	TYPE
Headlamps	
HI Beam (4–Door Sedan)	9005 HB3
LO Beam (4–Door Sedan)	9006 HB4
HI/LO Beam (2+2 Coupe)	Sealed Beam 2E1
Front Park/Turn Lamps	
4–Door Sedan	3057NA
2+2 Coupe	2057
Front Side Marker Lamps	
4–Door Sedan	24
2+2 Coupe	194
Stop/Tail Lamps	2057
Rear Turn Lamps	1156
Rear Side Marker Lamps	
4–Door Sedan	194
2+2 Coupe	168
Back–Up Lamps	
4–Door Sedan	2057
2+2 Coupe	1156
License Lamps	24
Center High–Mounted Stop Lamps	175

INTERIOR LAMPS

Ashtray Lamp	161
Cigar lamper Ring Lamp	73
Dome Lamp	562
Heater & A/C Control Lamps	NEO Wedge Bulb
Instrument Panel Warning Lamps	GE74
Instrument Cluster Illumination Lamps	GE73
Luggage Compartment Lamp	906
MAP Lamp	562
Panel Dimmer Lamp	GE73
PRND32 Shift Indicator Lamp	161

84196137

TRAILER WIRING

Wiring the vehicle for towing is fairly easy. There are a number of good wiring kits available and these should be used, rather than trying to design your own.

All trailers will need brake lights and turn signals as well as tail lights and side marker lights. Most areas require extra marker lights for overwide trailers. Also, most areas have recently required back-up lights for trailers, and most trailer manufacturers have been building trailers with back-up lights for several years.

Additionally, some Class I, most Class II and just about all Class III trailers will have electric brakes. Add to this number an accessories wire, to operate trailer internal equipment or to charge the trailer's battery, and you can have as many as seven wires in the harness.

Determine the equipment on your trailer and buy the wiring kit necessary. The kit will contain all the wires needed, plus a plug adapter set which includes the female plug, mounted on the bumper or hitch, and the male plug, wired into, or plugged into the trailer harness.

When installing the kit, follow the manufacturer's instructions. The color coding of the wires is usually standard throughout the industry. One point to note: some domestic vehicles, and most imported vehicles, have separate turn signals. On many domestic vehicles, however, the brake lights and rear turn signals operate with the same bulb. For those vehicles without separate turn signals, you can purchase an isolation unit so that the brake lights won't blink whenever the turn signals are operated.

One, final point: the best kits are those with a spring loaded cover on the vehicle mounted socket. This cover prevents dirt and moisture from corroding the terminals. Never let the vehicle socket hang loosely; always mount it securely to the bumper or hitch.

CIRCUIT PROTECTION

Fuses and Relays

There are 2 fuse or junction blocks in your Saturn vehicle. Both contain mini-fuses (small fuses with ratings from 5–30 amps), maxi-fuses (large fuses with ratings from 20–60 amps) and/or relays. The Under Hood Junction Block (UHJB) is located on the driver's side of the vehicle, near the battery. The Instrument Panel Junction Block (IPJB) is located at the base of the instrument panel center console, behind a trim piece on the passenger's side.

REPLACEMENT

Under Hood Junction Block (UHJB)

▶ See Figures 109, 110 and 111

1. Disconnect the negative battery cable.
2. Loosen the retaining screw, then remove the UHJB cover.
3. Pull the fuse or relay from its terminals. A small plastic puller is provided to help secure a firm grip on the fuses.
4. Installation is the reverse of the removal procedure.

Fig. 109 Loosen the retaining screw . . .

Fig. 110 . . . then remove the UHJB cover

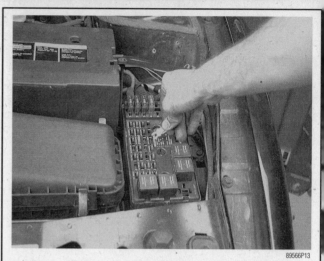

Fig. 111 Fuses can be removed with the fuse puller tool provided

Instrument Panel Junction Block (IPJB)

1. Remove the lower right trim panel from the instrument panel center column. Separate the Velcro™ fastener at the bottom of the trim panel and carefully unsnap the top corners of the panel from the fasteners.
2. Remove the applicable fuse or relay from the IPJB, based on the identification label located on the back of the trim panel. A plastic puller is also located on the trim panel to ease fuse removal.
3. Installation is the reverse of removal.

Circuit Breakers

RESETTING

Saturn vehicles use circuit breakers to protect devices which are subject to intermittent overloads, such as the power window motor. These breakers will automatically reset once the overload condition is removed (for example, when the window switch is released).

Flashers

REPLACEMENT

The turn signal flasher is located in the IPJB. Check the label on the back of the trim panel for the exact location on your vehicle.

To replace the flasher, simply grasp and pull it straight out. Align the contacts on the replacement flasher with its mounting position in the IPJB, then push it in place.

Troubleshooting Basic Turn Signal and Flasher Problems

Most problems in the turn signals or flasher system can be reduced to defective flashers or bulbs, which are easily replaced. Occasionally, problems in the turn signals are traced to the switch in the steering column, which will require professional service.

F = Front   R = Rear   ● = Lights off   ○ = Lights on

Problem		Solution
Turn signals light, but do not flash		• Replace the flasher
No turn signals light on either side		• Check the fuse. Replace if defective. • Check the flasher by substitution • Check for open circuit, short circuit or poor ground
Both turn signals on one side don't work		• Check for bad bulbs • Check for bad ground in both housings
One turn signal light on one side doesn't work		• Check and/or replace bulb • Check for corrosion in socket. Clean contacts. • Check for poor ground at socket
Turn signal flashes too fast or too slow		• Check any bulb on the side flashing too fast. A heavy-duty bulb is probably installed in place of a regular bulb. • Check the bulb flashing too slow. A standard bulb was probably installed in place of a heavy-duty bulb. • Check for loose connections or corrosion at the bulb socket
Indicator lights don't work in either direction		• Check if the turn signals are working • Check the dash indicator lights • Check the flasher by substitution
One indicator light doesn't light		• On systems with 1 dash indicator: See if the lights work on the same side. Often the filaments have been reversed in systems combining stoplights with taillights and turn signals. Check the flasher by substitution • On systems with 2 indicators: Check the bulbs on the same side Check the indicator light bulb Check the flasher by substitution

TCCA6C02

UNDERHOOD JUNCTION BLOCK CHARTS

MAXIFUSE/MINIFUSE FEEDS

This chart lists the maxifuses and the minifuses they control.

NO.	FUSE (CB)	COLOR (AMP)	MINI FUSE (AMP)
1	PWR CONVCE	Green (30)	Feeds Directly to Power Window Relay, Sun (15) (after Power Window Relay)
2	IGN 3	Green (30)	IGN 3 (7.5), Cruise (5), Recirc (10)
3	Spare		
4	Spare		
5	IGN 4	Green (30)	IGN 1 (10), Turn (7.5), B/U LP (10), EGR (10), EIS (7.5), PCM I (5), INJ (10), Wiper (25), Air Bag (10), Radio (10)
6	IP BATT	Green (30)	Body (10), F Pump (10), Locks (20), Chime (5), TRS 3/4 (10), TRS 2/TCC (10), TRS LP (7.5)
7	COOL FAN	Green (30)	Feeds Directly to Cooling Fan Relay
8	ABS	Green (30)	Feeds Directly to ABS Relay

89566G07

MAXIFUSE TO SYSTEM FEEDS

The following chart lists the maxifuses and the systems they control.

NO.	FUSE (CB)	COLOR (AMP)	SYSTEM
1	PWR CONVCE	Green (30)	Power Windows, Power Sunroof, Rear Wiper (Wagon)
2	IGN 3	Green (30)	Heater-Air Conditioning Blower Motor, Air Recirculation Door, Rear Defogger, Antilock Brakes, Power Windows, Air Conditioning, Cruise Control, Starter Solendoid, DRL Relay, Generator
3	Spare		
4	Spare		
5	IGN 4	Green (30)	Transaxle Shift Program Switch, Turn Signals, Brake Transaxle Shift Interlock, Backup Lamps, Linear EGR Solenoid, EVAP Canister Purge Solenoid, Electronic Ignition System, I/P Cluster Telltales, Lamps and Gauges, Powertrain Control Module, Fuel Injectors, Radio, Windshield Wipers/Washer, SIR, DRL Module, Chime, Remote Keyless Entry, Heated Oxygen Sensor, Coolant Level Switch
6	IP BATT	Green (30)	Chime, Wagon Hatch Ajar, Fuel System, Dome Lamps, Luggage Compartment Lamp, Outside Electric Mirror, Map Lamps, Radio/Clock Memory, Door Locks, Cigar Lighter, Automatic Transaxle Actuators, Remote Keyless Entry, Decklid/Liftgate Release
7	COOL FAN	Green (30)	Engine Cooling Fan
8	ABS	Green (30)	Antilock Brakes

89566G08

MINIFUSES CIRCUIT FEEDS

The following chart lists the minifuses, located in the underhood junction block, and the circuits they control.

FUSE	COLOR (AMP)	SYSTEM
TRS 2/TCC	Red (10)	Automatic Transmission 2nd Gear and Torque Converter Clutch Actuators
LH DR	Red (10)	Left Hand Headlamp Door Motor
HORN	Red (10)	Horn
BRAKE	Blue (15)	Stop Lamps
TRS 3/4	Red (10)	Automatic Transmission 3rd and 4th Gear Actuators
TRS LP	Brown (7.5)	Automatic Transmission Line Pressure Actuator
A/C	Brown (7.5)	Air Conditioning System
RR DFG	Natural (25)	Rear Window Defogger
HAZARD	Red (10)	Hazard Flasher, Headlamp HI Beam Indicator DRL Relay Center (SC2 Only), DRL Module
ABS	Tan (5)	Antilock Brake System Memory
ABS	Red (10)	ABS Left and Right Front Solenoids
EIS	Brown (7.5)	Electronic Ignition System
EGR	Red (10)	Linear EGR Solenoid, EVAP Canister Purge Solenoid, Heated Oxygen Sensor, PCM
RH DR	Red (10)	Right Hand Headlamp Door Motor
R HDLP	Red (10)	Right Headlamp Lights
B/U LP	Red (10)	Backup Lamps, Coolant Level Switch
INJ	Red (10)	Fuel Injectors
PCM 1	Tan (5)	Powertrain Control Module
PCM B	Tan (5)	Powertrain Control Module Memory
L HDLP	Red (10)	Left Headlamp Lights
FOG LP	Blue (15)	Fog Lamps
PARK	Red (10)	Park Lamps, Tail Lamps, Marker Lamps, Radio Display Lamp, IP Cluster Lamps, Cigar Lighter Ring Lamp, Ashtray Lamp, HVAC Control Head Lamps, I/P Dimmer, PRNDL Lamp, License Lamps

89566G09

RELAY CIRCUIT FEEDS

The following chart lists the relays, located in the underhood junction block, and the systems they control.

RELAY	MAXIMUM LOAD	SYSTEM
A/C	35 AMP	Air Conditioning System
COOL FAN	35 AMP	Engine Cooling Fan
ABS	N/A	Antilock Brake System (Non interchangeable with other relays)
FOG LAMP	35 AMP	Fog Lamps
DRL	20 AMP – Normally Open Contact 10 AMP – Normally Closed Contact	Daytime Running Lamps (Sedan, Wagon and Canadian SC1)

889566G10

INSTRUMENT PANEL JUNCTION BLOCK CHARTS

MINIFUSES CIRCUIT FEEDS

The following chart lists the minifuses, located in the instrument panel junction block, and the systems they control.

NOTICE: Fuses are numbered in instrument panel junction block. A label is provided on the trim cover for fuse system description.

NO.	FUSE	COLOR (AMP)	SYSTEM
1	–	–	Spare 1
2	RADIO	Red (10)	Radio
3	WIPER	Natural (25)	Windshield Wipers and Washers
4	IGN 1	Red (10)	Transaxle Shift Program Switch, I/P Cluster Telltale and Gauges, DRL Module, Chime, Remote Keyless Entry
5	TURN	Brown (7.5)	Turn Signals, Brake Transaxle Shift Interlock
6	AIR BAG	Red (10)	Sensing and Diagnostic Module (SDM)
7	–	–	Spare
8	–	–	Spare
9	IGN 3	Brown (7.5)	Antilock Brakes, Power Windows, Air Conditioning, Traction Switch, DRL, Chime, Rear Defogger, Blower Motor, Generator
10	CRUISE	Tan (5)	Cruise Control Module, Cruise Control Switch, Brake Switch
11	RECIRC	Red (10)	HVAC Recirculation Motor
12	LOCKS	Yellow (20)	Power Door Locks, Cigar Lighter
13	–	–	Spare
14	CHIME	Tan (5)	Radio/Clock Memory, Chime Module, Cargo Lamp (Wagon), Liftgate Ajar Indicator Light (Wagon)
15	BODY	Red (10)	Dome Lamp, Luggage Compartment Lamp (SL/SC), Outside Remote Mirror, Map Lamps, Liftgate/Decklid Release Switch, Door Lock Switches, Remote Keyless Entry
16	–	–	Spare
17	SUN	Blue (15)	Power Sunroof Rear Wiper/Washer (Wagon)
18	F PUMP	Red (10)	Fuel Pump

89566G11

RELAY/MODULE CIRCUIT FEEDS

The following chart lists the relays, located in the instrument panel junction block, and the systems they control.

RELAY	MAXIMUM LOAD	SYSTEM
WINDOW	35 AMP	Power Windows, Sunroof, Rear Wiper/Washer (Wagon)
FUEL PUMP	35 AMP – Normally Open Contact 20 AMP – Normally Closed Contact	Fuel Pump, Transaxle Actuators
FLASHER MODULE	6 Bulbs	Turn Signal/Hazard Flashers, (Back side of IPJB)
RR DEFOG	35 AMP	Rear Defogger
UNLOCK	20 AMP – Normally Closed Contact 10 AMP – Normally Closed Contact	Door Locks (Unlock)
LOCK	20 AMP – Normally Closed Contact 10 AMP – Normally Closed Contact	Door Locks (Lock)
CHIME MODULE	—	Chime, Rear Defogger Control, Seatbelt Telltale

89566G12

INDEX OF WIRING DIAGRAMS

89566W01

SAMPLE DIAGRAM: HOW TO READ & INTERPRET WIRING DIAGRAMS

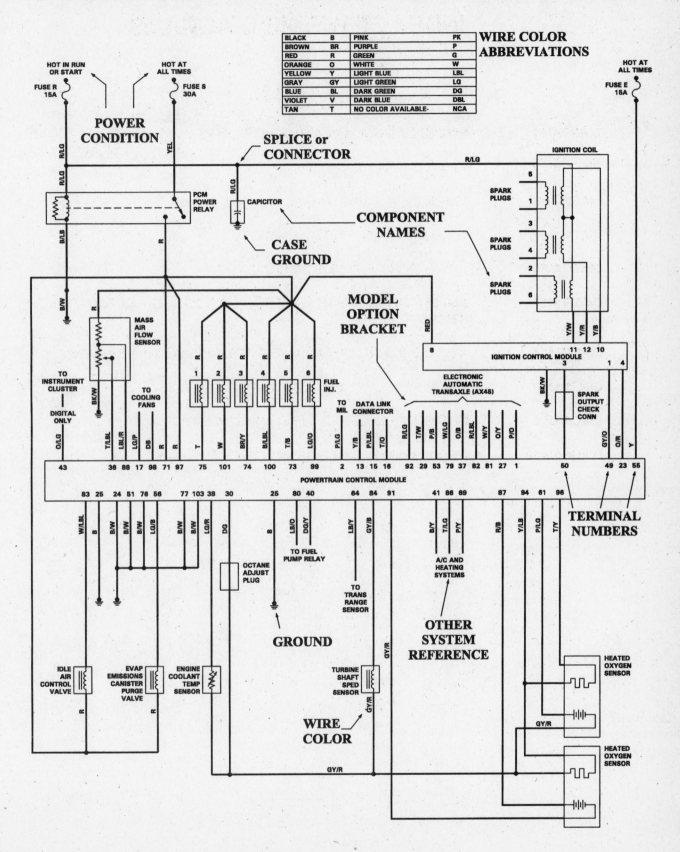

DIAGRAM 1

TCCA6W01

WIRING DIAGRAM SYMBOLS

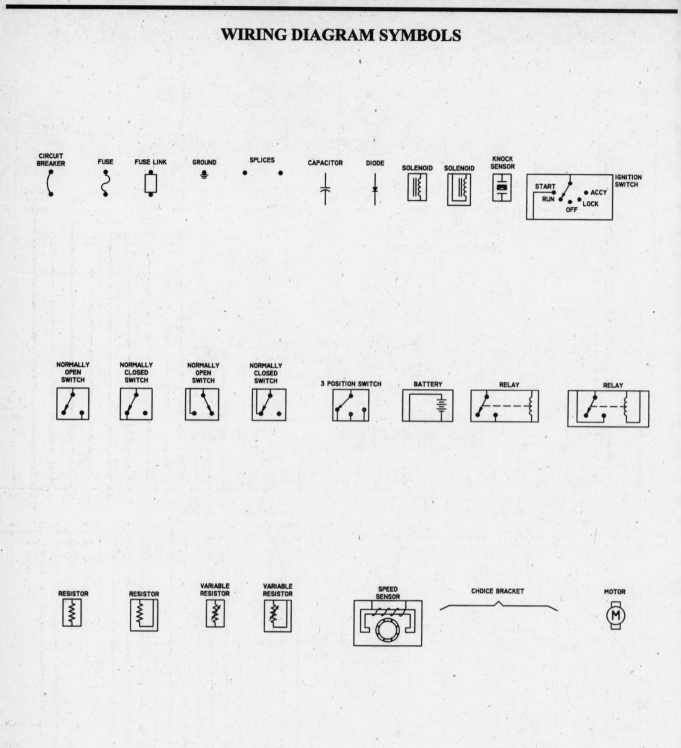

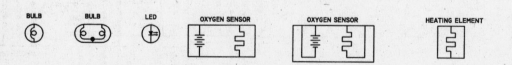

DIAGRAM 2

TCCA6W02

1996-98 ENGINE SCHEMATIC

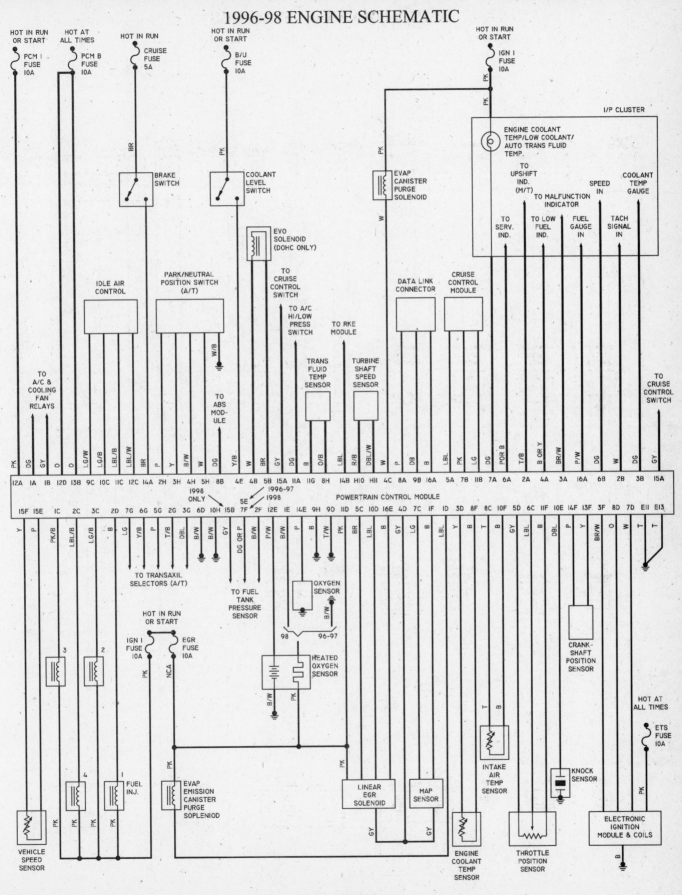

DIAGRAM 3

89566E01

1991-95 (DOHC) ENGINE SCHEMATIC

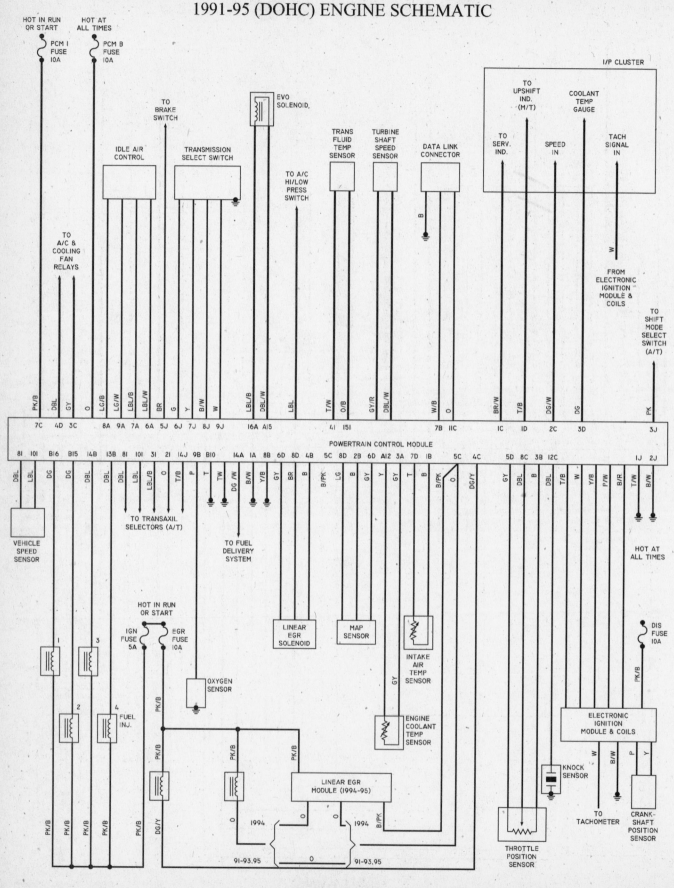

DIAGRAM 4

89566E02

1991-95 (SOHC) ENGINE SCHEMATIC

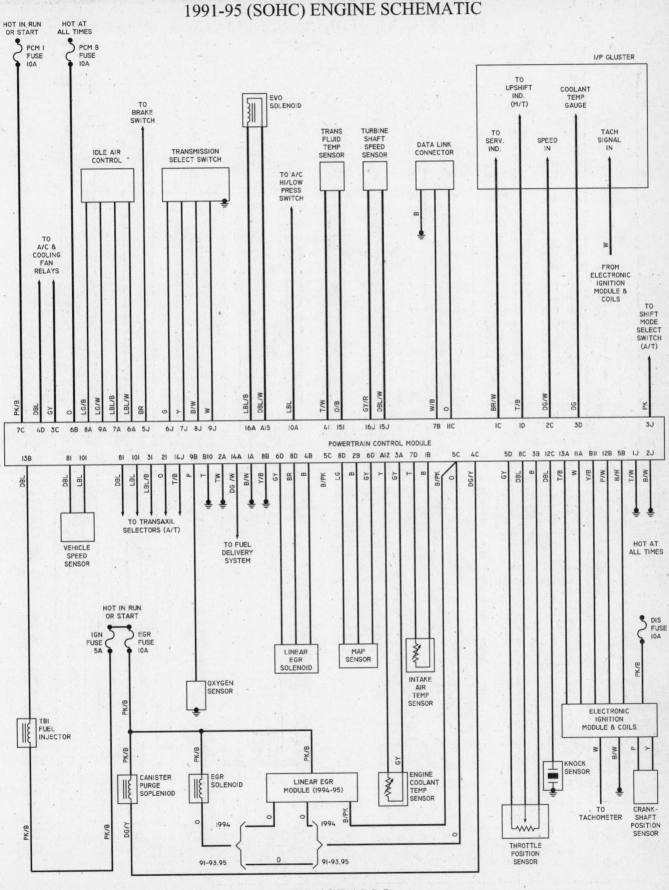

DIAGRAM 5

89566E03

1991-98 CHASSIS SCHEMATICS

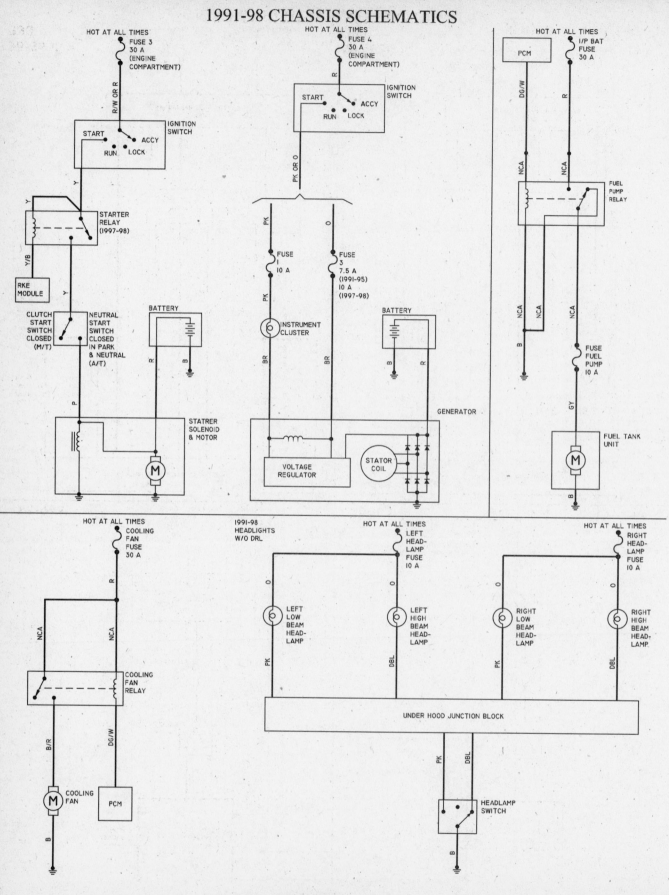

DIAGRAM 6

89566B01

1991-98 CHASSIS SCHEMATICS

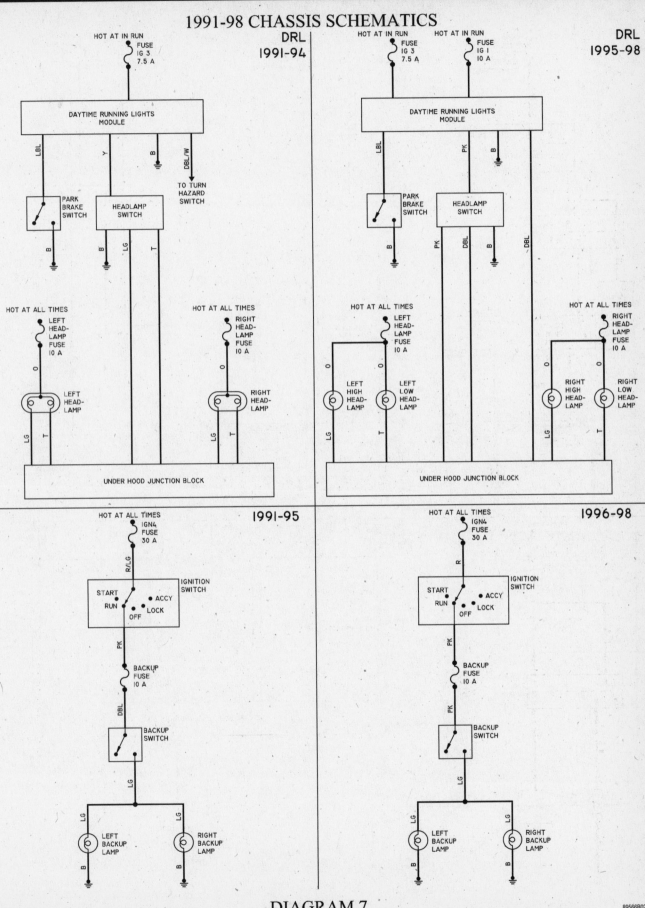

DIAGRAM 7

89566B02

1991-94 CHASSIS SCHEMATICS

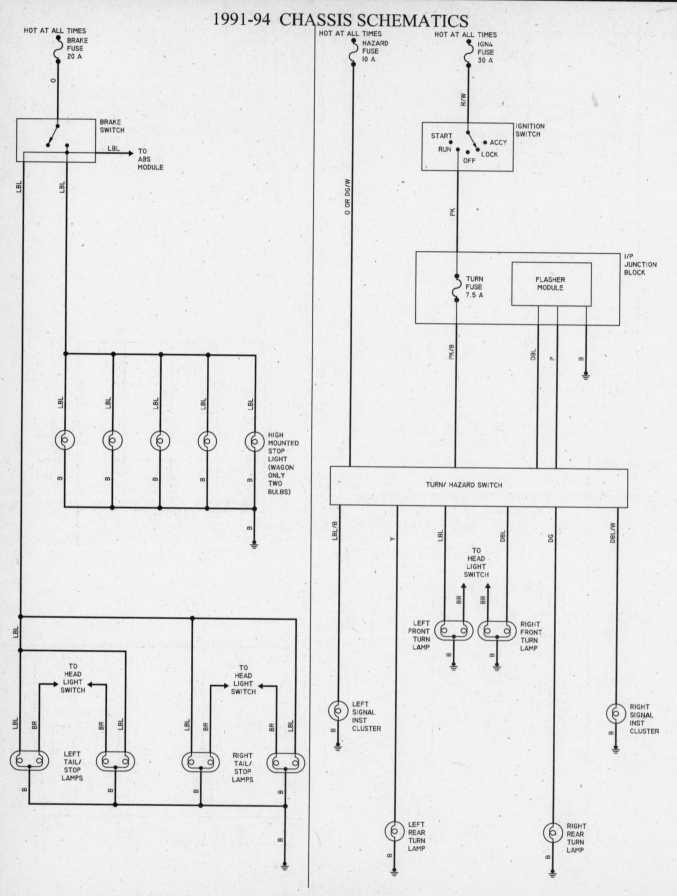

DIAGRAM 8

89566B03

1995-98 CHASSIS SCHEMATICS

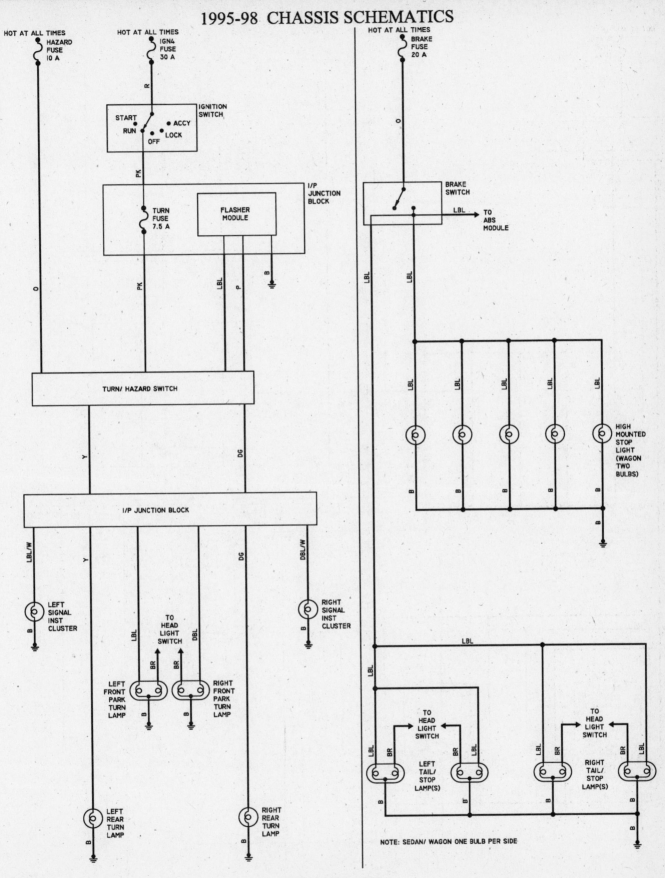

DIAGRAM 9

89566B04

1991-98 CHASSIS SCHEMATICS

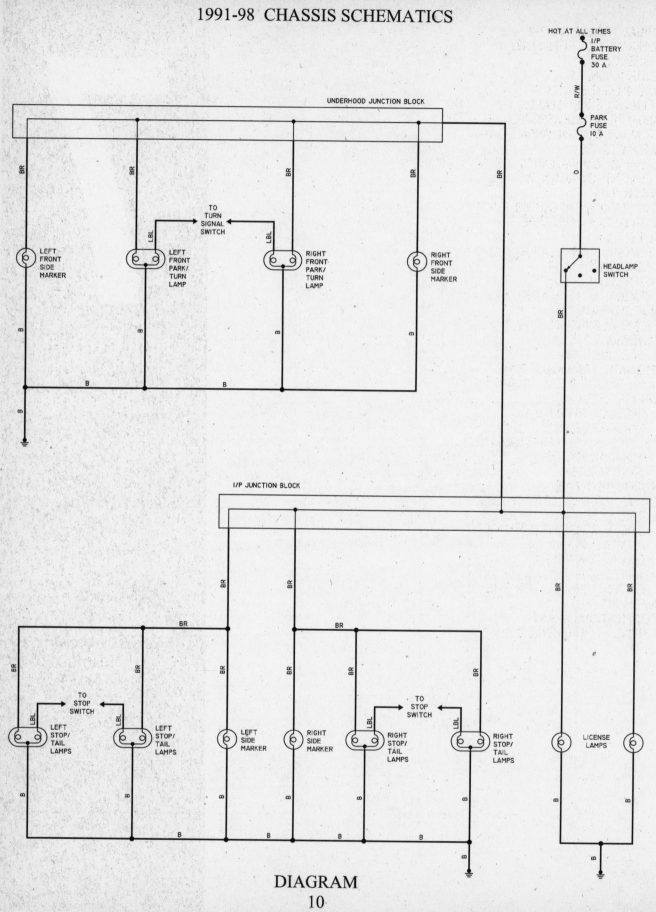

DIAGRAM

10

89566B05

7

DRIVE TRAIN

MANUAL TRANSAXLE

Understanding the Manual Transaxle

Because of the way an internal combustion engine breathes, it can produce torque, or twisting force, only within a narrow speed range. Most modern, overhead valve pushrod engines must turn at about 2500 rpm to produce their peak torque. By 4500 rpm they are producing so little torque that continued increases in engine speed produce no power increases. The torque peak on overhead camshaft engines is generally much higher, but much narrower.

The manual transaxle and clutch are employed to vary the relationship between engine speed and the speed of the wheels so that adequate engine power can be produced under all circumstances. The clutch allows engine torque to be applied to the transaxle input shaft gradually, due to mechanical slippage. Consequently, the vehicle may be started smoothly from a full stop. The transaxle changes the ratio between the rotating speeds of the engine and the wheels by the use of gears. The gear ratios allow full engine power to be applied to the wheels during acceleration at low speeds and at highway/passing speeds.

In a front wheel drive transaxle, power is usually transmitted from the input shaft to a mainshaft or output shaft located slightly beneath and to the side of the input shaft. The gears of the mainshaft mesh with gears on the input shaft, allowing power to be carried from one to the other. All forward gears are in constant mesh and are free from rotating with the shaft unless the synchronizer and clutch is engaged. Shifting from one gear to the next causes one of the gears to be freed from rotating with the shaft and locks another to it. Gears are locked and unlocked by internal dog clutches which slide between the center of the gear and the shaft. The forward gears employ synchronizers; friction members which smoothly bring gear and shaft to the same speed before the toothed dog clutches are engaged.

Identification

▶ See Figures 1 and 2

There are two 5-speed manual transaxles used in all Saturn vehicles. They are virtually the same by design and differ internally almost only by gear ratio. The 2 transaxles are equipped with different ratio/toothed gears in order to best utilize the engines to which they are mated. The MP2 transaxle is found only on vehicles with the SOHC engine, while the MP3 was available only with the DOHC engine.

If there is a question as to whether the transaxle in your vehicle is the original component, an identification code is stamped on the top of the transaxle clutch housing. The code includes information such as model year, transaxle

	MP2 (Base) Transaxle Teeth	MP3 (Performance) Transaxle Teeth		MP2 (Base) Transaxle Ratios	MP3 (Performance) Transaxle Ratios
1st Drive:	13	12	1st Gear:	3.077	3.250
1st Driven:	40	39	2nd Gear:	1.810	2.056
2nd Drive:	21	18	3rd Gear:	1.207	1.423
2nd Driven:	38	37	4th Gear:	0.861	1.033
3rd Drive:	29	26	5th Gear:	0.643	0.730
3rd Driven:	35	37	Reverse Gear:	2.923	2.923
4th Drive:	36	31	Final Drive Ratio:	4.063	4.063
4th Driven:	31	32			
5th Drive:	42	37			
5th Driven:	27	27			
Reverse Drive:	13	13			
Reverse Idler:	31	31			
Reverse Driven:	38	38			

84197002

Fig. 2 Manual transaxle gear specifications

type (MP2 or MP3), where the transaxle was built and when the unit was built. Once disassembled, there is another way to identify the transaxle. All gears from the MP3 have an identification groove on the outer diameter of the gear teeth. No such groove will appear on the gears of the MP2.

Back-up Light Switch

REMOVAL & INSTALLATION

▶ See Figure 3

1. Disconnect the negative battery cable.
2. For easier access, it may be necessary to remove the air cleaner and intake tube assembly.
3. Unplug the switch electrical connector.
4. Loosen the switch and remove it from the transaxle shift housing.
5. Installation is the reverse of removal. However, before installing, apply a light coating of Teflon® sealant, P/N 21485278 or equivalent, to the threads of the switch.
6. Tighten the switch to 28 ft. lbs. (38 Nm).

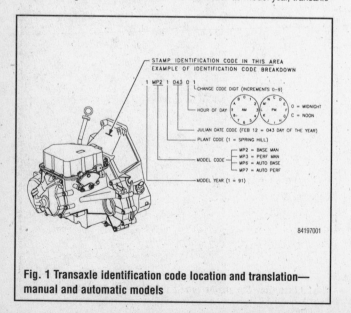

84197001

Fig. 1 Transaxle identification code location and translation—manual and automatic models

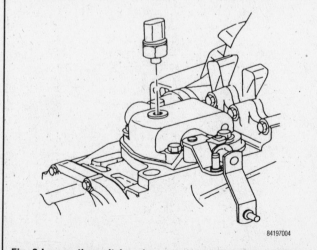

84197004

Fig. 3 Loosen the switch and remove it from the transaxle shift housing

Manual Transaxle Assembly

REMOVAL & INSTALLATION

▶ **See Figures 4 thru 10**

1. Properly disable the SIR system, if equipped, and disconnect the negative, then positive battery cables.
2. Remove the air inlet duct fasteners, detach the air temperature sensor connector and remove the air cleaner/inlet duct assembly.
3. Unfasten the battery hold-down, then remove the battery from the vehicle. Place the battery in a safe location, but do not store on a concrete surface or discharge will result. Next, unfasten and remove the battery tray. Note that one battery tray bolt is located in the fender well area.
4. For 1992–98 vehicles, remove the transaxle strut-to-cradle bracket through-bolt located on the radiator side of the transaxle. Flip the transaxle strut out of the way.
5. Disengage the back-up light switch and vehicle speed sensor electrical connectors from the transaxle. Remove the vent tube retaining clip and discard if damaged.
6. Remove the 2 ground terminals from the top 2 clutch housing studs.

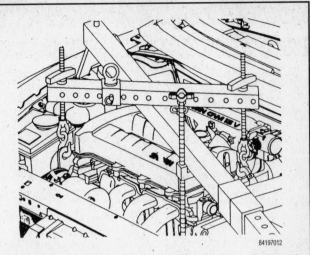
Fig. 6 When installing an engine support, position the bar hooks to the engine support brackets

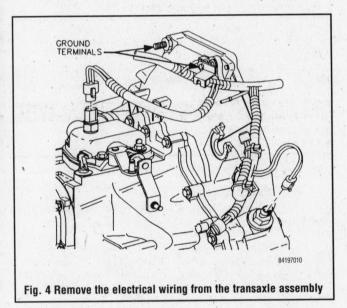
Fig. 4 Remove the electrical wiring from the transaxle assembly

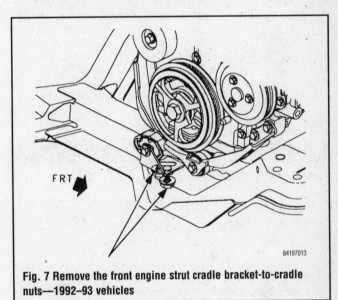
Fig. 7 Remove the front engine strut cradle bracket-to-cradle nuts—1992–93 vehicles

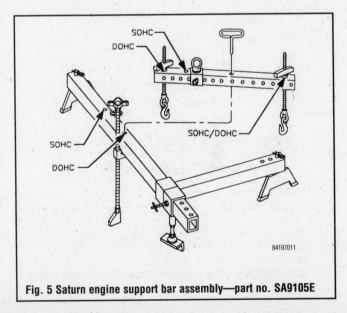
Fig. 5 Saturn engine support bar assembly—part no. SA9105E

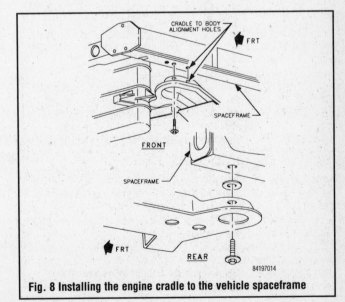
Fig. 8 Installing the engine cradle to the vehicle spaceframe

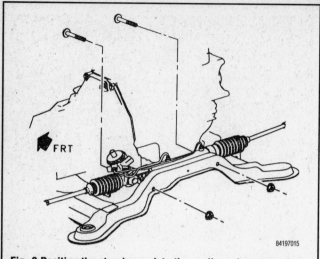

Fig. 9 Position the steering rack to the cradle and secure using the fasteners

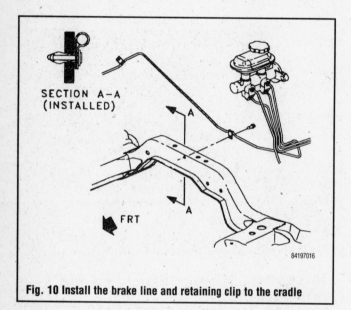

Fig. 10 Install the brake line and retaining clip to the cradle

7. On 1991 vehicles, unclip the oxygen sensor wire from the clutch housing.

8. On 1991 DOHC engines, remove the nut from the cradle-to-front engine cable bracket at the front right corner of the cradle, below the water pump.

9. Remove the top 2 clutch housing studs.

10. Remove the electronic ignition coil module. Discard the old coil retaining bolts and replace with new bolts upon installation.

11. For 1991 vehicles, loosen the 2 front transaxle mount-to-transaxle bolts.

12. Remove the shifter cables from the shift arms and clutch housing, taking care not to damage the cable bolt.

13. Rotate the clutch slave cylinder ¼ turn counterclockwise while pushing into the clutch housing, then remove the cylinder from the housing. Remove the 2 clutch hydraulic damper-to-clutch housing bolts, then wire the hydraulic assembly to the upper radiator hose.

14. Wire the radiator to the upper radiator support to hold the assembly in place when the cradle is removed.

15. Install engine support bar assembly tool SA9105E or an equivalent. Make sure the support feet are positioned on the outer edge of the shock tower, the bar hooks are connected to the engine bracket and the stabilizer foot is on the engine block, to the right of the engine oil dipstick.

16. Raise the front of the vehicle sufficiently to lower the transaxle and engine cradle. Support the vehicle safely using jackstands. Make sure the jackstands are not positioned under the engine cradle. Remove the drain plug from the lower center of the housing and drain the transaxle fluid into a clean container.

17. Remove the front wheels and the left, right and center splash shields from the vehicle. For coupes, remove the left and right lower facia braces.

18. For 1992–98 vehicles, remove the front engine strut cradle bracket-to-cradle nuts from below the cradle.

19. Remove the transaxle mount-to-cradle nut from under the cradle.

20. Remove the front exhaust pipe nuts at the manifold, then disconnect the pipe from the support bracket.

21. Remove the front pipe-to-catalytic converter bolts and lower the pipe from the vehicle.

22. Remove the engine-to-transaxle stiffening bracket bolts and remove the bracket. For 1991 vehicles, remove the rear transaxle mount bracket-to-transaxle bolt, then loosen the mount bolt and allow the bracket to hang out of the way.

23. Remove the clutch housing dust cover. Remove the steering rack-to-cradle bolts and wire the gear for support when cradle is removed. Remove the brake line and retainer from the cradle.

24. Remove and discard the cotter pin from the lower ball joints. Back off the ball joint nut until the top of the nut is even with the top of the threads.

25. Use ball joint separator tool SA9132S or equivalent to separate the ball joint from the lower control arm, then remove the nut. Do not use a wedge tool or seal damage may occur.

✳✳ WARNING

The outer CV-joint for vehicles equipped with ABS contains a speed sensor ring. Use of an incorrect tool to separate the control arm from the knuckle may result in damage to and loss of the ABS system.

26. For 1991 vehicles, remove the front transaxle mount-to-cradle nuts and the engine lower mount-to-cradle nuts. These nuts are all located under the front or side cradle members.

27. Position two 4 in. **x** 4 in. **x** 36 in. pieces of wood onto a powertrain support dolly, then position the dolly under the vehicle and against the cradle.

28. Remove the 4 cradle-to-body bolts and carefully lower the cradle from the vehicle with the support dolly. Tape or wire the 2 large washers from the rear cradle-to-body attachments in position to prevent loss.

29. Support the transaxle securely with a suitable jack.

30. Use an appropriate prybar to separate the left side halfshaft from the transaxle. Remove the halfshaft sufficiently to install seal protector tool SA91112T or equivalent around the shaft and into the seal to prevent cuts by the shaft spline.

31. Remove the 2 bottom clutch housing-to-engine bolts and install a guide bolt into the bottom rear clutch housing bolt hole from the side of the engine block.

32. Carefully separate the transaxle from the engine enough to clear the intermediate shaft (on the right side) and lower the transaxle from the vehicle.

To install:

33. Installation of the manual transaxle assembly is the reverse of removal. However, please note the following important steps:

34. Use a 6 **x** 1.0mm tap to clean the sealant from the ignition coil module mounting holes.

35. Place the transaxle assembly securely onto the jack and position under the vehicle for installation. Install axle seal protectors into the seals on both sides.

36. Tighten the 2 lower clutch housing-to-engine bolts to 103 ft. lbs. (140 Nm). Do not use the bolts to draw the transaxle to the engine.

37. Be sure to remove the seal protectors before installing the halfshaft. Push the halfshaft all the way into the transaxle and install the snapring. Remove the transaxle jack.

38. Clean and lubricate the ball joint threads before installing the ball joints into the knuckles.

39. Verify the correct positioning of the lower control arm ball studs to the knuckles, the cooling module support bushings, the engine strut bracket and the transaxle mount.

40. Insert ⁹⁄₁₆ inch round steel rods into the cradle-to-body alignment holes, near the front cradle-to-body fastener holes.

41. Tighten the 4 cradle-to-body mounting bolts to 155 ft. lbs. (210 Nm). Remove the support dolly, then the engine support bar assembly.

42. For 1992–98 vehicles, tighten the transaxle strut-to-cradle bracket through-bolt and nut to 52 ft. lbs. (70 Nm).

43. Remove the radiator assembly support wire.

44. Tighten the new ignition module bolts, with sealant, to 71 inch lbs. (8 Nm) and verify that the bolt heads are properly seated on the ignition module.

45. Tighten the 2 top clutch housing-to-engine studs to 74 ft. lbs. (100 Nm). Tighten the 2 ground terminals to 19 ft. lbs. (25 Nm).

46. Install the 2 slave cylinder-to-clutch housing nuts and tighten to 18 ft. lbs. (25 Nm).

47. Tighten the battery tray assembly bolts to 89 inch lbs. (10 Nm).

48. For 1991 vehicles, tighten the front transaxle mount-to-transaxle bolts and install the front transaxle mount cradle nuts. Tighten the fasteners to 35 ft. lbs. (48 Nm). Then, install the right side mount-to-cradle nuts and tighten to 40 ft. lbs. (54 Nm).

49. For 1992–98 vehicles, tighten the transaxle lower mount-to-cradle fastener and the 2 engine strut cradle bracket-to-cradle nuts to 37 ft. lbs. (50 Nm).

50. Tighten the steering gear bolts and nuts to 37 ft. lbs. (50 Nm).

51. Tighten the clutch housing dust cover fasteners to 97 inch lbs. (11 Nm).

52. Tighten the powertrain stiffening bracket bolts to 40 ft. lbs. (54 Nm).

53. Tighten the exhaust manifold-to-exhaust pipe retaining nuts in a crosswise pattern to 23 ft. lbs. (31 Nm). Tighten the front pipe-to-catalytic converter flange bolts to 18 ft. lbs. (25 Nm). Finally, tighten the front pipe-to-engine support bracket fasteners to 23 ft. lbs. (31 Nm).

➡ **If the converter flange threads are damaged, use the Saturn 21010753 converter fastener kit in place of the self-tapping screws to provide proper clamp load and prevent exhaust leaks.**

54. Tighten the ball joint nuts to 55 ft. lbs. (75 Nm). Continue to tighten the nuts as necessary and install new cotter pins.

55. For coupes, install the right left lower facia braces, J-nuts and fasteners. Tighten the fasteners to 89 inch lbs. (10 Nm).

56. Install the tire and wheel assemblies, then remove the jackstands and lower the vehicle.

57. Throroughly inspect the transaxle and engine compartment area to be sure that all wires, hoses and lines have been connected. Also inspect to make sure that all components and fasteners have been properly installed.

58. Connect the positive battery cable first, then the negative cable, and fill the transaxle to the proper level using Saturn transaxle fluid or equivalent.

59. Properly enable the SIR system, if equipped.

60. Warm the engine and check the transaxle fluid. Check and adjust front end alignment, as necessary.

Halfshafts

REMOVAL & INSTALLATION

▶ **See Figures 11 thru 23**

1. Remove the wheel cover or the center cap for access to the halfshaft nut. Have an assistant depress the brake pedal and loosen the front halfshaft nut. If no one is available, pump the brakes a few times so the calipers grab and hold the brake pads, then loosen the nut.

2. Raise the front of the vehicle and support it safely using jackstands, then remove the corresponding wheel and splash shield.

Fig. 11 With the vehicle firmly on the ground, have an assistant depress the brake pedal and loosen the front halfshaft nut

Fig. 12 After removing the wheel, remove the inner fender splash shield

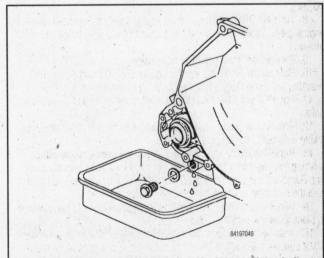

Fig. 13 If removing the left halfshaft or the intermediate shaft, first drain the transaxle fluid

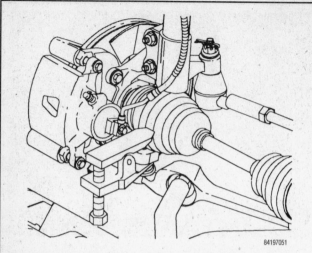

Fig. 14 Use tool SA9132S or equivalent to separate the ball joint from the lower control arm

Fig. 17 Check the tie rod end and rubber boot for rips or other damage

Fig. 15 Once the ball joint has been separated from the steering knuckle, remove the tie rod end cotter pin and castle nut

Fig. 18 Use a prybar to apply leverage and separate the control arm ball joint from the steering knuckle

Fig. 16 Use a separator tool to remove the tie rod end from the steering knuckle

Fig. 19 If difficulty is encountered, tap on the end of the half-shaft using a block of wood and a hammer

3. If removing the left side halfshaft or right side intermediate shaft, loosen the plug and drain the transaxle fluid into a clean container.

4. Remove the halfshaft nut and washer.

5. Remove and discard the cotter pin from the lower control arm ball joints. Back off the ball joint nut until the top of the nut is even with the top of the threads.

6. Use tool SA9132S or equivalent to loosen the lower control arm ball joint in the steering knuckle, then remove the nut. Do not use a wedge tool or seal damage may occur. The ball joint will be completely removed from the knuckle after the tie rod ball joint is separated.

❊❊ WARNING

The outer CV-joint for vehicles equipped with ABS contains a speed sensor ring. Use of an incorrect tool to separate the control arm from the knuckle may result in damage and loss of the ABS system.

7. Remove the tie rod cotter pin and loosen the castle nut, then separate the tie rod end from the knuckle using Tie Rod Separator SA91100C or equivalent. Once loosened sufficiently, remove the castle nut and separate the components completely. Do not use a wedge-type tool or the seal may be damaged.

8. You may wish to place a cloth over the sway bar to protect the surface, then use a prybar to leverage the knuckle and separate the lower control arm ball joint. Position a cloth at the prybar contact point with the cradle, then push down on the bar and separate the ball joint from the knuckle. Make sure the knuckle does not contact and damage the ball stud seal.

9. While pulling the knuckle/strut assembly away from the halfshaft, pull the end of the halfshaft from the wheel hub. If difficulty is encountered, tap on the end of the halfshaft using a block of wood and a hammer. Support the halfshaft assembly using a length of mechanic's wire or with a jack-stand.

10. If removing the right halfshaft, disconnect the halfshaft from the intermediate shaft by tapping the inner joint with a hammer and a block of wood. Remove the halfshaft from the vehicle.

11. If removing the left halfshaft, disconnect the halfshaft by inserting a large prybar into the space between the inner joint and transaxle. Pry the halfshaft from the transaxle, being careful not to contact and damage the transaxle oil seal. Remove the halfshaft from the vehicle.

To install:

12. If installing the left side halfshaft, install SA91112T or equivalent transaxle seal protector. Install the halfshaft into the transaxle; after the splines have safely passed the transaxle oil seal, remove the seal protector and fully seat the halfshaft.

13. If installing the right side halfshaft, insert the shaft onto the intermediate shaft and push firmly to engage the circlip. Install the shaft assembly into the transaxle.

14. Insert the outer end of the halfshaft into the wheel hub. Be careful not to damage the CV-joint boot.

15. Thoroughly clean and lubricate the ball joint stud threads of the lower control arm and tie rod end.

16. Install the lower control arm ball stud to the steering knuckle, then install the nut, but do not tighten at this time.

17. Attach the tie rod end to the steering knuckle, then install the nut. Tighten the nut to 33 ft. lbs. (45 Nm) and install a new cotter pin. If necessary, tighten the nut additionally, but do not back it off to insert the cotter pin.

18. Tighten the lower control arm ball stud nut to 55 ft. lbs. (75 Nm); tighten additionally if necessary to align the holes, then install a new cotter pin.

19. Install the washer and a new halfshaft nut, then tighten the nut to 148 ft. lbs. (200 Nm).

20. If equipped with ABS, inspect the wheel speed sensor signal for proper operation.

21. Install the inner splash shield and wheel.

22. Lower the vehicle and properly fill the transaxle.

23. Check and adjust the front end alignment as necessary.

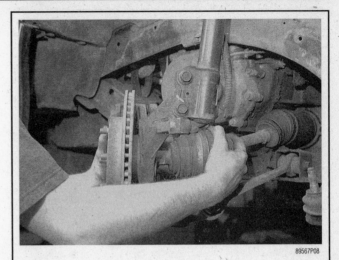

Fig. 20 While turning the steering knuckle outward, pull out the halfshaft

Fig. 21 Insert a prybar between the halfshaft and transaxle and carefully pry the shaft away

Fig. 22 Use both hands to support the joints of the halfshaft and carefully remove it from the vehicle

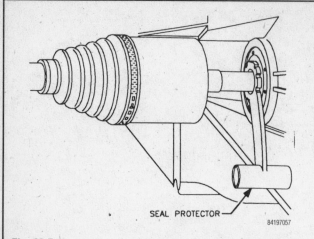

Fig. 23 Failure to use a seal protector may allow the halfshaft splines to damage the transaxle seal, causing a need for replacement

CV-JOINTS OVERHAUL

Disassembly

▶ **See Figures 24 thru 29**

1. Using soft metal or wood to protect the shaft, clamp the halfshaft to a workbench in a vise.
2. Remove and service the CV-joint from the end of the halfshaft, as follows:

  a. If the halfshaft has a damaged deflector ring, use a brass drift and hammer to remove the damaged component from the CV outer race.

  b. Either cut the outer seal retaining clamps using a side cutter or use a flat bladed screwdriver to disengage the outer band from the inner band at the retaining peg, then discard the clamp.

  c. Separate the joint seal from the CV-joint race at the large diameter, then slide the seal away from the joint, along the halfshaft. Wipe any excess grease from the CV inner race.

  d. Use a suitable pair of snapring pliers to spread the ears of the inner race retaining ring, then remove the CV-joint assembly from the shaft.

  e. Remove the seal from the halfshaft.

  f. Use a brass drift and gently tap on the cage until it is cocked and the first ball may be removed. Repeat this to remove the remaining balls.

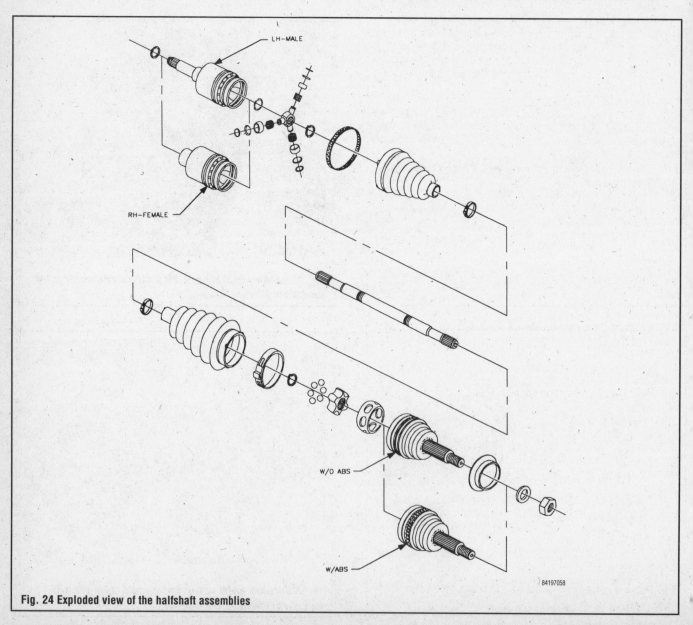

Fig. 24 Exploded view of the halfshaft assemblies

g. When the balls are removed, pivot the cage and inner race at a 90 degree angle to the center line of the outer race. The cage windows should align with the lands of the outer race, then lift the cage and inner race.

h. Rotate the inner race up and out of the cage.

i. Thoroughly clean and degrease all CV-joint parts and allow them to dry before assembly.

3. Remove and service the tri-pot joint and joint seal from the halfshaft's other end, as follows:

a. Cut the eared seal retaining clamp on the tri-pot seal using a side cutter, then discard the clamp.

b. Remove and discard the earless clamp using a small, flat bladed screwdriver.

c. Separate the seal from the tri-pot housing at the large diameter, then slide the seal away from the joint along the halfshaft. Wipe away excess grease from the face of the tri-pot spider and the inside of the housing.

d. Remove the tri-pot housing from the spider and shaft.

e. Spread the spacer ring using tool SA9198C or equivalent, and slide the spacer ring, along with the tri-pot spider, back on the halfshaft as shown.

f. Carefully remove the spider retaining ring from the halfshaft groove, and slide the spider assembly off the shaft. Use care not to lose

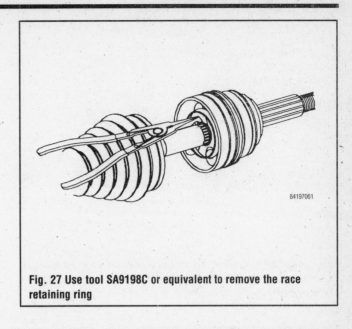

Fig. 27 Use tool SA9198C or equivalent to remove the race retaining ring

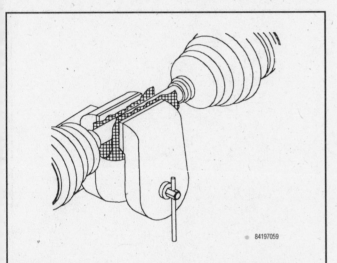

Fig. 25 Use soft metal or wood to protect the halfshaft from the vise

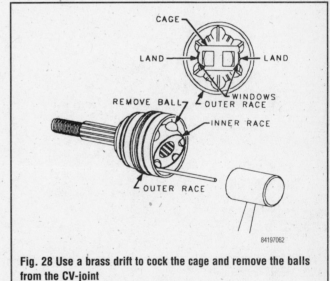

Fig. 28 Use a brass drift to cock the cage and remove the balls from the CV-joint

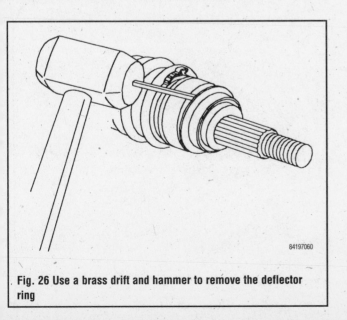

Fig. 26 Use a brass drift and hammer to remove the deflector ring

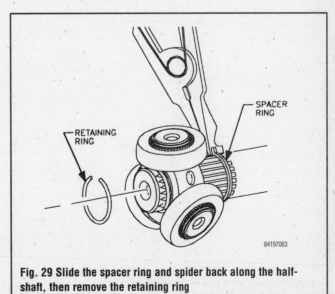

Fig. 29 Slide the spacer ring and spider back along the halfshaft, then remove the retaining ring

the tri-pot balls and needle rollers which may separate from the spider trunions.

  g. Remove the seal from the halfshaft.

  h. Thoroughly clean and degrease the housing and allow it to dry prior to assembly.

Assembly

▸ **See Figures 30, 31, 32 and 33**

1. Install the tri-pot joint and seal, as follows:

  a. Inspect the tri-pot joint components for unusual wear, cracks or damage and replace, as necessary. Clean the shaft; if rust is present in the seal mounting grooves, use a wire brush.

  b. Install the small seal retaining clamp on the neck of the seal, but do not crimp.

  c. Slide the seal onto the shaft and locate the neck of the seal on the halfshaft seal groove.

  d. Crimp the retaining clamp as illustrated, using tool SA9203C or equivalent. Measure the dimension of Section A and recrimp, if necessary.

  e. Install the spacer ring on the shaft and position it beyond the retaining groove, then install the spider, also past the retaining groove.

  f. Using SA9198 or an equivalent tool, install the retaining ring to the halfshaft ring groove, then slide the tri-pot spider towards the end of the shaft and seat the spacer ring in the axle groove.

  g. Place about ½ of the grease from the kit inside the seal and use the remainder to repack the tri-pot housing.

  h. Install the convolute retainer over the seal. The retainer must be in position when the joint is assembled or seal damage may result.

  i. Position the retaining clamp around the large diameter of the seal, then slide the tri-pot housing over the tri-pot assembly on the shaft.

  j. Slide the large diameter of the seal over the outside of the tri-pot housing and locate the seal lip in the housing groove. The seal must not be dimpled or stretched. If necessary, carefully insert a thin, flat and blunt tool between the large seal opening and outer race to equalize pressure, then seat the seal by hand and remove the tool.

  k. Position the tri-pot assembly at the proper dimensions, as illustrated, then install the large seal retaining clamp around the seal. Close the clamp using tool SA9161C or equivalent.

2. Assemble and install the outer CV-joint and seal, as follows:

  a. Inspect the CV-joint parts for any signs of unusual wear, cracks or damage and replace the joint assembly, if necessary.

  b. Put a light coat of grease on the inner and outer race grooves, then insert and rotate the inner race into the cage.

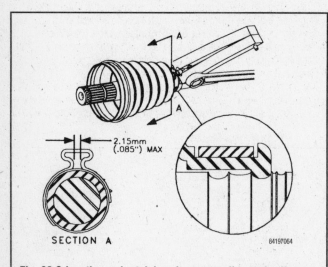

Fig. 30 Crimp the seal retaining clamp according to the dimensions shown in Section A

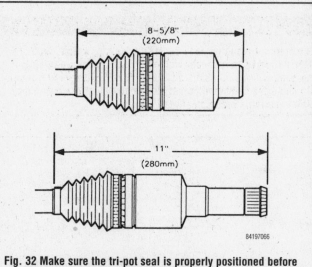

Fig. 32 Make sure the tri-pot seal is properly positioned before installing the large seal retaining clamp

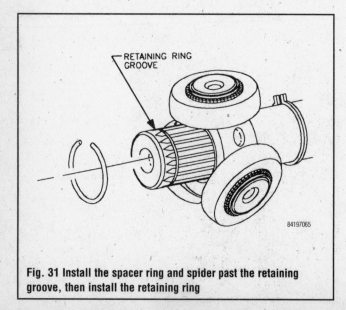

Fig. 31 Install the spacer ring and spider past the retaining groove, then install the retaining ring

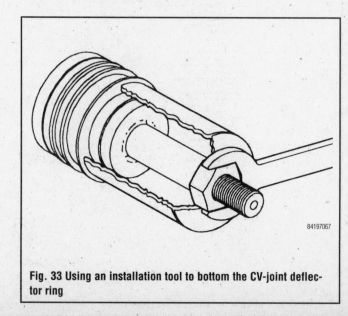

Fig. 33 Using an installation tool to bottom the CV-joint deflector ring

c. Install the cage and inner race into the outer race with the windows of the cage aligned with the lines of the outer race.

d. Use a brass drift to gently cock the cage/race and install the balls.

e. If removed, install the race retaining ring into the inner race.

f. Pack the assembled joint using the premeasured amount of grease from the service kit.

g. Install the small retaining clamp on the neck of the new seal, then slide the seal onto the shaft and locate the seal neck in the shaft seal groove.

h. Crimp the seal retaining ring using SA9203C or an equivalent crimping tool. A proper crimp will share the same dimensions of the tripot seal retaining ring crimp shown earlier.

i. Place about ½ of the provided grease inside the seal, then repack the CV-joint using the remaining grease.

j. Position the large seal retaining clamp around the seal.

k. Make sure the retaining ring side of the inner race is facing the halfshaft, then push the CV-joint onto the shaft until the ring is seated in the shaft groove.

l. Slide the seal large diameter over the outside of the CV-joint race and locate the lip of the seal in the housing groove. The seal must not be dimpled or stretched. If necessary, carefully insert a thin, flat and blunt tool between the large seal opening and outer race to equalize pressure, then seat the seal by hand and remove the tool.

m. Crimp the retaining clamp using tool SA9203C or equivalent and compare the crimp dimensions to the clamp crimped earlier. Recrimp if necessary to achieve the proper dimension.

n. If removed, position the deflecting ring at the CV-joint outer race. Use tool SA9160 or equivalent, along with a M20 x 1.0 nut, to tighten the tool until the deflector bottoms against the shoulder of the CV-joint outer race.

3. Remove the shaft assembly from the vise and install it in the vehicle.

CLUTCH

Understanding the Clutch

◆ See Figures 34 and 35

❊❊ CAUTION

The clutch driven disc may contain asbestos, which has been determined to be a cancer causing agent. Never clean clutch surfaces with compressed air! Avoid inhaling any dust from any clutch surface! When cleaning clutch surfaces, use a commercially available brake cleaning fluid.

The purpose of the clutch is to disconnect and connect engine power at the transaxle. A vehicle at rest requires a lot of engine torque to get all that weight moving. An internal combustion engine does not develop a high starting torque (unlike steam engines), so it must be allowed to operate without any load until it builds up enough torque to move the vehicle. Torque increases with engine rpm. The clutch allows the engine to build up torque by physically disconnecting the engine from the transaxle, relieving the engine of any load or resistance.

The transfer of engine power to the transaxle (the load) must be smooth and gradual; if it weren't, drive line components would wear out or break quickly. This gradual power transfer is made possible by gradually releas-

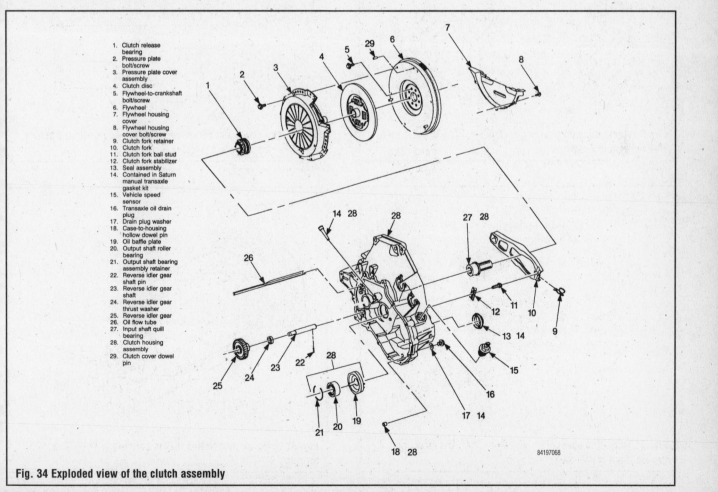

1. Clutch release bearing
2. Pressure plate bolt/screw
3. Pressure plate cover assembly
4. Clutch disc
5. Flywheel-to-crankshaft bolt/screw
6. Flywheel
7. Flywheel housing cover
8. Flywheel housing cover bolt/screw
9. Clutch fork retainer
10. Clutch fork
11. Clutch fork ball stud
12. Clutch fork stabilizer
13. Seal assembly
14. Contained in Saturn manual transaxle gasket kit
15. Vehicle speed sensor
16. Transaxle oil drain plug
17. Drain plug washer
18. Case-to-housing hollow dowel pin
19. Oil baffle plate
20. Output shaft roller bearing
21. Output shaft bearing assembly retainer
22. Reverse idler gear shaft pin
23. Reverse idler gear shaft
24. Reverse idler gear thrust washer
25. Reverse idler gear
26. Oil flow tube
27. Input shaft quill bearing
28. Clutch housing assembly
29. Clutch cover dowel pin

84197068

Fig. 34 Exploded view of the clutch assembly

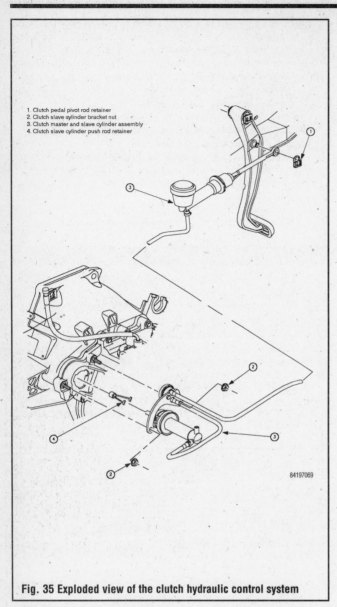

1. Clutch pedal pivot rod retainer
2. Clutch slave cylinder bracket nut
3. Clutch master and slave cylinder assembly
4. Clutch slave cylinder push rod retainer

84197069

Fig. 35 Exploded view of the clutch hydraulic control system

ing the clutch pedal. The clutch disc and pressure plate are the connecting link between the engine and transaxle. When the clutch pedal is released, the disc and plate contact each other (the clutch is engaged), physically joining the engine and transaxle. When the pedal is pushed inward, the disc and plate separate (the clutch is disengaged), disconnecting the engine from the transaxle.

The Saturn clutch is a single plate, dry friction disc with a diaphragm-style spring pressure plate. The clutch disc has a splined hub which attaches the disc to the input shaft. The disc has friction material where it contacts the flywheel and pressure plate. Torsion springs on the disc help absorb engine torque pulses. The pressure plate applies pressure to the clutch disc, holding it tight against the surface of the flywheel. The clutch operating mechanism consists of a release bearing, fork and cylinder assembly.

The release fork and actuating linkage transfer pedal motion to the release bearing. In the engaged position (pedal released), the diaphragm spring holds the pressure plate against the clutch disc, so engine torque is transmitted to the input shaft. When the clutch pedal is depressed, the release bearing pushes the diaphragm spring center toward the flywheel. The diaphragm spring pivots the fulcrum, relieving the load on the pressure plate. Steel spring straps riveted to the clutch cover lift the pressure plate from the clutch disc, disengaging the engine drive from the transaxle and enabling the gears to be changed.

The clutch is operating properly if:
• It will stall the engine when released with the vehicle held stationary
• The shift lever can be moved freely between 1st and Reverse gears with the vehicle stationary and the clutch disengaged

Driven Disc and Pressure Plate

REMOVAL & INSTALLATION

▶ **See Figures 36 thru 42**

1. Properly disable the SIR system, if equipped, and disconnect the negative battery cable.
2. Remove the transaxle from the vehicle.
3. Unsnap the release fork from the ball stud, then remove the fork and bearing from the vehicle. Slide the bearing from the fork. The bearing should be checked for excessive play and for minimal bearing drag. It should be replaced if no/little drag or excessive play is found.

➡**The release bearing is packed with grease and should not be washed with solvent.**

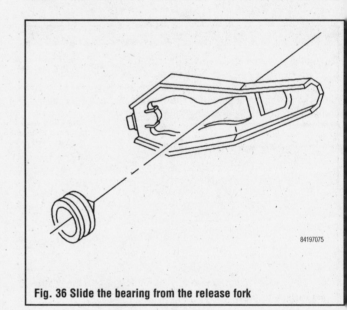

84197075

Fig. 36 Slide the bearing from the release fork

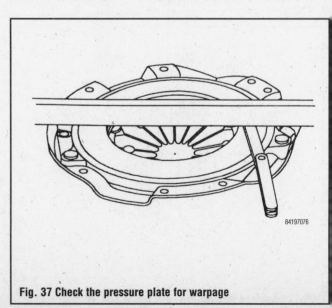

84197076

Fig. 37 Check the pressure plate for warpage

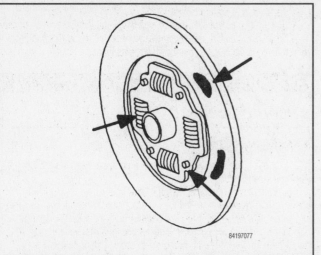

Fig. 38 Check the clutch disc for oil or burnt spots and check for loose springs, hub or rivets

4. Using a feeler gauge, measure the distance between the pressure plate and flywheel surfaces in order to determine clutch face thickness. Replace the clutch disc if it is not within specification, 0.205–0.287 in. (5.2–7.3mm).

5. Remove the pressure plate-to-flywheel bolts in a progressive criss-cross pattern to prevent warping the cover, then remove the pressure plate and clutch disc.

6. Inspect the pressure plate, as follows:

   a. Check for excessive wear, chatter marks, cracks or overheating (indicated by a blue discoloration). Black random spots on the friction surface of the pressure plate is normal.

   b. Check the plate for warpage using a straightedge and a feeler gauge; the maximum allowable warpage is 0.006 in. (0.15mm).

   c. Replace the plate, if necessary.

7. Inspect the clutch disc, as follows:

   a. Check the disc face for oil or burnt spots.

   b. Check the disc for loose damper springs, hub or rivets.

   c. Replace the disc, if necessary.

8. Check the flywheel, as follows:

   a. Check the ring gear for wear or damage.

   b. Check the friction surface for excessive wear, chatter marks, cracks or overheating (indicated by a blue discoloration). Black random spots on the friction surface of the pressure plate is normal.

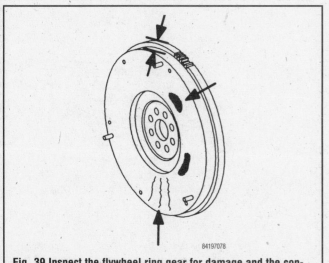

Fig. 39 Inspect the flywheel ring gear for damage and the contact surface for wear

Fig. 41 If removed, install the flywheel and tighten the mounting bolts in a crisscross pattern to specifications

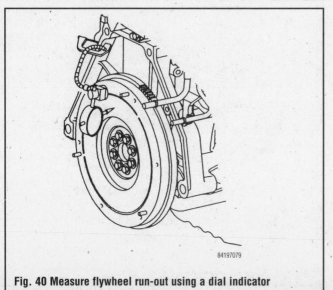

Fig. 40 Measure flywheel run-out using a dial indicator

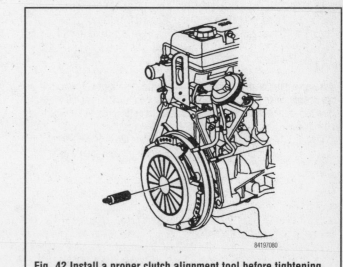

Fig. 42 Install a proper clutch alignment tool before tightening the pressure plate retaining bolts

c. Check flywheel thickness; the minimum allowable is 1.102 in. (28mm).

d. Measure flywheel run-out using a dial indicator, positioned for at least 2 flywheel revolutions. Push the crankshaft forward to take up thrust bearing clearance. Maximum flywheel run-out is 0.006 in. (0.15mm).

e. Check the flywheel for warpage using a straightedge and a feeler gauge; the maximum allowable warpage is 0.006 in. (0.15mm).

f. Replace the flywheel, if necessary.

9. If necessary, remove the flywheel retaining bolts and remove the flywheel from the crankshaft.

To install:

10. If removed, install the flywheel and tighten the bolts in a crisscross sequence to 59 ft. lbs. (80 Nm).

11. Install the clutch disc and pressure plate with the yellow dot on the pressure plate aligned as close as possible to the mark on the flywheel. The clutch disc is labeled FLYWHEEL SIDE in order to help correctly position the disc. Start the pressure plate bolts.

12. Install clutch alignment tool SA9145T or equivalent in the clutch disc, and push in until it bottoms out in the crankshaft.

13. Tighten the pressure plate bolts using multiple passes of a crisscross sequence to 18 ft. lbs. (25 Nm) and remove the alignment tool.

14. Lubricate the fork pivot point with high temperature grease and install the release bearing to the fork. Do not lubricate the release bearing or bearing quill.

15. Snap the release bearing and fork onto the ball stud.

16. Lubricate the splines of the input shaft lightly with a high temperature grease.

17. Install the transaxle assembly.

18. Connect the negative battery cable and, if equipped, properly enable the SIR system.

ADJUSTMENTS

Pedal Height/Travel Diagnosis

▶ **See Figures 43, 44 and 45**

The hydraulic clutch system is self-adjusting; therefore, no manual clutch pedal adjustments are necessary or possible. However, because the pedal travel is directly related to the clutch fork travel, the operating condition of the hydraulic system may be checked using clutch pedal travel.

1. Use a straightedge horizontally positioned from the center of the clutch pedal to the driver's seat, then depress the clutch pedal and measure pedal travel. The clutch pedal travel should be 5.3–6.2 in. (135–156mm). If the pedal travel is insufficient, look for an obvious cause, such as carpet or a floor mat blocking the pedal or a faulty/damaged pedal.

2. Through the access hole on the side of the transaxle (immediately to the right of the slave cylinder), use a caliper or depth gauge to measure travel of the clutch fork with the pedal in the full up and full down positions. Subtract the full down measurement from the full up figure to determine fork travel.

➡**If no caliper or depth gauge is available, use a round wire rod in the access hole and mark the pedal up/down positions. Then measure the distance between the 2 marks to determine fork travel.**

3. Compare the fork and pedal travel measurements using the chart.

4. If fork travel is less than the minimum allowable, check the following. (Conditions a, b and e require replacement of the master/slave cylinder assembly:

a. Fluid leaks in the hydraulic system.

b. Air in the system.

c. Improper installation of the master/slave cylinder.

d. Damaged master or slave cylinder.

e. Damage to the front of the dashboard.

5. If fork travel is acceptable and the hydraulics are working properly, check for a bent fork or damaged pressure plate, which may cause the improper pedal travel.

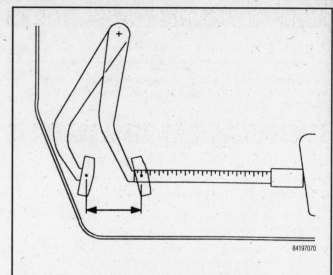

Fig. 43 Use a straightedge to measure pedal travel

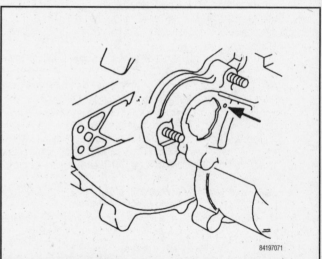

Fig. 44 An access hole is provided in the transaxle housing, in order to measure fork travel

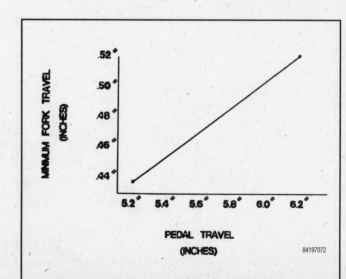

Fig. 45 Compare fork and pedal travel using the chart

Master Cylinder

REMOVAL & INSTALLATION

➥The master cylinder, pipes and slave cylinder are a complete assembly and must be replaced as a single unit. Refer to the following procedure.

Slave Cylinder

REMOVAL & INSTALLATION

◆ See Figures 46, 47, 48 and 49

➥The master cylinder, pipes and slave cylinder are a complete assembly and must be replaced as a single unit.

1. Block the clutch pedal to prevent it from being depressed while the slave cylinder is removed from the transaxle.
2. Remove the air cleaner/intake duct assembly.
3. Disconnect the negative battery cable first, then the positive cable.
4. Remove the battery hold-down retainer.
5. Remove the battery from the vehicle. Store the battery in a safe location, but do not place it on a concrete surface for any long period of time or it will discharge.
6. Remove the battery tray. Note that one battery tray mounting fastener is only accessible through the fender well.
7. Rotate the slave cylinder about ¼ turn counterclockwise while pushing toward the bell housing, in order to disengage the connector and remove the cylinder from the clutch housing. Remove the slave cylinder bracket retaining nuts and pull the assembly from the studs.
8. If equipped with ABS, remove the brake master cylinder-to-power booster mounting nuts. Move the master cylinder off of the mounting studs and slightly toward the engine, being careful not to bend or kink the brake lines.
9. Remove the master cylinder pushrod retaining clip from the clutch pedal pin and disconnect the pushrod from the pedal.
10. Rotate the clutch cylinder about ⅛ turn clockwise and remove it from the instrument panel. Remove the hydraulic assembly from the vehicle.

To install:

11. Install the clutch master cylinder into the dashboard with the reservoir leaning toward the driver's fender. Rotate the cylinder about ⅛ turn counterclockwise to lock it in position.

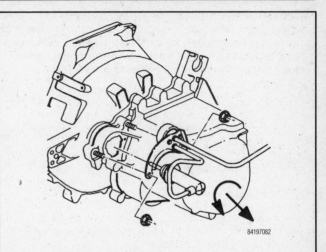

Fig. 47 To remove, rotate the slave cylinder about ¼ turn counterclockwise while pushing toward the bell housing

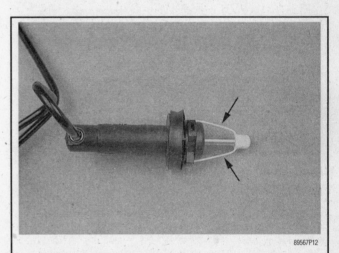

Fig. 48 This plastic retaining strap must remain on the slave cylinder to ensure proper seating of the actuator rod against the release fork

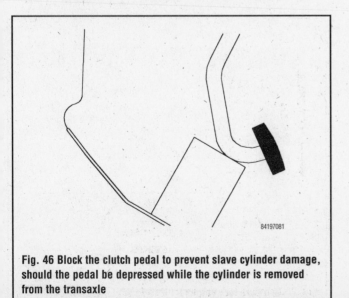

Fig. 46 Block the clutch pedal to prevent slave cylinder damage, should the pedal be depressed while the cylinder is removed from the transaxle

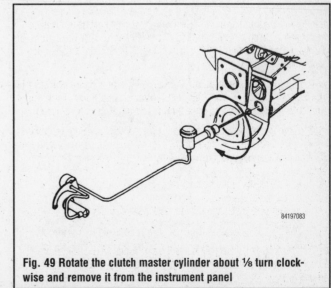

Fig. 49 Rotate the clutch master cylinder about ⅛ turn clockwise and remove it from the instrument panel

➥When installing a new assembly, the plastic retainer straps should remain in place on the slave cylinder, to ensure that the actuator rod seats on the release fork pocket upon installation. If reinstalling an assembly, be sure to position a new plastic retainer strap onto the end of the pushrod and attach the straps to the cylinder.

12. If equipped with ABS, place the brake master cylinder into position on the power booster mounting studs. Install and tighten the master cylinder-to-power booster mounting nuts to 20 ft. lbs. (27 Nm).

13. Slide the slave cylinder onto the clutch housing studs, then install the nuts and tighten to 18 ft. lbs. (25 Nm).

14. Insert the slave cylinder into the clutch housing with the hydraulic line facing down and rotate about ¼ turn clockwise while pushing it into the housing.

15. Lubricate the clutch pedal pin with silicone grease, then connect the pushrod to the clutch pedal and install the retaining clip.

16. Install the battery tray. Tighten the battery tray mounting fasteners to 89 inch lbs. (10 Nm).

17. Place the battery into the battery tray. Be careful that the battery terminals do not short against any metal during the installation.

18. Install the battery hold-down retainer.

19. Connect the positive battery cable first, then the negative cable.

20. Install the air cleaner/intake duct assembly.

21. Remove the block from behind the clutch pedal and, if equipped, properly enable the SIR system.

22. Start the engine and check the pedal for proper operation.

HYDRAULIC SYSTEM BLEEDING

The clutch hydraulic assembly has been filled with fluid and bled of air at the factory. Do not attempt to bleed the hydraulic system. While the unit does not require periodic checking, it must be serviced, when necessary, as a complete assembly. The system is full when the reservoir is half full.

Only DOT 3 brake fluid should be added to the system. If the fluid level drops, inspect the system, including the slave cylinder, for leakage. A slight wetting of the slave cylinder surface is normal. Fill the clutch master cylinder reservoir with brake fluid. Be careful not to spill brake fluid on the painted surface of the vehicle.

AUTOMATIC TRANSAXLE

Understanding the Automatic Transaxle

The automatic transaxle allows engine torque and power to be transmitted to the front wheels within a narrow range of engine operating speeds. It will allow the engine to turn fast enough to produce plenty of power and torque at very low speeds, while keeping it at a sensible rpm at high vehicle speeds (and it does this job without driver assistance). The transaxle uses a light fluid as the medium for the transmission of power. This fluid also works in the operation of various hydraulic control circuits and as a lubricant. Because the transaxle fluid performs all of these functions, trouble within the unit can easily travel from one part to another. For this reason, and because of the complexity and unusual operating principles of the transaxle, a very sound understanding of the basic principles of operation will simplify troubleshooting.

TORQUE CONVERTER

▶ See Figure 50

The torque converter replaces the conventional clutch. It has three functions:

1. It allows the engine to idle with the vehicle at a standstill, even with the transaxle in gear.

2. It allows the tranaxle to shift from range-to-range smoothly, without requiring that the driver close the throttle during the shift.

3. It multiplies engine torque to an increasing extent as vehicle speed drops and throttle opening is increased. This has the effect of making the transaxle more responsive and reduces the amount of shifting required.

The torque converter is a metal case which is shaped like a sphere that has been flattened on opposite sides. It is bolted to the rear end of the engine's crankshaft. Generally, the entire metal case rotates at engine speed and serves as the engine's flywheel.

The case contains three sets of blades. One set is attached directly to the case. This set forms the torus or pump. Another set is directly connected to the output shaft, and forms the turbine. The third set is mounted on a hub which, in turn, is mounted on a stationary shaft through a one-way clutch. This third set is known as the stator.

A pump, which is driven by the converter hub at engine speed, keeps the torque converter full of transmission fluid at all times. Fluid flows continuously through the unit to provide cooling.

Under low speed acceleration, the torque converter functions as follows:

The torus is turning faster than the turbine. It picks up fluid at the center of the converter and, through centrifugal force, slings it outward. Since the outer edge of the converter moves faster than the portions at the center, the fluid picks up speed.

The fluid then enters the outer edge of the turbine blades. It then travels back toward the center of the converter case along the turbine blades. In impinging upon the turbine blades, the fluid loses the energy picked up in the torus.

If the fluid was now returned directly into the torus, both halves of the converter would have to turn at approximately the same speed at all times, and torque input and output would both be the same.

In flowing through the torus and turbine, the fluid picks up two types of flow, or flow in two separate directions. It flows through the turbine blades, and it spins with the engine. The stator, whose blades are stationary when the vehicle is being accelerated at low speeds, converts one type of flow into another. Instead of allowing the fluid to flow straight back into the torus, the stator's curved blades turn the fluid almost 90° toward the direction of rotation of the engine. Thus the fluid does not flow as fast toward the torus, but is already spinning when the torus picks it up. This has the effect of allowing the torus to turn much faster than the turbine. This difference in speed may be compared to the difference in speed between the smaller and larger gears in any gear train. The result is that engine power output is higher, and engine torque is multiplied.

As the speed of the turbine increases, the fluid spins faster and faster in the direction of engine rotation. As a result, the ability of the stator to redirect the fluid flow is reduced. Under cruising conditions, the stator is eventually forced to rotate on its one-way clutch in the direction of engine rotation.

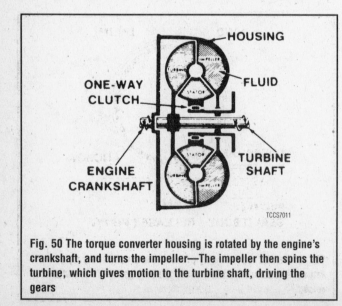

Fig. 50 The torque converter housing is rotated by the engine's crankshaft, and turns the impeller—The impeller then spins the turbine, which gives motion to the turbine shaft, driving the gears

Under these conditions, the torque converter begins to behave almost like a solid shaft, with the torus and turbine speeds being almost equal.

PLANETARY GEARBOX

▶ **See Figures 51, 52 and 53**

The ability of the torque converter to multiply engine torque is limited. Also, the unit tends to be more efficient when the turbine is rotating at relatively high speeds. Therefore, a planetary gearbox is used to carry the power output of the turbine to the driveshaft.

Planetary gears function very similarly to conventional transmission gears. However, their construction is different in that three elements make up one gear system, and, in that all three elements are different from one another. The three elements are: an outer gear that is shaped like a hoop, with teeth cut into the inner surface; a sun gear, mounted on a shaft and located at the very center of the outer gear; and a set of three planet gears, held by pins in a ring-like planet carrier, meshing with both the sun gear and the outer gear. Either the outer gear or the sun gear may be held stationary, providing more than one possible torque multiplication factor for each set of gears. Also, if all three gears are forced to rotate at the same speed, the gearset forms, in effect, a solid shaft.

Fig. 53 Planetary gears in the minimum reduction (drive) range. The ring gear is allowed to revolve, providing a higher gear ratio

Most automatics use the planetary gears to provide various reductions ratios. Bands and clutches are used to hold various portions of the gearsets to the transaxle case or to the shaft on which they are mounted. Shifting is accomplished, then, by changing the portion of each planetary gearset which is held to the transaxle case or to the shaft.

SERVOS & ACCUMULATORS

▶ **See Figure 54**

The servos are hydraulic pistons and cylinders. They resemble the hydraulic actuators used on many other machines, such as bulldozers. Hydraulic fluid enters the cylinder, under pressure, and forces the piston to move to engage the band or clutches.

The accumulators are used to cushion the engagement of the servos. The transmission fluid must pass through the accumulator on the way to the servo. The accumulator housing contains a thin piston which is sprung away from the discharge passage of the accumulator. When fluid passes through the accumulator on the way to the servo, it must move the piston against spring pressure, and this action smooths out the action of the servo.

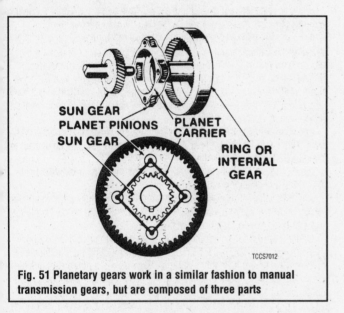

Fig. 51 Planetary gears work in a similar fashion to manual transmission gears, but are composed of three parts

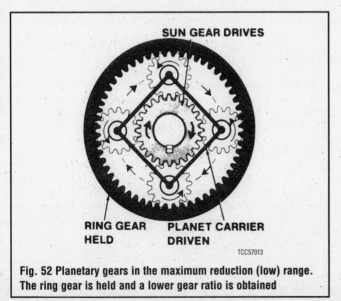

Fig. 52 Planetary gears in the maximum reduction (low) range. The ring gear is held and a lower gear ratio is obtained

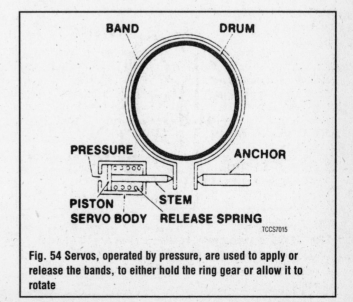

Fig. 54 Servos, operated by pressure, are used to apply or release the bands, to either hold the ring gear or allow it to rotate

HYDRAULIC CONTROL SYSTEM

The hydraulic pressure used to operate the servos comes from the main transaxle oil pump. This fluid is channeled to the various servos through the shift valves. There is generally a manual shift valve which is operated by the transaxle selector lever and an automatic shift valve for each automatic upshift the transaxle provides.

➡**Many new transaxles are electronically controlled. On these models, electrical solenoids are used to better control the hydraulic fluid. Usually, the solenoids are regulated by an electronic control module.**

There are two pressures which affect the operation of these valves. One is the governor pressure which is effected by vehicle speed. The other is the modulator pressure which is effected by intake manifold vacuum or throttle position. Governor pressure rises with an increase in vehicle speed, and modulator pressure rises as the throttle is opened wider. By responding to these two pressures, the shift valves cause the upshift points to be delayed with increased throttle opening to make the best use of the engine's power output.

Most transaxles also make use of an auxiliary circuit for downshifting. This circuit may be actuated by the throttle linkage the vacuum line which actuates the modulator, by a cable or by a solenoid. It applies pressure to a special downshift surface on the shift valve or valves.

The transaxle modulator also governs the line pressure, used to actuate the servos. In this way, the clutches and bands will be actuated with a force matching the torque output of the engine.

Identification

◆ **See Figure 1**

There are two 4-speed automatic transaxles used in all Saturn vehicles. They are virtually the same by design and differ internally almost only by gear ratio. The two transaxles are equipped with different gears, in order to best utilize the engines to which they are mated. The MP6 transaxle is found only on vehicles with the SOHC engine, while the MP7 was available only with the DOHC engine.

If there is a question as to whether the transaxle in your vehicle is the original component, an identification code is stamped on the top of the transaxle clutch housing. The code includes information such as model year, transaxle type (MP6 or MP7), where the transaxle was built and when the unit was built. An interpretation of the entire identification code can be found under the Manual Transaxle portion of this section.

The Saturn automatic transaxle utilizes four multiple disc clutches, a four-element torque converter with a lock-up clutch, 1st gear sprag clutch and a servo actuated dog clutch. The unit is controlled by the PCM through five electro-hydraulic actuators, in order to provide shift timing, shift feel and on-board diagnosis. A POWER/NORMAL button in the center console can be used to communicate with the PCM to alter shift points for performance/economy.

The PCM is capable of performing self-diagnostic functions on the various electronic transaxle control circuits. The SHIFT TO D2 light will illuminate and remain on if certain trouble codes are present in memory. Codes and flags are extracted using the ALDL, as described in Section 4 of this manual.

Beginning with the 1993 model year, the PCM transaxle controller was changed in order to produce smoother and more efficient shifts. This new controller also contained additional trouble codes and flags.

Neutral Safety/Selector/Reverse Light Switch

REMOVAL & INSTALLATION

◆ **See Figure 55**

1. Disconnect the negative battery cable.
2. Remove the air cleaner/induction tube assembly.
3. Unplug the electrical connectors from the switch and disconnect the shifter cable from the control lever.

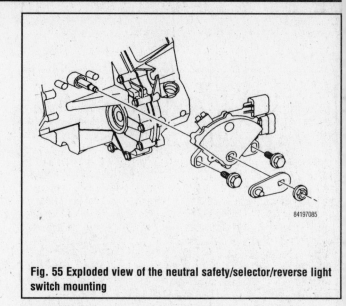

Fig. 55 Exploded view of the neutral safety/selector/reverse light switch mounting

4. For assembly purposes, note the position of the control lever shaft and manual lever, then remove the retaining nut from the control lever shaft and remove the manual lever. Be careful not to rotate the manual shift shaft excessively, or internal damage to the switch may occur.
5. Remove the 2 switch retaining bolts, then remove the switch from the transaxle.

To install:

6. Install the switch to the transaxle and loosely install the retaining bolts.
7. Position the manual lever as noted during removal and install the nut. Tighten the nut to 106 inch lbs. (12 Nm).
8. Adjust the switch and tighten the retaining bolts.
9. Install the cable to the control lever and adjust as necessary. Refer to the procedure which follows.
10. Attach the switch's electrical connectors and install the air induction tube.
11. Connect the negative battery cable and verify proper switch operation.

ADJUSTMENT

◆ **See Figures 56 and 57**

1. Place the transaxle selector in the **D** position. Use an ohmmeter or continuity tester to check for continuity across the switch terminals.

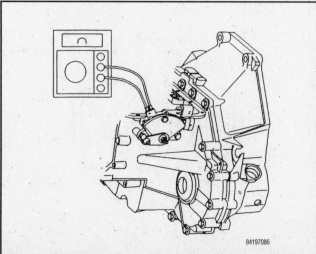

Fig. 56 Use an ohmmeter to properly adjust the neutral safety switch

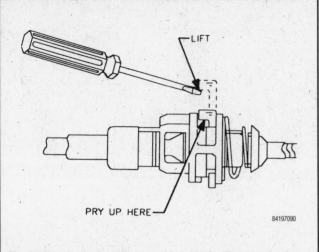

Fig. 57 To release the selector cable locktab, lift upward using a screwdriver or small prybar

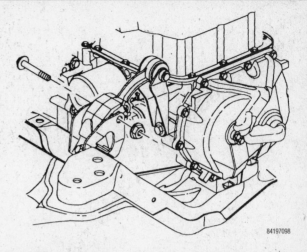

Fig. 58 Remove the transaxle strut-to-cradle bracket through-bolt—1992–98 vehicles

2. If no continuity exists, loosen the bolts and rotate the switch in the direction of the engine; otherwise, a Diagnostic Trouble Code (DTC) may result.

3. Tighten the switch bolts to 124 inch lbs. (14 Nm) and recheck continuity.

4. Connect the shift cable to the control lever of the switch. Perform the shift cable adjustment procedure as follows:

   a. Place the shifter in the **P** position.

   b. Raise and safely support the vehicle.

   c. Disengage the control cable from the lever on the switch assembly.

   d. Place the control lever assembly in the **P** position by moving the lever clockwise until it clicks and cannot be moved any further.

   e. Release the cable adjustment locktab using a screwdriver to pry the tab upward, then lift the tab by hand.

   f. Connect the cable to the shift transaxle lever and install the retainer.

   g. Move the cable housing back and forth in the adjuster to note end-play.

   h. Center the cable housing in the middle of the end-play.

   i. Press in the locktab.

   j. Lower the vehicle and verify shift cable operation.

5. Plug in the electrical connectors.

6. Install the air cleaner/induction tube assembly.

Automatic Transaxle Assembly

REMOVAL & INSTALLATION

▶ **See Figures 58 thru 69**

1. Properly disable the SIR system, if equipped, and disconnect the negative battery cable.

2. Remove the air inlet duct fasteners, unplug the air temperature sensor connector and remove the air cleaner/inlet duct assembly.

3. For 1992–98 vehicles, remove the transaxle strut-to-cradle bracket through-bolt located on the radiator side of the transaxle.

4. Unplug the vehicle and turbine speed sensor, transaxle temperature sensor, selector switch and actuator connectors from the transaxle.

5. Remove the 2 ground terminals from the top 2 converter housing studs.

6. Remove the ground wire from the neutral (selector) switch and unclip the oxygen sensor wire retainer from the converter housing.

7. Remove the top 2 converter housing studs.

8. Remove the 4 DIS coil-to-converter housing bolts, then wire the coil (module) to the cylinder head coolant outlet. Discard the old module's retaining bolts and replace with new bolts upon installation.

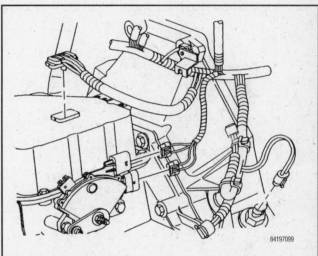

Fig. 59 Unplug the electrical connectors from these transaxle points

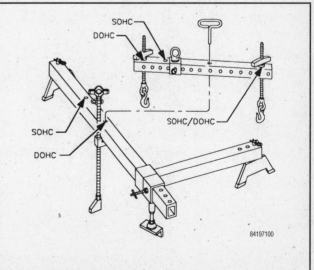

Fig. 60 Saturn engine support bar assembly—part no. SA9105E

9. For 1991 vehicles, loosen the 2 front transaxle mount-to-transaxle fasteners.

10. Wire the radiator to the upper radiator support, in order to hold the assembly in place when the cradle is removed.

11. Install engine support bar assembly tool SA9105E or equivalent. Make sure the support feet are positioned on the outer edge of the shock tower, the bar hooks are connected to the engine bracket and the stabilizer foot is on the engine block (to the right of the oil dipstick).

12. Raise the front of the vehicle sufficiently to lower the transaxle and the engine cradle. Support the vehicle safely using jackstands. Make sure the jackstands are not positioned under the engine cradle.

13. Remove the drain plug from the transaxle housing and drain the transaxle fluid. The drain plug is on the lower cowl side of the housing and is inserted from the engine side of the vehicle.

14. Remove the front wheels and engine splash shields from the vehicle. For coupes, remove the left and right lower facia braces.

15. For 1992–98 vehicles, remove the front engine strut cradle bracket-to-cradle nuts from below the cradle.

16. Remove the transaxle mount-to-cradle nut from under the cradle.

17. Remove and discard the cotter pin from the lower ball joints. Back off the ball joint nut until the top of the nut is even with the top of the threads.

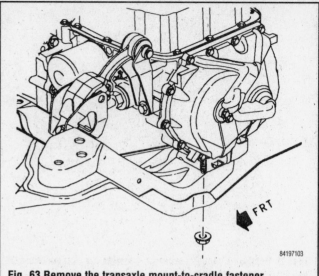

Fig. 63 Remove the transaxle mount-to-cradle fastener

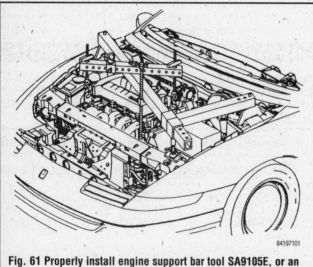

Fig. 61 Properly install engine support bar tool SA9105E, or an equivalent, to retain the engine when the cradle is removed

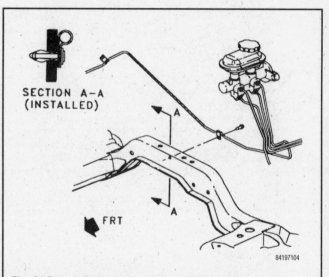

Fig. 64 Detach the brake line and retainer from the cradle

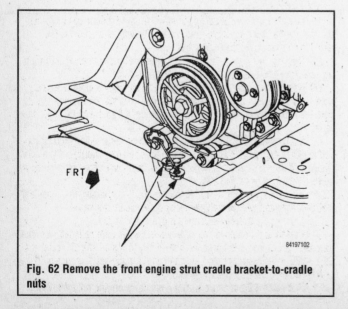

Fig. 62 Remove the front engine strut cradle bracket-to-cradle nuts

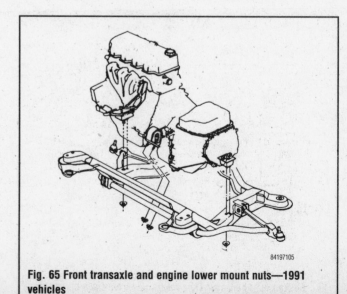

Fig. 65 Front transaxle and engine lower mount nuts—1991 vehicles

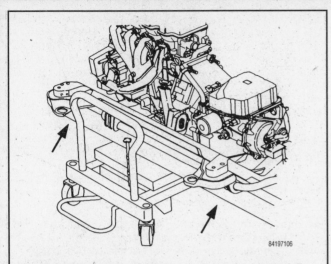

Fig. 66 Position a piece of wood on a powertrain support dolly to protect the cradle

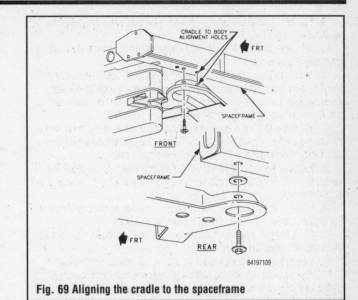

Fig. 69 Aligning the cradle to the spaceframe

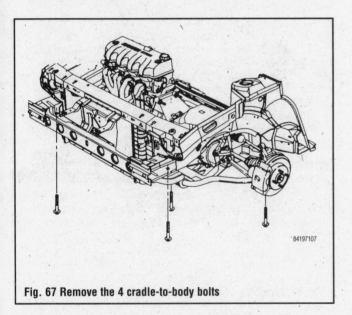

Fig. 67 Remove the 4 cradle-to-body bolts

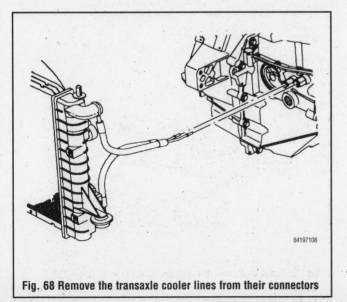

Fig. 68 Remove the transaxle cooler lines from their connectors

18. Use ball joint separator tool SA9132S or equivalent to separate the ball joint from the lower control arm, then remove the nut. Do not use a wedge tool or seal damage may occur.

✳✳ WARNING

The outer CV-joint for vehicles equipped with ABS contains a speed sensor ring. Use of an incorrect tool to separate the control arm from the knuckle may result in damage to and loss of the ABS system.

19. Remove the front exhaust pipe nuts at the manifold, then disconnect the pipe from the support bracket.

20. Remove the front pipe to catalytic converter bolts and lower the pipe from the vehicle.

21. Remove the engine-to-transaxle stiffening bracket bolts and remove the bracket. For 1991 vehicles, remove the rear transaxle mount bracket-to-transaxle bolts, then loosen the mount bolt and allow the bracket to hang out of the way.

22. Remove the steering rack-to-cradle bolts and wire the gear for support when the cradle is removed. Remove the brake line and retainer from the cradle.

23. Remove the converter housing dust cover, then remove the torque converter-to-flywheel bolts.

24. For 1991 vehicles, remove the front transaxle mount-to-cradle nuts and the engine lower mount-to-cradle nuts. These nuts are all located under the front or side cradle member.

25. Position two 4 in. **x** 4 in. **x** 36 in. pieces of wood onto a powertrain support dolly and position the dolly under the vehicle.

26. Remove the 4 cradle-to-body bolts and carefully lower the cradle from the vehicle with the support dolly. Tape or wire the 2 large washers from the rear cradle-to-body attachments in position to prevent losing these parts.

27. Squeeze the plastic tabs at the transaxle cooler line connectors and pull the lines out of the connectors. The plastic retainer should remain on the lines. Connect one end of a ⅜ in. rubber hose over each cooler line to prevent fluid loss or contamination.

28. If necessary for the transaxle to clear the body, lower the transaxle side of the assembly until the valve body cover clears the frame.

29. Support the transaxle securely with a suitable jack.

30. Use an appropriate prybar to separate the left side halfshaft from the transaxle. Only remove the halfshaft sufficiently to install seal protector tool SA91112T or equivalent around the shaft and into the seal, to prevent the seal from being cut by the shaft spline.

31. Remove the 2 bottom converter housing-to-engine bolts and lower

the transaxle sufficiently to reach the shifter cable. Separate the halfshaft from the right side, also using a seal protector tool.

32. Disconnect the transaxle shifter cable and remove the cable from the converter housing.

33. Carefully lower the transaxle from the vehicle. Use transaxle cooler cleaning tool SA9165T or an equivalent to clean the cooler and lines.

To install:

34. Installation of the automatic transaxle assembly is the reverse of removal. However, please note the following important steps:

35. Use a 6 **x** 1.0mm tap to clean the sealant from the ignition module's mounting holes.

36. Place the transaxle assembly securely onto the jack and position it under the vehicle for installation. Install axle seal protectors into the seals on both sides.

37. Tighten the 2 lower converter housing-to-engine bolts to 96 ft. lbs. (130 Nm). Do not use the bolts to draw the transaxle to the engine.

38. Be sure to remove the seal protectors before installing the halfshafts. Push the halfshaft all the way into the transaxle and install the snapring. Remove the transaxle jack.

39. Clean and lubricate the ball joint threads before installing the ball joints into the knuckles. Verify the correct positioning of the lower control arm ball studs to the knuckles, the cooling module support bushings, engine strut bracket and transaxle mount.

40. Insert 9/16 in. round steel rods into the cradle-to-body alignment holes near the front cradle-to-body fastener holes.

41. Tighten the 4 cradle mounting bolts to 155 ft. lbs. (210 Nm). Remove the support dolly, then the engine support bar assembly.

42. For 1992–98 vehicles, tighten the transaxle strut-to-cradle bracket through-bolt and nut to 52 ft. lbs. (70 Nm).

43. Remove the radiator assembly support wire.

44. Tighten the new ignition module bolts, with sealant, to 71 inch lbs. (8 Nm) and verify that the bolt heads are properly seated on the ignition module.

45. For 1991 vehicles, tighten the 2 front transaxle mount-to-transaxle fasteners to 35 ft. lbs. (48 Nm).

46. Tighten the 2 top converter housing-to-engine studs to 74 ft. lbs. (100 Nm). Tighten the 2 ground terminals to 19 ft. lbs. (25 Nm).

47. Adjust the shifter cable according to the procedure given in this section under Neutral Safety/Selector/Reverse Light Switch Adjustment.

48. For 1992–98 vehicles, tighten the transaxle mount-to-cradle nut, the transaxle strut cradle bracket-to-cradle nut and the 2 engine strut cradle bracket-to-cradle nuts to 52 ft. lbs. (70 Nm).

49. For 1991 vehicles, tighten the front transaxle mount cradle nuts to 35 ft. lbs. (48 Nm). Then, tighten the right side mount-to-cradle nuts to 40 ft. lbs. (54 Nm).

50. Tighten the steering gear bolts and nuts to 40 ft. lbs. (54 Nm).

51. Tighten the torque converter-to-flexplate bolts to 52 ft. lbs. (70 Nm). Tighten the converter housing dust cover bolts to 97 inch lbs. (11 Nm).

52. For 1991 vehicles, tighten the rear pitch restrictor-to-transaxle bolts to 40 ft. lbs. (54 Nm).

53. Tighten the powertrain stiffening bracket bolts to 40 ft. lbs. (54 Nm).

54. Tighten the exhaust manifold-to-exhaust pipe retaining nuts in a crosswise pattern to 23 ft. lbs. (31 Nm). Tighten the front pipe-to-catalytic converter flange bolts to 18 ft. lbs. (25 Nm). Finally, tighten the front pipe-to-engine support bracket fasteners to 23 ft. lbs. (31 Nm).

➡**If the converter flange threads are damaged, use the Saturn part no. 21010753 converter fastener kit in place of the self-tapping screws, to provide proper clamp load and prevent exhaust leaks.**

55. Tighten the ball joint nuts to 55 ft. lbs. (75 Nm). Continue to tighten the nuts as necessary and install new cotter pins.

56. For coupes, install the right and left lower facia braces, J-nuts and fasteners. Tighten the fasteners to 89 inch lbs. (10 Nm).

57. Install the tire and wheel assemblies, then remove the jackstands and lower the vehicle.

58. Thoroughly inspect the transaxle and engine compartment area to be sure that all wires, hoses and lines have been connected. Also inspect to make sure that all components and fasteners have been properly installed.

59. Connect the negative battery cable and fill the transaxle with Dexron®II or equivalent fluid.

60. Properly enable the SIR system, if equipped.

61. Warm the engine and check the transaxle fluid. Check and adjust front end alignment, as necessary.

Halfshafts

The Halfshaft removal, installation and overhaul procedures for the automatic transaxle are identical to those for the manual transaxle. Refer to the Halfshaft procedures under the manual transaxle portion of this section.

TORQUE SPECIFICATIONS

System	Component	Ft. Lbs.	Nm
Manual transaxle			
	Back-up light switch	28	38
	Manual transaxle assembly		
	Lower clutch housing-to-engine bolts	103	140
	Engine cradle-to-body bolts	155	210
	Ignition module mounting bolts	71 inch lbs.	8
	Top clutch housing-to-engine studs	74	100
	Ground terminal studs	19	25
	Slave cylinder-to-clutch housing nuts	19	25
	Powertrain stiffening bracket-to-powertrain bolts	40	54
	Clutch housing dust cover fasteners	97 inch lbs.	11
	Steering gear to cradle bolts and nuts	37	50
	Ball joint stud nuts	55	75
	Battery tray fasteners	89 inch lbs.	10
	1991 models		
	Front transaxle mount-to-transaxle bolts	35	48
	Front transaxle mount cradle nuts	35	48
	Right side mount-to-cradle nuts	40	54
	1992-98 models		
	Transaxle strut-to-cradle bracket through bolt and nut	52	70
	Transaxle lower mount-to-cradle fastener	37	50
	Engine strut cradle bracket-to-cradle nut	37	50
	Exhaust		
	Exhaust manifold-to-exhaust pipe retaining nuts	23	31
	Front pipe-to-catalytic converter flange bolts	18	25
	Front pipe-to-engine support bracket fasteners	23	31
	Coupes		
	Right left lower facia braces, J-nuts and fasteners	89 inch lbs.	10
Halfshafts			
	Tie rod end to steering knuckle nut	33	45
	Lower control arm ball joint nut	55	75
	Halfshaft nut	148	200
Clutch			
	Pressure plate bolts	18	25
	Flywheel bolts	59	80
	Hydraulic Clutch Master (Slave) Cylinder		
	Slave cylinder-to-clutch housing nuts	18	25
	Master cylinder-to-power booster mounting nuts	20	27
	Battery tray mounting fasteners	89 inch lbs.	10
Automatic transaxle			
	Neutral Safety/Selector/Reverse Light Switch		
	Manual lever nut	106 inch lbs.	12
	Switch bolts	124 inch lbs.	14
	Automatic transaxle assembly		
	Lower converter housing-to-engine bolts	96	130
	Engine cradle mounting bolts	155	210
	Top converter housing-to-engine fasteners	74	100
	Torque converter-to-flexplate bolts	52	70
	Powertrain stiffening bracket bolts	40	54
	Ignition module bolts	71 inch lbs.	8
	Ball joint stud nuts	55	75
	Steering gear bolts and nuts	40	54
	Ground terminal nuts	19	25

TORQUE SPECIFICATIONS

System	Component	Ft. Lbs.	Nm
Automatic transaxle			
	Automatic transaxle assembly		
	Converter housing dust cover bolts	97 inch lbs.	11
	1991 models		
	Rear pitch restrictor-to-transaxle bolts	40	55
	Front transaxle mount cradle nuts	35	48
	Right side mount-to-cradle nuts	40	54
	Front transaxle mount-to-transaxle fasteners	35	48
	1992-98 models		
	Transaxle mount-to-cradle nut	52	70
	Transaxle strut cradle bracket-to-cradle nut	52	70
	Engine strut cradle bracket-to-cradle nuts	52	70
	Transaxle strut-to-cradle bracket through bolt and nut	52	70
	Exhaust		
	Exhaust manifold-to-exhaust pipe retaining nuts	23	31
	Front pipe-to-catalytic converter flange bolts	18	25
	Front pipe-to-engine support bracket fasteners	23	31
	Coupes		
	Right left lower facia braces, J-nuts and fasteners	89 inch lbs.	10

89567C02

8

SUSPENSION AND STEERING

WHEELS

Wheels

REMOVAL & INSTALLATION

▶ See Figures 1 thru 7

1. Park the vehicle on a level surface.
2. Remove the jack, tire iron and, if necessary, the spare tire from their storage compartments.
3. Check the owner's manual or refer to Section 1 of this manual for the jacking points on your vehicle. Then, place the jack in the proper position.
4. If equipped with lug nut trim caps, remove them by unscrewing them from the lug nuts, using a socket. Consult the owner's manual, if necessary.
5. If equipped with a wheel cover or hub cap, insert the tapered end of the tire iron in the groove and pry off the cover.
6. Apply the parking brake and block the diagonally opposite wheel with a wheel chock or two.

➡Wheel chocks may be purchased at your local auto parts store, or a block of wood cut into wedges may be used. If possible, keep one or two of the chocks in your tire storage compartment, in case any of the tires has to be removed on the side of the road.

7. If equipped with an automatic transaxle, place the selector lever in **P** or Park; with a manual transaxle, place the shifter in Reverse.
8. With the tires still on the ground, use the tire iron/wrench to break the lug nuts loose.

➡If a nut is stuck, never use heat to loosen it or damage to the wheel and bearings may occur. If the nuts are seized, one or two heavy hammer blows directly on the end of the bolt usually loosens the rust. Be careful, as continued pounding will likely damage the brake drum or rotor.

9. Using the jack, raise the vehicle until the tire is clear of the ground. Support the vehicle safely using jackstands.
10. Remove the lug nuts, then remove the tire and wheel assembly.

To install:

11. Make sure the wheel and hub mating surfaces, as well as the wheel lug studs, are clean and free of all foreign material. Always remove rust from the wheel mounting surface and the brake rotor or drum. Failure to do so may cause the lug nuts to loosen in service.

TCCA8P00

Fig. 1 Place the jack at the proper lifting point on your vehicle

TCCA8P02

Fig. 3 With the vehicle still on the ground, break the lug nuts loose using the wrench end of the tire iron

TCCA8P01

Fig. 2 Before jacking the vehicle, block the diagonally opposite wheel with one or, preferably, two chocks

TCCA8P03

Fig. 4 After the lug nuts have been loosened, raise the vehicle using the jack until the tire is clear of the ground

12. Install the tire and wheel assembly and hand-tighten the lug nuts.
13. Using the tire wrench, tighten all the lug nuts, in a crisscross pattern, until they are snug.
14. Raise the vehicle and withdraw the jackstand, then lower the vehicle.
15. Using a torque wrench, tighten the lug nuts in a crisscross pattern to 103 ft. lbs. (140 Nm).

✳✳ WARNING

Do not overtighten the lug nuts, as this may cause the wheel studs to stretch or the brake disc (rotor) to warp.

16. If so equipped, install the wheel cover or hub cap. Make sure the valve stem protrudes through the proper opening before tapping the wheel cover into position.
17. If equipped, install the lug nut trim caps by screwing them on hand tight using a socket. Once tightened, use the ratchet to tighten the nuts an additional ¼ turn for steel wheels or an additional ½ turn for aluminum wheels.
18. Place the jack and tire iron/wrench in their storage compartments. Remove the wheel chock(s).
19. If you have removed a flat or damaged tire, place it in the storage compartment of the vehicle and take it to your local repair station to have it fixed or replaced as soon as possible.

INSPECTION

▶ See Figures 8 and 9

Inspect the tires for lacerations, puncture marks, nails and other sharp objects. Repair or replace as necessary. Also check the tires for treadwear and air pressure as outlined in Section 1 of this manual.

Check the wheel assemblies for dents, cracks, rust and metal fatigue. Repair or replace as necess-ary.

If replacement is necessary, the factory installed Saturn wheels have identification information in the wheel castings. This information includes size and, in the case of aluminum wheels, the part number.

Wheel Lug Studs

REPLACEMENT

With Disc Brakes
▶ See Figure 10

1. Raise the vehicle and safely support it with jackstands, then remove the wheel.

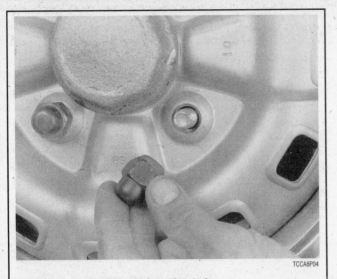
Fig. 5 Remove the lug nuts from the studs

Fig. 6 Remove the wheel and tire assembly from the vehicle

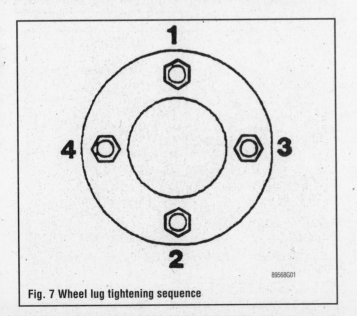
Fig. 7 Wheel lug tightening sequence

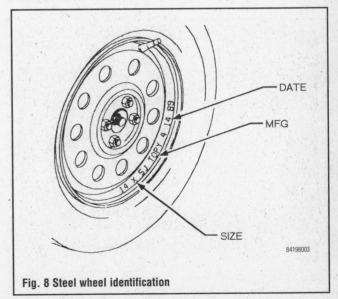
Fig. 8 Steel wheel identification

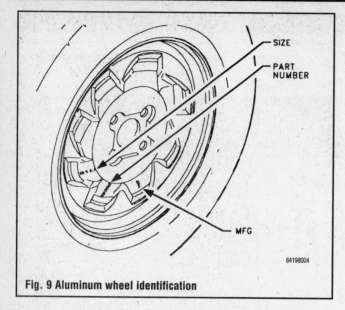

Fig. 9 Aluminum wheel identification

2. Remove the brake pads and caliper. Support the caliper aside using wire or a coat hanger. For details, please refer to Section 9 of this manual.

3. Remove the brake rotor.

4. Using wheel stud removal tool SA91107NE or equivalent, press the stud from the axle hub.

To install:

5. Clean the stud hole with a wire brush. Coat the serrated part of the stud with liquid soap and place it into the hole.

6. Position about 4 flat washers over the stud and thread the lug nut with the flat side of the nut facing the washers. Hold the flange while tightening the lug nut, and the stud should be drawn into position. MAKE SURE THE STUD IS FULLY SEATED, then remove the lug nut and washers.

7. Install the brake rotor.

8. Install the brake caliper and pads.

9. Install the wheel and lug nuts, then remove the jackstands and carefully lower the vehicle.

10. Tighten the lug nuts, in a crisscross pattern to 103 ft. lbs. (140 Nm).

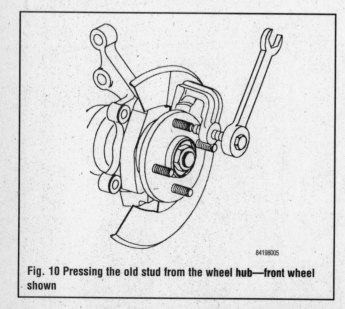

Fig. 10 Pressing the old stud from the wheel hub—front wheel shown

With Drum Brakes

▶ **See Figures 11 and 12**

1. Raise the vehicle and safely support it with jackstands, then remove the wheel.

2. Remove the brake drum.

3. If necessary to provide clearance, remove the brake shoes, as outlined in Section 9 of this manual.

4. Using wheel stud removal tool SA91107NE or equivalent, press the stud from the axle hub.

To install:

5. Coat the serrated part of the stud with liquid soap and place it into the hole.

6. Position about 4 flat washers over the stud and thread the lug nut with the flat side of the nut facing the washers. Hold the flange while tightening the lug nut, and the stud should be drawn into position. MAKE SURE THE STUD IS FULLY SEATED, then remove the lug nut and washers.

7. If applicable, install the brake shoes.

8. Install the brake drum.

9. Install the wheel and lug nuts, then remove the jackstands and carefully lower the vehicle.

10. Tighten the lug nuts in a crisscross pattern to 103 ft. lbs. (140 Nm).

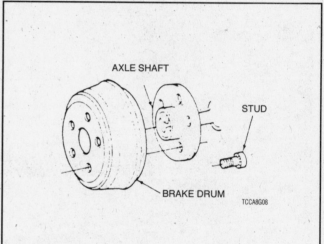

Fig. 11 Exploded view of the drum, axle flange and stud

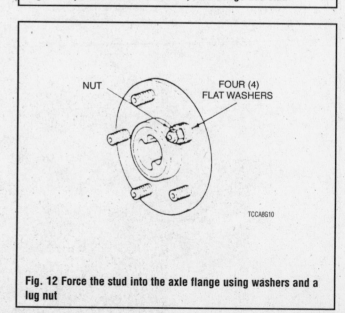

Fig. 12 Force the stud into the axle flange using washers and a lug nut

FRONT SUSPENSION

FRONT SUSPENSION ASSEMBLY COMPONENTS

1. MacPherson struts
2. Lower control arm
3. Sway bar
4. Steering knuckle
5. Lower ball joint (end of lower control arm)
6. Hub and bearing assembly (mounted on steering knuckle)

89568P15

The front suspension consists of 4 major components: MacPherson struts, lower control arms, steering knuckle assemblies and the stabilizer or sway bar. Strut towers located in the wheel wells locate the upper ends of the MacPherson struts. The lower end of the strut is attached to the steering knuckle and the control arm. The control arm provides side-to-side stability, while body lean on turns is controlled by a sway bar that connects to both lower control arms. The strut assembly, which consists of a coil spring and a strut, provides both functions that a spring and a shock absorber would.

The front suspension components are lubricated for life and require no routine greasing or lubrication. However, they should be periodically checked for damage or wear.

MacPherson Struts

REMOVAL & INSTALLATION

▶ See Figures 13 thru 21

✳✳ CAUTION

The MacPherson strut is under extreme spring pressure. Do not remove the strut shaft center support nut at the top without using an approved spring compressor. Personal injury may result if this caution is not followed.

1. If equipped with an Anti-lock Brake System (ABS), disconnect the negative battery cable and then raise the front of the vehicle. Support the vehicle safely using jackstands. Make sure the vehicle is at a height where underhood access is still possible.
2. Remove the front wheel.
3. If equipped, unplug and disconnect the ABS wire from the strut wiring bracket. Note the wiring position for assembly purposes, then place the ABS wiring out of the way to prevent damage. If the strut is being replaced, drill the rivet head retaining the ABS wiring bracket to the strut and remove the bracket.
4. Loosen the 2 steering knuckle-to-strut housing bolts, but do not remove them at this time. For reassembly purposes, matchmark the strut position to the steering knuckle.
5. Remove and discard the 3 upper strut-to-body nuts.
6. Place a rag over the CV-joint seal to protect it from damage, then remove the 2 steering knuckle-to-strut housing bolts.
7. Remove the strut assembly from the vehicle.

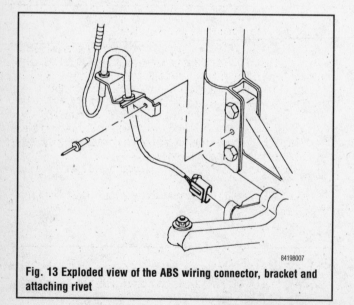

Fig. 13 Exploded view of the ABS wiring connector, bracket and attaching rivet

Fig. 14 View of the ABS wiring bracket and the strut-to-knuckle bolts

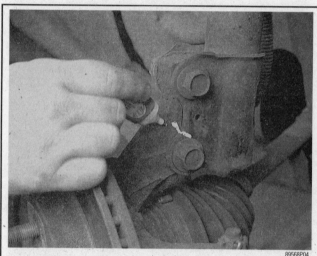

Fig. 15 Matchmark the strut-to-knuckle position for reassembly purposes

Fig. 16 Loosen, but do not remove, the strut-to-steering knuckle bolts

Fig. 17 Loosen the strut's 3 upper fasteners . . .

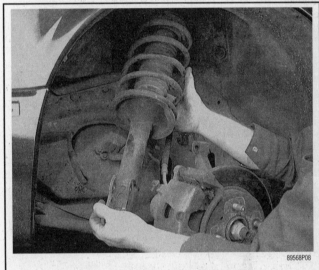

Fig. 20 Using a prying tool, separate the lower strut mount from the steering knuckle

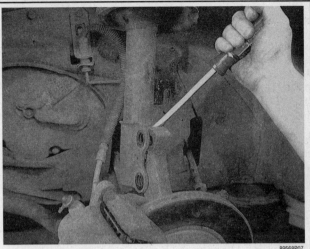

Fig. 18 . . . then remove them from the shock tower

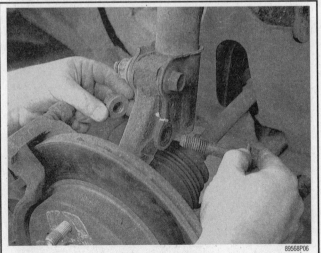

Fig. 21 Remove the strut assembly from the vehicle

Fig. 19 Remove the strut's 2 lower fasteners from the strut and knuckle

To install:

8. Position the strut in the vehicle and install 3 new upper mount nuts. New nuts must be used because the torque retention of the old nut may be insufficient. Tighten the nuts to 21 ft. lbs. (29 Nm).

9. Install the knuckle bolts, also using new nuts. Push the bottom of the strut inward while tightening the fasteners to 126 ft. lbs. (170 Nm).

10. If equipped with ABS and the strut was replaced, install the ABS wiring bracket to the strut using a new rivet. Connect the ABS wiring to the bracket and install the wiring to the speed sensor connector. Make sure the wiring is positioned as noted during removal.

11. Install the wheel assembly, remove the supports and lower the vehicle.

12. Connect the negative battery cable, then check and adjust the alignment as necessary.

OVERHAUL

◆ **See Figures 22 thru 27**

1. Mount the strut in a bench vise, then attach a suitable spring compressor/holding fixture. Be sure that the strut component is firmly secured.

2. Compress the spring sufficiently to completely unload the upper strut mount.

3. Remove the strut shaft nut while holding the strut stationary with a Torx® head socket wrench.

4. Remove the upper mount assembly and inspect the rubber for cracks or deterioration. Rotate the support bearing by hand and check for smooth operation.

5. Remove the spring from the strut and inspect the spring for damage.

6. Remove the dust shield assembly and inspect for cracks or deterioration.

7. Remove the strut from the vise or applicable holding fixture and retract the strut shaft, checking for smooth, even resistance.

8. If replacing the coil spring, carefully release the spring compressor.

To assemble:

9. Secure the strut in the bench vise, or applicable holding fixture.

10. Extend the strut shaft to the limit of its travel.

11. Install the dust shield assembly onto the strut, then install the spring with the compressor tool installed.

12. Install the spring isolator and the strut mount to the top of the assembly.

13. Guide the strut shaft through the upper strut mount assembly. Compress the coil until the washer and shaft nut can be installed to the end of the shaft, but do not overcompress and damage the spring.

14. Tighten the shaft to the nut using a Torx® head socket wrench and a torque wrench, while holding the nut steady with an open end wrench. Tighten the fastener to 37 ft. lbs. (50 Nm).

15. Release the spring compressor tool and remove the strut from the fixture.

89568P10

Fig. 23 Loosen the strut shaft nut while holding the strut stationary with a Torx® head socket wrench . . .

89568P11

Fig. 24 . . . then remove the strut shaft nut and washer

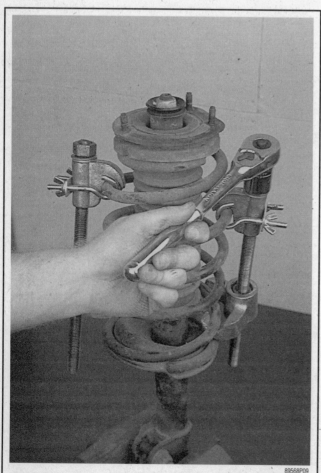

89568P09

Fig. 22 Mount the strut in a bench vise with a spring compressor tool or equivalent

89568P12

Fig. 25 Inspect the bearing plate and rubber bushing for wear or deterioration

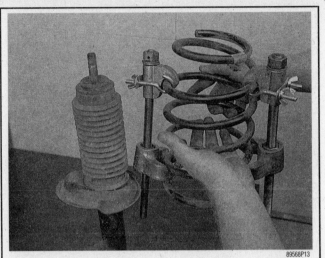

Fig. 26 Remove the coil spring and detach it from the compressor tool, if necessary

Fig. 27 Remove the strut shaft dust boot

Lower Ball Joint

INSPECTION

Raise and safely support the vehicle until the front wheel is clear of the floor. Try to rock the wheel up and down. If any play is felt, have an assistant rock the wheel while observing the lower ball joint. If any movement is seen between the steering knuckle and control arm, the ball joint is bad; if no movement is detected, the wheel play indicates wheel bearing wear.

REMOVAL & INSTALLATION

The lower ball joint is an integral part of the lower control arm. If the ball joint needs replacement, the entire lower arm must be replaced as an assembly.

Stabilizer (Sway) Bar

REMOVAL & INSTALLATION

▶ See Figures 28 thru 33

1. Place the steering wheel in the unlocked position, then raise the front of the vehicle and support safely it using jackstands.
2. Remove the left wheel and splash shield.
3. Remove and discard the cotter pin from the left lower control arm ball joint stud. Back off the ball joint nut until the top of the nut is even with the top of the threads.
4. Use ball joint separator tool SA9132S or equivalent to separate the ball joint from the lower control arm, then remove the nut. Do not use a wedge tool or seal damage may occur.

✳✳ WARNING

The outer CV-joint for vehicles equipped with an Anti-lock Brake System (ABS) contains a speed sensor ring. Use of an incorrect tool to separate the control arm from the knuckle may result in damage to and loss of the ABS.

5. Remove the left lower control arm-to-cradle fastener.
6. Turn the steering wheel to the left to access and remove the right stabilizer bar nut and washer.
7. Remove both stabilizer bar-to-cradle mounting brackets. If a cradle nut is damaged, cross-threaded or broken loose from the cradle, replace the nut as follows:
   a. If the nut is damaged but not broken loose, and the bolt can be removed, distort the bolt threads sufficiently to lock the bolt into the nut. Insert the bolt and tighten with an air impact wrench. Continue to turn the bolt until the nut breaks free of the cradle.
   b. If the nut was already broken loose and the bolt could not be extracted, or if the bolt was used to break the nut loose, cut the bolt head off.
   c. Retrieve the bolt shank and nut from the cradle cavity.
   d. Install a new bolt and nut (Saturn Part Nos. 21010823 and 21006321, or equivalents).
8. Remove the stabilizer bar with the left control arm from the vehicle. If necessary, remove the nut and left control arm from the bar.

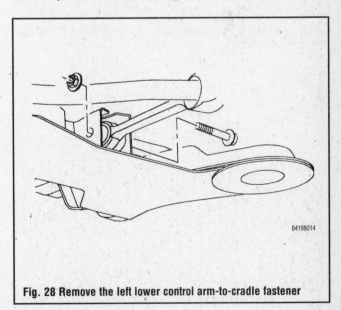

Fig. 28 Remove the left lower control arm-to-cradle fastener

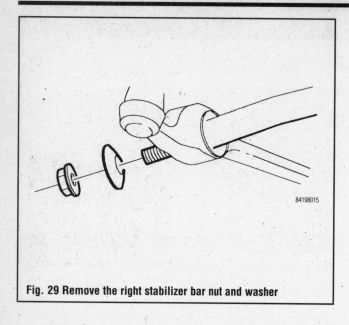

Fig. 29 Remove the right stabilizer bar nut and washer

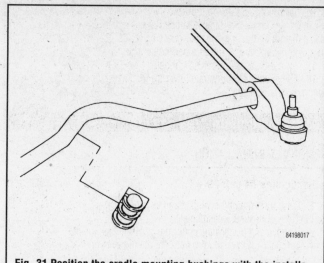

Fig. 31 Position the cradle mounting bushings with the installation slits facing the front of the vehicle

To install:

9. If removed, position the left control arm to the stabilizer bar, but do not tighten the fastener at this time.

10. Install the mounting bushings onto the bar with the installation slits facing the front of the vehicle.

11. Position the right end of the bar into the right control arm (still on the vehicle), then position the left control arm into the cradle. Do not install fasteners at this time.

12. Using new bolts or a suitable threadlock such as Loctite® 242, install the mounting brackets and tighten the bolts to 103 ft. lbs. (140 Nm). Then, tighten the fasteners a second time to 103 ft. lbs. (140 Nm).

13. Temporarily install the left wheel assembly onto the vehicle and have an assistant push the bottom of the left wheel toward the vehicle to allow lower control arm-to-cradle bolt installation.

14. Tighten the lower control arm cradle bolt to 92 ft. lbs. (125 Nm), then tighten the nut to 74 ft. lbs. (100 Nm). Remove the wheel assembly from the vehicle.

15. Install new nuts onto the right and left stabilizer bar to control arm studs and tighten the right nut, then the left nut to 106 ft. lbs. (144 Nm).

➡**New nuts must be used, because the torque retention of the old nuts may be insufficient.**

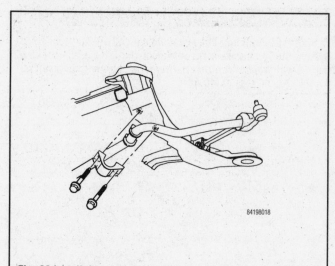

Fig. 32 Install the mounting brackets and tighten the bolts twice to ensure proper torque

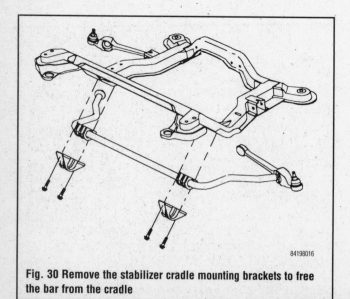

Fig. 30 Remove the stabilizer cradle mounting brackets to free the bar from the cradle

Fig. 33 View of the stabilizer bar, where it is mounted to the lower control arm

16. Thoroughly clean and lubricate the left ball joint stud threads, then install the left lower control arm ball stud into the left steering knuckle. Install the nut and tighten the stud nut to 55 ft. lbs. (75 Nm), tighten additionally if necessary and install a new cotter pin.

17. Install the splash shield, then install the wheel assembly.

18. Remove the supports and lower the vehicle. Check and adjust the alignment as necessary.

Lower Control Arm

REMOVAL & INSTALLATION

▶ See Figures 34 thru 39

1. Raise the front of the vehicle and support it safely using jackstands. Remove the wheel and splash shield.

2. Remove and discard the cotter pin from the lower control arm ball joint stud. Back off the ball joint nut until the top of the nut is even with the top of the threads.

3. Use ball joint separator tool SA9132S or equivalent to separate the lower control arm from the steering knuckle, then remove the nut. Do not use a wedge tool or seal damage may occur.

❊❊ WARNING

The outer CV-joint for vehicles equipped with an Anti-lock Brake System (ABS) contains a speed sensor ring. Use of an incorrect tool to separate the control arm from the knuckle may result in damage to and loss of the ABS.

4. Remove the control arm-to-cradle bolt and nut, then remove the sway bar-to-control arm nut and remove the control arm from the vehicle.

To install:

5. Position the control arm and install the arm onto the sway bar without the fastener, then place the end of the arm into the cradle. Install the cradle nut and bolt. Tighten the cradle bolt to 92 ft. lbs. (125 Nm), then tighten the cradle nut to 74 ft. lbs. (100 Nm).

6. Install the sway bar nut and tighten to 106 ft. lbs. (144 Nm).

7. Thoroughly clean and lubricate the ball joint stud threads, then install the lower control arm ball stud into the steering knuckle. Install the nut and tighten the lower control arm ball stud nut to 55 ft. lbs. (75 Nm); tighten additionally, if necessary to align the holes, and install a new cotter pin.

8. Install the splash shield and the wheel assembly.

9. Remove the supports and lower the vehicle. Check and adjust the front end alignment, as necessary.

Fig. 35 Loosen the lower control arm-to-cradle fasteners . . .

Fig. 36 . . . and remove the fasteners from the vehicle

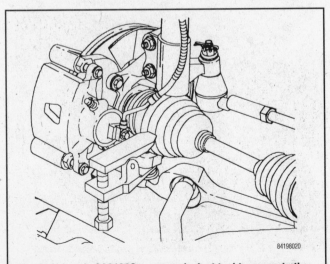

Fig. 34 Use only SA9132S or an equivalent tool to separate the lower control arm ball joint from the steering knuckle

Fig. 37 Loosen the sway bar-to-lower control arm retaining nut . . .

Fig. 38 . . . then remove the sway bar nut and washer

Fig. 39 Remove the lower control arm from the vehicle

Steering Knuckle

REMOVAL & INSTALLATION

▶ **See Figures 40 thru 51**

1. If equipped with an Anti-lock Brake System (ABS), disconnect the negative battery cable.

2. Have an assistant depress the brake pedal and loosen the front half-shaft nut, then raise the front of the vehicle and support it safely using jackstands.

3. Remove the wheel assembly.

4. Remove the brake caliper mounting bracket bolts and suspend the assembly from the strut spring with mechanic's wire or a coat hanger. Make sure the brake line is positioned so that it is not under stress and is not damaged.

5. Loosen the strut-to-knuckle bolts, but do not remove at this time.

6. Remove the rotor; if difficulty is encountered, use two M8 x 1.25 self-tapping bolts in the provided holes to drive the rotor from the hub. Remove the axle nut and washer.

7. Remove and discard the cotter pin from the lower control arm ball joint stud. Back off the ball joint nut until the top of the nut is even with the top of the threads.

8. Use ball joint separator tool SA9132S or equivalent to separate the lower control arm from the steering knuckle, then remove the nut. Do not use a wedge tool or seal damage may occur.

❋❋ WARNING

The outer CV-joint for vehicles equipped with ABS contains a speed sensor ring. Use of an incorrect tool to separate the control arm from the knuckle may result in damage and loss of the ABS.

9. Remove the tie rod cotter pin and castle nut, then separate the tie rod end from the knuckle, using tie rod separator SA91100C or equivalent. Do not use a wedge-type tool.

10. If equipped, unplug the ABS wheel speed sensor electrical connector.

11. Remove the strut-to-knuckle fasteners at this time.

12. Suspend the halfshaft from the body with wire, then remove the knuckle/hub fasteners and remove the knuckle/hub assembly from the vehicle.

13. If difficulty is encountered removing the knuckle, position a block of wood on the end of the halfshaft and tap on the wood with a hammer to free the hub assembly.

To install:

14. Thoroughly clean and lubricate the ball joint stud threads of the lower control arm and tie rod end.

15. Install the knuckle/hub assembly onto the axle shaft. Then, install the washer with a new nut, but do not tighten the nut at this time.

➡**A new nut must be used, because the torque retention of the old nut may not be sufficient.**

16. Install the lower control arm ball stud and install the nut, but do not tighten at this time.

17. Install the steering knuckle-to-strut fasteners, but do not tighten at this time.

18. Install the tie rod end and nut. Tighten the nut to 33 ft. lbs. (45 Nm) and install a new cotter pin. If necessary, tighten the nut additionally, but do not back it off to insert the cotter pin.

19. Push inward on the bottom of the strut and tighten the knuckle fasteners to 126 ft. lbs. (170 Nm).

20. Tighten the lower control arm ball stud nut to 55 ft. lbs. (75 Nm); tighten additionally only as necessary to install a new cotter pin.

Fig. 40 With the wheels on the ground and the brakes applied, loosen the halfshaft nut

Fig. 41 Remove the brake caliper and mounting bracket from the knuckle, and suspend the assembly from the strut spring with a wire or coat hanger

Fig. 44 Back off the ball joint nut until the top of the nut is even with the top of the threads

Fig. 42 After removing the brake rotor, remove the axle nut and washer

Fig. 45 Break the lower control arm loose from the steering knuckle using a ball joint separator tool . . .

Fig. 43 Remove and discard the cotter pin from the lower control arm ball joint stud

Fig. 46 . . . then remove the retaining nut from the ball joint stud

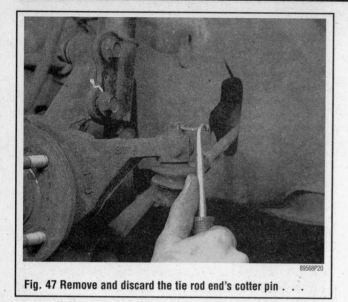

Fig. 47 Remove and discard the tie rod end's cotter pin . . .

Fig. 48 . . . then, after loosening the castle nut with a wrench, remove it from the tie rod end ball stud

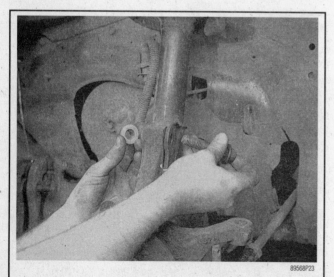

Fig. 49 Remove the strut-to-steering knuckle bolts

Fig. 50 Tie the halfshaft to the body with wire, then remove the steering knuckle assembly from the vehicle

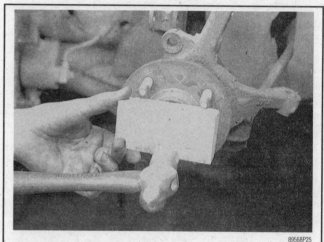

Fig. 51 A hammer can be used to separate the halfshaft and the knuckle, but ONLY if a block of wood is positioned to protect components

21. Install the rotor onto the hub and the caliper mount bracket onto the knuckle. Tighten the mount bracket assembly bolts to 81 ft. lbs. (110 Nm).

22. If equipped, engage the ABS electrical connector to the wheel speed sensor.

23. Have an assistant depress the brake pedal and tighten the halfshaft nut to 148 ft. lbs. (200 Nm).

24. Install the wheel assembly, remove the supports and lower the vehicle.

25. Connect the negative battery cable. Check and adjust the front end alignment, as necessary.

Front Hub and Bearing

Any time the hub or bearing is removed from the knuckle, a new bearing must be used during assembly.

BEARING REPLACEMENT

▶ See Figures 52 thru 58

1. Remove the steering knuckle assembly from the vehicle, as described earlier in this section.

2. For 1991 vehicles, remove the 3 dust shield fasteners and separate the shield from the assembly.

3. If equipped, remove the Anti-lock Brake System (ABS) wheel speed sensor from the knuckle.

4. Assemble wheel bearing removing tool SA9159S or equivalent using the hub driver, hub driver screw, bridge retainer and bridge. Install the tool to the knuckle and secure the assembly in a vise using the bridge as the vise contact point.

5. Hold the hub driver with a wrench and tighten the hub driver screw to remove the hub. If the inner bearing race is pulled out with the hub, remove the race with a bearing race remover. The service tool can be used by assembling the inner race puller, 2 bridge retainer plates, 2 bolts and 2 flat washers.

6. Inspect the hub at the bearing location for pitting, scoring or wear, and replace if necessary.

7. Remove the assembly from the vise and remove the wheel hub removal tools.

8. Remove the bearing retainer snapring.

9. Position the knuckle in a shop press on the knuckle support tube, and press the bearing from the knuckle with a small driver.

10. Inspect the knuckle bore for pitting, scoring, wear or corrosion. If damage cannot be easily cleaned with light sanding, the knuckle must be replaced.

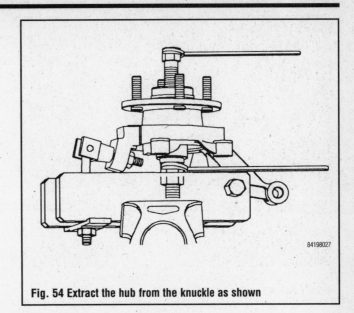

Fig. 54 Extract the hub from the knuckle as shown

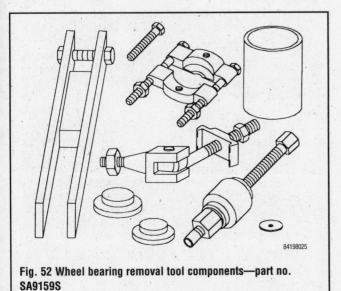

Fig. 52 Wheel bearing removal tool components—part no. SA9159S

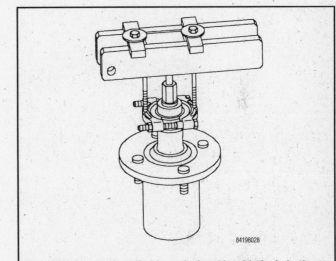

Fig. 55 If the race is pulled from the knuckle with the hub, the removal tool can be used to withdraw the race from the hub

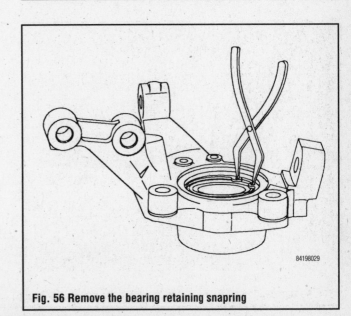

Fig. 53 Assemble the removal tool and install to the knuckle as shown

Fig. 56 Remove the bearing retaining snapring

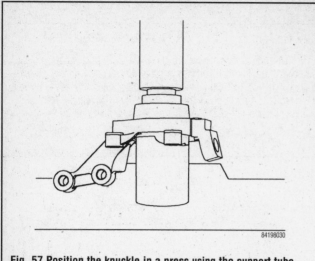

Fig. 57 Position the knuckle in a press using the support tube and a small driver

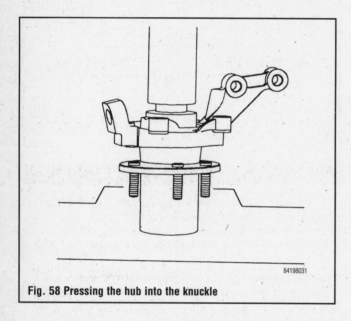

Fig. 58 Pressing the hub into the knuckle

To install:

11. Use the large driver from the service tool kit and properly position the knuckle into the press. (The knuckle should be inverted from its bearing removal position.) Using the driver, press in the new bearing until seated.

12. Use the small driver and knuckle support tube to press in the hub assembly. The small driver must be used to support the bearing inner race with its small (pilot) side facing toward the press and away from the bearing.

13. Install the bearing retainer snapring.

14. If equipped, install the ABS wheel speed sensor into the knuckle and tighten the fastener to 6 ft. lbs. (8 Nm).

15. For 1991 vehicles, install the brake dust shield and tighten the fasteners to 18 ft. lbs. (25 Nm).

16. Install the knuckle assembly to the vehicle.

Wheel Alignment

If the tires are worn unevenly, if the vehicle is not stable on the highway or if the handling seems uneven in spirited driving, the wheel alignment should be checked. If an alignment problem is suspected, first check for improper tire inflation and other possible causes, such as worn suspension or steering components, accident damage or even unmatched tires. If any worn or damaged components are found, they must be replaced before the wheels can be properly aligned. Wheel alignment requires very expensive equipment and involves minute adjustments which must be accurate; it should only be performed by a trained technician. Take your vehicle to a properly equipped shop.

Following is a description of the alignment angles which are adjustable on most vehicles and how they affect vehicle handling. Although these angles can apply to both the front and rear wheels, usually only the front suspension is adjustable.

CASTER

▶ **See Figure 59**

Looking at a vehicle from the side, caster angle describes the steering axis rather than a wheel angle. The steering knuckle is attached to a control arm or strut at the top and a control arm at the bottom. The wheel pivots around the line between these points to steer the vehicle. When the upper point is tilted back, this is described as positive caster. Having a positive caster tends to make the wheels self-centering, increasing directional stability. Excessive positive caster makes the wheels hard to steer, while an uneven caster will cause a pull to one side. Overloading the vehicle or sagging rear springs will affect caster, as will raising the rear of the vehicle. If the rear of the vehicle is lower than normal, the caster becomes more positive.

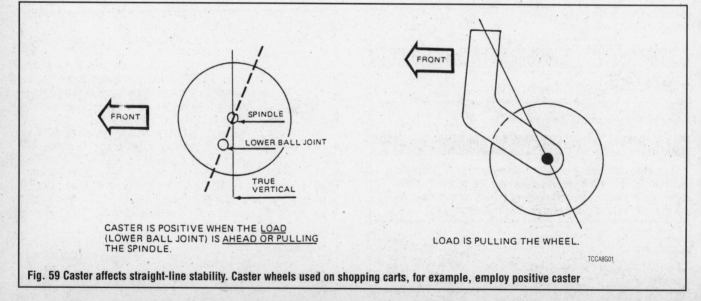

CASTER IS POSITIVE WHEN THE <u>LOAD</u> (LOWER BALL JOINT) IS <u>AHEAD OR PULLING</u> THE SPINDLE.

LOAD IS PULLING THE WHEEL.

Fig. 59 Caster affects straight-line stability. Caster wheels used on shopping carts, for example, employ positive caster

CAMBER

▶ **See Figure 60**

Looking from the front of the vehicle, camber is the inward or outward tilt of the top of wheels. When the tops of the wheels are tilted in, this is negative camber; if they are tilted out, it is positive. In a turn, a slight amount of negative camber helps maximize contact of the tire with the road. However, too much negative camber compromises straight-line stability, increases bump steer and torque steer.

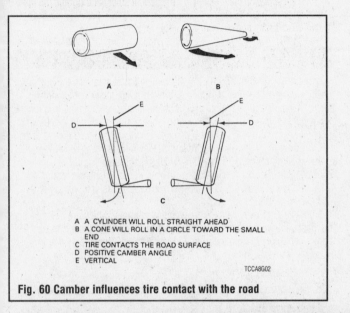

A A CYLINDER WILL ROLL STRAIGHT AHEAD
B A CONE WILL ROLL IN A CIRCLE TOWARD THE SMALL END
C TIRE CONTACTS THE ROAD SURFACE
D POSITIVE CAMBER ANGLE
E VERTICAL

TCCA8G02

Fig. 60 Camber influences tire contact with the road

TOE

▶ **See Figure 61**

Looking down at the wheels from above the vehicle, toe angle is the distance between the front of the wheels, relative to the distance between the back of the wheels. If the wheels are closer at the front, they are said to be toed-in or to have negative toe. A small amount of negative toe enhances directional stability and provides a smoother ride on the highway.

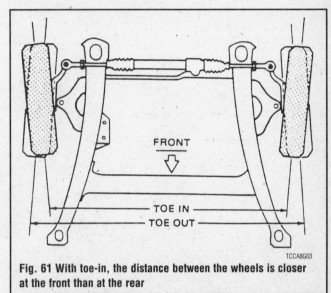

FRONT

TOE IN
TOE OUT

TCCA8G03

Fig. 61 With toe-in, the distance between the wheels is closer at the front than at the rear

REAR SUSPENSION

The major components of the rear suspension are: the MacPherson struts, rear crossmember, knuckles, stabilizer bar (sway bar), lateral links and trailing arms. The upper ends of the MacPherson struts are attached to the body by fasteners. The lower end of the strut is attached to the crossmember through the knuckle, lateral links and trailing arm. The lateral links provide side-to-side stability, while the trailing arm provides front-to-rear stability. The strut assembly, which consists of a coil spring and a strut provides both functions that a spring and a shock absorber would.

The rear suspension components are lubricated for life and require no routine greasing or lubrication. However, they should be periodically checked for damage or wear.

MacPherson Struts

REMOVAL & INSTALLATION

▶ **See Figures 62 thru 68**

❊❊ CAUTION

The MacPherson strut is under extreme spring pressure. Do not remove the strut shaft support nut at the top center of the assembly without using an approved spring compressor. Personal injury may result if this caution is not followed.

1. On coupes, remove the rear seat cushion bottom, left or right rocker panel interior moldings and the left or right rear sail interior panels.
2. On sedans, remove the left or right C-pillar interior molding.
3. Fold down the rear seat backs and remove the rear seat side bolsters from the vehicle, if equipped.
4. On coupes, remove the rear deck package shelf retainers that attach the shelf to the side of the cargo area.
5. If equipped, remove the speaker grill fasteners and grills from the shelf, then remove the seat belt bezel and separate the seat belts from the shelf. Remove the rear package shelf cover.
6. If equipped with an Anti-lock Brake System (ABS), disconnect the negative battery cable.
7. Raise the rear of the vehicle and support it safely using jackstands, then remove the appropriate rear wheel.
8. If equipped, disconnect the ABS wiring from the strut wiring bracket. If the strut is being replaced, drill out the rivet head, if required, retaining the ABS wiring bracket to the strut and remove the bracket. In either case, note the position of the wiring and move the wiring to prevent damage. If necessary, unplug the wheel speed sensor connector.
9. For reassembly purposes, matchmark the strut position to the knuckle. Then, loosen the 2 strut-to-knuckle bolts, but do not remove at this time.
10. Position a floor jack under the rear knuckle, then raise the jack only enough to support the knuckle. If a 2nd floor jack is not available, position a jackstand under the knuckle to support the strut.
11. Remove the 3 strut-to-body mounting nuts.

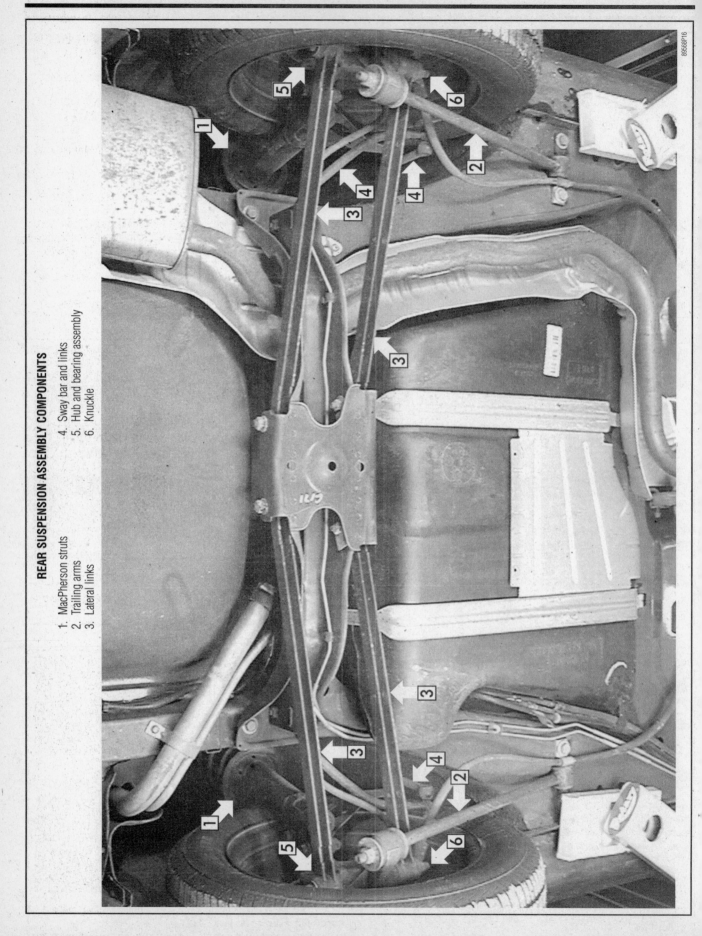

REAR SUSPENSION ASSEMBLY COMPONENTS

1. MacPherson struts
2. Trailing arms
3. Lateral links
4. Sway bar and links
5. Hub and bearing assembly
6. Knuckle

89568P16

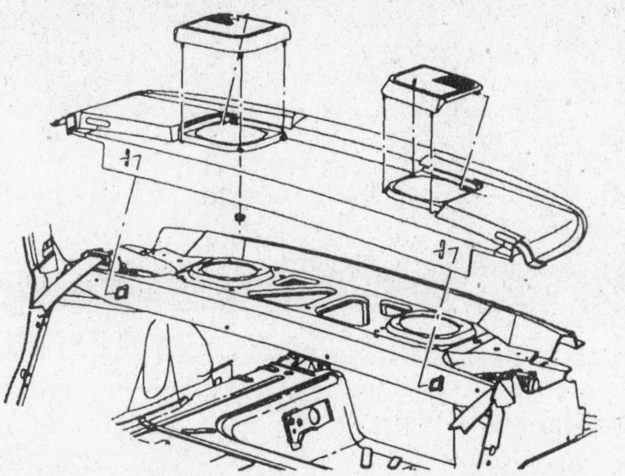

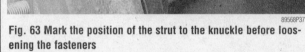

Fig. 62 Remove the speaker grilles, if equipped, and the rear deck package shelf cover

12. Slowly raise the vehicle using another jack, lowering the strut from the body. If another jack is not available, lower the jack holding the strut, but do so carefully.

13. Unfasten the strut-to-knuckle bolts and remove the strut assembly from the vehicle.

To install:

14. Install 3 new strut-to-upper mount nuts and tighten the nuts to 21 ft. lbs. (29 Nm).

➡ **New nuts must be used, because the torque retention of the old fasteners may not be sufficient.**

15. Install the knuckle bolts with new nuts, then push the bottom of the strut inward, aligning the marks made earlier. Tighten the fasteners to 126 ft. lbs. (170 Nm).

16. If the strut was replaced, install the ABS wiring bracket to the strut using a new rivet if necessary. Connect the ABS wiring to the bracket and, if unplugged, connect the wiring harness to the speed sensor. Make sure the wiring is positioned as noted during removal to prevent damage.

17. Install the wheel assembly, remove the supports and lower the vehicle.

18. Install the interior components.

19. Connect the negative battery cable, then check and adjust the rear alignment as necessary.

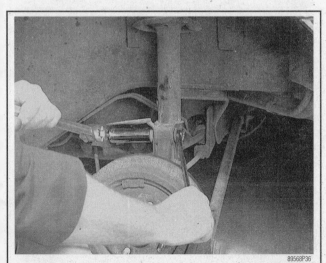

Fig. 63 Mark the position of the strut to the knuckle before loosening the fasteners

Fig. 65 Position a floor jack to support the strut when the upper fasteners are removed

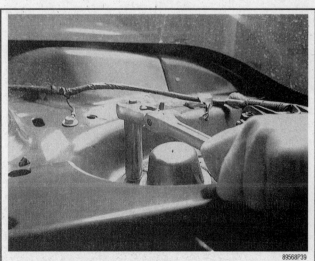

Fig. 64 Loosen the knuckle-to-strut fasteners, but do not remove at this time

Fig. 66 Remove the upper strut fasteners (before slowly raising the vehicle and/or lowering the strut)

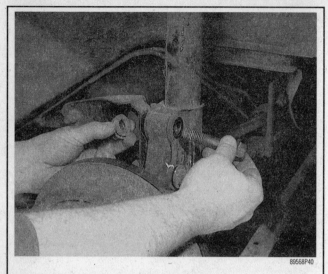

Fig. 67 Remove the strut-to-knuckle fasteners . . .

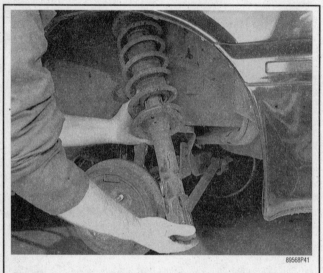

Fig. 68 . . . then remove the strut assembly from the vehicle

OVERHAUL

▶ See Figures 22, 23 and 24

1. Mount the strut in a bench vise, then attach a suitable spring compressor/holding fixture; be sure that the strut component is firmly secured.
2. Compress the spring sufficiently to completely unload the upper strut mount.
3. Remove the strut shaft nut while holding the strut stationary with a Torx® head socket wrench.
4. Remove the upper spring support and inspect the rubber for cracks or deterioration.
5. Remove the spring from the strut and inspect the spring for damage.
6. Remove the dust shield assembly and inspect for cracks or deterioration.
7. Remove the strut from the vise or applicable holding fixture and retract the strut shaft, checking for smooth, even resistance.
8. If replacing the coil spring, carefully release the spring compressor.

To assemble:
9. Secure the strut in the bench vise, or applicable holding fixture.
10. Extend the strut shaft to the limit of its travel.
11. Install the dust shield assembly onto the strut, then install the spring with the compressor tool installed.

12. Install the spring isolator and the strut mount to the top of the assembly.
13. Guide the strut shaft through the upper strut mount assembly. Compress the coil until the washer and shaft nut can be installed to the end of the shaft, but do not overcompress and damage the spring.
14. Tighten the shaft to the nut using a Torx® head socket wrench and a torque wrench, while holding the nut steady with an open end wrench. Tighten the fastener to 37 ft. lbs. (50 Nm).
15. Release the spring compressor tool and remove the strut from the fixture.

Lateral Links

REMOVAL & INSTALLATION

▶ See Figure 69

Whenever a front lateral link is to be replaced on 1991–94 models, the fuel tank must be removed. Also, 1991–92 vehicles were built with a rear assembly containing a special washer at the lateral link-to-crossmember attachment locations. For 1993, this washer was omitted, which necessitated the use of a special truss head bolt. The truss head bolt can be used on any model year, but the hex head bolt from the 1991–92 vehicles should not be used on rear suspensions without the built-in special washer.

1. Raise the rear of the vehicle and support it safely using jackstands, then remove the rear wheels.
2. If removing the front lateral link on 1991–94 models, remove the fuel tank. Refer to Section 5 of this manual.
3. If removing the front lateral link on 1995–98 models, perform the following steps:
    a. Remove the attaching nut that secures the rear brake lines to the floor pan.
    b. Place a jack under the rear crossmember for support. Remove the four crossmember-to-body mounting bolts.
    c. Lower the crossmember to gain access to the inboard lateral link fasteners.
4. Remove the inboard lateral link fasteners.
5. Remove the lateral link-to-knuckle fasteners.
6. Remove the link or links from the vehicle.

To install:
7. Install the link(s) into the crossmember with the fastener(s), but do not tighten at this time. If the front link is being replaced, remove the anti-rotation tab on the fastener nut.
8. Install the link(s) to the knuckle with the knuckle-to-link bolt, but do not tighten at this time.

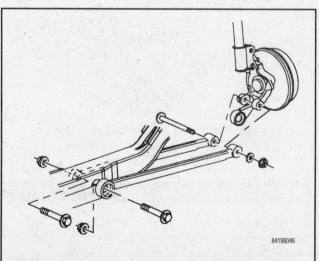

Fig. 69 The lateral links connect the rear crossmember and knuckles

9. Place the crossmember into proper position under the body. Align the crossmember to the body using a ⅜ in. rod at the alignment hole and slot. Tighten the crossmember mounting bolts to 89 ft. lbs. (120 Nm).

10. On 1991–94 models, tighten the front link bolt to 126 ft. lbs. (170 Nm) and/or the rear link bolt to 89 ft. lbs. (120 Nm).

11. On 1995–98 models, use front lateral link tool SA9411C, or equivalent, to prevent Torx® bolt rotation while tightening. Tighten the front and rear lateral link-to-crossmember fasteners to 89 ft. lbs. (120 Nm).

12. Tighten the lateral link-to-knuckle fasteners to 122 ft. lbs. (165 Nm).

13. On 1995–98 models, secure the rear brake lines to the floor pan and install the retaining nut.

14. On 1991–94 models, install the fuel tank, if removed.

15. Install the rear wheel(s), remove the supports and lower the vehicle. Check and adjust the rear alignment as necessary.

Trailing Arm

REMOVAL & INSTALLATION

▶ See Figure 70

1. Raise the rear of the vehicle and support it safely using jackstands, then remove the rear wheel(s).
2. Remove the trailing arm-to-knuckle nut.
3. Remove the trailing arm body bolts.
4. Slide the trailing arm from the knuckle.
To install:
5. Install the trailing arm into the knuckle and hand-tighten the fastener nut.
6. Position the arm to the body, then install and tighten the body bolts to 89 ft. lbs. (120 Nm).
7. Tighten the trailing arm-to-knuckle nut to 74 ft. lbs. (100 Nm).
8. Install the rear wheel(s).
9. Remove the supports and lower the vehicle.

Fig. 70 A view of the trailing arm and the 2 lateral links mounted to the knuckle of an ABS-equipped vehicle

Stabilizer (Sway) Bar and Links

REMOVAL & INSTALLATION

▶ See Figures 71, 72 and 73

1. Raise the rear of the vehicle and support it safely using jackstands, then remove the rear wheels.

2. Position a drain pan under the left rear brake line, at the brake line hose junction, and disconnect the line from the hose. Plug the brake line and hose to prevent excessive fluid loss and to avoid the possibility of contamination.

3. Remove the right and left stabilizer bar link-to-bracket fasteners.

4. Remove the stabilizer bar-to-crossmember fasteners.

5. Loosen, but do not remove, the left lateral link-to-knuckle fastener.

6. Remove the left trailing arm-to-knuckle nut, then remove the body fasteners. Slide the arm from the knuckle and remove it from the vehicle.

7. Remove the left lateral link fastener and pivot the links downward, away from the knuckle.

8. Note the position of the brake line to the crossmember, then unfasten the crossbody brake line by unsnapping the fasteners from the crossmember.

❊❊ WARNING

Attempting to remove the stabilizer bar from the crossmember without first loosening the brake lines, may result in a bent/damaged line which must be replaced.

9. Remove the stabilizer bar from the crossmember and the vehicle.
To install:
10. Install the stabilizer bar to the crossmember and secure using the bracket fasteners. Tighten the retaining bolts to 41 ft. lbs. (55 Nm).

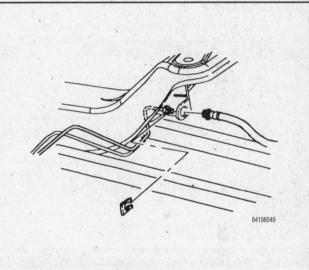

Fig. 71 Disconnect the left brake line from the brake hose

Fig. 72 View of the stabilizer bar end and the stabilizer link— model with rear disc brakes shown

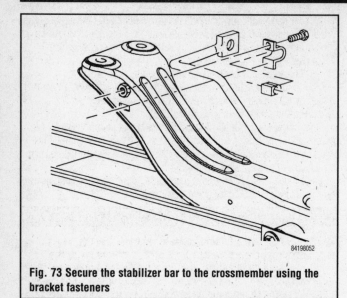

Fig. 73 Secure the stabilizer bar to the crossmember using the bracket fasteners

11. Pivot the lateral links into the knuckle and loosely install the fastener, but do not tighten at this time.

12. Slide the trailing arm into the knuckle and loosely install the fastener grommets, washer and nut. Install the arm-to-body fasteners and tighten to 89 ft. lbs. (120 Nm), then tighten the nut at the knuckle to 74 ft. lbs. (100 Nm).

13. Tighten the lateral link-to-knuckle fastener to 122 ft. lbs. (165 Nm).

14. Remove the plugs from the brake line and hose, then connect the fitting and tighten to 14 ft. lbs. (19 Nm).

15. Install the crossbody brake line to the crossmember by snapping the fasteners into the member. Make sure the line is positioned as noted during removal.

16. Install the rear stabilizer bar links-to-bracket fasteners and tighten to 30 ft. lbs. (40 Nm).

17. Properly bleed the brake system. Refer to the procedure in Section No. 9 of this manual.

18. Install the wheel.

19. Remove the supports and lower the vehicle.

Rear Wheel Hub/Bearing Assembly

The rear hub/bearing assembly is a sealed assembly, which requires no periodic maintenance and cannot be serviced. If the hub/bearing assembly becomes worn or damaged, the entire unit must be replaced.

STEERING

Steering Wheel

REMOVAL & INSTALLATION

Without Air Bag

▶ **See Figures 75 and 76**

1. Disconnect the negative battery cable.

2. Lift the horn pad from the steering wheel by pulling on the edge of the pad firmly, then disconnect the wires and remove the horn pad from the vehicle.

3. Remove the clip on the end of the steering column shaft and remove the retaining nut.

REMOVAL & INSTALLATION

▶ **See Figure 74**

1. Raise the rear of the vehicle and support safely using jackstands, then remove the rear wheel.

2. If equipped with an Anti-lock Brake System (ABS), disconnect the negative battery cable, then unplug the ABS speed sensor connector.

3. If equipped with rear disc brakes, remove the 2 caliper-to-knuckle mounting bolts from the rear of the knuckle, then use a length of wire to suspend the caliper from the strut, clear of the knuckle.

4. Remove the brake drum or rotor, as applicable.

5. Remove the 4 hub/bearing-to-knuckle bolts and remove the assembly from the vehicle.

To install:

6. Position the brake backing plate and the hub/bearing assembly, then secure the components to the knuckle using the retaining bolts. Tighten the bolts to 63 ft. lbs. (85 Nm).

7. Install the brake drum or the rotor and caliper. If applicable, tighten the caliper retaining bolts to 63 ft. lbs. (85 Nm).

8. If equipped, install the ABS wiring harness to the speed sensor.

9. Install the wheel assembly, remove the supports and carefully lower the vehicle. If applicable, connect the negative battery cable.

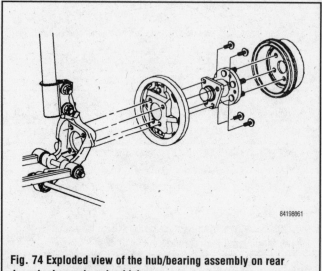

Fig. 74 Exploded view of the hub/bearing assembly on rear drum brake equipped vehicles

4. Note the position of the steering wheel locating notch for reassembly purposes.

5. Install a suitable steering wheel puller and remove the steering wheel from the steering column.

To install:

6. Route the wires through the wheel and position the steering wheel, making sure to properly align the locating notch. If the locating notch is not properly positioned, any attempt to install the steering wheel will damage the wheel and column beyond repair.

7. Install a new steering wheel nut. Tighten the nut to 30 ft. lbs. (40 Nm) and install a new clip on the end of the column.

8. Connect the wires to the horn pad and press the pad firmly into position on the wheel.

9. Connect the negative battery cable.

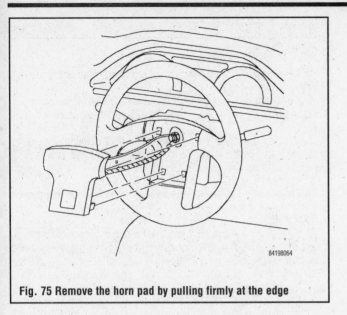

Fig. 75 Remove the horn pad by pulling firmly at the edge

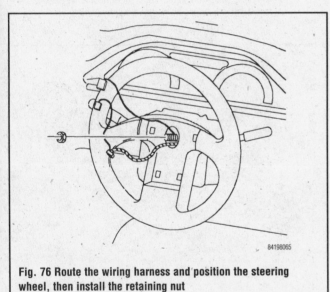

Fig. 76 Route the wiring harness and position the steering wheel, then install the retaining nut

With Air Bag
▶ See Figures 77 thru 89

※※ CAUTION

If your vehicle is equipped with a Supplemental Inflatable Restraint (SIR) system, follow the recommended disarming procedures before performing any work on or around the system. Failure to do so may result in possible deployment of the air bag and/or personal injury.

1. Disable the SIR system as follows:
   a. Align the steering wheel so the tires are in the straight-ahead position, then turn the ignition to the **OFF** or **LOCK** position and remove the key.
   b. Remove the SIR fuse from the Instrument Panel Junction Block (IPJB).
   c. Remove the Connector Position Assurance (CPA) device, then detach the yellow 2-way SIR connector at the base of the steering column.
2. If equipped with a passenger side air bag, perform the same SIR system disabling procedure with the addition of the following steps:
   a. Using a small, flat bladed screwdriver, pry off the upper instrument panel pad screw caps. Remove the trim panel retaining screws.
   b. Disengage the rear trim panel clips by lifting up. Then, disengage the front trim panel clips by lifting up at each outer corner and pulling rearward. Be careful not to damage the trim panel during its removal from the vehicle.
   c. Remove the upper trim panel insulator.
   d. Remove the CPA device, then detach the yellow 2-way SIR connector on the pigtail from the passenger side air bag module.
3. Disconnect the negative battery cable.
4. If necessary, remove the screw caps from behind the steering wheel by carefully prying with a small, flat bladed screwdriver. Loosen the 4 fasteners from the back of the steering wheel and lift the inflator module from the steering wheel.
5. Remove the CPA device and unplug the wiring harness from the module, then remove the inflator module from the steering wheel.

※※ CAUTION

When carrying a live inflator module, ensure that the bag and trim cover are pointed away from the body. Never carry the inflator module by the wires or connector on the underside of the module. This will minimize the chance of injury should the module accidentally deploy. When placing a live inflator module on a bench or other surface, always place the bag and trim cover up, away from the surface. This is necessary so a free space is provided to allow for air bag expansion in the unlikely event of accidental deployment.

6. Unplug the horn connector and, if equipped, unplug the cruise control switch connector.
7. Remove the clip on the end of the steering column shaft, if equipped, then remove the retaining nut.
8. Note the position of the steering wheel locating notch for reassembly purposes.
9. Install a suitable steering wheel puller and remove the steering wheel from the steering column.
10. Install a yellow retaining tab into the SIR coil assembly to keep it from rotating. If a retaining tab is not available, tape the coil in position to prevent coil damage.

To install:

11. Route the SIR wire and other electrical connections through the wheel, then position the steering wheel, making sure to properly align the locating notch. If the locating notch is not properly positioned, any attempt to install the steering wheel will damage the wheel and column beyond repair.
12. Remove the yellow retaining tape or the tape from the SIR coil assembly.
13. Install a new steering wheel nut. Tighten the nut to 30 ft. lbs. (40 Nm) and install a new clip on the end of the column, if equipped.
14. Connect the wiring harness to the horn and, if equipped, to the cruise control switch.
15. Position the inflator module and connect the SIR wiring harness. Install the CPA device to the connector, then seat the module on the steering wheel.
16. Secure the module using NEW fasteners, then tighten the new fasteners to 89 inch lbs. (10 Nm). Install the screw caps, if equipped.
17. Connect the negative battery cable.
18. Enable the SIR system as follows:
    a. Verify the ignition switch is in the **OFF** or **LOCK** position, then attach the SIR electrical connector at the base of the steering column. Install the CPA device to the connector.
    b. If equipped with a passenger side air bag, connect the yellow 2-way inflator module pigtail and attach the CPA device to the connector.
    c. If equipped with a passenger side air bag, install the insulator and upper trim panel assembly. Be sure that all flaps are tucked and clips engaged. Install and tighten the retaining screws to 20 inch lbs. (2.3 Nm). Install the trim panel screw caps.
    d. Install the SIR fuse to the IPJB and install the junction block cover.
    e. Turn the ignition **ON** and verify that the AIR BAG indicator lamp flashes 7–9 times, then extinguishes. If the light does not flash as indicated, inspect the system for malfunction.

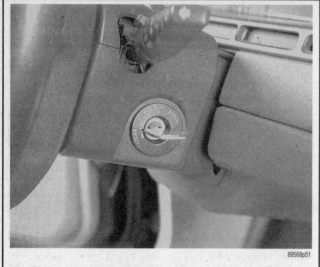

Fig. 77 Turn the ignition key to the ACC or LOCK position

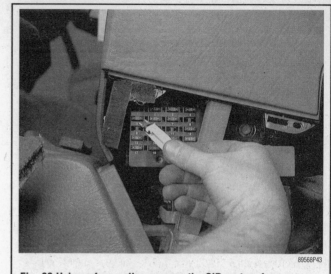

Fig. 80 Using a fuse puller, remove the SIR system fuse

Fig. 78 Remove the right side lower console trim panel . . .

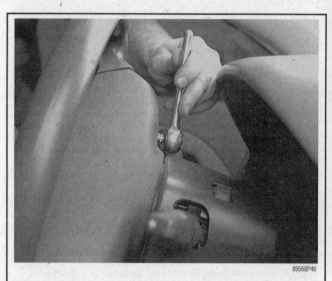

Fig. 81 Loosen the 4 air bag inflator module mounting fasteners

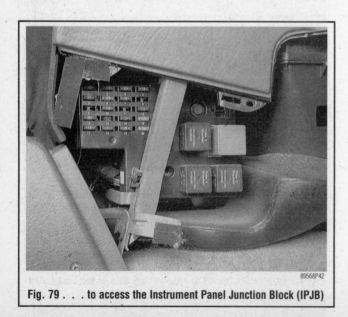

Fig. 79 . . . to access the Instrument Panel Junction Block (IPJB)

Fig. 82 Lift the inflator module away from the steering wheel

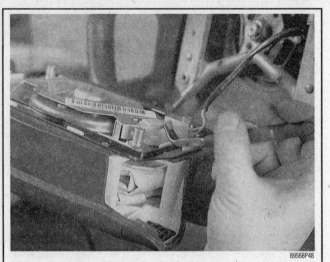

Fig. 83 Remove the CPA device from the inflator module connector . . .

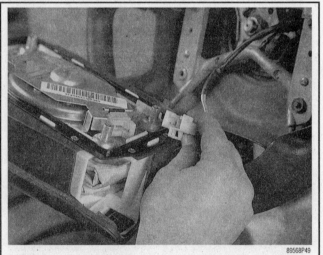

Fig. 84 . . . and unplug the wiring harness from the module, then remove the module

Fig. 85 Always place the air bag with the trim cover up, away from the surface, providing a free space in the unlikely event of accidental deployment

Fig. 86 Disengage the horn and, if equipped, cruise control wiring connectors

Fig. 87 Loosen the steering wheel retaining nut

Fig. 88 Remove the retaining nut, making note of the location of the steering wheel notch

Fig. 89 Using a steering wheel puller tool, remove the wheel from the column shaft

Turn Signal (Combination) Switch

REMOVAL & INSTALLATION

▶ See Figures 90, 91 and 92

✳✳ CAUTION

If your vehicle is equipped with a Supplemental Inflatable Restraint (SIR) system, follow the recommended disarming procedures before performing any work on or around the system. Failure to do so may result in possible deployment of the air bag and/or personal injury.

1. If equipped, disable the SIR system as follows:
   a. Align the steering wheel so the tires are in the straight-ahead position, then turn the ignition to the **OFF** or **LOCK** position and remove the key.
   b. Remove the SIR fuse from the Instrument Panel Junction Block (IPJB).
   c. Remove the Connector Position Assurance (CPA) device, then detach the yellow 2-way SIR connector at the base of the steering column.
2. If equipped with a passenger side air bag, perform the same SIR system disabling procedure with the addition of the following steps:
   a. Using a small, flat bladed screwdriver, pry off the upper instrument panel pad screw caps. Remove the trim panel retaining screws.
   b. Disengage the rear trim panel clips by lifting up. Then, disengage the front trim panel clips by lifting up at each outer corner and pulling rearward. Be careful not to damage the trim panel during its removal from the vehicle.
   c. Remove the upper trim panel insulator.
   d. Remove the CPA device, then detach the yellow 2-way SIR connector on the pigtail from the passenger side air bag module.
3. Disconnect the negative battery cable.
4. Remove the steering wheel. Refer to the procedure earlier in this section.
5. Remove the 2 retaining screws and the upper steering column cover from the steering column.
6. Remove the ignition lock bezel, then remove the 2 retaining screws and the lower steering column cover.
7. If equipped with an SIR system, remove the coil from the combination meter.
8. Remove the CPA device and disconnect the wires from the lever control (combination) switch.

9. Unfasten the retaining bolts and remove the combination switch assembly from the steering column. It may be necessary to remove the instrument panel trim covers and center console trim in order to access the switch fasteners; if this is necessary, refer to the procedures in Section 10.

To install:

10. Install the combination switch and retaining bolts. Tighten the lower mounting bolt first to assure proper location and seating, then attach the switch electrical connectors and insert the CPA device.
11. If applicable, install the SIR coil to the combination switch assembly.
12. If removed, install the instrument panel and center console trim panels.
13. Install the steering column covers and the ignition bezel.
14. Install the steering wheel.
15. Connect the negative battery cable.
16. Enable the SIR system as follows:
    a. Verify that the ignition switch is in the **OFF** or **LOCK** position, then attach the SIR electrical connector at the base of the steering column. Install the CPA device to the connector.
    b. If equipped with a passenger side air bag, connect the yellow 2-way inflator module pigtail and install the CPA device to the connector.

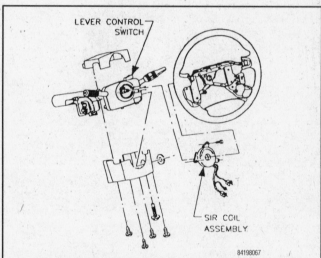

Fig. 90 Exploded view of the upper steering column and combination switch—SIR equipped vehicles

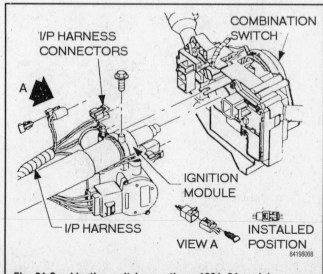

Fig. 91 Combination switch mounting—1991–94 models

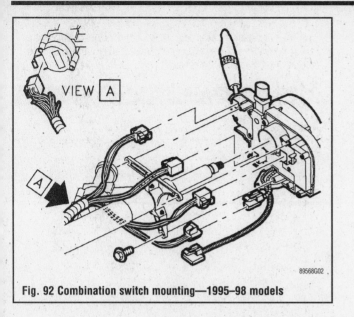

Fig. 92 Combination switch mounting—1995–98 models

c. If equipped with a passenger side air bag, install the insulator and upper trim panel assembly. Be sure that all flaps are tucked and clips engaged. Install and tighten the retaining screws to 20 inch lbs. (2.3 Nm). Install the trim panel screw caps.

d. Install the SIR fuse to the IPJB and install the junction block cover.

e. Turn the ignition **ON** and verify that the AIR BAG indicator lamp flashes 7–9 times, then extinguishes. If the light does not flash as indicated, inspect the system for malfunction.

Ignition Switch

REMOVAL & INSTALLATION

▶ See Figure 93

✳✳ CAUTION

If your vehicle is equipped with a Supplemental Inflatable Restraint (SIR) system, follow the recommended disarming procedures before performing any work on or around the system. Failure to do so may result in possible deployment of the air bag and/or personal injury.

1. If equipped, disable the SIR system as follows:
   a. Align the steering wheel so the tires are in the straight-ahead position, then turn the ignition to the **OFF** or **LOCK** position and remove the key.
   b. Remove the SIR fuse from the Instrument Panel Junction Block (IPJB).
   c. Remove the Connector Position Assurance (CPA) device, then detach the yellow 2-way SIR connector at the base of the steering column.
2. If equipped with a passenger side air bag, perform the same SIR system disabling procedure with the addition of the following steps:
   a. Using a small, flat bladed screwdriver, pry off the upper instrument panel pad screw caps. Remove the trim panel retaining screws.
   b. Disengage the rear trim panel clips by lifting up. Then, disengage the front trim panel clips by lifting up at each outer corner and pulling rearward. Be careful not to damage the trim panel during its removal from the vehicle.
   c. Remove the upper trim panel insulator.
   d. Remove the CPA device, then detach the yellow 2-way SIR connector on the pigtail from the passenger side air bag module.
3. Disconnect the negative battery cable.
4. Remove the steering wheel. Refer to the procedure earlier in this section.
5. Remove the combination switch. Refer to the procedure earlier in this section.

6. Unplug the ignition switch electrical connector and remove the 2 retaining screws, then disconnect the switch from the ignition module and remove the switch from the vehicle.

To install:

7. Install the combination switch and retaining bolts. Tighten the lower mounting bolt first to assure proper location and seating.

8. Connect the wiring harness to the ignition switch.

9. Install the ignition switch and retaining bolts to the ignition module. Tighten the ignition switch retaining bolts to 18–27 inch lbs. (2–3 Nm).

10. If applicable, install the SIR coil to the combination switch assembly.

11. If removed, install the instrument cluster and center console trim panels.

12. Install the steering column covers and the ignition bezel.

13. Install the steering wheel.

14. Connect the negative battery cable.

15. Enable the SIR system as follows:
    a. Verify that the ignition switch is in the **OFF** or **LOCK** position, then attach the SIR electrical connector at the base of the steering column. Install the CPA device to the connector.
    b. If equipped with a passenger side air bag, connect the yellow 2-way inflator module pigtail and install the CPA device to the connector.
    c. If equipped with a passenger side air bag, install the insulator and upper trim panel assembly. Be sure that all flaps are tucked and clips engaged. Install and tighten the retaining screws to 20 inch lbs. (2.3 Nm). Install the trim panel screw caps.
    d. Install the SIR fuse to the IPJB and install the junction block cover.
    e. Turn the ignition **ON** and verify that the AIR BAG indicator lamp flashes 7–9 times, then extinguishes. If the light does not flash as indicated, inspect the system for malfunction.

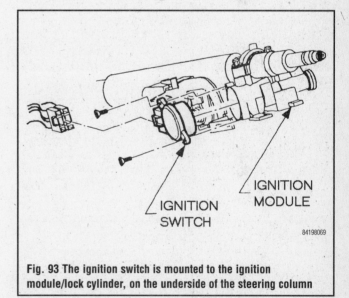

Fig. 93 The ignition switch is mounted to the ignition module/lock cylinder, on the underside of the steering column

Ignition Lock Cylinder

REMOVAL & INSTALLATION

▶ See Figure 94

✳✳ CAUTION

If your vehicle is equipped with a Supplemental Inflatable Restraint (SIR) system, follow the recommended disarming procedures before performing any work on or around the system. Failure to do so may result in possible deployment of the air bag and/or personal injury.

1. If equipped, disable the SIR system as follows:

a. Align the steering wheel so the tires are in the straight-ahead position, then turn the ignition to the **OFF** or **LOCK** position and remove the key.

b. Remove the SIR fuse from the Instrument Panel Junction Block (IPJB).

c. Remove the Connector Position Assurance (CPA) device, then detach the yellow 2-way SIR connector at the base of the steering column.

2. If equipped with a passenger side air bag, perform the same SIR system disabling procedure with the addition of the following steps:

a. Using a small, flat bladed screwdriver, pry off the upper instrument panel pad screw caps. Remove the trim panel retaining screws.

b. Disengage the rear trim panel clips by lifting up. Then, disengage the front trim panel clips by lifting up at each outer corner and pulling rearward. Be careful not to damage the trim panel during its removal from the vehicle.

c. Remove the upper trim panel insulator.

d. Remove the CPA device, then detach the yellow 2-way SIR connector on the pigtail from the passenger side air bag module.

3. Disconnect the negative battery cable.

4. Remove the steering wheel. Refer to the procedure earlier in this section.

5. Remove the combination switch. Refer to the procedure earlier in this section.

6. Insert the ignition key into the lock cylinder and turn it to the **ACC** position.

7. Depress the square locking button on the top of the ignition lock cylinder module. Slide the ignition lock cylinder assembly from the cylinder module.

To install:

8. Slide the ignition lock cylinder into the cylinder module and be sure that the square locking button is seated correctly.

9. Remove the key from the ignition lock cylinder.

10. Install the combination switch.

11. Install the steering wheel.

12. Connect the negative battery cable.

13. Enable the SIR system as follows:

a. Verify that the ignition switch is in the **OFF** or **LOCK** position, then attach the SIR electrical connector at the base of the steering column. Install the CPA device to the connector.

b. If equipped with a passenger side air bag, connect the yellow 2-way inflator module pigtail and install the CPA device to the connector.

c. If equipped with a passenger side air bag, install the insulator and upper trim panel assembly. Be sure that all flaps are tucked and clips engaged. Install and tighten the retaining screws to 20 inch lbs. (2.3 Nm). Install the trim panel screw caps.

d. Install the SIR fuse to the IPJB and install the junction block cover.

e. Turn the ignition **ON** and verify that the AIR BAG indicator lamp flashes 7–9 times, then extinguishes. If the light does not flash as indicated, inspect the system for malfunction.

Tie Rod Ends

REMOVAL & INSTALLATION

♦ See Figures 95, 96 and 97

1. Raise the front of the vehicle and support it safely using jackstands.

2. Remove the front wheel and, if necessary for access, remove the splash shield:

3. Remove the cotter pin and nut from the tie rod end.

4. Separate the tie rod end from the knuckle with separator tool SA91100C or equivalent. Do not use a wedge-type tool or the seal may be damaged.

5. Mark the threaded portion of the tie rod for installation purposes.

6. Loosen the tie rod jam nut and unthread the tie rod end from the tie rod. Count the number of turns necessary to remove.

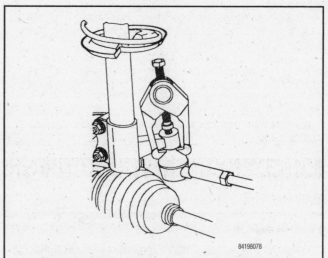

84198078

Fig. 95 Use SA91100C or an equivalent jawed separator tool to remove the tie rod end from the steering knuckle

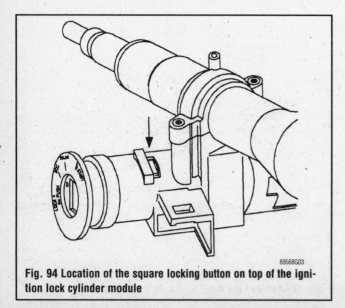

89568G03

Fig. 94 Location of the square locking button on top of the ignition lock cylinder module

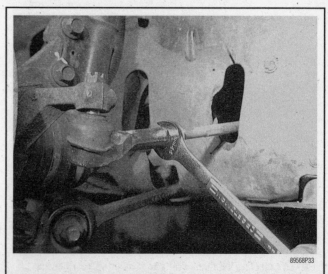

89568P33

Fig. 96 Loosen the tie rod jam nut

Fig. 97 Mark the tie rod end's position in relation to the threads, then unscrew the tie rod end from the tie rod

To install:

7. Clean and lubricate the threads before installation.

8. Install the tie rod end, using the same number of turns as counted during removal. Align the tie rod end to the marked location.

9. Install the tie rod end to the knuckle. Install the castle nut and tighten to 33 ft. lbs. (45 Nm), then install a new cotter pin. If necessary, tighten the nut additionally, but do not back it off, to insert the cotter pin.

10. Tighten the tie rod jam nut.

11. Install the splash shield, if removed, then install the wheel assembly.

12. Remove the supports and lower the vehicle, then check and adjust toe as necessary.

Manual Rack and Pinion

BEARING PRELOAD ADJUSTMENT

▶ **See Figure 98**

1. Center the steering wheel, then raise the front of the vehicle and support it safely using jackstands.

2. Loosen the steering gear adjuster plug locknut, then turn the adjuster plug clockwise until it bottoms in the gear housing.

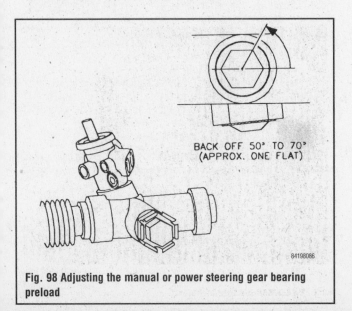

Fig. 98 Adjusting the manual or power steering gear bearing preload

3. Tighten the plug to 106 inch lbs. (12 Nm).

4. Back off the adjuster plug 50–70 degrees (about 1 flat of the nut).

5. While holding the plug steady, tighten the locknut with a crow's foot wrench to 52 ft. lbs. (70 Nm).

6. Check the steering for returnability, binding or difficulty in turning.

REMOVAL & INSTALLATION

▶ **See Figures 99 and 100**

1. Disconnect the negative battery cable, then raise the front of the vehicle and support it safely using jackstands.

2. Remove the front wheels and the left inner splash shield.

3. Remove and discard the tie rod ends' cotter pins, then remove the castle nuts. Disconnect the tie rod ends using SA91100C or an equivalent separator tool. Do not use a wedge-type tool or seal damage may occur.

4. Loosen the intermediate shaft cover from the steering gear and move it up enough to access the pinch bolt. Remove the pinch bolt.

5. Remove the steering gear fasteners and remove the gear through the left fenderwell.

To install:

6. Install the steering gear and tighten the steering gear-to-cradle bolts to 37 ft. lbs. (50 Nm).

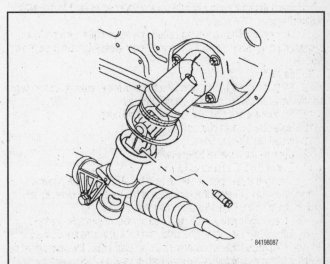

Fig. 99 Move the intermediate shaft cover away from the gear for access to the pinch bolt

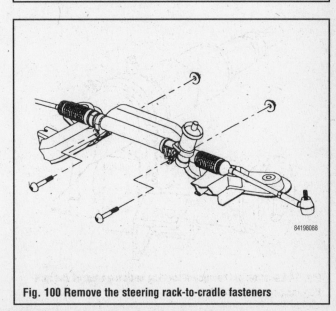

Fig. 100 Remove the steering rack-to-cradle fasteners

7. Apply Loctite® 242 Threadlocker, or equivalent, to the intermediate shaft pinch bolt threads. Install the intermediate steering shaft to the gear and tighten the pinch bolt to 35 ft. lbs. (47 Nm). Position the shaft cover.

8. Thoroughly clean and lubricate the threads of the tie rod ends, then install the ends into the steering knuckles. Install the castle nuts and tighten to 33 ft. lbs. (45 Nm). Install new cotter pins. If necessary, tighten the nuts additionally to install the pins, but do not loosen them.

9. Attach the left inner splash shield and install the front wheels.

10. Remove the supports and carefully lower the vehicle, then connect the negative battery cable.

11. Check front end alignment and adjust vehicle toe, as necessary.

Power Rack and Pinion

BEARING PRELOAD ADJUSTMENT

Bearing preload adjustment is the same for both the manual and power rack and pinion assemblies. Refer to the Manual Rack and Pinion adjustment procedure, earlier in this section.

REMOVAL & INSTALLATION

⬥ See Figures 101 and 102

1. Disconnect the negative battery cable, then raise the front of the vehicle and support it safely using jackstands.

2. Remove the front wheels and the left inner splash shield.

3. Remove and discard the tie rod ends' cotter pins, then remove the castle nuts. Disconnect the tie rod ends using SA91100C or an equivalent separator tool. Do not use a wedge-type tool or seal damage may occur.

4. Loosen the intermediate shaft cover from the steering gear and move it up enough to access the pinch bolt. Remove the pinch bolt.

5. For 1991 vehicles, unplug the power steering pressure switch electrical connector at the steering gear.

6. Place a clean container under the pressure and return hoses at the steering assembly. Disconnect the lines from the steering gear and allow the system to drain.

7. Remove the steering gear fasteners and remove the gear through the left fenderwell.

To install:

8. Install the steering gear and tighten the steering gear-to-cradle bolts to 37 ft. lbs. (50 Nm).

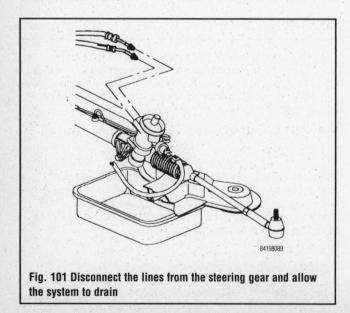

Fig. 101 Disconnect the lines from the steering gear and allow the system to drain

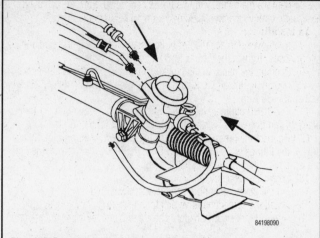

Fig. 102 Connect the pressure and return hoses, then for 1991 vehicles, connect the wiring harness to the power steering pressure switch

9. Apply Loctite® 242 Threadlocker, or equivalent, to the intermediate shaft pinch bolt threads. Install the intermediate steering shaft to the gear and tighten the pinch bolt to 35 ft. lbs. (47 Nm). Position the shaft cover.

10. Connect the pressure and return hoses, then tighten the fittings to 20 ft. lbs. (27.5 Nm).

11. For 1991 vehicles, connect the wiring harness to the power steering pressure switch.

12. Thoroughly clean and lubricate the threads of the tie rod ends, then install the ends into the steering knuckles. Install the castle nuts and tighten to 33 ft. lbs. (45 Nm). Install new cotter pins. If necessary, tighten the nuts additionally to install the pins, but do not loosen them.

13. Install the left inner splash shield and install the front wheels.

14. Remove the supports and lower the vehicle, then connect the negative battery cable.

15. Check front end alignment and adjust vehicle toe, as necessary.

16. Check and add power steering fluid as necessary. Bleed the power steering system. Refer to the procedure later in this section.

Power Steering Pump

REMOVAL & INSTALLATION

⬥ See Figures 103, 104, 105 and 106

1. Disconnect the negative battery cable.

2. Remove the reservoir fill cap.

3. Raise the front of the vehicle and support it safely using jackstands. Make sure the vehicle is raised sufficiently to allow both undervehicle and underhood access.

4. Place a clean container under the power steering hoses, then remove the hoses from the steering gear and allow the system to drain.

➡**Once the hoses have been removed from the steering gear, do not rotate the steering wheel, or additional fluid will be forced from the gear, creating even more of a mess.**

5. Use a box end wrench to relieve spring tension from the accessory drive belt tensioner and remove the belt from the steering pump pulley.

6. For DOHC engines, remove the pump-to-intake manifold and pump-to-engine fasteners and bracket.

7. Remove the 3 pump bracket-to-block bolts and raise the pump far enough to unplug the electrical connector from the pump's Electronic Variable Orifice (EVO) solenoid, if equipped.

8. Remove the pump, with the hoses connected, from the vehicle. If necessary, remove the pressure and return hoses from the pump.

To install:

9. If removed, replace the O-ring seals and install the pressure and return hoses. Tighten the fittings to 20 ft. lbs. (27.5 Nm).

10. Position the pump to the block and connect the wiring harness to the EVO solenoid. Install the 3 pump retaining bolts and tighten to 22–28 ft. lbs. (30–38 Nm).

11. For DOHC engines, install the pump-to-intake and pump-to-block brackets and fasteners. Tighten the fasteners to 22 ft. lbs. (30 Nm).

12. Connect the pressure and return hoses to the steering gear, then tighten the fittings to 20 ft. lbs. (27.5 Nm). Route the return hose, then the pressure hose into the retaining clip.

13. Lower the vehicle.

14. Fill the power steering reservoir with clean fluid.

15. Install the serpentine drive belt to the steering pump pulley.

16. Connect the negative battery cable and bleed the power steering system.

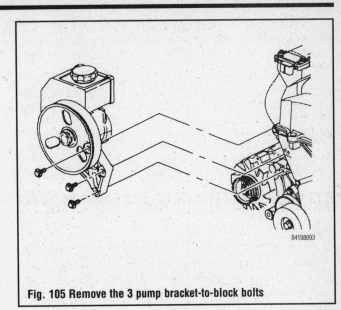

Fig. 105 Remove the 3 pump bracket-to-block bolts

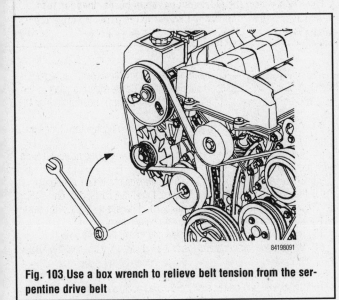

Fig. 103 Use a box wrench to relieve belt tension from the serpentine drive belt

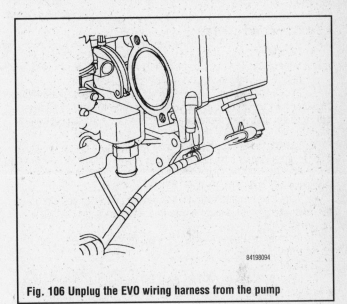

Fig. 106 Unplug the EVO wiring harness from the pump

SYSTEM BLEEDING

1. Fill the reservoir with the recommended fluid to the FULL mark. Refer to 2. Raise the front of the vehicle just sufficiently so that the drive wheels are off the ground, then safely support the vehicle using jackstands.

3. Bleed the system by turning the wheels from side-to-side without hitting the stops. It may take several cycles to bleed the system.

➡**Maintain the reservoir at the FULL mark during this procedure.**

4. Lower the vehicle.

5. Start the engine and, with the transaxle in **P** (A/T) or **N** (M/T), check the fluid level with the engine idling. If necessary, add fluid to bring the level to the FULL mark.

6. Road test the vehicle and check for proper operation. Recheck the fluid level and make sure it is at or slightly above the FULL mark after the system has stabilized at normal operating temperature.

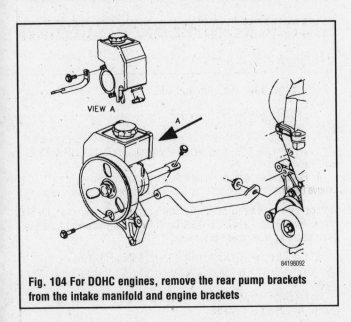

Fig. 104 For DOHC engines, remove the rear pump brackets from the intake manifold and engine brackets

Troubleshooting Basic Steering and Suspension Problems

Problem	Cause	Solution
Hard steering (steering wheel is hard to turn)	• Low or uneven tire pressure • Loose power steering pump drive belt • Low or incorrect power steering fluid • Incorrect front end alignment • Defective power steering pump • Bent or poorly lubricated front end parts	• Inflate tires to correct pressure • Adjust belt • Add fluid as necessary • Have front end alignment checked/adjusted • Check pump • Lubricate and/or replace defective parts
Loose steering (too much play in the steering wheel)	• Loose wheel bearings • Loose or worn steering linkage • Faulty shocks • Worn ball joints	• Adjust wheel bearings • Replace worn parts • Replace shocks • Replace ball joints
Car veers or wanders (car pulls to one side with hands off the steering wheel)	• Incorrect tire pressure • Improper front end alignment • Loose wheel bearings • Loose or bent front end components • Faulty shocks	• Inflate tires to correct pressure • Have front end alignment checked/adjusted • Adjust wheel bearings • Replace worn components • Replace shocks
Wheel oscillation or vibration transmitted through steering wheel	• Improper tire pressures • Tires out of balance • Loose wheel bearings • Improper front end alignment • Worn or bent front end components	• Inflate tires to correct pressure • Have tires balanced • Adjust wheel bearings • Have front end alignment checked/adjusted • Replace worn parts
Uneven tire wear	• Incorrect tire pressure • Front end out of alignment • Tires out of balance	• Inflate tires to correct pressure • Have front end alignment checked/adjusted • Have tires balanced

TCCA8C01

TORQUE SPECIFICATIONS

System / Component	Ft. Lbs.	Nm
Wheels		
Wheel lug nuts	103	140
Front suspension		
MacPherson struts		
Upper strut mount fasteners	21	29
Strut-to-knuckle fasteners	126	170
Upper strut shaft nut	37	50
Stabilizer bar (Sway bar)		
Mounting bracket bolts	103	140
Stabilizer-to-control arm nuts	106	144
Lower control arm		
Control arm-to-cradle bolt	92	125
Control arm-to-cradle nut	74	100
Lower control arm stud-to-steering knuckle nut	55	75
Steering knuckle assembly		
Strut-to-knuckle fasteners	126	170
Brake caliper mounting bracket-to-knuckle fasteners	81	110
Halfshaft-to-hub retaining nut	148	200
Wheel bearing and hub assembly		
Brake dust shield fasteners		
1991 models only	18	25
Rear suspension		
MacPherson struts		
Upper strut mount fasteners	21	29
Strut-to-knuckle fasteners	126	170
Upper strut shaft nut	37	50
Lateral links		
Crossmember-to-body mounting bolts	89	120
Rear lateral link-to-crossmember fasteners	89	120
Front lateral link-to-crossmember fasteners		
1991-94 models	126	170
1995-98 models	89	120
Lateral link-to-knuckle fasteners	122	165
Trailing arm		
Trailing arm-to-body bolts	89	120
Trailing arm-to-knuckle nut	74	100
Stabilizer bar (Sway bar) and links		
Stabilizer bar-to-crossmember bracket fasteners	41	55
Rear stabilizer links-to-bracket fasteners	30	40
Rear brake line fitting	14	19
Rear wheel hub and bearing assembly		
Hub/bearing assembly-to-brake backing plate bolts	63	85
Brake caliper retaining bolts (if equipped)	63	85
Steering		
Steering wheel		
Steering wheel retaining nut	30	40
Air bag inflator module fasteners (if equipped)	89 inch lbs.	10
Upper trim panel retaining screws	20 inch lbs.	2.3
Ignition switch		
Switch retaining bolts	18-27 inch lbs.	2-3

89568C01

TORQUE SPECIFICATIONS

System	Component	Ft. Lbs.	Nm
Steering			
Tie rod ends			
	Tie rod-to-steering knuckle nut	33	45
Rack and pinion			
	Steering gear adjuster plug	106 inch lbs.	12
	Adjuster plug locknut	52	70
	Steering gear-to-cradle mounting bolts	37	50
	Intermediate steering shaft-to-steering gear pinch bolt	35	47
	Power steering pressure and return hose fittings	20	27.5
Power steering pump			
	Electronic variable orifice (EVO) actuator retaining fasteners	22-28	30-38
	DOHC engines only		
	Pump-to-intake manifold fasteners	22	30
	Pump-to-engine block bracket fasteners	22	30

89568C02

9

BRAKES

BRAKE OPERATING SYSTEM

Basic Operating Principles

Hydraulic systems are used to actuate the brakes of all modern automobiles. The system transports the power required to force the frictional surfaces of the braking system together from the pedal to the individual brake units at each wheel. A hydraulic system is used for two reasons:

First, fluid under pressure can be carried to all parts of an automobile by small pipes and flexible hoses without taking up a significant amount of room or posing routing problems.

Second, a great mechanical advantage can be given to the brake pedal end of the system, and the foot pressure required to actuate the brakes can be reduced by making the surface area of the master cylinder pistons smaller than that of any of the pistons in the wheel cylinders or calipers.

The master cylinder consists of a fluid reservoir along with a double cylinder and piston assembly. Double type master cylinders are designed to separate the front and rear braking systems hydraulically in case of a leak. The master cylinder coverts mechanical motion from the pedal into hydraulic pressure within the lines. This pressure is translated back into mechanical motion at the wheels by either the wheel cylinder (drum brakes) or the caliper (disc brakes).

Steel lines carry the brake fluid to a point on the vehicle's frame near each of the vehicle's wheels. The fluid is then carried to the calipers and wheel cylinders by flexible tubes in order to allow for suspension and steering movements.

In drum brake systems, each wheel cylinder contains two pistons, one at either end, which push outward in opposite directions and force the brake shoe into contact with the drum.

In disc brake systems, the cylinders are part of the calipers. At least one cylinder in each caliper is used to force the brake pads against the disc.

All pistons employ some type of seal, usually made of rubber, to minimize fluid leakage. A rubber dust boot seals the outer end of the cylinder against dust and dirt. The boot fits around the outer end of the piston on disc brake calipers, and around the brake actuating rod on wheel cylinders.

The hydraulic system operates as follows: When at rest, the entire system, from the piston(s) in the master cylinder to those in the wheel cylinders or calipers, is full of brake fluid. Upon application of the brake pedal, fluid trapped in front of the master cylinder piston(s) is forced through the lines to the wheel cylinders. Here, it forces the pistons outward, in the case of drum brakes, and inward toward the disc, in the case of disc brakes. The motion of the pistons is opposed by return springs mounted outside the cylinders in drum brakes, and by spring seals, in disc brakes.

Upon release of the brake pedal, a spring located inside the master cylinder immediately returns the master cylinder pistons to the normal position. The pistons contain check valves and the master cylinder has compensating ports drilled in it. These are uncovered as the pistons reach their normal position. The piston check valves allow fluid to flow toward the wheel cylinders or calipers as the pistons withdraw. Then, as the return springs force the brake pads or shoes into the released position, the excess fluid reservoir through the compensating ports. It is during the time the pedal is in the released position that any fluid that has leaked out of the system will be replaced through the compensating ports.

Dual circuit master cylinders employ two pistons, located one behind the other, in the same cylinder. The primary piston is actuated directly by mechanical linkage from the brake pedal through the power booster. The secondary piston is actuated by fluid trapped between the two pistons. If a leak develops in front of the secondary piston, it moves forward until it bottoms against the front of the master cylinder, and the fluid trapped between the pistons will operate the rear brakes. If the rear brakes develop a leak, the primary piston will move forward until direct contact with the secondary piston takes place, and it will force the secondary piston to actuate the front brakes. In either case, the brake pedal moves farther when the brakes are applied, and less braking power is available.

All dual circuit systems use a switch to warn the driver when only half of the brake system is operational. This switch is usually located in a valve body which is mounted on the firewall or the frame below the master cylinder. A hydraulic piston receives pressure from both circuits, each circuit's pressure being applied to one end of the piston. When the pressures are in balance, the piston remains stationary. When one circuit has a leak, however, the greater pressure in that circuit during application of the brakes will push the piston to one side, closing the switch and activating the brake warning light.

In disc brake systems, this valve body also contains a metering valve and, in some cases, a proportioning valve. The metering valve keeps pressure from traveling to the disc brakes on the front wheels until the brake shoes on the rear wheels have contacted the drums, ensuring that the front brakes will never be used alone. The proportioning valve controls the pressure to the rear brakes to lessen the chance of rear wheel lock-up during very hard braking.

Warning lights may be tested by depressing the brake pedal and holding it while opening one of the wheel cylinder bleeder screws. If this does not cause the light to go on, substitute a new lamp, make continuity checks, and, finally, replace the switch as necessary.

The hydraulic system may be checked for leaks by applying pressure to the pedal gradually and steadily. If the pedal sinks very slowly to the floor, the system has a leak. This is not to be confused with a springy or spongy feel due to the compression of air within the lines. If the system leaks, there will be a gradual change in the position of the pedal with a constant pressure.

Check for leaks along all lines and at wheel cylinders. If no external leaks are apparent, the problem is inside the master cylinder.

DISC BRAKES

Instead of the traditional expanding brakes that press outward against a circular drum, disc brake systems utilize a disc (rotor) with brake pads positioned on either side of it. An easily-seen analogy is the hand brake arrangement on a bicycle. The pads squeeze onto the rim of the bike wheel, slowing its motion. Automobile disc brakes use the identical principle but apply the braking effort to a separate disc instead of the wheel.

The disc (rotor) is a casting, usually equipped with cooling fins between the two braking surfaces. This enables air to circulate between the braking surfaces making them less sensitive to heat buildup and more resistant to fade. Dirt and water do not drastically affect braking action since contaminants are thrown off by the centrifugal action of the rotor or scraped off the by the pads. Also, the equal clamping action of the two brake pads tends to ensure uniform, straight line stops. Disc brakes are inherently self-adjusting. There are three general types of disc brake:

- Fixed caliper
- Floating caliper
- Sliding caliper

The fixed caliper design uses two pistons mounted on either side of the rotor (in each side of the caliper). The caliper is mounted rigidly and does not move.

The sliding and floating designs are quite similar. In fact, these two types are often lumped together. In both designs, the pad on the inside of the rotor is moved into contact with the rotor by hydraulic force. The caliper, which is not held in a fixed position, moves slightly, bringing the outside pad into contact with the rotor. There are various methods of attaching floating calipers. Some pivot at the bottom or top, and some slide on mounting bolts. In any event, the end result is the same.

DRUM BRAKES

Drum brakes employ two brake shoes mounted on a stationary backing plate. These shoes are positioned inside a circular drum which rotates with the wheel assembly. The shoes are held in place by springs. This allows them to slide toward the drums (when they are applied) while keeping the linings and drums in alignment. The shoes are actuated by a wheel cylinder which is mounted at the top of the backing plate. When the brakes are applied, hydraulic pressure forces the wheel cylinder's actuating links outward. Since these links bear directly against the top of the brake shoes, the tops of the shoes are then forced against the inner side of the drum. This action forces the bottoms of the two shoes to contact the brake drum by rotating the entire assembly slightly (known as servo action). When pressure within the wheel cylinder is relaxed, return springs pull the shoes back away from the drum.

Most modern drum brakes are designed to self-adjust themselves during application when the vehicle is moving in reverse. This motion causes both shoes to rotate very slightly with the drum, rocking an adjusting lever, thereby causing rotation of the adjusting screw. Some drum brake systems are designed to self-adjust during application whenever the brakes are applied. This on-board adjustment system reduces the need for maintenance adjustments and keeps both the brake function and pedal feel satisfactory.

POWER BOOSTERS

Virtually all modern vehicles use a vacuum assisted power brake system to multiply the braking force and reduce pedal effort. Since vacuum is always available when the engine is operating, the system is simple and efficient. A vacuum diaphragm is located on the front of the master cylinder and assists the driver in applying the brakes, reducing both the effort and travel he must put into moving the brake pedal.

The vacuum diaphragm housing is normally connected to the intake manifold by a vacuum hose. A check valve is placed at the point where the hose enters the diaphragm housing, so that during periods of low manifold vacuum brakes assist will not be lost.

Depressing the brake pedal closes off the vacuum source and allows atmospheric pressure to enter on one side of the diaphragm. This causes the master cylinder pistons to move and apply the brakes. When the brake pedal is released, vacuum is applied to both sides of the diaphragm and springs return the diaphragm and master cylinder pistons to the released position.

If the vacuum supply fails, the brake pedal rod will contact the end of the master cylinder actuator rod and the system will apply the brakes without any power assistance. The driver will notice that much higher pedal effort is needed to stop the car and that the pedal feels harder than usual.

Vacuum Leak Test

1. Operate the engine at idle without touching the brake pedal for at least one minute.
2. Turn off the engine and wait one minute.
3. Test for the presence of assist vacuum by depressing the brake pedal and releasing it several times. If vacuum is present in the system, light application will produce less and less pedal travel. If there is no vacuum, air is leaking into the system.

System Operation Test

1. With the engine **OFF**, pump the brake pedal until the supply vacuum is entirely gone.
2. Put light, steady pressure on the brake pedal.
3. Start the engine and let it idle. If the system is operating correctly, the brake pedal should fall toward the floor if the constant pressure is maintained.

Power brake systems may be tested for hydraulic leaks just as ordinary systems are tested.

✳✳ WARNING

Clean, high quality brake fluid is essential to the safe and proper operation of the brake system. You should always buy the highest quality brake fluid that is available. If the brake fluid becomes contaminated, drain and flush the system, then refill the master cylinder with new fluid. Never reuse any brake fluid. Any brake fluid that is removed from the system should be discarded.

Brake Light Switch

For removal, installation and adjustment of the brake light switch, refer to Section 6 of this manual.

Master Cylinder

REMOVAL & INSTALLATION

Without Anti-Lock Brakes

▶ See Figures 1, 2, 3, 4 and 5

1. Disconnect the negative battery cable, then unplug the fluid level sensor connector at the reservoir.
2. Position a rag to catch any leaking fluid, then remove the hydraulic pipes from the master cylinder using a suitable wrench. Plug the pipes and cylinder bores to prevent fluid contamination or loss.

✳✳ WARNING

Do not allow brake fluid to spill on or come in contact with the vehicle's finish, as it will remove the paint. In case of a spill, immediately flush the area with water.

3. Remove the 2 master cylinder retaining nuts, then remove the master cylinder from the vehicle.

Fig. 1 Disengage the brake fluid level sensor connector

Fig. 2 Loosen the master cylinder brake line fittings

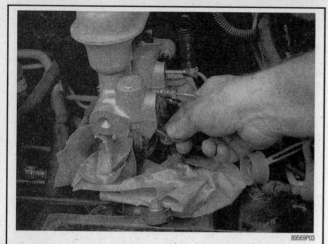

Fig. 3 Place shop rags or paper towels below the fittings to catch any fluid spill and disconnect the hydraulic lines from the master cylinder

Fig. 4 Unfasten the master cylinder-to-brake booster mounting nuts . . .

Fig. 5 . . . then remove the master cylinder from the power brake booster

To install:

4. Install the master cylinder onto the brake booster studs, then install the retaining nuts and tighten to 20 ft. lbs. (27 Nm).

5. Remove the protective caps or line plugs, then connect the hydraulic brake pipes to the master cylinder and tighten the fittings to 24 ft. lbs. (32 Nm).

6. Properly bleed the hydraulic brake system. Refer to the procedure later in this section.

7. Install the wiring harness connector to the brake fluid level sensor.

8. Connect the negative battery cable and start the engine. Have an assistant depress the brake pedal while you check for leaks.

With Anti-Lock Brakes

The master cylinder for ABS equipped vehicles is part of the ABS control assembly. Refer to the removal and installation procedure later in this section.

Power Brake Booster

REMOVAL & INSTALLATION

▶ **See Figures 6 and 7**

1. Disconnect the negative, then the positive battery cable. Remove the air cleaner assembly.

2. Remove the battery, battery hold-down device and tray.

3. Remove the master cylinder or, if equipped, ABS control assembly mounting nuts, then remove and support the master cylinder or control assembly aside to allow for booster removal. Be careful not to bend or damage the brake lines.

4. Disconnect the vacuum line at the power booster check valve.

5. From under the dashboard, remove the retainer and washer from the booster pushrod, then remove the rod from the pedal pin.

6. Remove the 4 retaining nuts, then remove the booster from the vehicle.

To install:

7. Position the booster in the vehicle, but do not install the fasteners at this time.

8. Connect the booster pushrod to the brake pedal pin using the retainer and washer, then tighten the 4 booster retaining nuts to 20 ft. lbs. (27 Nm).

9. Connect the vacuum line to the check valve.

10. Remove the support and reposition the master cylinder or ABS control assembly to the brake booster studs, then install and tighten the retaining nuts to 20 ft. lbs. (27 Nm).

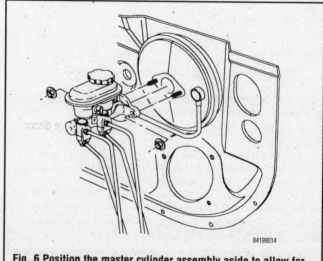

Fig. 6 Position the master cylinder assembly aside to allow for booster removal

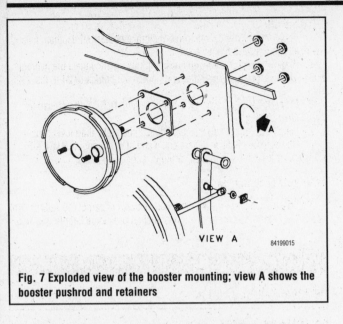

Fig. 7 Exploded view of the booster mounting; view A shows the booster pushrod and retainers

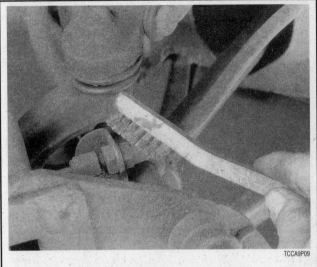

Fig. 8 Use a brush to clean the fittings of any debris

11. Install the battery tray, hold-down device and battery.
12. Install the air cleaner assembly. Connect the positive, then the negative battery cable.

Proportioning Valve

The proportioning valves on all Saturn models are integral components of the master cylinder assembly. If any brake malfunction is attributed to the proportioning valves, the master cylinder must be replaced. Master cylinder removal and installation procedures can be found earlier in this section.

Brake Hoses and Lines

Metal lines and rubber brake hoses should be checked frequently for leaks and external damage. Metal lines are particularly prone to crushing and kinking under the vehicle. Any such deformation can restrict the proper flow of fluid and, therefore, impair braking at the wheels. Rubber hoses should be checked for cracking or scraping; such damage can create a weak spot in the hose and it could fail under pressure.

Any time the lines are removed or disconnected, extreme cleanliness must be observed. Clean all joints and connections before disassembly (use a stiff bristle brush and clean brake fluid); be sure to plug the lines and ports as soon as they are opened. New lines and hoses should be flushed clean with brake fluid before installation to remove any contamination.

REMOVAL & INSTALLATION

▶ See Figures 8, 9, 10 and 11

1. Disconnect the negative battery cable.
2. Raise and safely support the vehicle on jackstands.
3. Remove any wheel and tire assemblies necessary for access to the particular line you are removing.
4. Thoroughly clean the surrounding area at the joints to be disconnected.
5. Place a suitable catch pan under the joint to be disconnected.
6. Using two wrenches (one to hold the joint and one to turn the fitting), disconnect the hose or line to be replaced.
7. Disconnect the other end of the line or hose, moving the drain pan if necessary. Always use a back-up wrench to avoid damaging the fitting.
8. Disconnect any retaining clips or brackets holding the line and remove the line from the vehicle.

➡If the brake system is to remain open for more time than it takes to swap lines, tape or plug each remaining clip and port to keep contaminants out and fluid in.

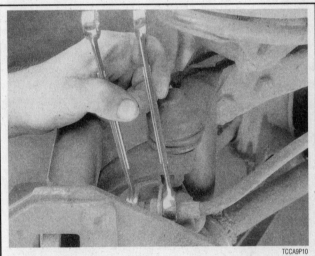

Fig. 9 Use two wrenches to loosen the fitting. If available, use flare nut type wrenches

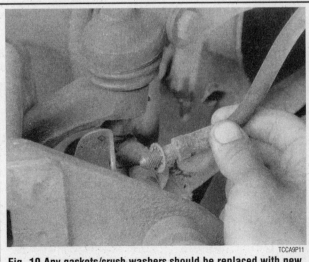

Fig. 10 Any gaskets/crush washers should be replaced with new ones during installation

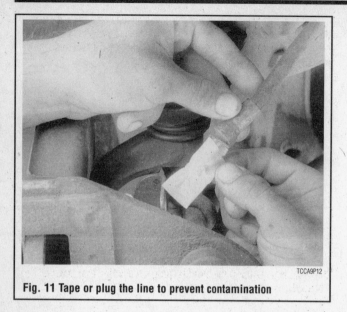

Fig. 11 Tape or plug the line to prevent contamination

To install:

9. Install the new line or hose, starting with the end farthest from the master cylinder. Connect the other end, then confirm that both fittings are correctly threaded and turn smoothly using finger pressure. Make sure the new line will not rub against any other part. Brake lines must be at least ½ in. (13mm) from the steering column and other moving parts. Any protective shielding or insulators must be reinstalled in the original location.

✳✳ **WARNING**

Make sure the hose is NOT kinked or touching any part of the frame or suspension after installation. These conditions may cause the hose to fail prematurely.

10. Using two wrenches as before, tighten each fitting.
11. Install any retaining clips or brackets on the lines.
12. If removed, install the wheel and tire assemblies, then carefully lower the vehicle to the ground.
13. Refill the brake master cylinder reservoir with clean, fresh brake fluid, meeting DOT 3 specifications. Properly bleed the brake system.
14. Connect the negative battery cable.

Bleeding the Brake System

▶ See Figures 12, 13 and 14

The brake system bleeding procedure differs for ABS and non-ABS vehicles. The following procedure pertains only to non-ABS vehicles. For details on bleeding ABS equipped vehicles, refer to the ABS procedures later in this section.

✳✳ **WARNING**

Make sure the master cylinder contains clean DOT 3 brake fluid at all times during the procedure.

1. The master cylinder must be bled first if it is suspected of containing air. Bleed the master cylinder as follows:
   a. Position a container under the master cylinder to catch the brake fluid.
   b. Loosen the left front brake line (front upper port) at the master cylinder and allow the fluid to flow from the front port.
   c. Connect the line and tighten to 24 ft. lbs. (32 Nm).
   d. Have an assistant depress the brake pedal slowly one time and hold it down, while you loosen the front line to expel air from the master cylinder. Tighten the line, then release the brake pedal. Repeat until all air is removed from the master cylinder.

   e. Tighten the brake line to 24 ft. lbs. (32 Nm) when finished.
   f. Repeat these steps for the right front brake line (rear upper port) at the master cylinder.

✳✳ **WARNING**

Do not allow brake fluid to spill on or come in contact with the vehicle's finish, as it will remove the paint. In case of a spill, immediately flush the area with water.

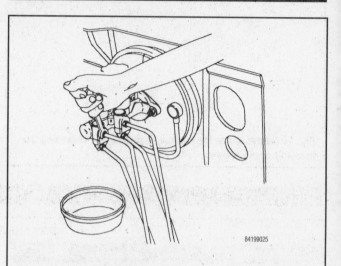

Fig. 12 Loosen the front brake line in order to bleed the master cylinder

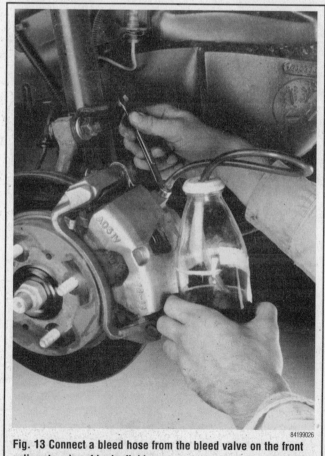

Fig. 13 Connect a bleed hose from the bleed valve on the front caliper to a jar of brake fluid

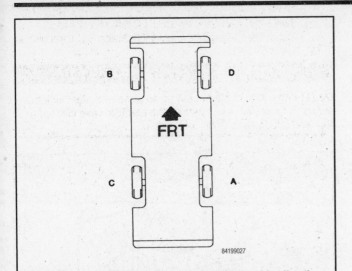

Fig. 14 Always follow the lettered sequence when bleeding the hydraulic brake system

2. If a single line or fitting was the only hydraulic line disconnected, then only the caliper(s) or wheel cylinder(s) affected by that line must be bled. If the master cylinder required bleeding, then all calipers and wheel cylinders must be bled in the proper sequence:

    a. Right rear

    b. Left front

    c. Left rear

    d. Right front

3. Bleed the individual calipers or wheel cylinders as follows:

    a. Place a suitable wrench over the bleeder screw and attach a clear plastic hose over the screw end.

    b. Submerge the other end in a transparent container of brake fluid.

    c. Loosen the bleed screw, then have an assistant apply the brake pedal slowly and hold it down. Close the bleed screw, then release the brake pedal. Repeat the sequence until all air is expelled from the caliper or cylinder.

    d. When finished, tighten the bleed screw to 97 inch lbs. (11 Nm) for the front, or 66 inch lbs. (7.5 Nm) for the rear.

4. Check the pedal for a hard feeling with the engine not running. If the pedal is soft, repeat the bleeding procedure until a firm pedal is obtained.

DISC BRAKES

DISC BRAKE COMPONENTS

1. Brake caliper
2. Brake caliper support bracket
3. Brake pads
4. Brake rotor

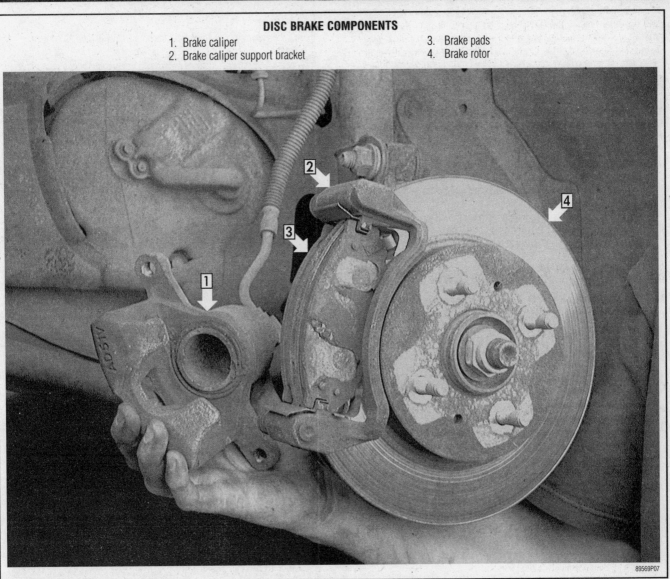

❊❊ CAUTION

Older brake pads may contain asbestos, which has been determined to be a cancer-causing agent. Never clean brake surfaces with compressed air! Avoid inhaling any dust from any brake surface! When cleaning brake surfaces, use a commercially available brake cleaning fluid.

Brake Pads

REMOVAL & INSTALLATION

Front Disc Brakes

◗ See Figures 16 thru 23

➥Always replace the brake pads in sets, meaning BOTH front wheels.

1. Raise the front of the vehicle and support it safely using jackstands, then remove the front wheels.
2. Remove the lower caliper lock pin.

➥The Saturn brake pad replacement procedure states that after removing the lower caliper pin, you can simply pivot the caliper to a vertical position in order to remove the brake pads. This procedure, when attempted on a 1992 SC with ABS, looked as if it would place excessive stress on the brake line. No changes in steering position seemed to alleviate this tension, so a full pivot for pad removal was not attempted. Another option is to remove the upper caliper pin, but not the lower pin, and simply pivot the caliper down for possible access. However, to prevent damage to the brake line you may wish to remove both guide pins and support the caliper assembly from the strut using a coat hanger or length of wire.

3. Pivot the caliper up or down on either guide pin or remove both guide pins and support the caliper from the strut using a coat hanger or length of wire.
4. Remove the 2 brake pads and the pad clips from the caliper support Discard the old pad clips.
5. Check the caliper pins, pin boots and the piston boot for deterioration or damage.

To install:

6. Either by hand or using a C-clamp, bottom the piston all the way into the caliper bore.
7. Carefully lift the inner edge of the piston boot by hand to release any trapped air.
8. Install new pad clips into the caliper support.

Fig. 17 Remove the lock pin, being careful not to damage the boot

Fig. 18 Loosen and remove the upper guide pin from the caliper

Fig. 16 Loosen the caliper lower lock pin

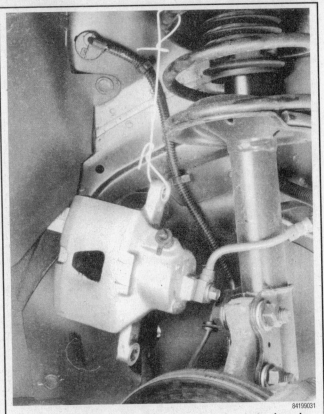

Fig. 19 If removed, support the caliper from the strut using wire or a coat hanger

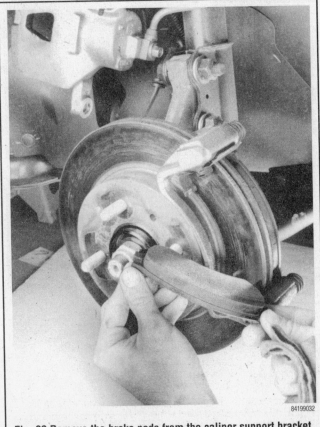

Fig. 20 Remove the brake pads from the caliper support bracket

Fig. 21 Remove and discard the old brake pad retaining clips from the support bracket

Fig. 22 If necessary, use a C-clamp to bottom the caliper piston

Fig. 23 With the new brake pads properly installed, guide the caliper over the pads and onto the support bracket

9. Install the inner and outer brake pads into the support. If installed, remove the temporary support wire from the caliper.

10. Pivot or position the caliper body on the upper guide pin into position. Compress the boots by hand as the caliper is positioned onto the support.

11. Lubricate the smooth ends of the lock and guide pins with silicone grease, then install the pin and tighten to 27 ft. lbs. (36 Nm). Do not get grease on the pin threads.

12. Repeat the procedure for the opposite side brake pads.

13. Install the front wheels, then remove the supports and carefully lower the vehicle.

14. Prior to operating the vehicle, depress the brake pedal a few times until the brake pads are seated against the rotor and a firm pedal is felt. For safety's sake, DO NOT ATTEMPT TO MOVE THE VEHICLE UNTIL THIS STEP IS PERFORMED.

Rear Disc Brakes

▶ **See Figures 24, 25, 26 and 27**

➡**Always replace the brake pads in sets, meaning BOTH rear wheels.**

1. Raise the rear of the vehicle and support it safely using jackstands, then remove the rear wheels.

2. Remove the caliper lock and guide pins.

3. Remove the caliper from the support, being careful not to damage the pin boots and suspend the caliper from the strut using wire or a coat hanger.

4. Remove the brake pads, then remove and discard the old brake pad clips from the caliper support.

5. Inspect the piston and pin boots for deterioration and the pins for corrosion. Damaged boots must be replaced; if the piston boot is damaged, the caliper must be overhauled. Corroded pins must be replaced; do not attempt to remove the corrosion.

To install:

6. Using SA91110NE or an equivalent piston driver tool, bottom the piston by rotating it clockwise into the caliper bore. Do NOT use a C-clamp to press the piston into the bore; otherwise, the piston and/or caliper will be damaged.

7. Align the piston slots so the slots run perpendicular to a line drawn through the center of the pin holes in the support, as shown in the illustration.

8. Carefully lift the inner edge of the piston boot by hand to release any trapped air. Do not use a sharp tool or damage the piston boot; otherwise, caliper overhaul will be necessary.

9. Install new pad clips into the caliper support.

10. Install the inner and outer brake pads into the clips on the support. The pad with the wear sensor should be located outboard. The piston indentation slots should be positioned to correctly accept the brake pads.

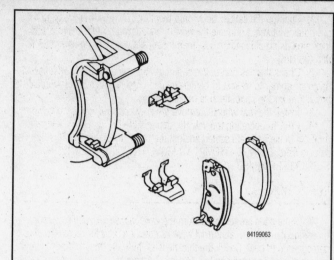

Fig. 25 Remove and discard the old shoes and clips from the caliper support

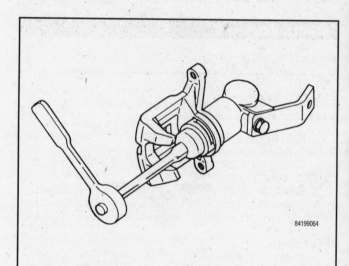

Fig. 26 Use a suitable brake piston driver tool to screw the piston back into the caliper

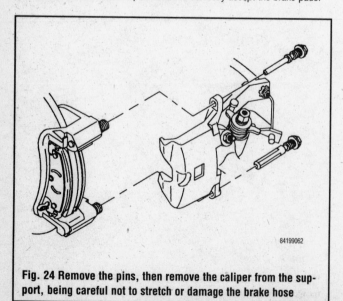

Fig. 24 Remove the pins, then remove the caliper from the support, being careful not to stretch or damage the brake hose

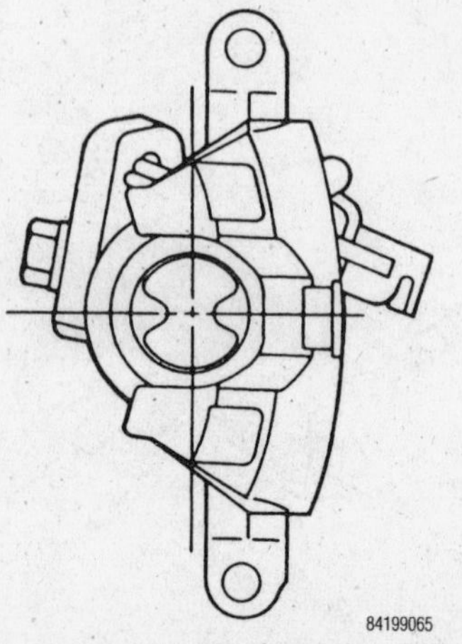

Fig. 27 Once properly threaded, the piston indentation slots will align as shown

11. Position the caliper body onto the support while compressing the pin boots by hand. Lubricate the non-threaded portion of the guide and lock pins using silicone grease, then install the pins and tighten to 27 ft. lbs. (36 Nm).

12. Check the position of the pad clips. If necessary, use a small screwdriver or prybar to re-seat or center the pad clips on the support. Repeat the procedure for the opposite side brake pads.

13. Install the rear wheels, remove the supports and lower the vehicle.

14. Prior to operating the vehicle, depress the brake pedal a few times until the brake pads are seated against the rotor and a firm pedal is felt. For safety's sake, DO NOT ATTEMPT TO MOVE THE VEHICLE UNTIL THIS STEP IS PERFORMED.

INSPECTION

When the pads are removed from the caliper, inspect them for oil or grease contamination, abnormal wear or cracking, and for deterioration or damage due to heat. Check the thickness of the pads; the minimum allowable thickness is approximately $\frac{3}{32}$ in. (2.4mm).

Most brake pads are equipped with a wear indicator that will make a squealing noise when the pads are worn. This alerts you to the need for brake service before any rotor damage occurs.

Brake Caliper

REMOVAL & INSTALLATION

Front Disc Brakes

▶ See Figures 28, 29, 30 and 31

1. Raise the front of the vehicle and support it safely using jackstands, then remove the front wheel.

2. Disconnect the brake hose from the caliper. Plug the openings to prevent excessive fluid contamination or loss.

3. Remove the lock pin and guide pin from the caliper and support.

4. Remove the caliper from the support, being careful not to damage the pin boots. Remove the pin boots from the caliper support and inspect for damage.

To install:

5. Make sure the piston is bottomed in the bore. If necessary, bottom the piston by hand or by using a C-clamp.

6. If removed, install the brake pads to the caliper support.

7. Lubricate the pin boots and guide pins with silicone grease. If removed, install the pin boots into the caliper support.

Fig. 29 Remove the guide and lock pins, then remove the caliper from the support. Remove and inspect the pin boots

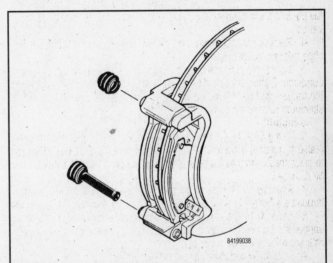

Fig. 30 Lubricate and install the pin boots into the caliper support

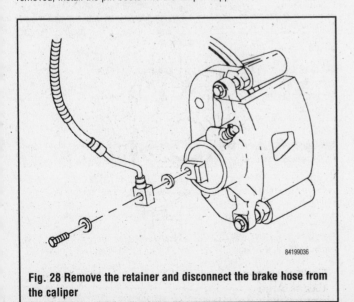

Fig. 28 Remove the retainer and disconnect the brake hose from the caliper

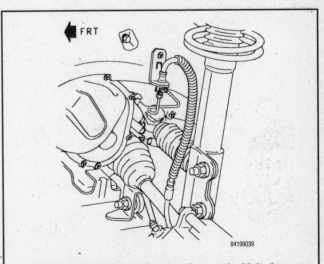

Fig. 31 The brake hose must be properly routed with its loop towards the rear of the vehicle

8. Position the caliper onto the support and over the brake pads, then lubricate the non-threaded portion of the guide and lock pins with silicone grease. Install the pins and tighten to 27 ft. lbs. (36 Nm).

9. Make sure the brake line is properly routed with a loop to the rear, then install the brake hose with new washers. Tighten the fitting to 36 ft. lbs. (49 Nm).

10. Properly bleed the hydraulic brake system.

11. Install the wheel, then remove the supports and carefully lower the vehicle.

Rear Disc Brakes

♦ See Figures 32, 33 and 34

1. Raise the rear of the vehicle and support it safely using jackstands, then remove the rear wheel.

2. Disconnect the brake hose from the caliper. Plug the openings to prevent fluid contamination or loss.

3. Slip the end of the parking brake cable off the parking brake lever, then remove the cable outer housing from the cable bracket with SA9151BR, or an equivalent cable release tool.

4. Remove the lock pin and guide pin.

5. Remove the caliper from the support, being careful not to damage the pin boots. If necessary, remove the pin boots from the caliper support.

6. Inspect the piston and pin boots for deterioration and the pins for corrosion. Damaged boots must be replaced; if the piston boot is damaged, the caliper must be overhauled. Corroded pins must be replaced; do not attempt to remove the corrosion.

To install:

7. Make sure the piston is bottomed in the bore. Do not compress the piston using a C-clamp; instead, the piston must be rotated into the caliper on its threads using a piston driver tool.

8. If removed, install the brake pads to the caliper support.

9. Lubricate the pin boots and guide pins with silicone grease. If removed, install the pin boots into the caliper support.

10. Position the caliper, then lubricate the non-threaded portion of the guide and lock pins. Install the pins and tighten to 27 ft. lbs. (36 Nm).

11. Install the brake hose with new washers, then tighten the fitting to 36 ft. lbs. (49 Nm).

12. Connect the parking brake cable.

13. Properly bleed the hydraulic brake system.

14. Install the wheel, then remove the supports and lower the vehicle.

15. Prior to operating the vehicle, depress the brake pedal a few times until the brake pads are seated against the rotor and a firm pedal is felt. For safety's sake, DO NOT ATTEMPT TO MOVE THE VEHICLE UNTIL THIS STEP IS PERFORMED.

Fig. 33 The parking brake cable attaches to the rear of the caliper/support assembly

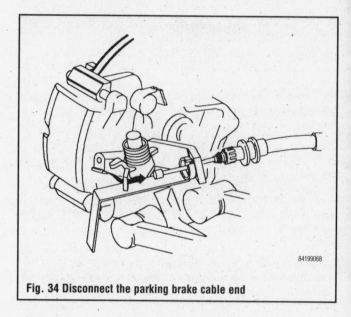

Fig. 34 Disconnect the parking brake cable end

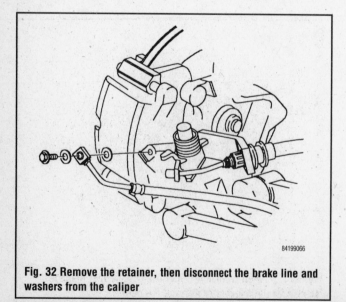

Fig. 32 Remove the retainer, then disconnect the brake line and washers from the caliper

OVERHAUL

♦ See Figures 35 thru 42

➡Some vehicles may be equipped with dual piston calipers. The procedure to overhaul the caliper is essentially the same with the exception of multiple pistons, O-rings and dust boots.

1. Remove the caliper from the vehicle and place on a clean workbench.

✷✷ CAUTION

NEVER place your fingers in front of the pistons in an attempt to catch or protect the pistons when applying compressed air. This could result in personal injury!

➡Depending upon the vehicle, there are two different ways to remove the piston from the caliper. Refer to the brake pad replacement procedure to make sure you have the correct procedure for your vehicle.

2. The first method is as follows:

a. Stuff a shop towel or a block of wood into the caliper to catch the piston.

b. Remove the caliper piston using compressed air applied into the caliper inlet hole. Inspect the piston for scoring, nicks, corrosion and/or worn or damaged chrome plating. The piston must be replaced if any of these conditions are found.

3. For the second method, you must rotate the piston to retract it from the caliper.

4. If equipped, remove the anti-rattle clip.

5. Use a prytool to remove the caliper boot, being careful not to scratch the housing bore.

6. Remove the piston seals from the groove in the caliper bore.

7. Carefully loosen the brake bleeder valve cap and valve from the caliper housing.

8. Inspect the caliper bores, pistons and mounting threads for scoring or excessive wear.

9. Use crocus cloth to polish out light corrosion from the piston and bore.

10. Clean all parts with denatured alcohol and dry with compressed air.

To assemble:

11. Lubricate and install the bleeder valve and cap.

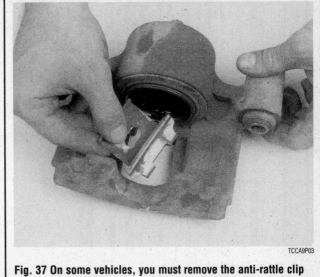

Fig. 37 On some vehicles, you must remove the anti-rattle clip

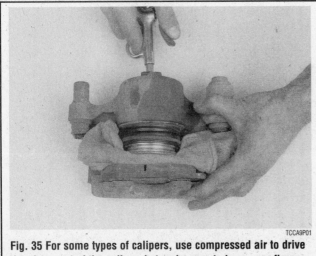

Fig. 35 For some types of calipers, use compressed air to drive the piston out of the caliper, but make sure to keep your fingers clear

Fig. 38 Use a prytool to carefully pry around the edge of the boot . . .

Fig. 36 Withdraw the piston from the caliper bore

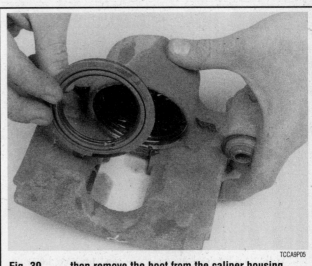

Fig. 39 . . . then remove the boot from the caliper housing, taking care not to score or damage the bore

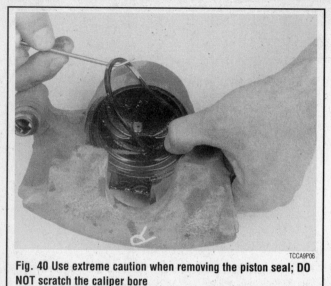

Fig. 40 Use extreme caution when removing the piston seal; DO NOT scratch the caliper bore

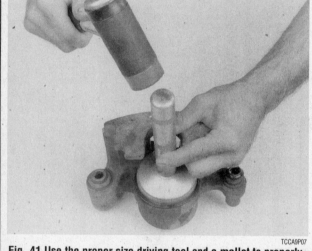

Fig. 41 Use the proper size driving tool and a mallet to properly seal the boot in the caliper housing

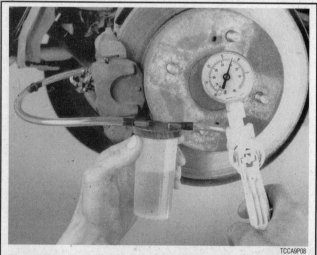

Fig. 42 There are tools, such as this Mighty-Vac, available to assist in proper brake system bleeding

12. Install the new seals into the caliper bore grooves, making sure they are not twisted.

13. Lubricate the piston bore.

14. Install the pistons and boots into the bores of the calipers and push to the bottom of the bores.

15. Use a suitable driving tool to seat the boots in the housing.

16. Install the caliper in the vehicle.

17. Install the wheel and tire assembly, then carefully lower the vehicle.

18. Properly bleed the brake system.

Brake Disc (Rotor)

REMOVAL & INSTALLATION

Front

▶ See Figures 43, 44, 45 and 46

1. Raise the front of the vehicle and support it safely using jackstands.
2. Remove the wheel.

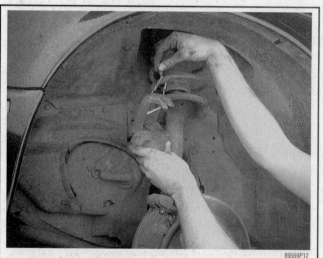

Fig. 43 Support the caliper from the strut with wire or a coat hanger to protect the brake line from damage

Fig. 44 Remove the 2 brake caliper support bracket-to-steering knuckle mounting bolts . . .

Fig. 45 . . . then remove the support bracket from the knuckle

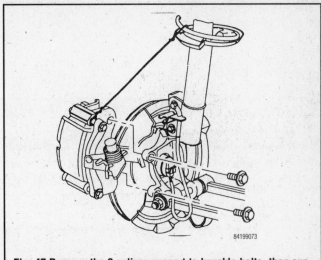

Fig. 47 Remove the 2 caliper support-to-knuckle bolts, then support the bracket from the strut with wire or a coat hanger

Fig. 46 Remove the brake rotor from the hub assembly

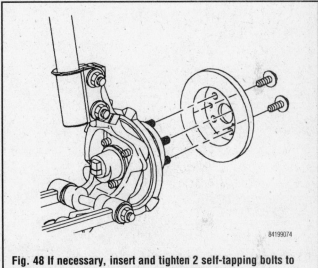

Fig. 48 If necessary, insert and tighten 2 self-tapping bolts to press the rotor from the hub and bearing assembly

3. Remove the brake caliper, then support it from the strut with wire or a coat hanger to protect the brake line from damage. Refer to the procedure earlier in this section.

4. Remove the 2 caliper support-to-knuckle bolts.

5. Remove the rotor from the vehicle. If difficulty is encountered, insert two M8 x 1.25 self-tapping bolts into the holes provided on the rotor to press it from the hub.

To install:

6. Position the rotor over the hub wheel studs.

7. Install the caliper support bracket assembly. Tighten the bolts to 81 ft. lbs. (110 Nm).

8. Install the brake caliper to the support bracket. Tighten the caliper bolts to 27 ft. lbs. (36 Nm).

9. Install the wheel.

10. Remove the supports and carefully lower the vehicle.

Rear

▶ See Figures 47 and 48

1. Raise the rear of the vehicle and support it safely using jackstands.

2. Remove the wheel.

3. Remove the 2 caliper support-to-knuckle bolts, then support the bracket from the strut with wire or a coat hanger to protect the brake line from damage.

4. Remove the rotor from the wheels studs on the hub and bearing assembly. If difficulty is encountered, insert two M8 x 1.25 self-tapping bolts into the holes provided on the rotor and press it from the hub.

To install:

5. Position the rotor over the hub wheel studs.

6. Unwire and install the caliper/support bracket assembly. Tighten the bolts to 63 ft. lbs. (85 Nm).

7. Install the wheel.

8. Remove the supports and carefully lower the vehicle.

INSPECTION

Check the disc brake rotor for scoring, cracks or other damage. Check the minimum thickness and rotor run-out.

Check the disc brake rotor thickness using a micrometer or caliper. The brake rotor minimum thickness is 0.625 in. (15.8mm) for the front, and 0.35 in. (9mm) for the rear. This is the thickness at which point the rotor becomes unsafe to use and must be discarded. The minimum thickness that the rotor can be machined to is 0.633 in. (16.1mm) for the front, and 0.37 in. (9.3mm) for the rear, as this thickness allows room for additional rotor wear after it has been machined and returned to service.

Check the rotor for run-out using a dial indicator. Install a few lug nuts to hold the rotor against the hub while checking. Mount the dial indicator on the strut and position the indicator foot on the outermost diameter of the contact surface of the disc pads. Make sure there is no wheel bearing free-play. Rotor run-out must not exceed 0.0005 in. (0.013mm). If the rotor run-out exceeds specification, machine the rotor if it is not below the minimum thickness specification for machining. If the rotor run-out is still excessive after machining, the rotor must be replaced.

DRUM BRAKES

✳✳ CAUTION

Older brake shoes may contain asbestos, which has been determined to be a cancer causing agent. Never clean the brake surfaces with compressed air! Avoid inhaling any dust from any brake surface! When cleaning brake surfaces, use a commercially available brake cleaning fluid.

Brake Drums

REMOVAL & INSTALLATION

▶ **See Figures 50 and 51**

1. Release the parking brake.
2. Raise the rear of the vehicle and support it safely using jackstands.
3. Remove the rear wheel, then remove the brake drum.

DRUM BRAKE COMPONENTS

1. Leading brake shoe
2. Parking brake shoe
3. Brake shoe hold-down cup and spring
4. Parking brake lever
5. Brake shoe adjuster assembly
6. Wheel cylinder
7. Lower return spring
8. Upper return spring
9. Adjuster spring

89569P26

Fig. 50 After removing the rear wheel, pull the drum from the hub and bearing assembly

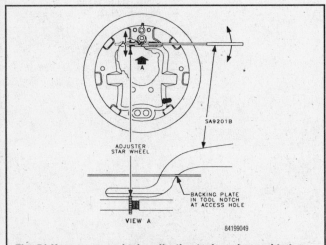

Fig. 51 If necessary, a brake adjusting tool can be used to turn the star wheel and loosen the brake shoes for easier drum removal

4. If difficulty is encountered, turn the star wheel of the brake adjuster assembly through the access hole in the rear of the brake backing plate using an adjusting tool. This will loosen the brake shoes and allow for drum removal. Do not pry between the drum and plate, or the backing plate will become bent and damaged.

INSPECTION

♦ See Figure 52

Clean all grease, brake fluid, and other contaminants from the brake drum using brake cleaner. Visually check the drum for scoring, cracks, or other damage and replace, if necessary.

Check the drum inner diameter using a brake shoe clearance gauge. The maximum allowable inner diameter is 7.93 in. (201.4mm). This is the diameter at which point the drum becomes unsafe to use and must be discarded. The maximum diameter to which the drum can be machined is 7.9 in. (200.6mm), as this distance allows room for additional wear after the drum has been machined and returned to service.

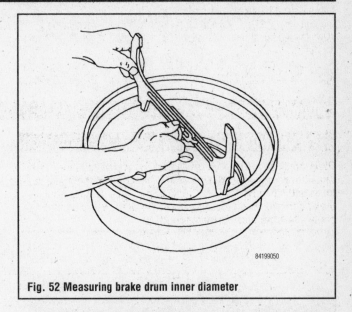

Fig. 52 Measuring brake drum inner diameter

Brake Shoes

INSPECTION

Inspect the brake shoes for peeling, cracking, or extremely uneven wear on the lining. Check the lining thickness using calipers. If the brake lining is less than approximately 3/32 in. (2.4mm), the shoes must be replaced. The shoes must also be replaced if the linings are contaminated with brake fluid or grease.

REMOVAL & INSTALLATION

♦ See Figures 53 thru 63

➡Brake shoes must be replaced as axle sets. Only disassemble one side at a time so that a complete assembly is available as a reference. If necessary, compare the component positions from the side you are currently installing to the other assembly, in order to verify correct positioning; just remember that the other side is a mirror image and is reversed from the side on which you are currently working.

1. Raise the rear of the vehicle and support it using jackstands, then remove the wheels and brake drums.
2. Remove the lower return and adjuster springs using a universal brake spring remover. Do not overextend the springs, or they will be damaged and will need to be replaced.
3. Compress the leading brake shoe hold-down cup and spring, then, from the rear of the backing plate, turn and remove the pin. Release spring compression, then remove the hold-down cup and spring.
4. Pull the leading shoe towards the front of the vehicle and remove the adjuster assembly and lever. It may be necessary to turn the adjuster star wheel to shorten the adjuster's length.
5. Remove the leading shoe by twisting it out of engagement with the upper return spring.
6. Remove the upper return spring from the parking (trailing) brake shoe, then remove the hold-down cup, spring and pin assembly.
7. Push the parking brake shoe lever into the cable spring, while disengaging the cable from the end lever, and remove the parking brake shoe, lever and cable spring from the vehicle.
8. Remove the retainer and wave washer, then remove the parking brake lever from the shoe.

9. Disassemble the brake adjuster socket, screw and nut, then clean the components in denatured alcohol. Inspect the assembly, making sure the screw threads smoothly into the adjusting nut over the full threaded length.

10. On the brake backing plate, inspect the wheel cylinder for signs of leakage and for cut or damaged boots. Do not attempt to repair a damaged cylinder; the assembly must be replaced. Refer to the procedure later in this section.

To install:

11. Lubricate the adjuster assembly, as well as the backing plate's 6 raised shoe contact pads, the brake lever pin and all surfaces which contact the brake shoe webs with brake lubricant.

12. Install the parking brake lever onto the pin on the brake shoe and secure with the wave washer and retainer clip. Crimp the ends of the retainer to secure the brake lever.

13. Install the cable spring into the cage on the parking brake lever, then install the cable through the spring and onto the lever.

14. Install the parking brake shoe with the hold-down cup and spring assembly; using a universal spring cup remover/installer tool. Make sure the shoe is correctly engaged into the wheel cylinder (top) and the anchor (bottom).

15. Install the long straight end of the upper return spring into the back hole in the parking brake shoe, then position the leading brake shoe and install the other end of the spring into the back of the leading shoe.

16. Pull the leading shoe toward the front of the vehicle and install the adjuster between the parking and leading brake shoes. Verify that the adjuster notches properly engage the brake shoe notches and that the shoe is properly aligned in the wheel cylinder and anchor.

17. Install the adjuster lever and adjuster spring. Make sure that the notch on the lever engages the pin on the parking brake shoe and the notch on the adjusting socket. The lower leg of the lever should engage the teeth of the star wheel adjuster assembly.

18. Secure the leading brake shoe using the hold-down cup assembly.

19. Install the adjuster spring to the upper side of the brake shoes with the short end to the leading shoe and the long end to the parking brake shoe. Then, install the lower return spring into the lower holes of the shoes.

20. Verify the correct location of all brake components; if necessary, use the other side's brake assembly for comparison.

21. Using a suitable drum clearance gauge, measure the inner diameter of the brake drum and adjust the outside diameter of the brake shoes to 0.02 in. (0.50mm) less than the inner diameter of the drum.

22. Repeat the procedure for the opposite brake shoes and install the brake drums.

23. If the wheel cylinders have been replaced, bleed the hydraulic brake system.

24. Install the rear wheels.

Fig. 54 . . . followed by the adjuster spring from the top of the brake assembly

Fig. 55 Remove the leading shoe hold-down cup and spring assembly

Fig. 53 Disconnect the lower return spring from the leading brake shoe . . .

Fig. 56 Pull out the leading brake shoe . . .

Fig. 57 . . . and remove the star wheel adjuster assembly

Fig. 58 After twisting the shoe to disengage the upper return spring, remove the leading brake shoe

Fig. 59 Remove the upper return spring from the parking (trailing) brake shoe

Fig. 60 After removing the hold-down cup and spring, pull the brake shoe/parking brake lever assembly from the backing plate, then disengage the cable

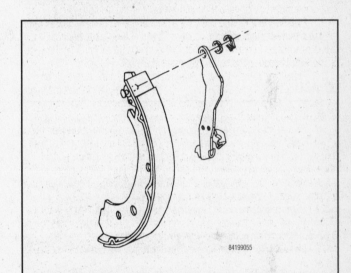

Fig. 61 Remove the retainer and wave washer, then remove the parking brake lever from the shoe

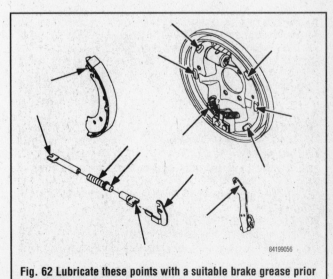
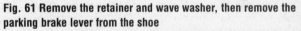

Fig. 62 Lubricate these points with a suitable brake grease prior to assembling the brake components

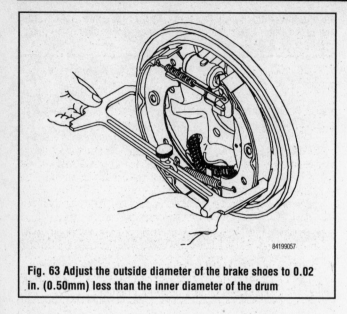

Fig. 63 Adjust the outside diameter of the brake shoes to 0.02 in. (0.50mm) less than the inner diameter of the drum

Fig. 64 Use a wrench to loosen the fitting . . .

Fig. 65 . . . then disconnect the hydraulic brake line from the rear of the wheel cylinder assembly

Fig. 66 Remove the retainers, then remove the wheel cylinder from the brake backing plate

25. Remove the supports and carefully lower the vehicle, then apply and release the brake pedal 20 times to allow the adjuster to properly position the brake shoes.

26. Check and adjust the parking brake cable, as necessary.

ADJUSTMENTS

The Saturn features a self-adjuster mechanism that is part of the brake assembly and compensates for normal brake lining wear. No external adjustment is necessary or possible.

When the brake shoes or components of the brake assembly are replaced, the inside diameter of the drum should be measured using a brake shoe clearance gauge. The outer diameter of the brake shoes should then be given the base setting of 0.020 in. (0.5mm) less than the drum's inner diameter. The automatic adjuster will take care of all adjustments from that point.

Wheel Cylinders

REMOVAL & INSTALLATION

▶ See Figures 64, 65 and 66

1. Raise the rear of the vehicle and support it safely using jackstands.
2. Remove the wheel and the brake drum.
3. Remove the brake shoes. Refer to the procedure earlier in this section.
4. Disconnect the hydraulic brake line and washer from the rear of the wheel cylinder. Plug the openings to prevent excessive fluid loss or contamination.
5. Remove the bleeder valve and cap.
6. Remove the wheel cylinder retaining bolts, then remove the cylinder from the backing plate.

To install:

7. Install the cylinder to the backing plate and tighten the retainers to 89 inch lbs. (10 Nm).
8. Connect the hydraulic brake line using NEW washers and tighten the fastener to 36 ft. lbs. (49 Nm).
9. Install the bleeder valve and tighten to 66 inch lbs. (7.5 Nm), then install the cap.
10. Install the brake shoes and drum.
11. Properly bleed the hydraulic brake system.
12. Install the wheel assembly, then remove the supports and carefully lower the vehicle.
13. Apply and release the brake pedal at least 20 times to make sure the shoes are properly seated in the drum.

OVERHAUL

◆ **See Figures 67 thru 76**

Wheel cylinder overhaul kits may be available, but often at little or no savings over a reconditioned wheel cylinder. It often makes sense with these components to substitute a new or reconditioned part instead of attempting an overhaul.

If no replacement is available, or you would prefer to overhaul your wheel cylinders, the following procedure may be used. When rebuilding and installing wheel cylinders, avoid getting any contaminants into the system. Always use clean, new, high quality brake fluid. If dirty or improper fluid has been used, it will be necessary to drain the entire system, flush the system with proper brake fluid, replace all rubber components, then refill and bleed the system.

✳✳ WARNING

Clean, high quality brake fluid is essential to the safe and proper operation of the brake system. You should always buy the highest quality brake fluid that is available. If the brake fluid becomes contaminated, drain and flush the system, then refill the master cylinder with new fluid. Never reuse any brake fluid. Any brake fluid that is removed from the system should be discarded.

1. Remove the wheel cylinder from the vehicle and place on a clean workbench.
2. First remove and discard the old rubber boots, then withdraw the pistons. Piston cylinders are equipped with seals and a spring assembly, all located behind the pistons in the cylinder bore.
3. Remove the remaining inner components, seals and spring assembly. Compressed air may be useful in removing these components. If no compressed air is available, be VERY careful not to score the wheel cylinder bore when removing parts from it. Discard all components for which replacements were supplied in the rebuild kit.
4. Wash the cylinder and metal parts in denatured alcohol or clean brake fluid.

✳✳ WARNING

Never use a mineral-based solvent such as gasoline, kerosene or paint thinner for cleaning purposes. These solvents will swell rubber components and quickly deteriorate them.

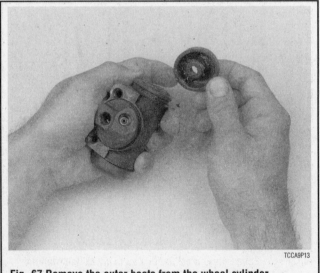

TCCA9P13

Fig. 67 Remove the outer boots from the wheel cylinder

TCCA9P15

Fig. 69 Remove the pistons, cup seals and spring from the cylinder

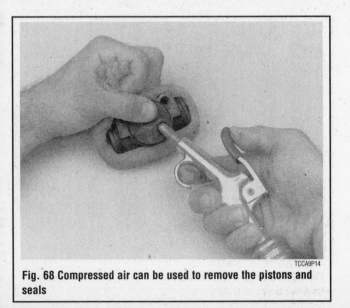

TCCA9P14

Fig. 68 Compressed air can be used to remove the pistons and seals

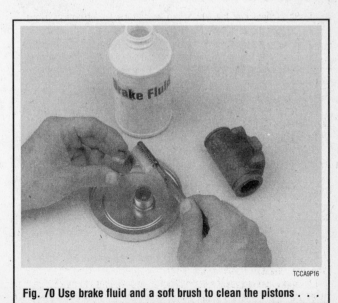

TCCA9P16

Fig. 70 Use brake fluid and a soft brush to clean the pistons . . .

Fig. 71 . . . and the bore of the wheel cylinder

Fig. 72 Once cleaned and inspected, the wheel cylinder is ready for assembly

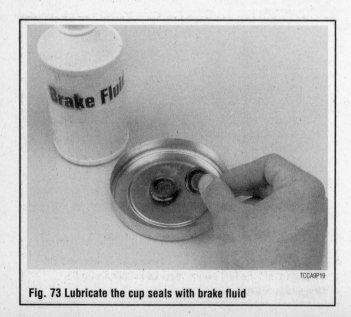

Fig. 73 Lubricate the cup seals with brake fluid

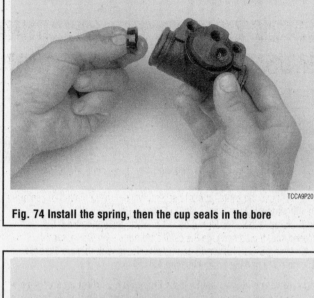

Fig. 74 Install the spring, then the cup seals in the bore

Fig. 75 Lightly lubricate the pistons, then install them

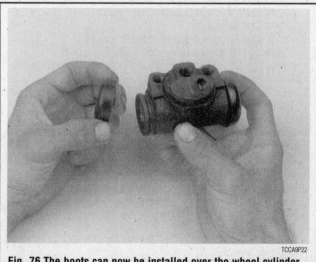

Fig. 76 The boots can now be installed over the wheel cylinder ends

5. Allow the parts to air dry or use compressed air. Do not use rags for cleaning, since lint will remain in the cylinder bore.

6. Inspect the piston and replace it if it shows scratches.

7. Lubricate the cylinder bore and seals using clean brake fluid.

8. Position the spring assembly.

PARKING BRAKE

Cables

REMOVAL & INSTALLATION

Except Equalizer Cable

▶ **See Figures 77 and 78**

1. Disconnect the negative battery cable. If equipped, disable the air bag system.

2. Remove the center console assembly from the vehicle. For center console removal procedures, refer to Section 10.

3. Remove the adjuster nut and, if present, the spring from the threaded rod, then disengage the cables from the equalizer assembly.

4. Using SA9151BR or an equivalent release tool, remove the brake cable from the parking brake base bracket.

➡ **Before removing the cable from the vehicle, tie a piece of wire or string to the console end of the cable. After removal, the string can be used to pull the new cable through the floor pan and into position.**

5. Remove the rear seat cushion and pull back the carpeting.

6. Raise the rear of the vehicle and support it safely using jackstands, then remove the rear wheels.

7. Remove the cable grommet/cable assembly from the floor pan. Then, remove the trailing arm/parking brake cable-to-body fasteners.

8. Cut and remove the brake cable retaining tie strap.

9. If equipped with drum brakes, proceed as follows:

   a. Remove the brake drum. Refer to the procedure earlier in this section.

   b. Remove the lower return spring and the adjuster spring.

   c. Remove the parking brake shoe hold-down cup assembly.

   d. Pull the parking brake shoe towards the rear of the vehicle and remove the adjuster assembly.

   e. Disconnect the upper return spring from the parking brake shoe.

   f. Remove the parking brake cable from the parking brake lever and remove the parking brake cable spring.

   g. Use SA9151BR or an equivalent cable release tool to remove the cable from the backing plate.

10. If equipped with disc brakes, proceed as follows:

   a. Remove the nut securing the cable to the floor pan stud.

   b. Use SA9151BR or an equivalent cable release tool to remove the cable from the caliper lever.

11. Remove the cable from the vehicle.

To install:

12. If equipped with disc brakes, proceed as follows:

   a. Install the parking brake cable into the caliper bracket and correctly seat the retaining fingers.

   b. Attach the cable to the caliper park brake lever.

   c. Install the trailing arm/parking brake cable-to-body fasteners, then tighten the fasteners to 89 ft. lbs. (120 Nm).

   d. Install the cable bracket to the floor pan stud. Install the nut and tighten to 25 inch lbs. (2.8 Nm).

13. If equipped with drum brakes, proceed as follows:

   a. Install the cable through the brake backing plate and correctly seat the retaining fingers.

   b. Install the cable spring onto the cable, then install the cable end into the actuator lever.

c. Connect the short end of the upper return spring to the leading brake shoe, then connect the other end of the spring to the parking brake shoe.

d. Pull the parking brake shoe toward the vehicle's rear and install the adjuster assembly between the shoes, then install the adjuster lever.

e. Install the adjuster spring with the short end into the leading shoe and the other end into the adjuster lever.

f. Install the lower return spring, then install the parking brake shoe hold-down cup assembly.

g. Adjust the brake shoe outer diameter to 0.02 in. (0.50mm) less than the inner diameter of the brake drum and install the brake drum.

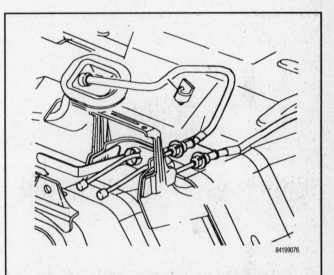

Fig. 77 Remove the brake cable from the console bracket

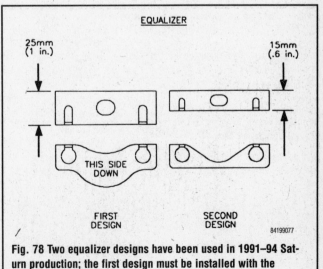

Fig. 78 Two equalizer designs have been used in 1991–94 Saturn production; the first design must be installed with the curved side downward

h. Install the trailing arm/parking brake cable-to-body fasteners, then tighten the fasteners to 89 ft. lbs. (120 Nm).

14. Pull the cable into the vehicle with the wire or string that was attached during disassembly.

15. Install the grommet into the floor pan.

16. Install the rear wheels.

17. Install the cables into the parking brake base bracket, then attach the cables to the equalizer.

➡On 1991–94 models, refer to the appropriate illustration for equalizers of the first production design; on these models, the cable entry holes (rounded side of the equalizer) must face downward. For the second design, the cable entry holes may face up or down.

18. Install the cable and equalizer assembly onto the threaded rod, then install the adjusting nut.

19. For rear drum brake equipped vehicles, apply and release the brake pedal 20 times to allow the drum brake adjuster to position the brake shoes.

20. Adjust the parking brake; refer to the procedure later in this section.

21. Place the carpet back to its original position. Install the center console and the rear seat cushion.

22. Remove the supports and carefully lower the vehicle to the ground.

23. Enable the air bag system and connect the negative battery cable.

Equalizer Cable

1995–98 MODELS

1. Disconnect the negative battery cable. Disable the air bag system, as described in Section 6.

2. Remove the center console assembly from the vehicle. For center console removal procedures, refer to Section 10.

3. Remove the cable adjuster nut located on the parking brake lever.

4. Disengage the rear cables from the equalizer assembly.

5. Pull up the parking brake lever to its highest position.

6. Move the equalizer cable over the parking brake indicator switch and swing it forward. Pull the cable down and out of the parking brake lever.

To install:

7. Route the equalizer cable into the parking brake lever and start the adjuster nut by hand.

8. Swing the equalizer cable rearward and lift it over the parking brake indicator switch.

9. Release the parking brake lever to the fully lowered position.

10. Install both of the rear cables into the equalizer.

11. Adjust the parking brake; refer to the procedure later in this section.

12. Install the center console assembly; refer to Section 10.

13. Enable the air bag system and connect the negative battery cable.

ADJUSTMENT

◆ See Figures 79 thru 85

➡If equipped with rear drum brakes that have been serviced, before performing parking brake adjustment procedures, apply and release the brake pedal 20 times. This allows the adjuster to position the brake shoes and prevents premature wear of the brake linings due to improper parking brake adjustment.

1. Disconnect the negative battery cable.

2. On 1991 models, remove the center console assembly; refer to Section 10 for removal procedures. On 1992–94 models, remove the center console rear ashtray. On 1995–98 models, lift up the brake lever, remove the cover retaining screw and slide the cover off.

3. Raise the rear of the vehicle and support it safely using jackstands.

4. On 1991–94 models, pull the brake lever up to the 3rd click from the released position. On 1995–98 models, pull the brake lever up to the 1st click from the released position.

5. Tighten the adjuster nut until a slight brake drag is felt at the rear wheels.

6. Apply and release the lever several times.

7. Pull the lever up to the 2nd click. There should be no brake drag at the rear wheels.

8. Pull the lever to the 3rd click and check for a slight drag at the rear wheels.

9. Pull the lever to the 4th click and verify that the rear wheels are locked (they cannot be turned by hand).

10. Loosen or tighten the adjuster nut as necessary until these conditions are met.

11. On 1991–94 models, if the proper adjustment cannot be obtained, perform the following:

a. Check measurement A, from the center of the pivot pin attachment to the end of the adjuster rod, as shown in the appropriate illustration.

b. If the measurement is 5.78 in. (147mm) or less, replace the brake lever assembly.

c. If the measurement is 6.33 in. or more, check for damage or incorrect installation of the drum brake assembly components, parking brake cables and/or parking brake lever assembly.

12. Install the center console assembly, ashtray or parking brake lever cover, if applicable.

13. Remove the supports, lower the vehicle and connect the negative battery cable.

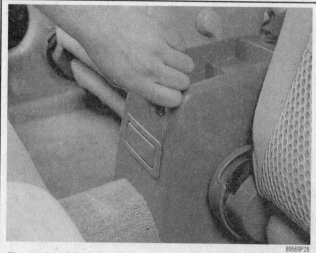

Fig. 79 On 1992–94 models, insert a small prytool between the console and ashtray . . .

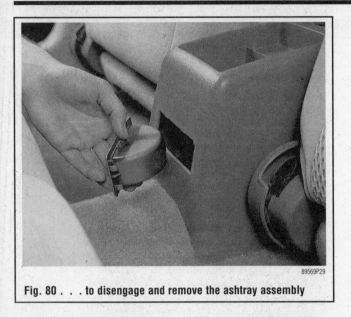

Fig. 80 . . . to disengage and remove the ashtray assembly

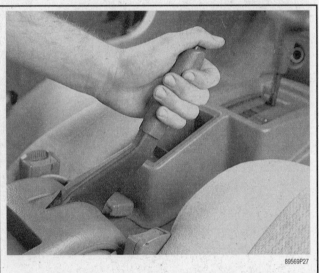

Fig. 81 Pull the brake lever up from the released position

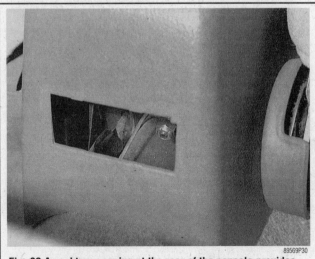

Fig. 82 An ashtray opening at the rear of the console provides convenient access to the parking brake adjustment

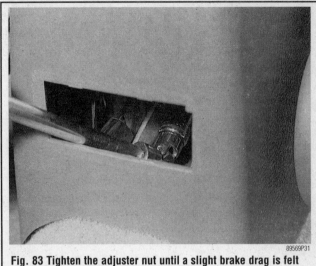

Fig. 83 Tighten the adjuster nut until a slight brake drag is felt at the rear wheels

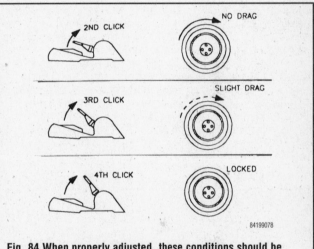

Fig. 84 When properly adjusted, these conditions should be observed while attempting to turn the rear wheel by hand, depending on parking brake lever position

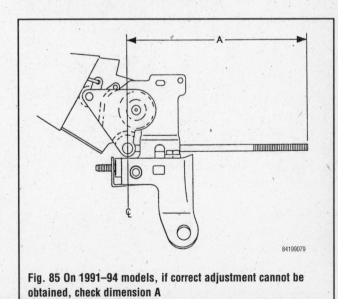

Fig. 85 On 1991–94 models, if correct adjustment cannot be obtained, check dimension A

ANTI-LOCK BRAKE SYSTEM

General Information

▶ **See Figure 86**

The Saturn Anti-Lock Brake System (ABS) is lightweight, compact and highly efficient at controlling wheel locking. The basic premise of ABS is that in order to slow a vehicle, the brake system needs tire traction (friction). When wheel lock-up occurs during braking (a tire skids rather than rolls), the friction between the tire and road surface decreases enormously. This is extremely undesirable since friction is necessary for both steering and slowing of the vehicle. The Saturn ABS prevents wheel lock-up, increasing braking and steering ability (under certain, otherwise adverse, conditions such as wet pavement) by monitoring wheel speed and activating to relieve brake pressure to the portion of the system which is approaching lock-up. In other words, the system monitors how fast each wheel is turning and, if it detects there is an "odd man out" (a tire is moving much slower than the rest) it will regulate the hydraulic brake pressure to that tire's caliper or piston and prevent wheel lock-up. Without ABS activation under these conditions, the system acts as a normal hydraulic brake system and does not regulate hydraulic pressure.

Many other ABS systems use only electric solenoids to control fluid flow. The solenoid is commanded to one of three positions, allowing line pressure to build, release or hold. On the Saturn system, quick-response electric motors are used to alter the position of displacement pistons. The pistons serve as movable valves, altering the line pressure in small increments. Expressed another way, the solenoid systems can be compared to digital circuits, being either on or off while the motor-operated valves in the Saturn system compare to an analog circuit with continuous modulation. This modulation of brake line pressure yields much smoother operation with less pedal vibration during operation.

The system uses input signals from each of 4 wheel sensors. Four output channels (2 front, 2 rear) are controlled by the control module, though only the 2 front channels can be considered independent. Each front chan-nel is provided with a separate piston, solenoid valve and motor, so they can be controlled separately, while the rear wheels share a single motor driving 2 pistons and are, therefore, activated in unison. When locking is detected at one wheel, a solenoid closes the hydraulic path from the master cylinder. The electric motor for that brake circuit cycles the piston up and down to modulate braking force. The system cannot increase the brake pressure above that developed in the master cylinder; the system cannot apply the brakes by itself.

During normal braking, the ABS system is transparent to the operator. Internally, each control piston is in the uppermost or home position allowing brake fluid pressure to pass to the wheels unrestricted. A small internal Expansion Spring Brake (ESB) device is applied to each piston, preventing it from being forced downward by the pressurized fluid passing above it.

When impending wheel lock is noted at one or more wheels, the Electronic Brake Control Module (EBCM) commands the system into ABS mode. Solenoids in each front wheel circuit close. The brakes on the pistons are released and the pistons are driven downward by the electric motors through a system of driven gears and against the spring pressure. The amount of current applied to the motors controls the speed and distance travelled. As the motors move backwards, the piston moves downward, allowing a check valve to seat. The brake pressure to the wheel is now a function of the controlled volume within the piston chamber.

To reduce pressure, the motor continues to drive the piston downward. If an increase in pressure is necessary, the piston is driven upward. Total pressure available is limited to the amount present when ABS was entered.

The rear brakes are controlled in similar fashion. Wheel speed signals are received from each rear wheel. The EBCM uses a Select Low strategy, controlling a single motor to control output to the rear brakes based on the wheel with the greatest tendency to lock.

Many of the service procedures, including troubleshooting and control assembly overhaul require the use the Saturn Portable Diagnostic Tool (PDT) or an equivalent scan tool. If a tool is not available for these procedures, the vehicle should be taken to a qualified repair shop.

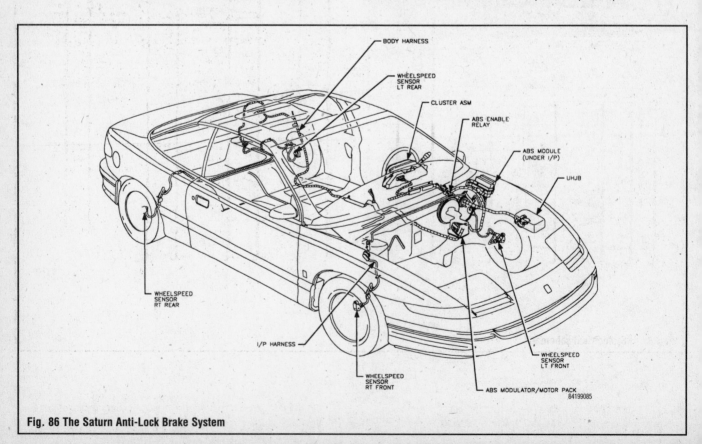

Fig. 86 The Saturn Anti-Lock Brake System

Troubleshooting the ABS System

▶ **See Figures 87, 88 and 89**

The Saturn ABS Electronic Brake Control Module (EBCM) is similar in many ways to the PCM described in Section 4 of this manual. It is an electronic controller of the ABS system which is capable of monitoring system function through the system circuits. It will constantly monitor the output from its various sensors to determine if a sensor or circuit is operating properly.

When the EBCM detects a problem, it will set a trouble code or flag. If the fault is serious enough to affect braking, the EBCM will illuminate the red ANTILOCK warning lamp. Whenever the lamp is lit, the ABS system is disabled and will NOT operate, but normal hydraulic braking will exist. In this way, the system is prevented from ever disabling or adversely affecting the vehicle's hydraulic brakes due to a partial or complete ABS failure.

As with the PCM system, trouble codes and flags are stored in both general information and malfunction history circuits of the control module. General information is cleared each time the ignition switch is turned **OFF**, but codes will reset when the ignition is turned **ON** again and the malfunction is still present. Data will remain in malfunction history for a series of 100 ignition cycles after a fault disappears, unless a scan tool is used to clear the memory earlier. Disconnecting the power will have no affect on data stored in the EBCM malfunction history circuits.

Unlike the PCM system, on which a jumper wire can activate a flash code readout, a scan tool MUST be used for fault code diagnosis and troubleshooting on the ABS EBCM. The scan tool is connected to the system through the Assembly Line Data Link (ALDL) connector which is under the dashboard, near the driver's door.

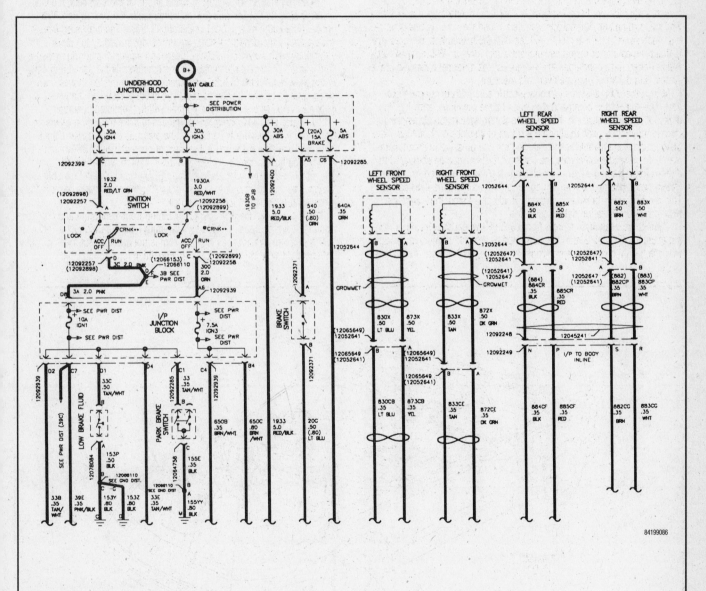

Fig. 87 ABS electrical schematic

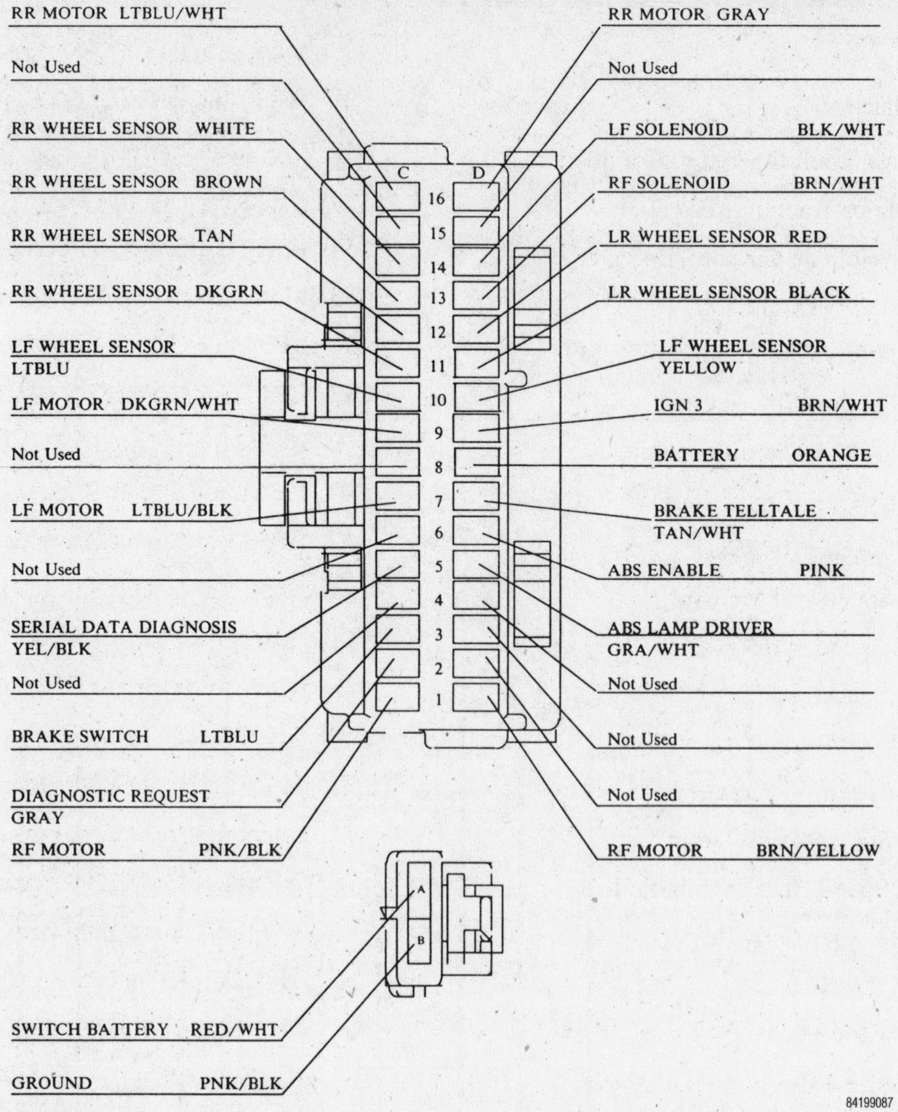
RR MOTOR LTBLU/WHT
Not Used
RR WHEEL SENSOR WHITE
RR WHEEL SENSOR BROWN
RR WHEEL SENSOR TAN
RR WHEEL SENSOR DKGRN
LF WHEEL SENSOR LTBLU
LF MOTOR DKGRN/WHT
Not Used
LF MOTOR LTBLU/BLK
Not Used
SERIAL DATA DIAGNOSIS YEL/BLK
Not Used
BRAKE SWITCH LTBLU
DIAGNOSTIC REQUEST GRAY
RF MOTOR PNK/BLK
RR MOTOR GRAY
Not Used
LF SOLENOID BLK/WHT
RF SOLENOID BRN/WHT
LR WHEEL SENSOR RED
LR WHEEL SENSOR BLACK
LF WHEEL SENSOR YELLOW
IGN 3 BRN/WHT
BATTERY ORANGE
BRAKE TELLTALE TAN/WHT
ABS ENABLE PINK
ABS LAMP DRIVER GRA/WHT
Not Used
Not Used
Not Used
RF MOTOR BRN/YELLOW
C
D
16
15
14
13
12
11
10
9
8
7
6
5
4
3
2
1
A
B
SWITCH BATTERY RED/WHT
GROUND PNK/BLK
84199087

Fig. 88 EBCM pinout—1991–95 models

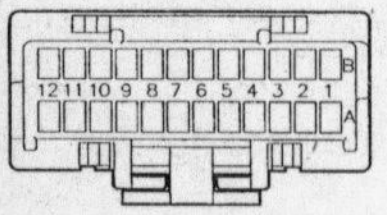
12 11 10 9 8 7 6 5 4 3 2 1
B
A

ABS CONTROL MODULE – 12110088				
CAVITY	WIRE COLOR	CIRCUIT NUMBER	WIRE SIZE	CIRCUIT FUNCTION
A1	PPL	1807B	.35	CLASS 2 DATA
A2	WHT	883CG	.35	RR + WHEEL SENSOR
A3	RED	885CF	.35	LR + WHEEL SENSOR
A4	TAN	833CE	.35	RF + WHEEL SENSOR
A5	LT BLU	830CB	.35	LF + WHEEL SENSOR
A6	LT GRN	1656	.35	ABS/TRAC ACTIVE TT CONTROL
A7	LT GRN/BLK	875	.35	ABS TT CONTROL
A8				
A9	BRN	641B	.35	IGN 3
A10	DK GRN	1288	.80	LF SOLENOID
A11	PNK	1632	.35	ABS ENABLE
A12				
B1	BRN	882CG	.35	RR – WHEEL SENSOR
B2	BLK	884CF	.35	LR – WHEEL SENSOR
B3	DK GRN	872CE	.35	RF – WHEEL SENSOR
B4	YEL	873CB	.35	LF – WHEEL SENSOR
B5	TAN/WHT	33E	.35	BRAKE TT CONTROL
B6	LT BLU	1289	.80	RF SOLENOID
B7				
B8				
B9	LT BLU	20G	.50	BRAKE SW
B10				
B11	PPL/WHT	1572	.35	TRACTION SWITCH ENABLE
B12	ORN	640A	.35	B+

89569G02

Fig. 89 ABS control module pinout—1996–98 models

DIAGNOSTIC CODES

➧ **See Figures 90 and 91**

ABS Control Module

REMOVAL & INSTALLATION

1991–95 Models

➧ **See Figure 92**

The Electronic Brake Control Module (EBCM) is located under the instrument panel, to the left of the steering column. The module is outboard of the Powertrain Control Module (PCM), closest to the left kick panel.

➡**If replacing the control module with a service replacement, the Electronically Erasable Programmable Read Only Memory (EEPROM) of the new module must be programmed by the Saturn Service Stall or equivalent system.**

I   No Scan Data

11 91–92 Only Circuit Open Or Grounded

12 ABS Telltale or Traction LED Fault

13 91–92 Only Telltale Circuit Shorted to B+

14 Switched Battery Circuit Open

15 Switched Battery Circuit Shorted To B+

16 Enable Relay Coil Circuit Open

17 Enable Relay Coil Circuit Grounded

18 Enable Relay Coil Circuit Shorted To B+

21 LF Wheel Speed = 0 MPH

22 RF Wheel Speed = 0 MPH

23 LR Wheel Speed = 0 MPH

24 RR Wheel Speed = 0 MPH

25 LF Wheel Speed Acceleration fault

26 RF Wheel Speed Acceleration fault

27 LR Wheel Speed Acceleration fault

28 RR Wheel Speed Acceleration fault

31 Any 2 Wheel Speeds = 0 MPH

36 System Voltage Low

37 ABS System Voltage High

38 Left Front ESB Does Not Hold Motor

41 Right Front ESB Does Not Hold Motor

42 Rear ESB Does Not Hold Motor

44 LF Motor Frozen

45 RF Motor Frozen

46 Rear Motor Frozen

47 LF Motor Circuit Current Low

48 RF Motor Circuit Current Low

51 Rear Motor Circuit Current Low

52 LF In Release Too Long

53 RF In Release Too Long

54 Rear In Release Too Long

55 Motor Circuit Fault Detected

56 LF Motor Circuit Open

57 LF Motor Circuit Grounded

58 LF Motor Circuit Shorted to B+

61 RF Motor Circuit Open

62 RF Motor Circuit Grounded

63 RF Motor Circuit Shorted to B+

64 Rear Motor Circuit Open

65 Rear Motor Circuit Grounded

66 Rear Motor Circuit Shorted To B+

76 Solenoid Circuit 1288 Open or Shorted to B+

77 Solenoid Circuit 1288 Grounded

78 Solenoid Circuit 1289 Open or Shorted to B+

81 Solenoid Circuit 1289 Grounded

82 ABS Calibration Fault

86 ABS turned on red *Brake* telltale

87 Red Brake Telltale Circuit Open

88 Red Brake Telltale Circuit Shorted To B+

91 Brake Switch Circuit Open During Normal Stop

92 Brake Switch Circuit Open During ABS Stop

93 Brake Switch Circuit Open On Initialization

94 Brake Switch Circuit Always Closed

95 Stop Lamp Circuit Open NO TAG

96 Stop Lamp Circuits Open or Grounds Open

89569G03

Fig. 90 ABS diagnostic trouble codes—1991–95 models

I No Scan Data

11 ABS Telltale – Circuit Fault

13 Traction Active Telltale – Circuit Fault

14 Enable Relay Circuit – Open

15 Enable Relay Circuit – Shorted to B+

16 Enable Relay Coil Circuit – Open

17 Enable Relay Coil Circuit – Grounded

18 Enable Relay Coil Circuit – Shorted To B+

21 LF Wheel Speed = Too Low (Vehicle Moving)

22 RF Wheel Speed = Too Low (Vehicle Moving)

23 LR Wheel Speed = Too Low (Vehicle Moving)

24 RR Wheel Speed = Too Low (Vehicle Moving)

25 LF Wheel Speed Acceleration Malfunction

26 RF Wheel Speed Acceleration Malfunction

27 LR Wheel Speed Acceleration Malfunction

28 RR Wheel Speed Acceleration Malfunction

32 LF Wheel Speed Circuit Malfunction

33 RF Wheel Speed Circuit Malfunction

34 LR Wheel Speed Circuit Malfunction

35 RR Wheel Speed Circuit Malfunction

36 ABS System – Voltage Low

37 ABS System – Voltage High

38 Left Front ESB Does Not Hold Motor

41 Right Front ESB Does Not Hold Motor

42 Rear ESB Does Not Hold Motor

44 LF Motor Frozen

45 RF Motor Frozen

46 Rear Motor Frozen

47 LF Motor Circuit – Current Low

48 RF Motor Circuit – Current Low

51 Rear Motor Circuit – Current Low

52 LF In Release Too Long

53 RF In Release Too Long

54 Rear In Release Too Long

55 Controller Fault

56 LF Motor Circuit – Open

57 LF Motor Circuit – Grounded

58 LF Motor Circuit – Shorted to B+

61 RF Motor Circuit – Open

62 RF Motor Circuit – Grounded

63 RF Motor Circuit – Shorted to B+

64 Rear Motor Circuit – Open

65 Rear Motor Circuit – Grounded

66 Rear Motor Circuit – Shorted To B+

75 Serial Communication Malfunction

76 Solenoid Circuit 1288 – Open or Grounded

77 Solenoid Circuit 1288 – Shorted To B+

78 Solenoid Circuit 1289 – Open or Grounded

81 Solenoid Circuit 1289 – Shorted to B+

82 ABS Calibration Malfunction

86 ABS turned on Red *Brake* Warning Lamp

87 Red Brake Warning Lamp Circuit – Open or Shorted To B+

91 Brake Switch Circuit – Open All the Time

92 Brake Switch Circuit – Open During ABS Stop

93 Brake Switch Circuit – Open On Key-Up

94 Brake Switch – Always Closed

95 Stop Lamp Circuit – Open

89569G04

Fig. 91 ABS diagnostic trouble codes—1996–98 models

1. Disconnect the negative battery cable.

2. Remove the CPA device and unplug the 2-way connector from the EBCM.

3. Unplug the 32-way electrical connector from the EBCM.

4. Rotate the module retaining screw ¼ turn and remove the module by pulling downward. Be careful not to snag the wiring.

To install:

5. Position the EBCM into the bracket, taking care not to snag the wiring.

6. Seat the retaining screw by pushing upward 2 clicks.

7. Connect the wiring and insert the CPA device onto the 2-way connector.

8. If necessary, program the EEPROM.

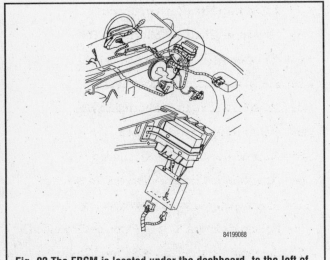

Fig. 92 The EBCM is located under the dashboard, to the left of the steering column and next to the PCM

1996–98 Models

▶ See Figure 93

The ABS control module is located under the instrument panel top cover, directly over the steering column. The ABS control module and PCM are mounted in the same bracket.

1. Disconnect the negative battery cable.

2. Using a small, flat bladed screwdriver, pry off the upper instrument panel pad screw caps. Remove the trim panel retaining screws.

3. Disengage the rear trim panel clips by lifting up. Then, disengage the front trim panel clips by lifting up at each outer corner and pulling rearward. Be careful not to damage the trim panel during its removal from the vehicle.

4. Remove the upper trim panel insulator.

5. Loosen the ABS module mounting nut at the crossbeam, and lift the control module upward towards the windshield to disengage it from the ABS/PCM mounting bracket.

6. Disengage the ABS module wiring connectors and remove the module from the vehicle.

To install:

7. Plug in the ABS module electrical connectors and lower the module into position in the mounting bracket. Guide the attachment stud into the slot in the crossbeam.

8. Install and tighten the mounting nut to 89 inch lbs. (10 Nm). It is extremely important to tighten the mounting nut to the correct torque specification, or damage to the ABS module can result.

9. Install the instrument panel insulator pad and upper trim panel assembly. Be sure that all flaps are tucked and clips engaged. Install and tighten the retaining screws to 20 inch lbs. (2.3 Nm). Install the trim panel screw caps.

10. Connect the negative battery cable.

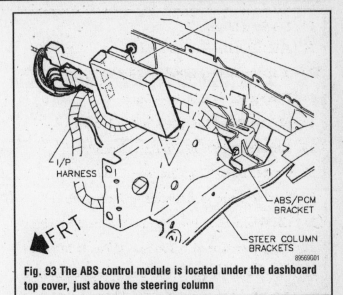

Fig. 93 The ABS control module is located under the dashboard top cover, just above the steering column

ABS Control Assembly

The ABS control assembly consists of the solenoid valves, master cylinder and hydraulic modulator.

REMOVAL & INSTALLATION

▶ See Figures 94 thru 99

✳✳ CAUTION

The ABS modulator pistons are normally in the HOME or top position. In this position, the modulator drive gears are under spring tension and will turn during disassembly if not unloaded. The sudden rotation of gears and release of tension could cause injury if done without extreme care. It is recommended that the pistons be run down and the tension released *prior* to removing the brake control assembly, if the unit is to be disassembled.

1. If removing the unit for disassembly, connect the Saturn Portable Diagnostic Tool (PDT), or an equivalent scan tool, and perform the RUN ABS MOTORS, PISTONS DOWN-REL test to run the modulator pistons down and release spring tension.

2. Remove the air cleaner/induction assembly.

3. Disconnect the negative battery cable, then remove the battery box and battery from the vehicle.

4. Tag and unplug the 2 electrical connectors from the solenoids and the connector from the brake fluid level sensor on the lower side of the fluid reservoir.

5. Remove the CPA device from the motor pack 6-way connector and disengage the connector from the bottom of the motor pack.

6. Position a cloth to catch any escaping fluid, then tag and disconnect the brake lines from the control assembly. Plug the openings to prevent excessive fluid contamination or loss.

✳✳ WARNING

Once the brake lines are disconnected, do not excessively pull or bend the lines away from the module, or the lines may be damaged.

7. Remove the 2 brake control assembly-to-brake booster retaining nuts and remove the assembly.

8. If necessary, disassemble the unit as follows:

a. Remove the 6 gear cover Torx® screws from the bottom of the assembly and remove the cover. If the entire motor pack assembly is being replaced, skip to Step C, and then to the assembly procedure.

b. Mark the location of the modulator drive gears for reassembly. Insert a small suitable prybar between the holes in the gears to keep them from moving and remove the 3 gear-to-driveshaft retaining nuts. Remove the gears from the modulator.

❄❄ WARNING

Do not allow the gears to turn while removing the driveshaft nuts, or the piston may hit the top of the modulator bore and damage the pistons.

c. Remove the motor pack-to-modulator Torx® screws and separate the motor pack from the modulator.

d. Remove the 2 modulator-to-master cylinder through-bolts and separate the master cylinder from the assembly.

e. Remove the 2 transfer tubes and O-rings from the master cylinder and modulator.

f. Remove the through-bolt O-rings from the master cylinder and modulator.

To install:

9. If necessary, assemble the ABS control unit as follows:

a. Lubricate the new transfer tube O-rings with clean brake fluid. Press the NEW tubes and O-rings into the modulator by hand until fully bottomed.

b. Lubricate the new through-bolt O-rings with clean fluid and install the rings into the master cylinder and modulator.

c. Install the master cylinder onto the modulator, and press the transfer tubes into position on the master cylinder.

d. Install the through-bolts and tighten to 146 inch lbs. (16.5 Nm).

e. Position the drive gears onto the driveshafts as noted earlier. Hold the gears from turning and tighten the retaining nuts to 75 inch lbs. (8.5 Nm).

f. Hold the modulator upside down with the gears facing you and carefully rotate each gear counterclockwise until movement stops. This will position the pistons close to the top or HOME position and simplify the bleeding procedure.

g. Position the motor pack to the modulator, aligning the gears, and install the 4 motor pack retaining screws. Tighten the screws to 40 inch lbs. (4.5 Nm).

h. Install the gear cover onto the modulator assembly with the 6 retaining screws, then tighten the screws to 20 inch lbs. (2.25 Nm).

10. Position the control assembly onto the brake booster studs, then install the retaining nuts and tighten to 20 ft. lbs. (27 Nm).

11. Position the brake lines into the control assembly as originally noted and tighten the fittings to 24 ft. lbs. (32 Nm).

➡From the front of the master cylinder moving rearward, the lines are: LF, RR, LR and RF.

12. Install the 6-way wiring harness connector and insert the CPA device. Install the wiring harness connectors to the brake fluid level and 2 solenoid valve electrical connectors.

13. Install the battery and hold-down device. Connect the positive battery cable first, then the negative battery cable.

14. Install the air cleaner/induction assembly.

15. If the ABS pistons were run down for unit disassembly, connect the scan tool and perform the RUN ABS MOTORS, PISTONS UP-HOME test to restore tension to the modulator gears before moving the vehicle.

❄❄ CAUTION

Never attempt to move the vehicle with the modulator pistons in the DOWN position! With the tension released and the piston free to travel, the master cylinder must first fill the piston bores (force the pistons downward with brake fluid) before effective braking can occur. This will cause excessive brake pedal travel and reduced braking until the system is initialized (at a speed greater than 3 MPH) and the pistons are run back to the HOME position.

16. Fill the reservoir with clean brake fluid and properly bleed the ABS system.

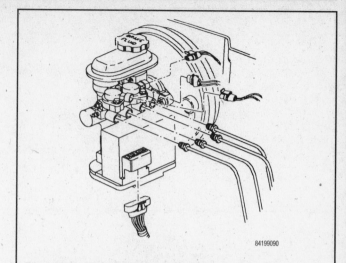

Fig. 94 Label and disconnect the wiring harnesses and brake lines from the control assembly

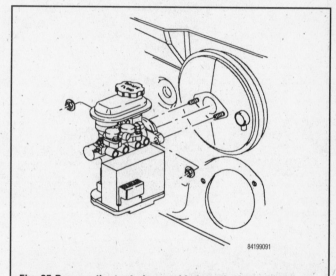

Fig. 95 Remove the control assembly from the brake booster

Fig. 96 View of the control assembly from the driver's side of the vehicle

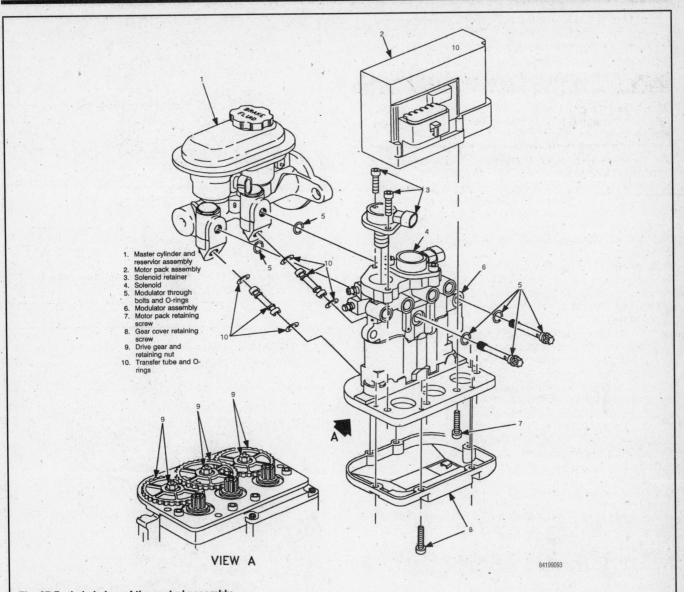

1. Master cylinder and reservoir assembly
2. Motor pack assembly
3. Solenoid retainer
4. Solenoid
5. Modulator through bolts and O-rings
6. Modulator assembly
7. Motor pack retaining screw
8. Gear cover retaining screw
9. Drive gear and retaining nut
10. Transfer tube and O-rings

VIEW A

84199093

Fig. 97 Exploded view of the control assembly

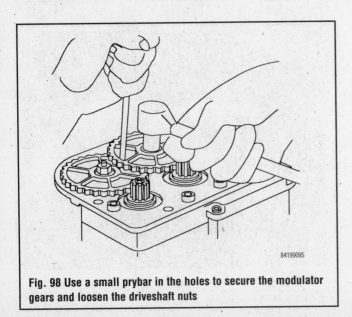

84199095

Fig. 98 Use a small prybar in the holes to secure the modulator gears and loosen the driveshaft nuts

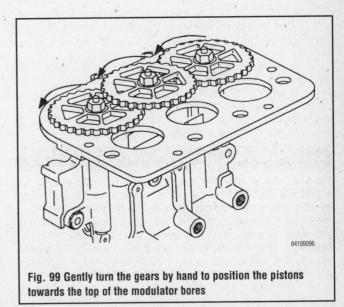

84199096

Fig. 99 Gently turn the gears by hand to position the pistons towards the top of the modulator bores

TESTING

> **See Figure 100**

1. Disengage the ABS motor pack connector.
2. Using a Digital Volt/Ohmmeter (DVOM), measure the voltage at circuit E, A or C of the motor pack connector with the key **ON**. Which circuit is being checked depends on which trouble code is set.
3. If battery voltage is present, the circuit is shorted to voltage. If battery voltage is not present, measure the voltage at circuit(s) F, B and/or D.
4. If battery voltage is not present, the ABS motor pack is at fault.
5. Turn the key **OFF** and disconnect the negative battery cable.
6. With the DVOM at the ohmmeter setting, measure the resistance to ground of the motor circuit to be tested.
7. If the resistance measures above 200 ohms, the ABS motor pack is at fault.

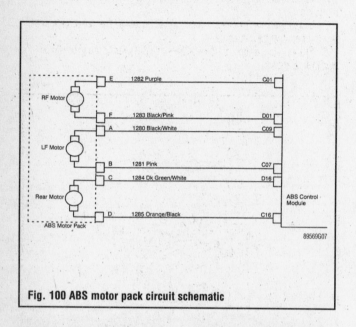

Fig. 100 ABS motor pack circuit schematic

ABS System Enable Relay

REMOVAL & INSTALLATION

The enable relay is located under the dashboard, above the steering column, taped to the wiring harness on 1991–94 Saturn models. On 1995–98 Saturn models, the enable relay is located in the underhood junction block.

1. Disconnect the negative battery cable.
2. On 1991–94 models, remove the enable relay from under the dashboard, above the steering column.
3. On 1995–98 models, remove the underhood junction block cover, located on the left side of the engine compartment. Remove the relay.
4. Installation is the reverse of removal.

Wheel Speed Sensors

REMOVAL & INSTALLATION

Front

> **See Figures 101 and 102**

1. Disconnect the negative battery cable, then raise the front of the vehicle and support using jackstands.

2. Unplug the electrical connector from the speed sensor.
3. Remove the Torx® head retaining screw from the top of the sensor.
4. Remove the wheel speed sensor from the vehicle.
5. Make sure the sensor locating pin was removed with the sensor. If the pin remained in the knuckle, pull it out with a pair of pliers. A pin which

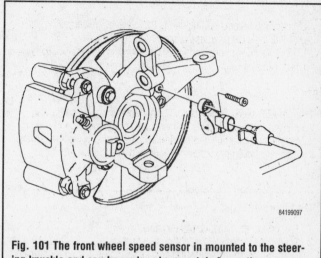

Fig. 101 The front wheel speed sensor in mounted to the steering knuckle and can be replaced separately from other components

Fig. 102 View of the knuckle-mounted speed sensor and the sensor ring which is part of the halfshaft/CV-joint assembly—front ABS brakes

is stuck in the knuckle must be drilled out with an 8mm drill bit. Be very careful not to enlarge the locating hole in the knuckle; incorrect sensor location would result in loss of ABS operation.

6. Use a small wooden dowel with sand paper wrapped around it to clean the locating hole of any corrosion and allow proper sensor seating.

To install:

7. Position the sensor to the steering knuckle, making sure that it is fully seated.

8. Install the fastener and tighten to 89 inch lbs. (10 Nm).

9. Install the wiring harness connector to the sensor.

10. Remove the supports and lower the vehicle, then connect the negative battery cable.

Rear

The rear ABS speed sensor is contained within the hub and bearing assembly and is not serviceable. If the sensor is damaged and in need of replacement, the entire hub and bearing assembly must be replaced. Refer to the rear suspension procedures in Section 8 of this manual.

TESTING

▶ **See Figure 103**

1. Disconnect the negative battery cable.

2. Disengage the wheel speed sensor connector.

3. Using an ohmmeter, measure the resistance circuit A and B of the wheel speed sensor connector.

4. If the resistance measures below 200 ohms, circuit B is shorted to ground. If the resistance measures above 200 ohms, replace the wheel speed sensor.

5. If necessary, repeat these steps for the other speed sensors.

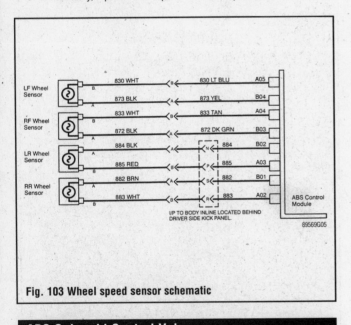

Fig. 103 Wheel speed sensor schematic

ABS Solenoid Control Valve

REMOVAL & INSTALLATION

▶ **See Figure 104**

The solenoid control valves are located on top of the ABS control assembly, which is mounted to the power brake booster. The valves are directly outboard of the master cylinder. There are 2 different valve types which can be identified by an 8-digit number etched on top of the valve and referenced in the Saturn parts catalog. Always be sure to use the correct replacement type.

1. Disconnect the negative battery cable, then clean the dirt from the area surrounding the solenoid valve.

2. Unplug the electrical connector from the valve.

3. Remove the 2 Torx® head retainers, then remove the solenoid.

4. Check to see if the O-ring remained in the valve groove and was lifted up from the modulator bore, as should occur if equipped with the type A solenoids. If using a type B solenoid, or if the type A O-ring remained in the bore, carefully extract the valve lip seal or O-ring from the modulator.

To install:

5. Lubricate the valve O-ring or lip seal with clean brake fluid and install onto the valve. O-rings should be installed into the groove provided on the solenoid. Lip seals should be inserted over the bottom of the valve with the lip or cupped side of the seal facing upward toward the solenoid top.

6. Position the solenoid into the modulator and push downward until the valve flange is fully seated. Install the 2 screws and tighten to 45 inch lbs. (5 Nm).

7. Install the wiring harness electrical connector to the valve, then connect the negative battery cable.

8. Properly bleed the anti-lock brake system. Refer to the procedure later in this section.

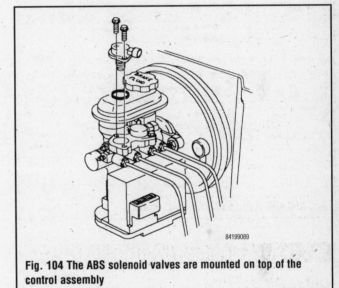

Fig. 104 The ABS solenoid valves are mounted on top of the control assembly

TESTING

▶ **See Figure 105**

1. Disconnect the negative battery cable.

2. Disengage the solenoid valve electrical connector.

3. Using an ohmmeter, measure the resistance between circuits A and B of the solenoid valve connector terminal.

4. If the resistance measures below 200 ohms, check for continuity of the power circuit. If the power circuit displays continuity, replace the solenoid valve.

5. If necessary, repeat these steps on the other solenoid valve.

Tone (Exciter) Ring

On all Saturn vehicles equipped with ABS, the tone ring is an integral component of the halfshaft assembly on the front and the hub/bearing assembly on the rear. If the vehicle requires replacement of the tone ring, the rear hub/bearing assembly and/or the front halfshaft(s) must be replaced. Refer to the rear suspension procedures in Section 8 of this manual.

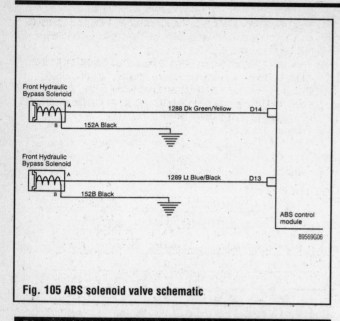

Fig. 105 ABS solenoid valve schematic

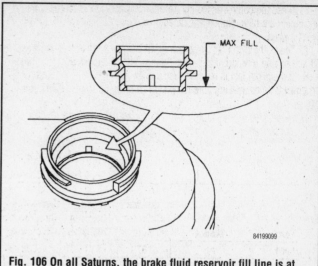

Fig. 106 On all Saturns, the brake fluid reservoir fill line is at the base of the fill neck

Bleeding the ABS System

▶ See Figures 106, 107 and 108

➡Prior to bleeding the rear brakes, the rear displacement cylinder pistons must be returned to the topmost or HOME position. To return the pistons to the HOME position, use a Scan tool to perform the special test, "RUN ABS PISTONS UP-HOME." This test will run the pistons to the top of their travel.

1. Fill the master cylinder with clean brake fluid and keep the reservoir at least ½ full during the bleeding operation. Install the fluid reservoir cap.

2. Prime the control assembly as follows:

   a. Attach a clear tube to the rear bleeder valve and submerge the tube in a transparent container of clean brake fluid.

   b. Slowly open the valve ½–¾ of a turn, then have an assistant depress the brake pedal.

   c. Hold the brake pedal until fluid begins to flow from the valve, then close the valve and release the pedal.

   d. Tighten the valve to 62 inch lbs. (7 Nm).

   e. Repeat the procedure at the front bleeder valve.

3. Once fluid flows from both control assembly valves, it may not be completely purged of air. To assure that the unit is free of air, bleed the calipers to remove air from the system's lowest points, then return and bleed the control assembly one last time.

4. Be sure to bleed the calipers in the proper order:

   a. Right rear
   b. Left rear
   c. Right front
   d. Left front

➡If, when performing the bleeding procedure on the rear calipers, brake fluid does not come out of the bleeder, the rear displacement pistons may not be at the HOME or topmost position.

5. Bleed each caliper, in the proper order, as follows:

   a. Attach a clear tube to the caliper bleeder valve and submerge the other end of the tube in a transparent container of clean brake fluid.

   b. Open the valve ½–¾ of a turn, then have an assistant slowly depress the brake pedal and hold.

   c. Watch for air bubbles as the fluid begins to flow from the valve, then close the valve and release the pedal.

   d. Wait 5 seconds and repeat until the pedal feels firm and no air is present in the brake line.

   e. Tighten the valve to 97 inch lbs. (11 Nm) for the calipers and 66 inch lbs. (7.5 Nm) for the wheel cylinders.

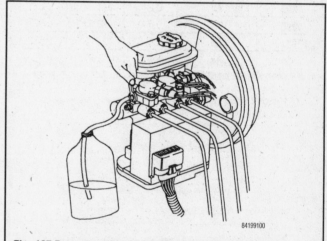

Fig. 107 Run a transparent tube from the ABS control assembly's front bleeder screw to a transparent container partially filled with clean brake fluid

Fig. 108 Attach a transparent tube to the caliper bleeder valve—rear brakes shown

6. Bleed the control assembly from the 2 valves in the same fashion as the calipers are bled. Tighten the bleeder valves to 62 inch lbs. (7 Nm) when finished.

7. Check the pedal for excessive travel both with the engine OFF and with the engine running. If pedal feel is firm, proceed to Step 11.

8. If the pedal feel is not firm, use a Scan tool to run the ABS motor up and down 2 times, making sure the pistons are run up to the HOME position.

9. Start the engine and let it run for 2 seconds after the ABS light goes out, then turn the engine OFF. Repeat this ignition cycle 9 more times.

10. Re-bleed the entire system.

11. With the engine running and brake applied, check the system for leaks.

12. Road test the vehicle and make several normal, non-ABS stops. Then, make 1–2 ABS stops from a higher speed (about 50 mph/80 kmh).

13. After road testing the vehicle, it is recommended that the entire system be bled and inspected one final time.

BRAKE SPECIFICATIONS

All measurements in inches unless noted

Year	Model	Master Cylinder Bore	Brake Disc			Brake Drum Diameter			Minimum Lining Thickness	
			Original Thickness	Minimum Thickness	Maximum Runout	Original Inside Diameter	Max. Wear Limit	Maximum Machine Diameter	Front	Rear
1991	All	0.870	0.710 ①	0.633 ②	0.0005	7.87	7.93	7.9	3/32	3/32
1992	All	0.870	0.710 ①	0.633 ②	0.0005	7.87	7.93	7.9	3/32	3/32
1993	All	0.870	0.710 ①	0.633 ②	0.0005	7.87	7.93	7.9	3/32	3/32
1994	All	0.870	0.710 ①	0.633 ②	0.0005	7.87	7.93	7.9	3/32	3/32
1995	All	0.870	0.710 ①	0.633 ②	0.0005	7.87	7.93	7.9	3/32	3/32
1996	All	0.870	0.710 ①	0.633 ②	0.0005	7.87	7.93	7.9	3/32	3/32
1997	All	0.870	0.710 ①	0.633 ②	0.0005	7.87	7.93	7.9	3/32	3/32
1998	All	0.870	0.710 ①	0.633 ②	0.0005	7.87	7.93	7.9	3/32	3/32

Note: Both front and rear disc specifications are minimums to which the rotors may be machined. The discard wear limit is 0.625 for front discs and 0.350 for rear discs.

① Rear Disc: 0.430
② Rear disc: 0.370

89569C01

Troubleshooting the Brake System

Problem	Cause	Solution
Low brake pedal (excessive pedal travel required for braking action.)	• Excessive clearance between rear linings and drums caused by inoperative automatic adjusters	• Make 10 to 15 alternate forward and reverse brake stops to adjust brakes. If brake pedal does not come up, repair or replace adjuster parts as necessary.
	• Worn rear brakelining	• Inspect and replace lining if worn beyond minimum thickness specification
	• Bent, distorted brakeshoes, front or rear	• Replace brakeshoes in axle sets
	• Air in hydraulic system	• Remove air from system. Refer to Brake Bleeding.
Low brake pedal (pedal may go to floor with steady pressure applied.)	• Fluid leak in hydraulic system	• Fill master cylinder to fill line; have helper apply brakes and check calipers, wheel cylinders, differential valve tubes, hoses and fittings for leaks. Repair or replace as necessary.
	• Air in hydraulic system	• Remove air from system. Refer to Brake Bleeding.
	• Incorrect or non-recommended brake fluid (fluid evaporates at below normal temp).	• Flush hydraulic system with clean brake fluid. Refill with correct-type fluid.
	• Master cylinder piston seals worn, or master cylinder bore is scored, worn or corroded	• Repair or replace master cylinder
Low brake pedal (pedal goes to floor on first application—o.k. on subsequent applications.)	• Disc brake pads sticking on abutment surfaces of anchor plate. Caused by a build-up of dirt, rust, or corrosion on abutment surfaces	• Clean abutment surfaces
Fading brake pedal (pedal height decreases with steady pressure applied.)	• Fluid leak in hydraulic system	• Fill master cylinder reservoirs to fill mark, have helper apply brakes, check calipers, wheel cylinders, differential valve, tubes, hoses, and fittings for fluid leaks. Repair or replace parts as necessary.
	• Master cylinder piston seals worn, or master cylinder bore is scored, worn or corroded	• Repair or replace master cylinder
Decreasing brake pedal travel (pedal travel required for braking action decreases and may be accompanied by a hard pedal.)	• Caliper or wheel cylinder pistons sticking or seized	• Repair or replace the calipers, or wheel cylinders
	• Master cylinder compensator ports blocked (preventing fluid return to reservoirs) or pistons sticking or seized in master cylinder bore	• Repair or replace the master cylinder
	• Power brake unit binding internally	• Test unit according to the following procedure: (a) Shift transmission into neutral and start engine (b) Increase engine speed to 1500 rpm, close throttle and fully depress brake pedal (c) Slow release brake pedal and stop engine (d) Have helper remove vacuum check valve and hose from power unit. Observe for backward movement of brake pedal. (e) If the pedal moves backward, the power unit has an internal bind—replace power unit

TCCA9C01

Troubleshooting the Brake System (cont.)

Problem	Cause	Solution
Spongy brake pedal (pedal has abnormally soft, springy, spongy feel when depressed.)	• Air in hydraulic system • Brakeshoes bent or distorted • Brakelining not yet seated with drums and rotors • Rear drum brakes not properly adjusted	• Remove air from system. Refer to Brake Bleeding. • Replace brakeshoes • Burnish brakes • Adjust brakes
Hard brake pedal (excessive pedal pressure required to stop vehicle. May be accompanied by brake fade.)	• Loose or leaking power brake unit vacuum hose • Incorrect or poor quality brakelining • Bent, broken, distorted brakeshoes • Calipers binding or dragging on mounting pins. Rear brakeshoes dragging on support plate. • Caliper, wheel cylinder, or master cylinder pistons sticking or seized • Power brake unit vacuum check valve malfunction • Power brake unit has internal bind • Master cylinder compensator ports (at bottom of reservoirs) blocked by dirt, scale, rust, or have small burrs (blocked ports prevent fluid return to reservoirs). • Brake hoses, tubes, fittings clogged or restricted • Brake fluid contaminated with improper fluids (motor oil, transmission fluid, causing rubber components to swell and stick in bores • Low engine vacuum	• Tighten connections or replace leaking hose • Replace with lining in axle sets • Replace brakeshoes • Replace mounting pins and bushings. Clean rust or burrs from rear brake support plate ledges and lubricate ledges with molydisulfide grease. **NOTE:** If ledges are deeply grooved or scored, do not attempt to sand or grind them smooth—replace support plate. • Repair or replace parts as necessary • Test valve according to the following procedure: (a) Start engine, increase engine speed to 1500 rpm, close throttle and immediately stop engine (b) Wait at least 90 seconds then depress brake pedal (c) If brakes are not vacuum assisted for 2 or more applications, check valve is faulty • Test unit according to the following procedure: (a) With engine stopped, apply brakes several times to exhaust all vacuum in system (b) Shift transmission into neutral, depress brake pedal and start engine (c) If pedal height decreases with foot pressure and less pressure is required to hold pedal in applied position, power unit vacuum system is operating normally. Test power unit. If power unit exhibits a bind condition, replace the power unit. • Repair or replace master cylinder **CAUTION:** Do not attempt to clean blocked ports with wire, pencils, or similar implements. Use compressed air only. • Use compressed air to check or unclog parts. Replace any damaged parts. • Replace all rubber components, combination valve and hoses. Flush entire brake system with DOT 3 brake fluid or equivalent. • Adjust or repair engine

TCCA9C02

Troubleshooting the Brake System (cont.)

Problem	Cause	Solution
Grabbing brakes (severe reaction to brake pedal pressure.)	• Brakelining(s) contaminated by grease or brake fluid	• Determine and correct cause of contamination and replace brakeshoes in axle sets
	• Parking brake cables incorrectly adjusted or seized	• Adjust cables. Replace seized cables.
	• Incorrect brakelining or lining loose on brakeshoes	• Replace brakeshoes in axle sets
	• Caliper anchor plate bolts loose	• Tighten bolts
	• Rear brakeshoes binding on support plate ledges	• Clean and lubricate ledges. Replace support plate(s) if ledges are deeply grooved. Do not attempt to smooth ledges by grinding.
	• Incorrect or missing power brake reaction disc	• Install correct disc
	• Rear brake support plates loose	• Tighten mounting bolts
Dragging brakes (slow or incomplete release of brakes)	• Brake pedal binding at pivot	• Loosen and lubricate
	• Power brake unit has internal bind	• Inspect for internal bind. Replace unit if internal bind exists.
	• Parking brake cables incorrrectly adjusted or seized	• Adjust cables. Replace seized cables.
	• Rear brakeshoe return springs weak or broken	• Replace return springs. Replace brakeshoe if necessary in axle sets.
	• Automatic adjusters malfunctioning	• Repair or replace adjuster parts as required
	• Caliper, wheel cylinder or master cylinder pistons sticking or seized	• Repair or replace parts as necessary
	• Master cylinder compensating ports blocked (fluid does not return to reservoirs).	• Use compressed air to clear ports. Do not use wire, pencils, or similar objects to open blocked ports.
Vehicle moves to one side when brakes are applied	• Incorrect front tire pressure	• Inflate to recommended cold (reduced load) inflation pressure
	• Worn or damaged wheel bearings	• Replace worn or damaged bearings
	• Brakelining on one side contaminated	• Determine and correct cause of contamination and replace brakelining in axle sets
	• Brakeshoes on one side bent, distorted, or lining loose on shoe	• Replace brakeshoes in axle sets
	• Support plate bent or loose on one side	• Tighten or replace support plate
	• Brakelining not yet seated with drums or rotors	• Burnish brakelining
	• Caliper anchor plate loose on one side	• Tighten anchor plate bolts
	• Caliper piston sticking or seized	• Repair or replace caliper
	• Brakelinings water soaked	• Drive vehicle with brakes lightly applied to dry linings
	• Loose suspension component attaching or mounting bolts	• Tighten suspension bolts. Replace worn suspension components.
	• Brake combination valve failure	• Replace combination valve
Chatter or shudder when brakes are applied (pedal pulsation and roughness may also occur.)	• Brakeshoes distorted, bent, contaminated, or worn	• Replace brakeshoes in axle sets
	• Caliper anchor plate or support plate loose	• Tighten mounting bolts
	• Excessive thickness variation of rotor(s)	• Refinish or replace rotors in axle sets

TCCA9C03

10

BODY

AND

TRIM

Doors

REMOVAL & INSTALLATION

♦ **See Figures 1 and 2**

➡**Two people are needed to replace the entire door assembly. Most components, including the hinges, can be replaced without removing the door from the vehicle and, thereby, eliminating the need for an assistant.**

1. For doors equipped with electrical components such as power windows or stereo speakers, disconnect the negative battery cable.

2. If applicable, disconnect the electrical wiring boot and unplug the wiring. Should the wiring connectors be unaccessible from the boot, the interior trim panel(s) must be removed so the wires can be disconnected. Refer to the procedure later in this section.

3. Remove the check link bolt(s) from the door or the body, whichever is accessible.

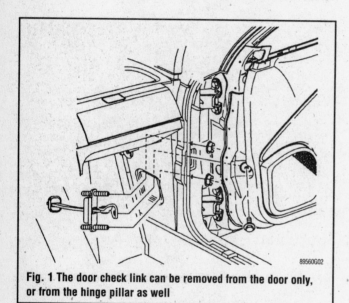

Fig. 1 The door check link can be removed from the door only, or from the hinge pillar as well

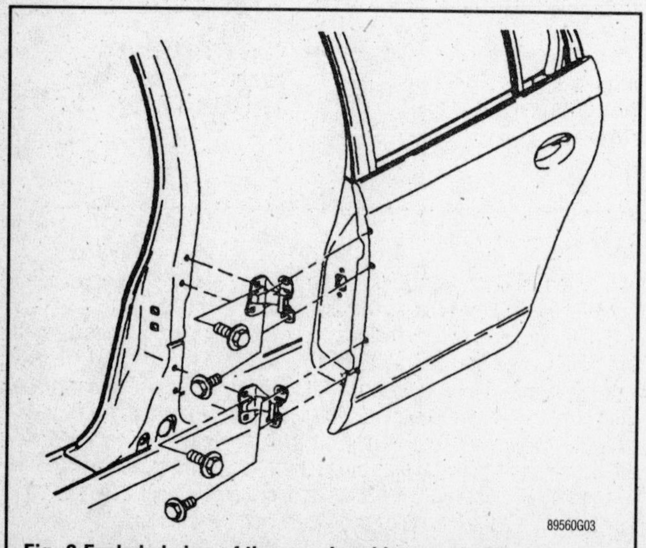

4. If the door is to be reinstalled, use a soft marker to outline the hinge positioning in relation to the door assembly.

5. While an assistant supports the door assembly, remove the bolts attaching the hinges to the door, then remove the door from the vehicle.

To install:

6. Have an assistant support the door assembly to the vehicle and secure using the bolts, but do not completely tighten at this time.

7. Adjust the door. Refer to the procedure later in this section. If reinstalling a door, use the marks made during removal as a starting point and adjust from there, as necessary. Then, tighten the hinge-to-door bolts to 26 ft. lbs. (35 Nm) on 1991–95 models or 22 ft. lbs. (30 Nm) on 1996–98 models.

8. Secure the check link to the door or body using the retaining bolt(s) and tighten to 12 ft. lbs. (16 Nm).

9. Connect the door wiring, then install the trim panel(s) and wiring boot, as applicable.

10. If removed, connect the negative battery cable.

ADJUSTMENT

♦ **See Figures 3, 4 and 5**

1. Determine if adjustment is necessary by opening and closing the door checking for proper fit and alignment. Make sure the door does not bind on adjacent panels.

2. If adjustment is necessary, the door striker must be removed from the body. Use a soft marker to scribe alignment marks, then remove the screws, striker and, if present, shim(s) to allow for proper door adjustment.

3. To adjust the door up or down, loosen the 4 hinge-to-door bolts and position the door as needed.

4. To adjust the door in or out, loosen the 4 hinge-to-door bolts, one hinge at a time, and position the door as needed.

5. To adjust the door fore or aft, loosen the 4 hinge-to-body bolts, one hinge at a time, and position as needed.

6. When the door has been properly adjusted, tighten the hinge bolts to 26 ft. lbs. (35 Nm).

7. Loosely install the striker, screws and, if present, the shims. Align the striker with the marks made during removal and tighten the retainers to 18 ft. lbs. (25 Nm). Check striker alignment by closing the door.

8. If necessary, adjust the striker alignment. The striker may be adjusted up and down or in and out by loosening the screws and repositioning the striker. Fore and aft adjustments are made by adding or removing shims from the behind the striker. When adjustments are made, tighten the screws to 18 ft. lbs. (25 Nm).

Fig. 2 Exploded view of the rear door hinge assembly

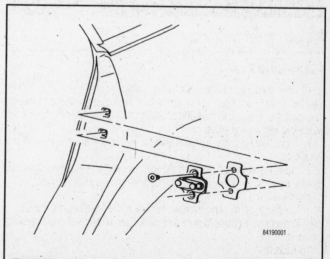

Fig. 3 The striker must be removed when adjusting the door assembly

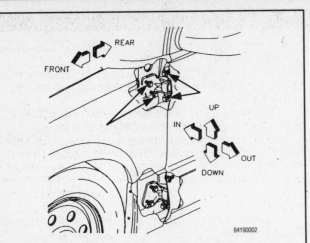

Fig. 4 Loosen the body bolts only to make front and rearward door adjustments; loosen the door bolts to make all other adjustments

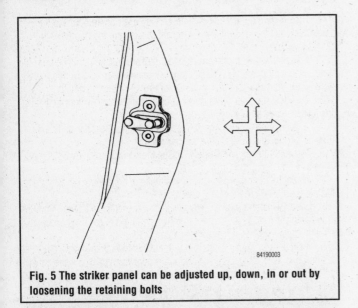

Fig. 5 The striker panel can be adjusted up, down, in or out by loosening the retaining bolts

Hood

REMOVAL & INSTALLATION

▶ See Figures 6 and 7

1. Open the hood and position fender covers to protect the paint. It may also be useful to position rags on top of the cowl panel, behind the hinges, to protect the vehicle and the hood should it slide back onto the cowl panel when the bolts are loosened.

2. Using a soft marker, scribe alignment lines of the hood's position on the hinges.

3. Have 2 assistants hold the hood and remove the 4 retaining bolts. Be careful not to let the hood slip backwards and onto the vehicle when the bolts are removed.

4. Carefully lift the hood from the vehicle. If the hood is to be stored standing, be sure to place rags or protective foam under the hood to protect the paint.

To install:

5. Have your assistants lift the hood and position it to the hinges. Make

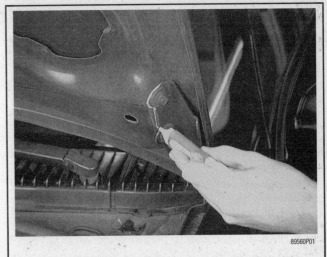

Fig. 6 Trace the position of the hinges on the underside of the hood before removal

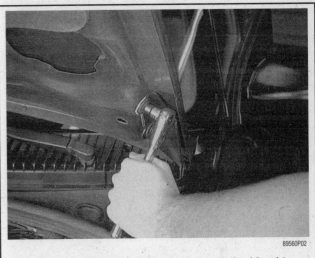

Fig. 7 While 2 assistants hold the hood, loosen the 4 hood-to-hinge retaining bolts (2 on each side)

sure the hood aligns with the marks made earlier, then install the retaining bolts.

6. Remove the fender covers, then check and adjust hood alignment, as necessary. Refer to the procedure later in this section. Once the hood is properly adjusted, tighten the retaining bolts to 19 ft. lbs. (25 Nm).

ALIGNMENT

▶ See Figure 8

1. To adjust the hood fore, aft and cross-car, loosen the hinge-to-body bolts and position the hood, as needed.

2. To adjust the hood up and down at the rear, loosen the hinge-to-hood bolts and position the hood, as needed.

3. To adjust the hood up and down at the front, the hood latch and/or the hood bumpers can be repositioned. The hood latch can be adjusted by loosening the retainers and repositioning the latch, as needed. The bumpers are adjusted by turning them to raise or lower their positions, as needed.

4. When all adjustments are completed, tighten the hinge bolts to 19 ft. lbs. (25 Nm) and the hood latch bolts to 89 inch lbs. (10 Nm).

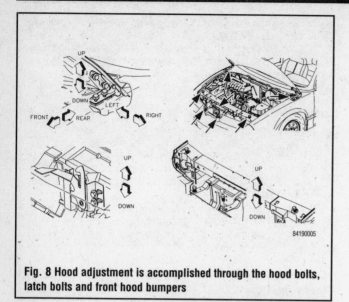

Fig. 8 Hood adjustment is accomplished through the hood bolts, latch bolts and front hood bumpers

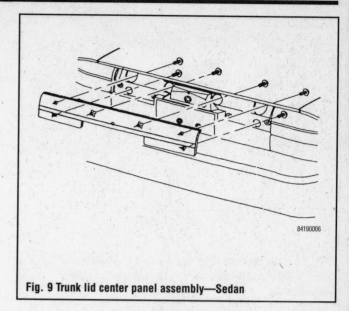

Fig. 9 Trunk lid center panel assembly—Sedan

Trunk Lid/Liftgate

REMOVAL & INSTALLATION

1991–96 Coupe and 1991–95 Sedan

▶ See Figures 9, 10, 11 and 12

1. Disconnect the negative battery cable.
2. For Sedans, remove the trunk lid center panel and lower outer panel assemblies, as follows:
    a. Remove the trunk center panel assembly bolts, then rotate the panel outboard at the bottom and remove the panel from the vehicle.
    b. Remove the license plate, then disconnect the lock rod from the trunk lock cylinder pawl.
    c. Remove the lower outer panel attachment bolts.
    d. Disengage the panel bottom flange from the inner panel by pulling the top of the panel rearward and the lower panel downward.
    e. Remove the wiring harness retainer clip, then disconnect the license plate bulb sockets from the lamps.
    f. Remove the trunk lid lower outer panel from the vehicle.
3. For Coupes, remove the trunk lid lower panel, as follows:
    a. Remove the retaining screws from the reflex panels located on either side of the trunk lid, then remove the panels from the lid.
    b. Remove the clip and disconnect the lock rod from the lock cylinder.
    c. Remove the screws and plastic retainers, then remove the lower panel from the trunk lid.
    d. If necessary, remove the sealing sleeve, lock cylinder retainer, reinforcement and the lock cylinder.
4. Remove the wiring harness grommet from the inner panel by pushing rearward. Pull the wiring harness inside the trunk lid through the grommet hole.
5. Disconnect the trunk lid release cable from the lock.
6. Unplug the center stop lamp wiring harness connector.
7. Open the wiring harness routing connectors, then remove the harness and lock cable from the rear of the trunk lid.
8. Remove the trunk lid-to-hinge attachment bolts, then, with the aid of an assistant, remove the trunk lid from the vehicle.

To install:

9. With the help of an assistant, lower the trunk lid onto the hinges and install the attaching bolts. Check trunk alignment and adjust, as necessary, then tighten the retaining bolts.
10. Route the wiring harness and lock cable through the connectors on the rear of the trunk lid, then lock the harness and cable into the connectors.

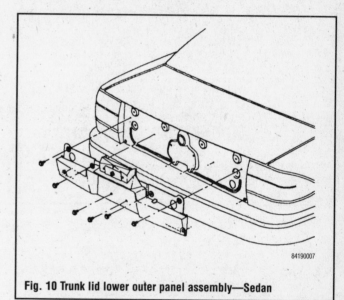

Fig. 10 Trunk lid lower outer panel assembly—Sedan

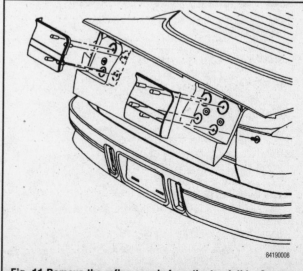

Fig. 11 Remove the reflex panels from the trunk lid—Coupe

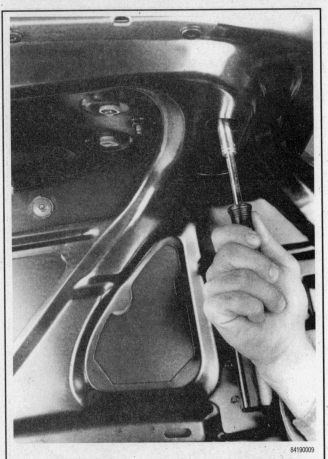

Fig. 12 It will be necessary to use a socket and extension to reach some of the reflex panel screws which are recessed in the trunk lid—Coupe

11. Install the wiring harness to the center stop lamp terminal.
12. Connect the lock release cable to the lock.
13. Install the wiring harness through the rear compartment lid inner panel, then seat the wiring harness grommet to the panel.
14. For Coupes, install the trunk lid lower panel, as follows:
    a. If removed, install the lock cylinder into the panel, then install the reinforcement and retainer to the lock cylinder. Install the sealing sleeve to the cylinder.
    b. Install the lower panel to the trunk lid and secure using the retaining screws. Tighten the screws slowly and evenly to 53 inch lbs. (6 Nm), using multiple passes.
    c. Connect the lock rod to the lock cylinder using the clip.
    d. Install the reflex panels using the retaining screws. Make sure the sealing feature of the screws and panels is adequate. If necessary, replace the seals or apply a body caulking compound to make sure no water or moisture will intrude.
15. For the Sedan, install the trunk lid lower outer panel and center panel assemblies, as follows:
    a. Position the lower outer panel to the trunk lid and connect the license plate bulb sockets to the license plate lamps.
    b. Connect the wiring harness retainer clip to the trunk lid lower outer panel.
    c. Install the bottom flange of the lower outer panel over the lower edge of the trunk lid inner panel, then install the attaching bolts and tighten to 53 inch lbs. (6 Nm).
    d. Connect the lock rod to the lock cylinder pawl.
    e. Position the upper edge of the center panel to the trunk lid, then rotate the bottom of the center panel inboard.
    f. Check to see that the sealing feature of the bolts is adequate and, if necessary, replace the bolts or seal using a body caulking compound to protect against water intrusion.
    g. Loosely install the attaching bolts and check the panel alignment. Adjust the alignment, as necessary, and tighten the bolts. Install the license plate.
16. Connect the negative battery cable.

1997–98 Coupe and 1996–98 Sedan

♦ See Figure 13

1. Disconnect the negative battery cable.
2. Open the trunk lid and pull back the left side carpet.
3. Disengage the trunk lid wiring harness from the body harness at the blue connector on the wheelhousing.
4. Remove the trunk lid-to-hinge attachment bolts, then with the aid of an assistant, remove the trunk lid from the vehicle.
5. If replacing the trunk lid, remove the license plate, lock cylinder, lock, center panel and wiring harness.
To install:
6. With the help of an assistant, lower the trunk lid onto the hinges and install the attaching bolts. Check trunk alignment and adjust, as necessary, then tighten the retaining bolts to 89 inch lbs. (10 Nm).
7. Connect the trunk lid wiring harness to the body connector and place the carpet back to its correct position.
8. Connect the negative battery cable.

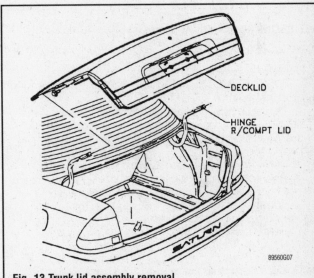

Fig. 13 Trunk lid assembly removal

Wagon

♦ See Figure 14

1. Disconnect the negative battery cable.
2. Remove the trim access plate from the inner liftgate trim panel.
3. Unplug the wiring connectors and the window washer hose.
4. Remove the rubber grommet between the liftgate and body, then retrieve the wires and washer hose from the liftgate.
5. Either support the liftgate using a prop or have an assistant hold the liftgate, then remove the liftgate struts.
6. Use a soft marker to scribe small alignment marks for installation purposes. With the help of an assistant, remove the hinge attachment bolts from the liftgate and remove the liftgate from the vehicle.
To install:
7. Raise the liftgate into position and align the marks made during removal, then install the gate to the hinges and tighten the bolts to 89 inch lbs. (10 Nm).
8. Install the struts.

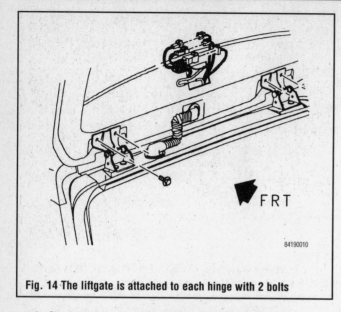

Fig. 14 The liftgate is attached to each hinge with 2 bolts

9. Check and adjust the liftgate alignment, as necessary.

10. Route the wiring harness and washer hose through the liftgate, then connect the harness to the terminals and hose to the window washer.

11. Install the trim panel access plate.

12. Install the rubber grommet between the liftgate and the body.

13. Connect the negative battery cable.

ALIGNMENT

Coupe and Sedan

▶ See Figure 15

1. To adjust the trunk lid cross-car, loosen the hinge-to-package shelf retainers and position the assembly, as needed.

2. To adjust the front of the trunk lid fore, aft, up or down, loosen the lid-to-hinge arm bolts and position the lid, as needed.

3. To adjust the rear of the trunk lid up and down, loosen the striker retainers and move the striker upward or downward, as needed.

4. When all adjustments are made, tighten the package shelf retainers to 19 ft. lbs. (25 Nm). Tighten the trunk lid bolts and/or striker bolts to 89 inch lbs. (10 Nm).

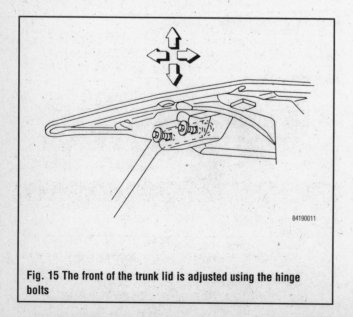

Fig. 15 The front of the trunk lid is adjusted using the hinge bolts

Wagon

The Wagon liftgate may be adjusted up, down or cross-car by loosening the hinge-to-liftgate bolts. Fore and aft adjustments are usually made using the hinge-to-body bolts.

Outside Mirrors

REMOVAL & INSTALLATION

▶ See Figures 16 thru 22

1. If equipped with a power right side mirror, disconnect the negative battery cable.

2. Remove the screw from the mirror trim panel.

3. Loosen the setscrew, then disconnect the cable from the trim panel on the manual mirror.

4. Remove the trim panel from the vehicle.

5. Remove the foam filler from the door frame.

6. If equipped, unplug the wiring harness connector on the power mirror.

7. Remove the nuts from the mirror assembly, then remove the mirror and seal from the vehicle.

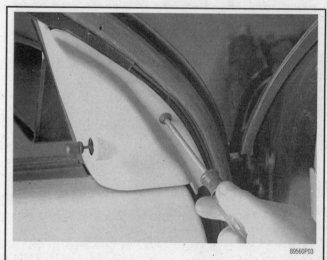

Fig. 16 Loosen the outside mirror's interior trim panel mounting screw . . .

Fig. 17 . . . and remove the trim panel from the side of the door

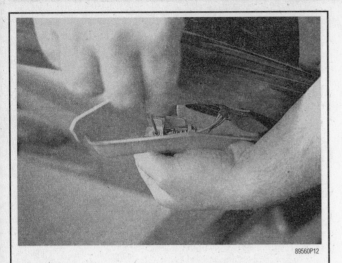

Fig. 18 Using a hex key, loosen the mirror cable retaining setscrew and . . .

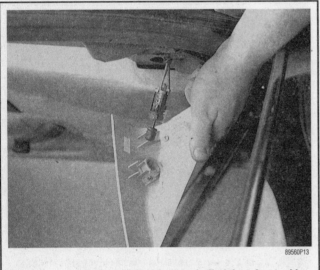

Fig. 19 . . . slide the interior trim panel off of the mirror cable

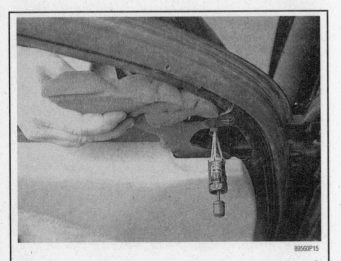

Fig. 20 Pull the foam filler insulation out, but do not damage or discard

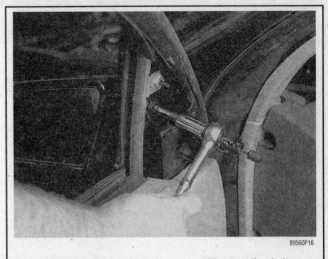

Fig. 21 Remove the outside mirror assembly mounting bolts from inside the door

Fig. 22 Pull the mirror assembly out from the mounting holes on the door

To install:

8. Install the seal onto the mirror assembly.

9. If equipped, connect the wiring harness to the right side power mirror.

10. Install the mirror assembly to the door, then tighten the retaining nuts to 53 inch lbs. (6 Nm) on 1991–95 models, or 89 inch lbs. (10 Nm) on 1996–98 models.

11. Install the foam filler to the door frame.

12. Install the mirror cable into the trim panel and tighten the setscrew to 27 inch lbs. (3 Nm), for the manual mirror.

13. Position the trim panel to the door frame, then secure using the screw.

14. If equipped with a power right side mirror, connect the negative battery cable.

Antenna

REPLACEMENT

1. Remove the right front fender. Refer to the procedure later in this section.

2. Remove the retaining bolts from the antenna bracket.

3. Unplug the antenna wiring, then remove the antenna from the vehicle.

To install:

4. Position the antenna to the vehicle and connect the wiring.

5. Install the antenna retaining screws.

6. Install the fender.

Fenders

REMOVAL & INSTALLATION

1991–95 Sedan and Wagon

▶ **See Figures 23, 24, 25, 26 and 27**

1. Remove the plastic retainers from the wheelhouse liner-to-lower fender extension.

2. Remove the plastic retainers from the bottom of the fender extension-to-fender support.

3. Detach the fender extension from the lower portion of the fender by exerting pressure at the bottom of the extension and carefully inserting a flat screwdriver into the groove. Using the screwdriver, depress the tab and pull the extension outward and from the vehicle.

4. Remove the plastic retainers from the bottom of the fender.

5. Remove the bolts at the fender upper-to-lower joint which thread into the support bracket on the front body hinge pillar. The bolts are driven vertically upward.

6. Remove the plastic retainers from the wheel liner-to-fender joint.

7. If the fender is being replaced, remove the rubber seals from the rear of the fender, the Saturn emblem and the metal bracket from the top of the rear fender.

8. Remove the fender-to-fascia stud plate at the horizontal joint by removing the rear bolt and the front nuts.

9. Remove the headlamp housing.

10. Remove the fender-to-fascia stud plate and reinforcement at the vertical joint.

11. Remove the bolt from the rear upper corner of the fender-to-front body hinge pillar support bracket.

12. Remove the bolts from the fender flange-to-upper engine compartment rail, then remove the fender from the vehicle.

To install:

13. If the fender is being replaced, install the metal bracket on the top rear of the fender, the Saturn emblem and the rubber seals to the fender rear flange.

14. Position the fender onto the engine compartment side rail.

15. Align the fender to the door and hood, then install the fender flange bolts using the sequence shown in the illustration.

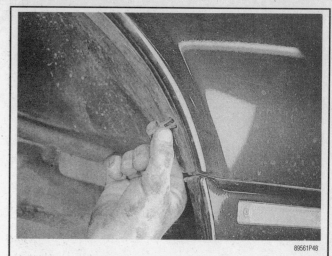

Fig. 24 . . . then pull out the pushpin retainers from the wheelhouse liner-to-lower fender extension

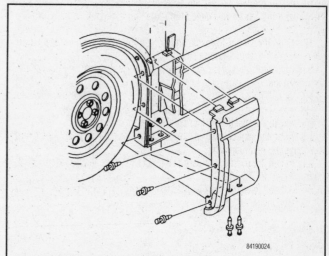

Fig. 25 The fender lower extension must be removed before the fender can be removed from the vehicle

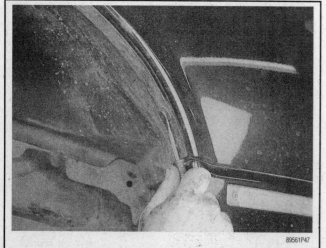

Fig. 23 Using a small prying tool, pull up on the center of the plastic pushpin retainers . . .

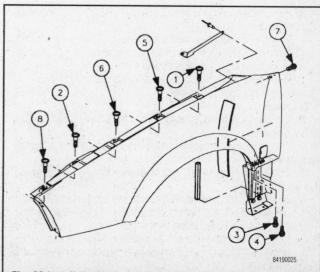

Fig. 26 Install the fender retaining bolts in the proper sequence

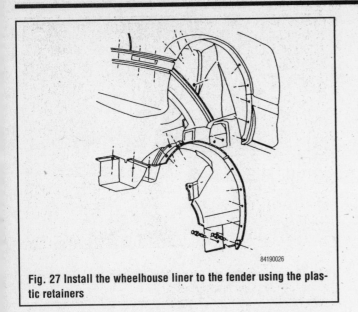

Fig. 27 Install the wheelhouse liner to the fender using the plastic retainers

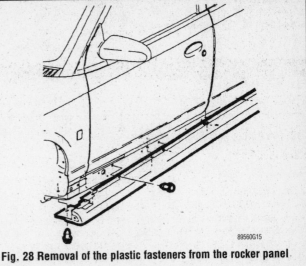

Fig. 28 Removal of the plastic fasteners from the rocker panel cover

16. Install the fender-to-front body hinge pillar support bracket bolt, then install the fender-to-fascia stud plate and vertical joint reinforcement.

17. Align the fender-to-fascia and tighten the nuts to the vertical stud plate. Install the headlamp housing.

18. Install the fender-to-fascia upper support bracket assembly along the horizontal joint using the nuts and bolts.

19. Install the wheelhouse liner to the fender using the plastic retainers.

20. Install the fender-to-body hinge pillar lower bracket bolts.

21. Install the plastic retainers at the bottom of the fender.

22. Align the fender lower extension to the fender and push it rearward to engage the retaining tabs.

23. Install the plastic retainers at the bottom of the fender extension-to-fender support.

24. Install the wheel liner-to-fender extension plastic retainers.

25. Check and adjust headlamp aim, as necessary.

1996–98 Sedan and Wagon

▶ **See Figures 28, 29 and 30**

1. Remove the plastic retainers from the outer wheelhouse liner.

2. Remove the front and middle mounting fasteners on the rocker panel cover and lower the front end of the cover. Remove the bottom rear fender mounting bolt.

3. Remove the windshield frame filler panel mounting bolt and slide the panel upward.

4. Slide the front fascia forward and disengage it from the fender bolts.

5. Remove the rear upper fender corner-to-front body hinge piller bolt.

6. Remove the top fender-to-spaceframe mounting bolts.

7. Remove the two fender-to-fender reinforcement bracket bolts. Remove the fender from the vehicle.

To install:

8. If the fender is being replaced, transfer the Saturn emblem and any rubber flange seals.

9. Place the fender in position on the upper spaceframe and align to the door and hood.

10. Install the fender-to-front body hinge pillar bolt and tighten to 89 inch lbs. (10 Nm).

11. Install all spaceframe side rail, bottom rear and fender-to-fender reinforcement bracket bolts. Tighten all the bolts to 89 inch lbs. (10 Nm).

12. Position the front fascia and slide it into place.

13. Install the windshield frame filler panel and mounting bolt.

14. Install the front and middle mounting fasteners which retain the front rocker panel cover.

15. Install the plastic retainers to the outer wheelhouse liner.

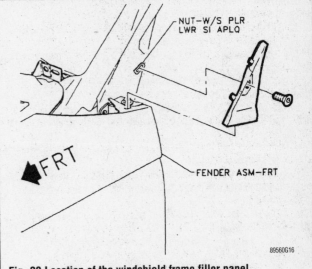

Fig. 29 Location of the windshield frame filler panel

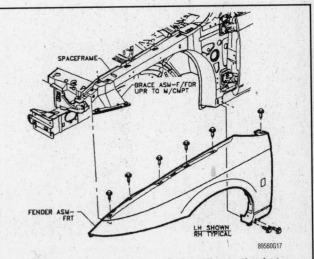

Fig. 30 Exploded view of the front fender and mounting fasteners

1991–96 Coupe

♦ See Figures 31 and 32

1. Remove the plastic retainers from the wheelhouse liner-to-fender extension.

2. Remove the plastic retainers from the bottom of the fender extension-to-fender support.

3. Disconnect the fender extension from the lower portion of the fender by exerting pressure at the bottom of the extension and carefully inserting a flat screwdriver into the groove. Using the screwdriver, depress the tab and pull the extension outward and from the vehicle.

4. Remove the wheelhouse liner-to-fascia and front fender fasteners.

5. Remove the retainers from the bottom of the fender.

6. Remove the bolts at the fender upper-to-lower joint which thread into the support bracket on the front body hinge pillar. The bolts are driven vertically upward.

7. Remove the nuts and reinforcement plate from the horizontal fender-to-fascia joint, then remove the stud plate.

8. Remove the bolt from the rear upper corner of the fender-to-front body hinge pillar support bracket.

9. Remove the fender-to-headlamp housing assembly and fender-to-engine compartment rail bolts, then remove the fender from the vehicle.

10. If replacing the fender, remove the Saturn emblem.

To install:

11. If the fender is being replaced, install the Saturn emblem.

12. Position the fender onto the engine compartment side rail, then align the fender with the door and hood.

13. Install the fender-to-engine compartment side rail fasteners, fender-to-headlamp housing assembly fasteners and the fender-to-front body hinge pillar support bracket bolts. Tighten all the bolts to 89 inch lbs. (10 Nm) using the sequence shown in the illustration.

14. Attach the fender-to-fascia by lowering the stud plate through the bracket and fender, then install the reinforcement plate and nuts. Align the fender-to-fascia and tighten the nuts.

15. Install the vertically driven fender bolts to the support bracket on the lower front body hinge pillar.

16. Install the retainers to the bottom of the fender.

17. Install the wheelhouse-to-front fascia and the front fender fasteners.

18. Align the fender lower extension to the fender and push it rearward to engage the retaining tabs.

19. Install the plastic retainers at the bottom of the fender extension-to-fender support.

20. Install the wheel liner-to-fender extension plastic retainers.

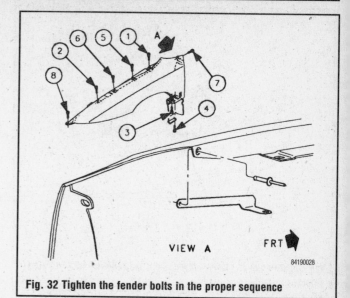

Fig. 32 Tighten the fender bolts in the proper sequence

1997–98 Coupe

♦ See Figures 33, 34 and 35

1. Remove the plastic retainers from the outer wheelhouse liner.

2. Remove the plastic pushpin fasteners from the rocker panel cover.

3. Slide the cover forward to disengage it from the front fender, rocker cover support and quarter panel. Remove the panel from the vehicle.

4. Remove the bottom rear fender mounting bolts.

5. Remove the top middle plastic pushpin retainers from the front fascia.

6. Remove the windshield frame filler panel mounting bolt and slide the panel upward.

7. Slide the front fascia forward and disengage it from the fender bolts.

8. Remove the front fender bracket shoulder bolts.

9. Remove one outer headlamp mounting bolt. Slide the outer side of the headlamp assembly forward to gain access to the forward front fender-to-spaceframe mounting bolt.

10. Remove the rear upper fender corner-to-front body hinge piller bolt.

11. Remove the top fender-to-spaceframe mounting bolts.

12. Remove the two fender-to-fender reinforcement bracket bolts. Remove the fender from the vehicle.

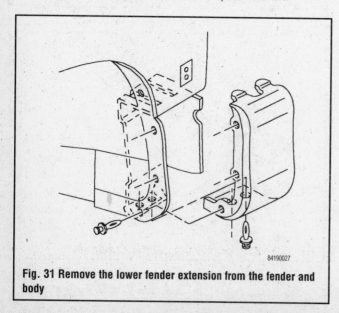

Fig. 31 Remove the lower fender extension from the fender and body

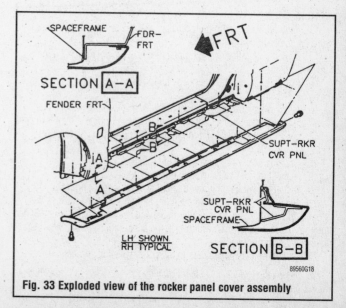

Fig. 33 Exploded view of the rocker panel cover assembly

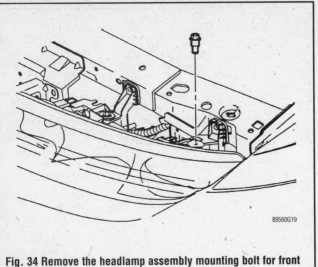

Fig. 34 Remove the headlamp assembly mounting bolt for front fender mounting bolt access

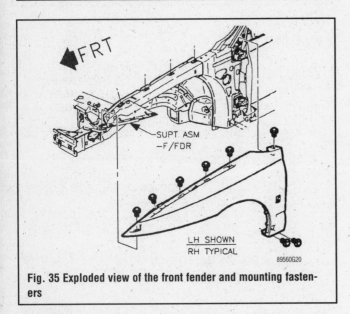

Fig. 35 Exploded view of the front fender and mounting fasteners

To install:

13. If the fender is being replaced, transfer the Saturn emblem and any rubber bumper or flange seals.

14. Place the fender in position on the upper spaceframe and align to the door and hood.

15. Install the fender-to-front body hinge pillar bolt and tighten to 89 inch lbs. (10 Nm).

16. Install all spaceframe side rail, bottom rear and fender-to-fender reinforcement bracket bolts. Tighten all of the bolts to 89 inch lbs. (10 Nm).

17. Install the headlamp mounting bolt and tighten to 61 inch lbs. (7 Nm).

18. Position the front fascia and slide it into place.

19. Install the front fascia plastic pushpin retainers.

20. Install the windshield frame filler panel and mounting bolt.

21. Place the rocker cover in position, aligning the panel tabs to the slots in the front fender, rear quarter panel and rocker cover support. Slide the rocker cover panel rearward to lock the panel into position, then install the plastic pushpin retainers.

22. Install the plastic retainers to the outer wheelhouse liner.

Power Sunroof

REMOVAL & INSTALLATION

▶ **See Figure 36**

Glass

▶ **See Figure 37**

1. Turn the ignition **ON** and set the sunroof to the VENT position, then turn the ignition **OFF**.

2. Remove the retaining screws that attach the sunroof glass to the guide assemblies, then carefully lift the glass from the roof opening.

✳✳ WARNING

Once the glass has been removed from the sunroof, DO NOT cycle the roof to the full closed position or the guide assemblies will be damaged.

To install:

3. With the guide assemblies still in the VENT position, lower the glass onto the guides and align the mounting holes.

4. Apply Loctite® 222 or an equivalent small screw threadlocker to the attaching screws, then loosely install the screws.

5. Turn the ignition **ON** and close the sunroof, then adjust the glass to 0.04 in. (1mm) below the roof surface at the front and rear of the glass, then tighten the glass attaching screws to 35–44 inch lbs. (4–5 Nm).

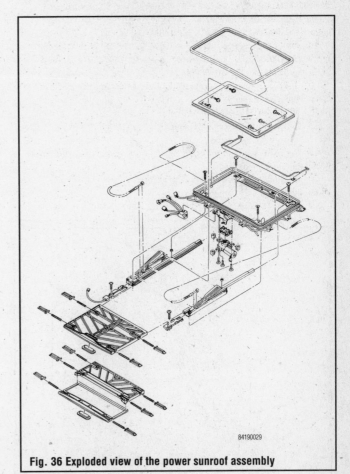

Fig. 36 Exploded view of the power sunroof assembly

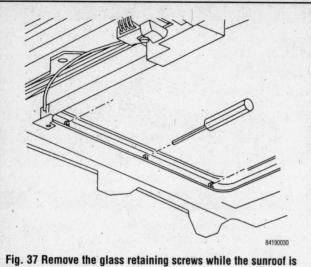

Fig. 37 Remove the glass retaining screws while the sunroof is in the vent position

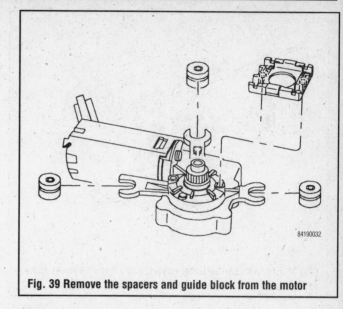

Fig. 39 Remove the spacers and guide block from the motor

Motor

▶ See Figures 38 and 39

1. Disconnect the negative battery cable.
2. Remove the headliner from the vehicle. Refer to the procedure later in this section.
3. Remove the Connector Position Assurance (CPA) device from the motor harness wire connector, then unplug the harness from the motor.
4. Remove the 3 retaining screws, then remove the sunroof motor, spacers and guide block.

To install:

5. Install the motor spacers onto the motor, then slide the guide assembly to the full rearward position in order to time the cables.
6. Position the guide block onto the motor, aligning it to the locator pins on the motor housing.
7. Install the sunroof motor using the 3 screws, then tighten the screws to 27–35 inch lbs. (3–4 Nm).
8. Install the wiring harness connector to the motor terminals, then install the CPA device.
9. Install the headliner to the vehicle.
10. Connect the negative battery cable.

Switch

▶ See Figures 40 and 41

1. Disconnect the negative battery cable.
2. Remove the map lamp lens and bezel assembly from the roof console assembly.
3. Disengage the sunroof opening windlace at the switch area.
4. Remove the sun visor inboard anchor clips.
5. Remove the 2 retaining screws, then lower the switch and roof console assembly.
6. Unplug the wiring harness connector from the switch, then depress the switch tabs at the console to remove the switch.

To install:

7. Push the switch into the roof console until the tabs engage, then install the wiring harness to the switch terminals.
8. Install the switch and console assembly using the retaining screws.
9. Install the sun visor inboard anchor clips and the sunroof opening windlace.
10. Install the map lens and bezel assembly to the roof console, then connect the negative battery cable.

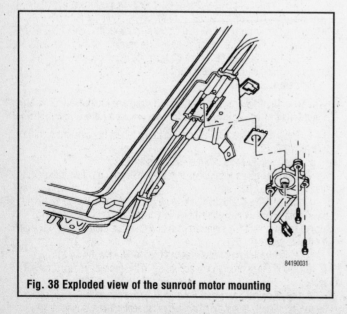

Fig. 38 Exploded view of the sunroof motor mounting

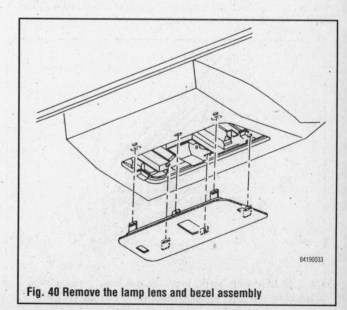

Fig. 40 Remove the lamp lens and bezel assembly

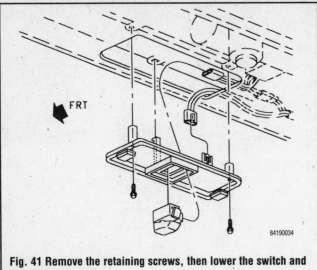

Fig. 41 Remove the retaining screws, then lower the switch and roof console assembly

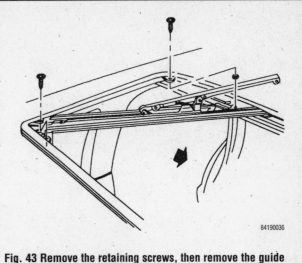

Fig. 43 Remove the retaining screws, then remove the guide assembly from the vehicle

Guide Assembly

▶ See Figures 42 and 43

1. Position the sunroof glass in the VENT position, then remove the glass according to the procedure earlier in this section.

2. If equipped, remove the wind deflector by pushing the spring ends down and forward, then slide the deflector rearward and out of the vehicle.

3. Remove the screw from each locator. Remove the locator from the right side, then remove the locator and microswitch from the left side. Each locator can be removed by inserting a screwdriver into the screw hole, tilting the screwdriver toward the center of the vehicle and then sliding it forward.

4. Remove the sunshade from the vehicle. Refer to the procedure later in this section.

5. Lift the drive cable end out of the guide assembly.

6. Remove the guide assembly screws, then remove the rear screw and spacer.

7. Slide the rear of the guide assembly towards the center of the vehicle and remove the guide assembly.

To install:

8. Position the front of the guide assembly at the front mount location with the guide's rear at the opposite corner of the sunroof opening. Slide the rear of the guide under the roof panel and toward the outboard side of the opening to the rear mounting position.

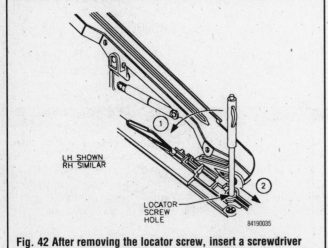

Fig. 42 After removing the locator screw, insert a screwdriver into the screw hole, tilt the screwdriver toward the vehicle's center and slide it forward to remove the locator

9. Install the guide rear spacer and guide screws, then tighten the screws to 35–44 inch lbs. (4–5 Nm).

10. Engage the drive cable into the guide assembly.

11. Install the sunshade to the vehicle.

12. Install each locator by positioning the locator into the guide assembly slightly ahead of its installed position. Insert a screwdriver into the locator screw hole, then push rearward and down on the locator to engage it with the guide.

13. Install each locator screw and tighten to 35 inch lbs. (4 Nm).

14. If equipped, install the wind deflector by positioning the deflector front tabs under the roof flange at the front and by engaging the spring ends into the brackets.

15. Install the sunroof glass. Refer to the procedure earlier in this section. Be sure to properly adjust the glass before tightening the retaining screws.

Cables

▶ See Figure 44

1. Position the sunroof glass in the VENT position, then remove the glass according to the procedure earlier in this section.

2. Remove the sunroof motor according to the procedure earlier in this section.

3. Remove the screw from each locator. Remove the locator from the right side, then remove the locator and microswitch from the left side. Each locator may be removed by inserting a screwdriver into the screw hole, tilting the screwdriver toward the center of the vehicle and then sliding it forward.

4. Lift the drive cable ends out of the guide assembly, then pull the cables out of the sunroof cable guide tubes.

To install:

➡Cables should always be replaced if they are kinked. Also, cables should always be replaced in pairs and greased before installation.

5. Push the greased cables into the guide tubes, but be careful not to kink the cables.

6. Engage the cables into the guide assemblies.

7. Slide the guide assemblies to the full rearward position in order to time the cables.

8. Install each locator by positioning the locator into the guide assembly slightly ahead of its installed position. Insert a screwdriver into the locator screw hole, then push rearward and down on the locator to engage it with the guide.

9. Install each locator screw and tighten to 35 inch lbs. (4 Nm).

10. Install the sunroof motor according to the procedure earlier in this section.

11. Install the sunroof glass. Refer to the procedure earlier in this section. Be sure to properly adjust the glass before tightening the retaining screws.

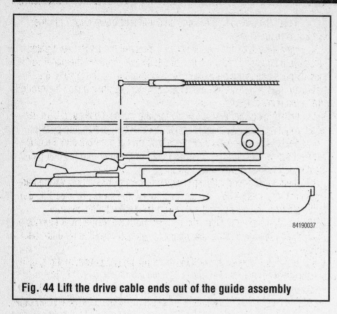

Fig. 44 Lift the drive cable ends out of the guide assembly

Sunshade

▶ See Figures 45 and 46

1. Position the sunroof glass in the VENT position, then remove the glass according to the procedure earlier in this section.
2. Push the sunshade slide blocks into the sunshade in order to release them from the sunroof guide assembly, as follows:
    a. For the Sedan, both slide blocks on one side of the vehicle should

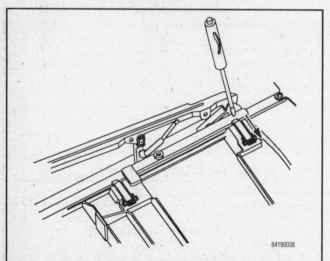

Fig. 45 Push the slide blocks into the sunshade in order to engage or disengage the shade from the guides

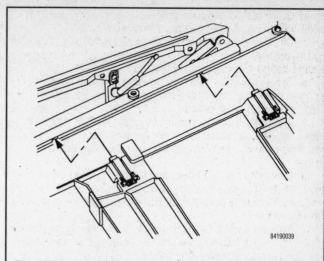

Fig. 46 Two slide blocks engage the guide assemblies on each side of the sunshade

be released first, then release both blocks on the other side and remove the shade through the sunroof opening.
    b. For the Coupe, release the rear slide blocks first and remove the rear section through the sunroof opening, then repeat the procedure at the front slide blocks and remove the shade from the vehicle.

To install:

3. For the Coupe:
    a. Install the front half of the sunshade first by inserting the slide blocks on the right side of the sunshade in the lower slide position of the right guide assembly. Then, push the slide blocks on the left side into the sunshade to allow the front half of the shade to drop into position. When positioned, engage the blocks into the lower channel of the left guide.
    b. Push the front half of the sunshade to the full forward position.
    c. Position the rear half of the shade so that the stop tabs are against the bumpers on the guide assembly. Engage the right slide blocks of the shade's upper half into the upper channel on the right guide assembly.
    d. Engage the slide blocks on the left side into the upper channel of the left guide assembly.
    e. Slide the sunshade back and forth to assure proper and smooth operation.
4. For the Sedan:
    a. Insert the slide blocks on the right side of the sunshade into the sunroof guide assembly.
    b. Push the sunshade slide blocks on the left side into the sunshade to allow the sunshade to drop into position. Once the sunshade is positioned, engage the slide blocks into the left guide assembly.
    c. Slide the sunshade back and forth to assure proper and smooth operation.
5. Install the sunroof glass. Refer to the procedure earlier in this section. Be sure to properly adjust the glass before tightening the retaining screws.

INTERIOR

Instrument Panel and Pad

REMOVAL & INSTALLATION

1991–94 Models

▶ See Figures 47, 48, 49, 50 and 51

1. Disconnect the negative battery cable.
2. Remove the left and right end cap/HVAC air outlet assemblies by

removing the attaching screws and pulling each assembly from the retaining clips.
3. For 1991 vehicles, remove the cigarette lighter trim bezel.
4. Carefully remove the center air outlet/trim panel by pulling outward at the clip locations. Begin at the bottom of the panel and work towards the upper clips, being careful not to score or damage the panel.
5. If equipped, remove the trim panel extension strip by pulling outward at the clip locations.
6. Remove the 2 dashboard upper trim panel screw caps (located near

the windshield at either end of the dashboard) by carefully prying with a small, flathead tool. Remove the upper trim panel screws.

7. Lift the upper trim panel to disengage the clips at the rear edge, then pull the panel rearward and out of the clips at the base of the windshield. Remove the upper trim panel from the vehicle.

8. Remove the screws from the instrument cluster trim panel, then pull the panel rearward sufficiently to disengage it from the retainers and to reach the electrical wiring behind it.

9. Remove the Connector Position Assurance (CPA) devices, then unplug the electrical connectors from the instrument panel light and the rear window defogger switches. Remove the cluster trim panel from the vehicle.

10. Remove the retaining screws, then pull the instrument cluster outward sufficiently to reach the wiring harness connectors. Unplug the wiring harness connectors by depressing the retainer legs, then remove the cluster from the vehicle.

11. Remove the wiring harness clips from the dashboard reinforcement.

12. Remove the Assembly Line Data Link (ALDL) screws, then remove the connector from the underside of the dashboard.

13. Remove the screws and lower the steering column filler panel from under the column. Be careful not to scratch the console with the mounting tabs when removing the panel.

14. Remove the hood release lever screw, located underneath and behind the lever.

15. Remove the radio and HVAC control panel. Refer to the procedures in Section 6 of this manual.

16. Unplug the cigarette lighter connector, then remove the lighter bulb holder by rotating counterclockwise and pulling it straight out.

17. Apply the parking brake, then remove the parking brake filler panel by carefully lifting at the rear edge. If the vehicle is equipped with a manual transaxle, remove the gear shift knob by pulling straight upward.

18. Remove the ashtray, then unclip the ashtray bulb holder. If equipped, remove the power window/mirror switch by sliding forward, then lifting at the rear edge. Remove the CPA devices and unplug the electrical connectors.

19. Remove the liner from the center console's rear storage compartment, then remove the console side and rear compartment screws. Lift the back of the console sufficiently to reach underneath and push out the seat belt bezels. Feed the seat belts though the cutouts and remove the console from the vehicle.

20. Remove the retaining screws, then carefully pull the Instrument Panel Junction Block (IPJB) away from the dashboard reinforcement. Disconnect the ground wire from the reinforcement.

21. Remove the screw and electrical connector from the rear of the IPJB. If equipped with an automatic transaxle, remove the CPA devices, then disconnect the 2-way instrument panel-to-body harness connector.

22. Remove the wiring harness and antenna hold-down clips from the dashboard reinforcement.

23. Loosen the floor shifter assembly. Remove the nuts and bolts, then lift the lower reinforcement bracket off of the studs and slide the bracket rearward.

24. If equipped with cruise control, remove the nuts from the module, then lower the module from the mounting bracket.

25. Remove the screws and nuts from the dashboard retainer and the reinforcement bracket.

26. Remove the bolts, then carefully lower the steering column assembly onto the driver's seat.

27. Carefully remove the instrument panel/dashboard assembly from the vehicle.

To install:

28. Install the instrument panel assembly. When positioning the panel, feed the fuse block and wiring harness through the lower reinforcement and into position. Install and tighten the screws and nuts on the dashboard retainer and reinforcement bracket.

29. Carefully raise the steering column assembly into position, then install the retaining bolts and tighten to 33 ft. lbs. (45 Nm).

30. Install the lower reinforcement bracket and nuts, then tighten the floor shifter assembly.

31. Install the wiring harness and antenna hold-down clips onto the dashboard reinforcement.

32. Install the electrical connector to the rear of the IPJB, then tighten the screw. If equipped with an automatic transaxle, connect the panel-to-body harness connector and install the lock pin. Install the IPJB to the dashboard reinforcement using the retaining screws, then attach the ground wire to the reinforcement.

33. Install the center console while feeding the seat belts and wire harness through the console cutouts. Snap the seat belt bezels into position.

34. Secure the console using the side screws at the front and the compartment screws at the top of the rear. Position the liner into the rear storage compartment and over the screws.

35. If equipped, connect the power window/mirror electrical wiring harnesses and install the CPA devices, then install the switch by inserting the front edge first and snapping the rear of the switch into the console.

36. Install the ashtray bulb holder, then install the ashtray to the console.

37. If equipped with a manual transaxle, push the shift knob onto the shifter.

38. Install the parking brake filler panel by pressing down at the clip locations.

39. Install the cigarette lighter bulb by inserting it into position and rotating counterclockwise to lock it in, then attach the lighter electrical connector.

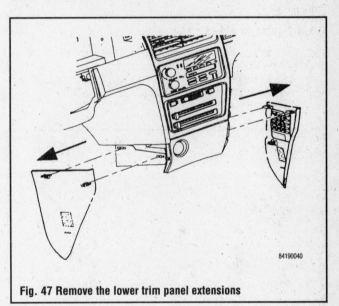

Fig. 47 Remove the lower trim panel extensions

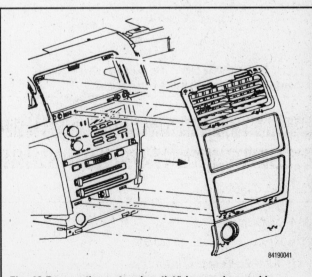

Fig. 48 Remove the center air outlet/trim panel assembly

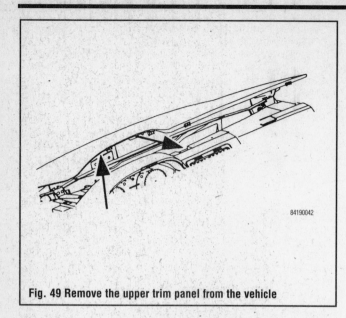

Fig. 49 Remove the upper trim panel from the vehicle

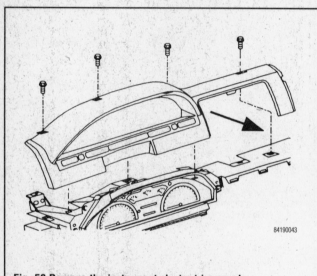

Fig. 50 Remove the instrument cluster trim panel

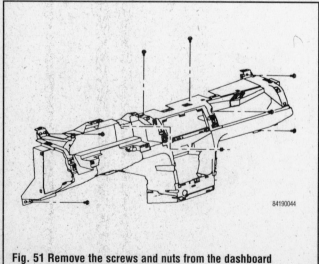

Fig. 51 Remove the screws and nuts from the dashboard retainer and reinforcement bracket

40. Install the HVAC control panel and adjust the cables accordingly. Refer to the appropriate procedures in Section 6 of this manual.

41. Install the radio assembly.

42. Connect the harness clips to the dashboard reinforcement. Connect the instrument cluster wiring harnesses, then position the cluster. Make sure the cluster lighting and rear window defogger harnesses are in position, then install and tighten the cluster retaining screws.

43. Install the hood release lever and screw.

44. If equipped with cruise control, position the module and secure using the retaining nuts.

45. Install the steering column filler panel and screws.

46. Install the ALDL connector and screws.

47. Position the instrument cluster trim panel and connect the wiring harness to the instrument light and rear window defogger switches. Install the CPA devices into the electrical connectors, then install the cluster trim panel into the retainers and secure using the screws.

48. Install the upper trim panel by inserting into the clips at the windshield, then snapping the rear of the panel into place. Install the upper trim panel screws, then snap the screw covers into position.

49. Install the center air outlet/trim panel, then install the lower left and right trim panel extensions. For 1991 vehicles, install the cigarette lighter trim bezel. If equipped, install the trim panel extension.

50. Install the left and right end cap assemblies, then connect the negative battery cable.

1995–98 Models

▶ See Figures 52, 53, 54, 55 and 56

1. Disconnect the negative battery cable. Disable the air bag system.

2. Remove the left and right end cap/HVAC air outlet assemblies by removing the attaching screws and pulling each assembly from the retaining clips.

3. Remove the center console assembly. Removal procedures for the center console can be found later in this section.

4. Remove the radio assembly. Removal procedures for the radio assembly can be found in Section 6.

5. Remove the HVAC control panel assembly. Removal procedures for the HVAC control panel assembly can be found in Section 6.

6. Remove the closeout seal by releasing it from the tabs.

7. Remove the screw, ground wire and wiring harness from the H-bracket.

8. Remove the retaining screw and unplug the electrical connector from the instrument panel junction block.

9. Remove the instrument panel junction block from its mounting pad by first removing the screw and then using a small, prying tool to release the locking tabs.

10. Pass the junction block through the H-bracket, towards the front of the vehicle.

11. Remove the H-bracket bolts.

12. Remove the instrument cluster assembly. Removal procedures for the instrument cluster assembly can be found in Section 6.

13. Remove the dimmer switch wiring harness from the instrument panel reinforcement and feed through, towards the front of the car.

14. Remove the passenger side air bag wiring harness from the cross-car beam and energy absorber.

15. Remove the glove box door stops by turning them 90 degrees, then allowing the door to hang down.

16. Unfasten the retaining screws and remove the glove box assembly by pulling rearward at the clip locations.

17. Disengage the radio antenna cable at the lower right side of the instrument panel reinforcement.

✳✳ CAUTION

When handling any air bag module, always hold the module with the outer trim surface facing up and away from your body. An air bag that accidentally deploys can cause severe physical injury if not handled properly.

18. Remove the mounting screws and passenger side air bag module.
19. Remove the instrument panel mounting fasteners.
20. Remove the mounting nuts, screws and instrument panel pad/reinforcement assembly. Place the panel pad/reinforcement on a clean surface to avoid any damage.
21. Remove the A/C outlets from the instrument panel by rotating them downward and pushing out through the back side.
22. Use a small, prying tool to disengage the locking tabs on both sides of the center outlet housing while pulling the unit rearward.

To install:

23. Install the center A/C outlet housing into instrument panel retainer until the tabs are flush with the sides of the unit. Snap the center outlets into the housing.
24. Install the instrument panel pad/reinforcement assembly into the vehicle. Make sure that all wiring harnesses, as well as the instrument panel junction block, are in position. Install and tighten the mounting nuts and screws to 89 inch lbs. (10 Nm).
25. Install the passenger side air bag module into the instrument panel assembly and feed the wiring harness into position. Install and tighten the mounting fasteners to 89 inch lbs. (10 Nm).
26. Connect the radio antenna cable at the lower right side of the instrument panel reinforcement.
27. Install the glove box and snap it in at the clip locations. Tighten the mounting screws to 20 inch lbs. (2.2 Nm). Install the glove box door stops and adjust the striker as necessary.
28. Attach the passenger side air bag wiring harness to the energy absorber and cross-car beam.
29. Install the dimmer switch wiring harness to the instrument panel reinforcement.
30. Install the instrument cluster assembly. Refer to Section 6 for installation procedures.
31. Install and tighten the H-bracket bolts to 19 ft. lbs. (25 Nm).
32. Route the instrument panel junction block through the H-bracket, towards the rear of the vehicle, and slide it onto the mounting pad to lock the tabs. Tighten the junction block screw to 20 inch lbs. (2.2 Nm).
33. Connect the wiring harness to the instrument panel junction block and tighten the retaining screw to 5 ft. lbs. (6.5 Nm).
34. Install the wiring harness, ground wire and screw to the H-bracket. Tighten the screw to 20 inch lbs. (2.2 Nm).
35. Install the closeout seal, making sure that the tabs engage the seal holes on both sides.
36. Install the HVAC control panel assembly. Refer to Section 6 for installation procedures.
37. Install the radio assembly. Refer to Section 6 for installation procedures.

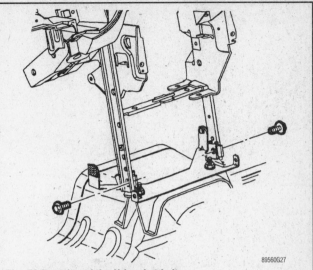
Fig. 53 Location of the H-bracket bolts

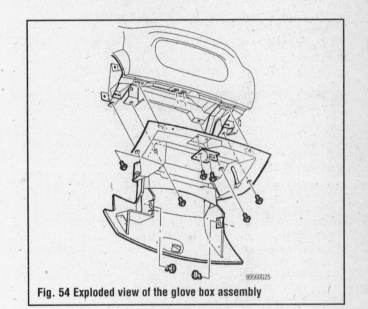
Fig. 54 Exploded view of the glove box assembly

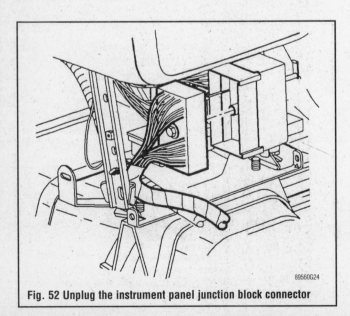
Fig. 52 Unplug the instrument panel junction block connector

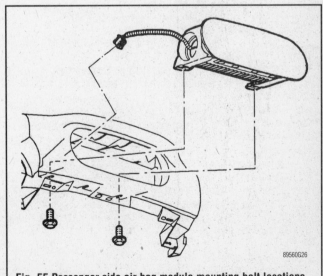
Fig. 55 Passenger side air bag module mounting bolt locations

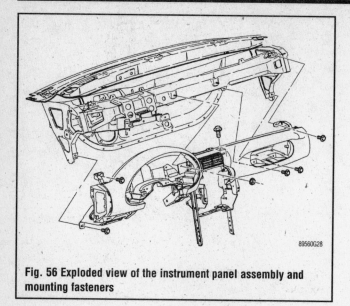

Fig. 56 Exploded view of the instrument panel assembly and mounting fasteners

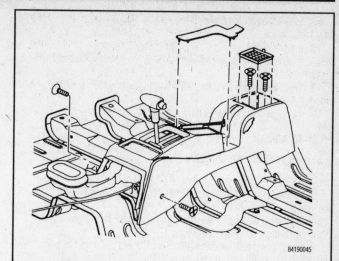

Fig. 57 Remove the rear and side screws, then remove the parking brake filler panel

38. Install the center console assembly. Installation procedures can be found later in this section.

39. Install the left and right end cap/HVAC air outlet assemblies. Tighten the mounting screws to 20 inch lbs. (2.2 Nm).

40. Enable the air bag system. Connect the negative battery cable.

Center Console

REMOVAL & INSTALLATION

1991–94 Models

▶ See Figures 57, 58 and 59

1. Disconnect the negative battery cable.
2. If equipped with an automatic transaxle, tape the shift button in on the shifter handle.
3. If the vehicle is equipped with a manual transaxle, remove the gear shift knob by pulling straight upward.
4. Remove the liner from the center console's rear storage compartment, then remove both rear compartment screws.
5. Remove the 2 screws from the front sides of the console.
6. Apply the parking brake, then remove the parking brake filler panel by carefully lifting at the rear edge.
7. Remove the ashtray, then unclip the ashtray bulb holder.
8. If equipped, remove the power window/mirror switch by sliding it forward, then lifting at the rear edge. Remove the CPA devices and unplug the electrical connectors.
9. Remove the left and right lower trim panel extensions by disconnecting the Velcro™ at the bottom of the panels and pulling them out of the upper retaining clips.
10. Lift the back of the console sufficiently to reach underneath and push out the seat belt bezels. Feed the seat belts though the cutouts and remove the console from the vehicle.

To install:

11. Install the console while feeding the seat belts and wire harness through the console cut outs. Snap the seat belt bezels into position.
12. If equipped, connect the power window/mirror electrical wiring harnesses and install the CPA devices, then install the switch by inserting the front edge first and snapping the rear of the switch into the console.
13. Install the ashtray bulb holder, then install the ashtray to the console.
14. Secure the console using the side screws at the front, and the compartment screws at the top rear of the console. Position the liner into the rear storage compartment and over the screws.

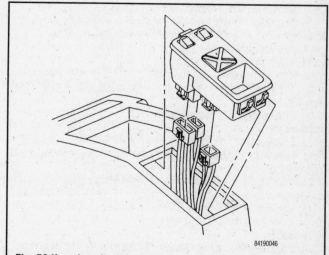

Fig. 58 If equipped, remove the power window/mirror switch assembly

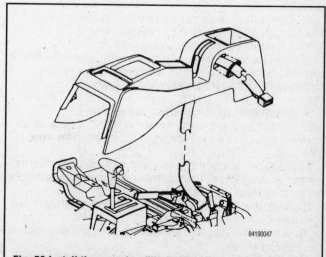

Fig. 59 Install the console while feeding the seat belts through the console cutouts

15. Install the parking brake filler panel by pressing down at the clip locations.

16. Install the left and right lower trim panel extension.

17. If equipped with a manual transaxle, push the shift knob onto the shifter.

18. If equipped with an automatic transaxle, remove the tape from the shifter button.

19. Connect the negative battery cable.

1995–98 Models

▶ See Figures 60, 61 and 62

1. Disconnect the negative battery cable. Disable the air bag system.

2. If equipped, remove the retaining fasteners and center armrest from the vehicle.

3. Apply the parking brake, then remove the mounting screw and lift the cover straight off of the brake lever.

4. If equipped with an automatic transaxle, place the shifter in the Neutral position. Lift the ashtray straight out of the cup holder.

5. Starting at the rear edge, pull upwards to remove the cup holder from the console.

6. Remove the ashtray bulb socket by lifting the tab and sliding it forward. Then, pull the socket straight out.

7. Remove the wiring harness from the cup holder.

8. If equipped, remove the power window/mirror switch by sliding forward, then lifting at the rear edge. Unplug the electrical connectors.

9. If not equipped with a center armrest, lift out the rear screw cover and remove both rear console screws.

10. Remove the left and right lower trim panel extensions by disconnecting the Velcro™ at the bottom of the panels and pulling them out of the retaining clips.

11. Remove the front console mounting screws and move the console unit rearward.

12. Disengage the cigarette lighter electrical connector.

13. Remove the cigarette lighter bulb socket by rotating it clockwise and pulling straight out.

14. If the vehicle is equipped with an automatic transaxle, tape the gear shift button in to clear the shifter opening.

15. Lift the back of the console. Move the console back and lift straight up to remove the console from the vehicle.

To install:

16. Place the shifter in the Neutral position. Lower the front of the console unit over the shifter.

17. Install the wire harness connectors through the console cutouts into the correct position.

18. Lower the rear console over the parking brake lever and onto the rear mounting pad.

19. Move the console rearward and install the cigarette lighter bulb socket through the opening, rotating it counterclockwise.

20. Connect the cigarette lighter electrical harness.

21. Move the console forward into position. Engage the front console upper tabs to the sheet metal on the lower instrument panel brace. Install and tighten the front console mounting screws to 11 inch lbs. (1.2 Nm).

22. Install the left and right lower trim panel extensions.

23. If not equipped with a center armrest, install both the rear console screws. Tighten the screws to 14 inch lbs. (1.6 Nm). Install the screw cover.

24. If equipped, connect the power window/mirror electrical wiring harnesses, then install the switch by inserting the front edge first and snapping the rear of the switch into the console.

25. Insert the bulb socket into the cup holder, pushing down to engage the locking tab. Install the wiring harness to the cup holder.

26. Install the cup holder into the console by pushing straight down, then install the ashtray.

27. Install the parking brake cover over the lever. Tighten the screw to 14 inch lbs. (1.6 Nm). Be very careful not to overtighten the screw or it will pull through the cover.

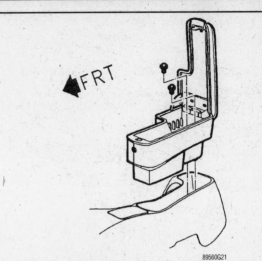

89560G21

Fig. 60 Remove the mounting screws and center armrest, if equipped

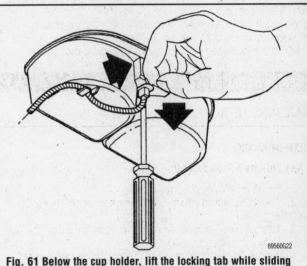

89560G22

Fig. 61 Below the cup holder, lift the locking tab while sliding the bulb socket forward, then pull straight out

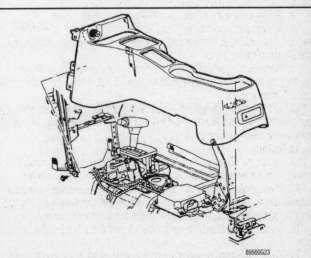

89560G23

Fig. 62 Install the console while feeding the wiring harnesses through the console cutouts

28. If equipped, install the center arm rest. Tighten the mounting screws to 14 inch lbs. (1.6 Nm).

29. If equipped with an automatic transaxle, remove the tape from the shifter button.

30. Enable the air bag system. Connect the negative battery cable.

Headliner

REMOVAL & INSTALLATION

Coupe and Sedan

▶ See Figures 63, 64, 65 and 66

1. Disconnect the negative battery cable.

2. Remove the bezel from the map lamp or lamp and sunroof switch console by prying at the notch on the outside edge.

3. Disconnect the dome lamp assembly form the headliner, then unplug the wiring harness and remove the lamp assembly from the vehicle.

4. Remove the screws, then remove the sun visor retainer/supports and the sun visors from the headliner.

5. Remove the screw covers from the assist straps, then remove the screws and the straps from the headliner. Be careful. If the front assist strap has different color screws, the black screw will have LEFT-HANDED threads.

6. Remove the coat hooks, then, for the Coupe, remove the rear upper garnish molding by pulling at the clip locations.

7. Remove the 2 dashboard upper trim panel screw caps (located near the windshield at either end of the dash or in the center of the dash toward the front) by carefully prying with a small flathead tool. Remove the upper trim panel screws.

8. Lift the upper trim panel to disengage the clips at the rear edge, then pull the panel rearward and out of the clips at the base of the windshield. Remove the upper trim panel from the vehicle.

9. If equipped with shoulder belt height adjusters:
    a. Unsnap the shoulder belt fastener cover and remove the shoulder belt mounting fastener.
    b. Using an awl, unsnap the shoulder belt adjuster knob locking clip.

10. On 1996–98 Sedan and 1997–98 Coupe models, remove the high mounted stop lamp cover by pulling down to disengage the mounting clips.

11. On 1991–95 Sedan and 1991–96 Coupe models, remove the 2 screws and instrument panel end cap/HVAC outlet assembly from the left side of the dashboard.

12. Check for any garnish molding mounting fasteners and remove. Remove the windshield garnish molding by firmly pulling at the attaching clip locations.

13. For the Sedan, pull the left rear seat bolster at the top to disengage the fastener, then remove the bolster from the vehicle. Remove the lower rear screw from the left rear garnish molding, then remove the molding by pulling at the clip locations.

14. If equipped, disengage the sunroof finish lace, then remove the lace from the headliner.

❋❋ WARNING

Be very careful when handling the headliner. Should the liner become bent it will break and leave a wrinkle which cannot be repaired.

15. Disengage the Velcro™ attachments at the rear of the headliner, then fully recline the front seats and carefully remove the headliner through the right or left front door of the vehicle.

To install:

16. Carefully install the headliner through the right door, then position the headliner to the dome lamp opening, sun visor attachment points and, if equipped, to the sunroof opening. Engage the Velcro™ attachments at the rear of the headliner.

17. Install the sun visors.

18. For the Sedan, install the left rear garnish molding by pushing at the clip locations, then install the rear retaining screw. Position the rear seat side bolster onto the outboard pivot pin, then secure the bolster by aligning and engaging the guide pin and upper fastener.

19. If equipped, install the sunroof lace.

20. Install the front upper garnish molding.

21. For the Coupe, install the rear upper garnish molding.

22. Install the coat hooks.

23. Install the assist strap and strap covers. If equipped, the black left-handed thread screw should always be installed in the rearward attachment of the assist strap.

24. Install the instrument panel end cap and screws.

25. Install the upper trim panel by inserting into the clips at the windshield, then snapping the rear of the panel into place. Install the upper trim panel screws, then snap the screw covers into position.

26. If equipped with shoulder belt height adjusters:
    a. Install the shoulder belt, being careful not to twist it. Tighten the shoulder belt mounting fastener to 37 ft. lbs. (50 Nm).
    b. Snap into place, the shoulder belt mounting fastener cover and install the adjuster knob.

27. On 1996–98 Sedan and 1997–98 Coupe models, install the high

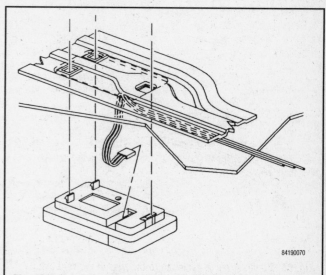

84190070

Fig. 63 Remove the dome lamp assembly from the headliner

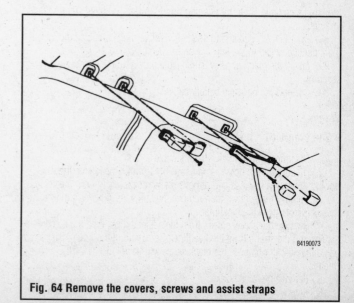

84190073

Fig. 64 Remove the covers, screws and assist straps

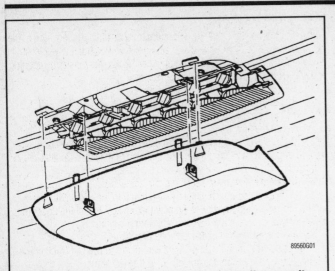

Fig. 65 Pull down on the high stop lamp cover to disengage its mounting clips

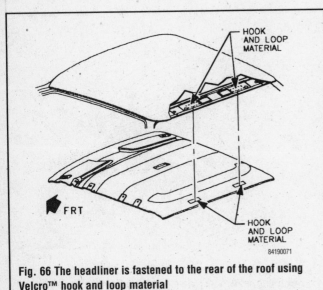

Fig. 66 The headliner is fastened to the rear of the roof using Velcro™ hook and loop material

mounted stop lamp cover by pushing firmly upward to engage the mounting clips.

28. On 1991–95 Sedan and 1991–96 Coupe models, install the instrument panel end cap to the left side of the dashboard.

29. Install the map lamp/switch console bezel and the dome lamp assembly.

30. Connect the negative battery cable.

Wagon

▶ See Figure 67

1. Disconnect the negative battery cable.

2. Disconnect the dome lamp assembly from the headliner, then unplug the wiring harness and remove the lamp assembly from the vehicle.

3. Remove the screws, then remove the sun visor retainer/supports and the sun visors from the headliner.

4. Remove the screw covers from the assist straps, then remove the screws and the straps from the headliner. Be careful. If the front assist strap has different color screws, the black screw will have LEFT-HANDED threads.

5. Remove the coat hooks.

6. Remove the 2 dashboard upper trim panel screw caps (located near

the windshield at either end of the dash or in the center of the dash toward the front) by carefully prying with a small flathead tool. Remove the upper trim panel screws.

7. Lift the upper trim panel to disengage the clips at the rear edge, then pull the panel rearward and out of the clips at the base of the windshield. Remove the upper trim panel from the vehicle.

8. Remove the 2 screws and instrument panel end cap/HVAC outlet assembly from the left side of the dashboard.

9. If equipped with shoulder belt height adjusters:

a. Unsnap the shoulder belt fastener cover and remove the shoulder belt mounting fastener.

b. Using an awl, unsnap the shoulder belt adjuster knob locking clip.

10. On 1991–95 Sedan and 1991–96 Coupe models, remove the 2 screws and instrument panel end cap/HVAC outlet assembly from the left side of the dashboard.

11. Check for any garnish molding mounting fasteners and remove. Remove the windshield garnish molding by firmly pulling at the attaching clip locations.

12. Pull the rear seat bolster at the top to disengage the fastener, then remove the bolster from the vehicle.

13. If necessary, remove the left rear shock tower cover.

14. Remove the rear headliner trim panel.

15. Remove the body rear corner trim panel assembly.

16. Remove the left latch pillar upper trim molding assembly.

❋❋ WARNING

Be very careful when handling the headliner. Should the liner become bent, it will break and leave a wrinkle which cannot be repaired.

17. Disengage the Velcro™ attachments at the rear of the headliner and from the right garnish moldings, then fully recline the front seats and carefully remove the headliner through the rear of the vehicle.

18. Disengage the attachments at the rear of the headliner, fully recline the front seats and carefully remove the headliner through the right or left front door of the vehicle.

To install:

19. With the aid of an assistant and with the seats reclined, carefully install the headliner through the rear of the vehicle, then position the headliner to the dome lamp opening and sun visor attachment points. Engage the attachments at the rear of the headliner.

20. Install all of the interior garnish moldings/trim panels that were removed for this procedure. Be sure to apply firm pressure at the mounting points in order to positively engage the attachment clips.

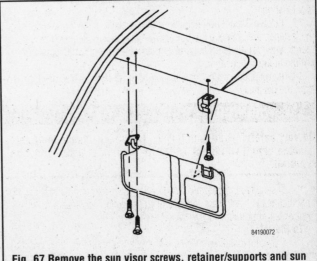

Fig. 67 Remove the sun visor screws, retainer/supports and sun visors

21. Install the rear header trim.
22. Install the left rear shock tower cover.
23. Install the left seat side bolster.
24. Install the left windshield garnish molding.
25. Install the instrument panel end cap and screws.
26. Install the upper trim panel by inserting into the clips at the windshield, then snapping the rear of the panel into place. Install the upper trim panel screws, then snap the screw covers into position.
27. If equipped with shoulder belt height adjusters:
   a. Install the shoulder belt, being careful not to twist it. Tighten the shoulder belt mounting fastener to 37 ft. lbs. (50 Nm).
   b. Snap into place, the shoulder belt mounting fastener cover and install the adjuster knob.
28. Install the assist strap and strap covers. If equipped, the black left-handed thread screw should always be installed in the rearward attachment of the assist strap.
29. Install the coat hooks. Install the sun visor retainer/supports and the sun visors.
30. Install the dome lamp assembly.
31. Connect the negative battery cable.

Door Panels

REMOVAL & INSTALLATION

▶ See Figures 16, 17 and 68 thru 75

1. If equipped with power windows, disconnect the negative battery cable.
2. For front door panels, remove the screws and the mirror trim panel. If equipped with a manual remote mirror, allow the panel to hang from the adjustment cable.
3. For rear door panels, remove the inner belt seal from the trim panel. If necessary, use a putty knife which has its blade covered with tape to gently pry the sealing strip from the flange.
4. If equipped, remove the power door lock switch by prying out at the top of the switch with a small, straight screwdriver. Make sure the screwdriver tip is covered with tape to protect the surfaces. Unplug the wiring harness and remove the switch.
5. Remove the screws, then slide the door handle assembly forward and pull the assembly outward. Use a long, thin screwdriver to disengage the lock and latch rods from the retainers, then remove the handle assembly from the vehicle.
6. If equipped with manual windows, disengage the regulator handle clips using a standard handle tool or by working a cloth back and forth between the handle and the bearing plate. Once disengaged, remove the handle and plate from the door.
7. Remove the screws and the door pull cup.
8. For front door panels, remove the inner belt sealing strip.
9. For the Coupe, remove the screw from the rear of the door trim panel.
10. Disengage the door trim panel retainers from the door assembly using a trim panel tool, then remove the panel from the vehicle.
To install:
11. Position the trim panel to the door assembly, then engage the retainers by pressing firmly at the panel retainer locations.
12. Install the door pull cup and screws, then tighten the screws to 22 inch lbs. (2.2 Nm).
13. If equipped, install the manual window regulator handle clip on the handle, then install the bearing plate and handle. The handle should be pointing in a forward and slightly upward position with the window closed.
14. Position the door handle assembly to the trim panel, then tip the handle outward at the top. Using a thin hook shaped tool, lift the latch and lock rods into the retainers and engage the clamps. Push the handle into the trim panel and slide it rearward to engage the tabs. Install and tighten the retaining screws.

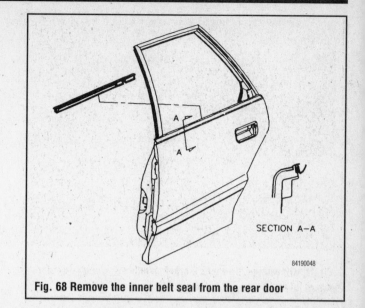

Fig. 68 Remove the inner belt seal from the rear door

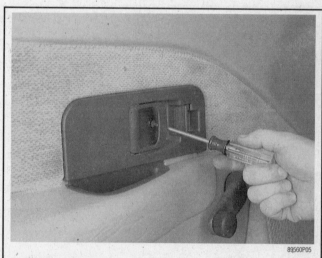

Fig. 69 Slide the door handle assembly forward after removing the mounting screw

Fig. 70 Remove the door handle assembly after disengaging the lock and latch rods from their retainers

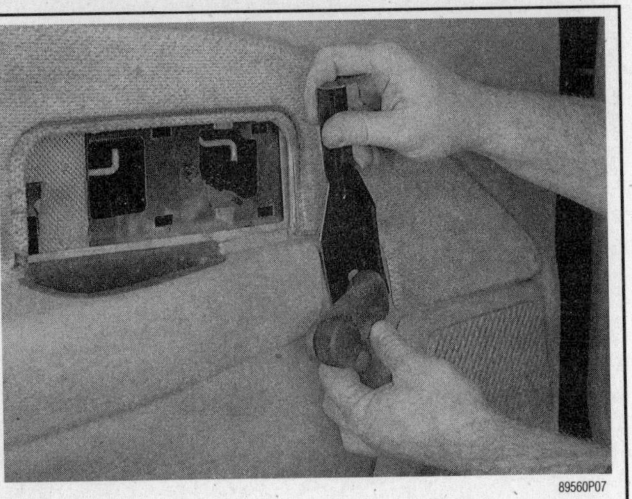

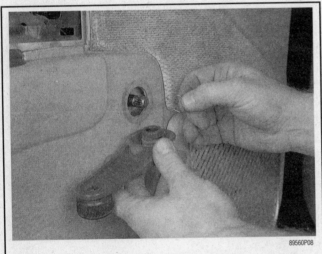

Fig. 71 Disengage the window crank handle retaining clip using a special tool

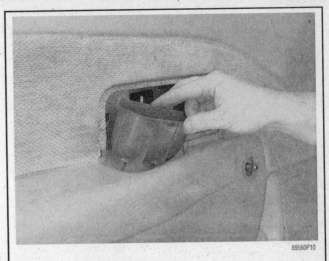

Fig. 74 . . . lift the door pull cup up and out of the inner door panel

Fig. 72 Once the retaining clip has been disengaged, slide the handle off of the window crank

Fig. 75 Disengage the trim panel retainers from the door assembly, then remove the panel

15. If equipped, connect the wiring harness to the power door lock switch and install the switch into the handle assembly.

16. Install the inner belt sealing strip.

17. For front doors, install the mirror trim panel and tighten the retaining screw(s).

18. If applicable, connect the negative battery cable.

Liftgate Trim Panel and Molding

REMOVAL & INSTALLATION

Inner Lower Trim Panel

▶ **See Figures 76 and 77**

1. Raise the liftgate.
2. Remove the wedge blocks from either end of the liftgate assembly.
3. Remove the lower fasteners from the liftgate lower trim panel, by pushing in the center of each pin approximately ⅛ inch until it clicks, then remove the fastener.
4. Insert a screwdriver or small prybar into the hole in the lower trim panel (located near the wiper pivot on the pivot hump) so that the tool sits on

Fig. 73 Loosen the door pull cup mounting screw and . . .

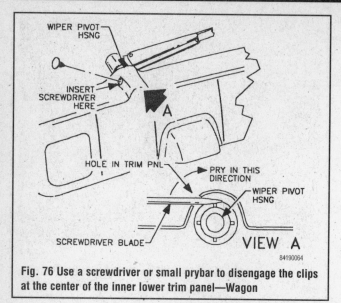

Fig. 76 Use a screwdriver or small prybar to disengage the clips at the center of the inner lower trim panel—Wagon

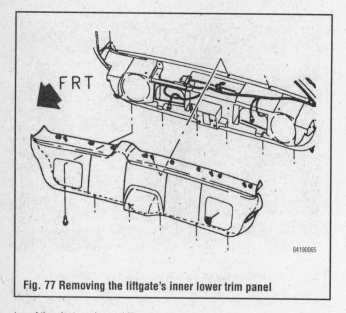

Fig. 77 Removing the liftgate's inner lower trim panel

top of the pivot as shown. Lift up on the tool handle to disengage the trim panel upper clips, then remove the inner lower trim panel from the vehicle.

To install:

5. Align the upper clips on the lower trim panel to the liftgate slots, then install the panel by pushing at the clip locations.

6. Reset the trim panel push-in fasteners by spreading the center pin tabs and moving the pin so that it sits approximately ¼ in. out of the fastener. Insert the fasteners into the bottom of the trim panel and push the center pin until flush.

7. Install the liftgate wedge blocks.

8. Close the liftgate.

Upper Garnish Window Molding

♦ **See Figure 78**

1. Raise the liftgate, then remove the access panel from the top of the liftgate molding.

2. Unplug the liftgate harness from the body wiring harness.

3. Remove the liftgate inner lower trim panel assembly. Refer to the procedure earlier in this section.

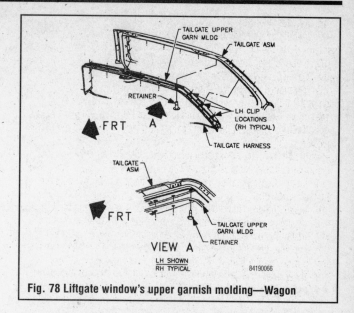

Fig. 78 Liftgate window's upper garnish molding—Wagon

4. Remove the fasteners from the top and lower end of the liftgate window upper garnish molding. Push the center pin of each fastener approximately ⅛ inch until it clicks, then remove the fastener.

5. Grasp the garnish molding at the lower ends and pull downward to disengage the clips on the molding upper sides. Disengage the wiring harness attachments from the trim panel, then remove the molding from the vehicle.

To install:

6. Attach the liftgate harness to the upper molding, then align the guide pins and clips on the molding with the holes in the liftgate. Push on the molding at the clip locations to secure the molding to the liftgate.

7. Install the liftgate inner lower trim panel assembly.

8. Connect the liftgate harness to the body wiring harness, then install the access panel.

9. Close the liftgate.

Door Locks

REMOVAL & INSTALLATION

♦ **See Figures 79 thru 93**

To remove the lock cylinder or the power lock actuator from the door assembly, the outer door panel must first be removed, as follows:

1. If working on a front door, remove the side mirror from the door assembly. Refer to the procedure earlier in this section.

2. Disengage the outer seal strip by pulling upward at the rear. If necessary, use a putty knife which has its blade covered with tape to gently pry the sealing strip from the flange.

3. Push the center portion of the handle pushpins through the pins, then pull the pins from the handle. Slide the handle off the pivot assembly and remove the handle from the vehicle.

4. Remove the door panel screws. Remove the door panel by first pulling the front, then the rear outward and finally the entire panel.

5. Inspect the foam tape fillers at the front edge of the door panel and replace, if damaged.

6. Once the outer door panel has been removed, partially remove the door insulator, if any. Disconnect the lock rod from the cylinder.

7. If equipped with power door locks, unplug the harness and the harness retainer.

8. Remove retainer, then remove the lock cylinder from the door assembly.

To install:

9. Install the lock cylinder and secure using the retainer. Place the door insulator back to proper position.

10. If applicable, connect the harness and harness retainer.

11. Connect the lock rod to the cylinder.

12. If removed or replacing, install the foam tape panels at the front edge of the outer door panels, as illustrated. Replacement fillers may be cut to the proper lengths using a suitable foam of the same thickness.

13. If removed, install the foam seal strip to the perimeter of the door structure.

14. Install the outer door panel alignment pins into net hole and slot in the door structure outer reinforcement.

15. Install the door panel screws, then tighten in the proper sequence to 53 inch lbs. (6 Nm).

16. Install the door handle by lifting the pivot assembly and sliding the handle on the pivot. Insert the pushpins to secure the handle assembly.

17. Install the outer sealing strip by aligning the location tab at the rear of the strip with the slot in the door panel. Fully seat the strip by pushing the strip down on the flange, working from the rear forward.

18. If applicable, install the side mirror to the door assembly. Refer to the procedure earlier in this section.

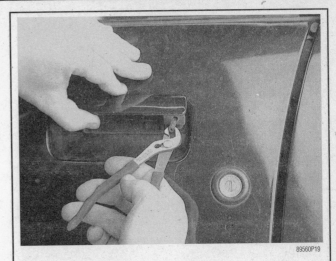

Fig. 81 Using small pliers, pull out the plastic door handle pushpins

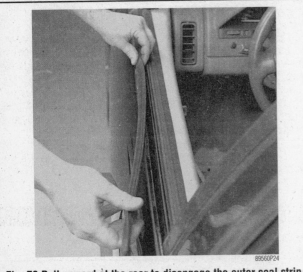

Fig. 79 Pull upward at the rear to disengage the outer seal strip

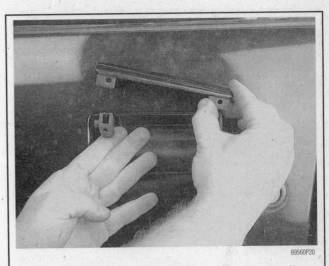

Fig. 82 With the pushpins removed, separate the door handle from the lever

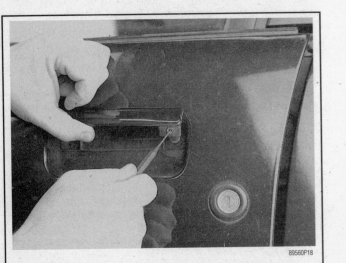

Fig. 80 Using a small punch, push in the center portion of the door handle pushpin

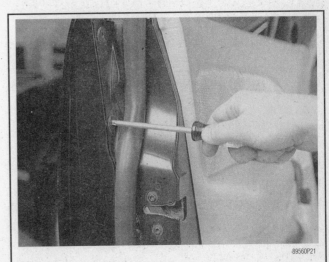

Fig. 83 Remove the outer door panel retaining screws using a Torx® head screwdriver

Fig. 84 Separate the outer panel from the door assembly

Fig. 85 Full view of an inner door assembly showing door lock location (1) and window regulator assembly (2)

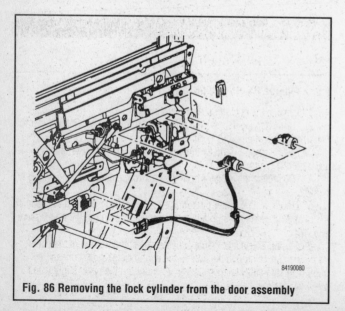

Fig. 86 Removing the lock cylinder from the door assembly

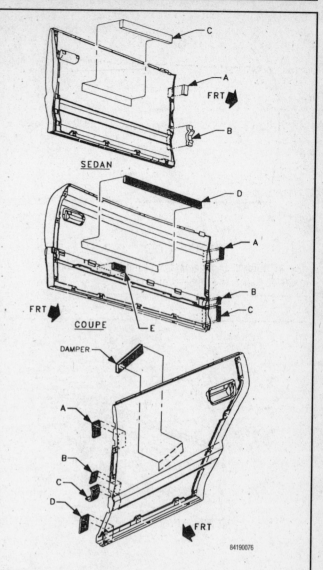

Fig. 87 Position foam tape fillers of proper length in these positions

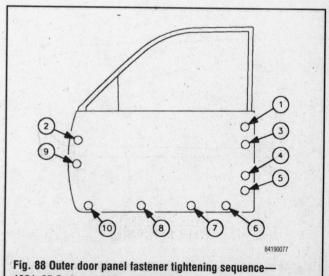

Fig. 88 Outer door panel fastener tightening sequence—1991-95 Sedan and Wagon front door

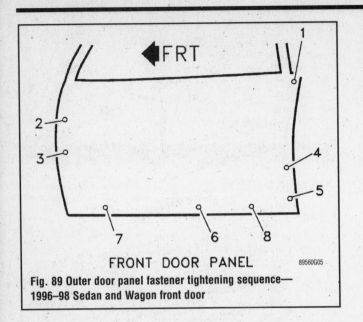

Fig. 89 Outer door panel fastener tightening sequence—
1996–98 Sedan and Wagon front door

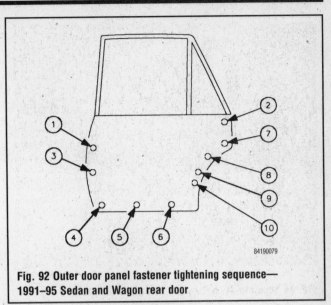

Fig. 92 Outer door panel fastener tightening sequence—
1991–95 Sedan and Wagon rear door

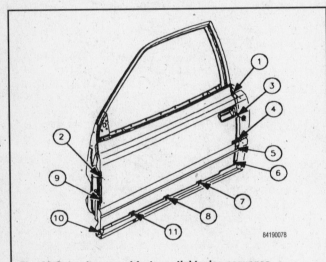

Fig. 90 Outer door panel fastener tightening sequence—
1991–96 Coupe

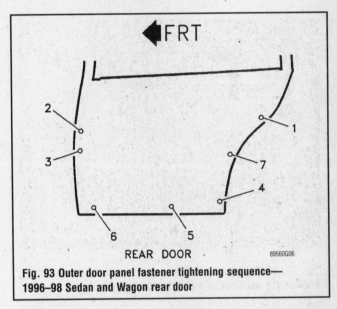

Fig. 93 Outer door panel fastener tightening sequence—
1996–98 Sedan and Wagon rear door

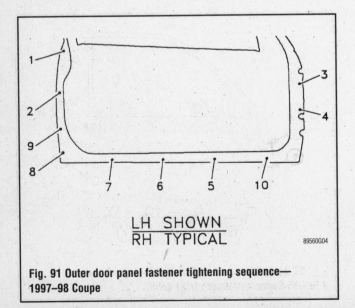

Fig. 91 Outer door panel fastener tightening sequence—
1997–98 Coupe

Power Door Lock Actuator

REMOVAL & INSTALLATION

▶ See Figures 94, 95 and 96

1. Remove the outer door panel assembly.
2. For the Coupe, remove the door inner trim panel assembly. Refer to the procedure earlier in this section.
3. For the Coupe, partially remove the rear ½ of the water shield for access to the rivets.
4. Disconnect the actuator from the lock rod, then unplug the wiring harness.
5. Drive the rivet center pin rearward sufficiently to drill out the head of the rivet. Drill out the rivet(s) and remove the actuator from the door assembly.
6. For the Sedan and Wagon, cut the remaining portion of the rivets short, then push through the hole in the door assembly. Disconnect the inner door trim panel lower fasteners, then pry out the panel slightly and allow the rivets to fall.

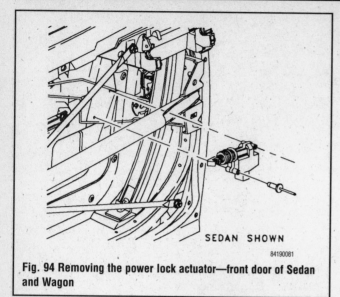

Fig. 94 Removing the power lock actuator—front door of Sedan and Wagon

Fig. 95 Removing the power lock actuator—Coupe

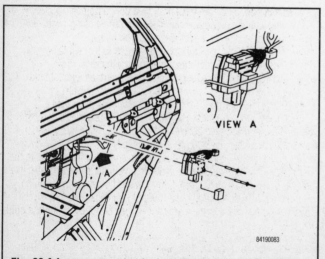

Fig. 96 A foam spacer must be installed on the rear door of the Sedan and Wagon when replacing the power lock actuator

To install:

7. For the Sedan and Wagon rear door, install the foam spacer. Failure to install a spacer will cause rattles.

8. Install the actuator using rivets, then connect the harness and actuator rod. For the Coupe, install the rear ½ of the water shield.

9. Reposition and fasten or install the door assembly trim panel, as applicable.

10. Install the outer door panel assembly.

Trunk Lock

REMOVAL & INSTALLATION

▶ See Figures 97, 98 and 99

1. For the Sedan, disconnect the negative battery cable, then remove the trunk lid center panel assembly or reflector.

2. For the Coupe, remove the reflex panels from both sides of the trunk lid.

3. Disconnect the lock rod from the cylinder.

4. For the Sedan, remove the license plate, then remove the inner and outer retainers securing the lower outer panel. Pull the panel outward, disconnect the bulb sockets from the lamps and remove the panel from the vehicle.

5. For the Coupe, remove the screws, plastic retainers and the lower panel from the trunk lid.

6. Remove the sealing sleeve, lock cylinder retainer, reinforcement and the lock cylinder from the lower panel.

To install:

7. Install the lock cylinder to the panel, then install the reinforcement and the retainer to the cylinder. Install the sealing sleeve.

8. Position the panel to the vehicle. For the Sedan, connect the bulb sockets.

9. Install the lower panel and tighten the retainers. For the Coupe, be sure to follow the proper sequence as illustrated, then install the plastic retainers.

10. Connect the lock rod to the cylinder using the retaining clip.

11. For the Sedan, install the center panel or reflector using the retaining nuts, then connect the negative battery cable.

12. For the Coupe, install the reflex panels. Make sure the fastener sealing feature is intact. If necessary, apply a body caulking compound to seal against water intrusion.

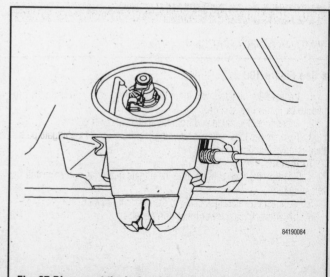

Fig. 97 Disconnect the lock rod from the lock cylinder

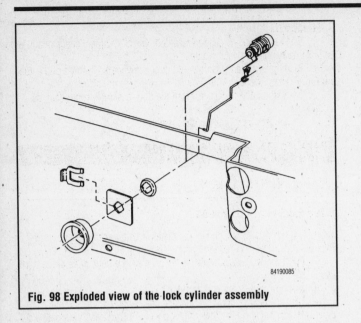

Fig. 98 Exploded view of the lock cylinder assembly

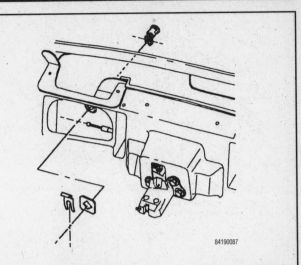

Fig. 100 Disconnect the cable end from the liftgate lock cylinder housing

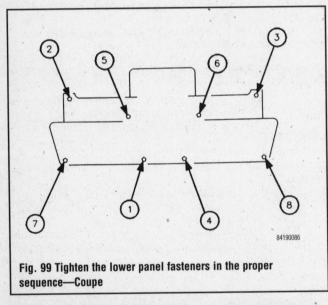

Fig. 99 Tighten the lower panel fasteners in the proper sequence—Coupe

Liftgate Lock

REMOVAL & INSTALLATION

▶ **See Figure 100**

1. Remove the liftgate lower inner trim panel assembly. Refer to the procedure earlier in this section.
2. Disconnect the cable end from the lock cylinder housing.
3. Remove the lock cylinder retaining clip from inside the liftgate panel, then remove the lock cylinder.

To install:

4. Position the lock cylinder into the liftgate outer panel, then install the retaining clip.
5. Connect the latch lock cable to the cylinder housing.
6. Install the liftgate lower inner trim panel.

Door Glass and Regulator

REMOVAL & INSTALLATION

▶ **See Figures 85, 101, 102, 103 and 104**

1. Remove the outer door panel assembly. Refer to the sub-procedure under Door Locks, earlier in this section.
2. For rear doors, loosen the bottom fastener of the rear door stationary window assembly, in order to provide clearance, then apply tape to the lower half of the outboard exposed portion of the stationary window channel. The tape must be applied in order to prevent damage to the paint surface.
3. Lower the door glass, then use SA9148B, or an equivalent tool, to remove the glass nuts. Push the glass upward and rotate slightly, then remove the glass from the vehicle.
4. If the regulator is also to be removed, proceed as follows:
   a. Remove the inner door trim panel. Refer to the procedure earlier in this section.
   b. If applicable, unplug the wiring harness from the power regulator.
   c. Except for the rear door, remove the regulator cam bolts.
   d. Drill out the regulator rivets and remove the regulator from the vehicle.

To install:

5. If removed, install the regulator to the door assembly:
   a. Position the regulator to the door and secure using rivets.
   b. Except for the rear door, install the cam bolts.
   c. If applicable, connect the wiring harness to the regulator.
   d. Install the inner door trim panel to the vehicle.
6. Install the door glass into the door and over the regulator studs.
7. Install the glass nuts using tool SA9148B, or equivalent.
8. Check the door glass for proper operation and alignment.
9. If necessary, adjust the glass alignment. For rear doors, fore/aft adjustments are made after loosening the "B" bolts, while in/out adjustments are made when the "A" bolts are loosened.
10. Except for the rear doors, glass which is not running parallel to the door header may be adjusted after loosening the inner panel cam attachment bolts and adjusting the cam. If the glass is floating in the opening, lower the glass ¾ downward, loosen the front guide fasteners, push the glass rearward to seat it in the rear channel, then bring the front glass guide

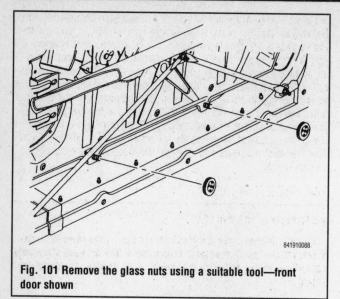

Fig. 101 Remove the glass nuts using a suitable tool—front door shown

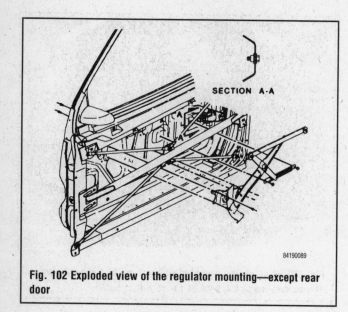

Fig. 102 Exploded view of the regulator mounting—except rear door

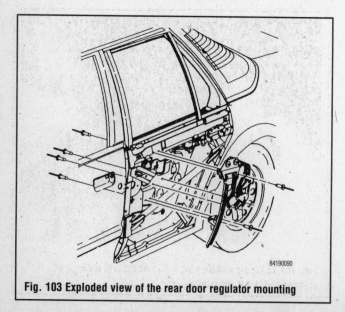

Fig. 103 Exploded view of the rear door regulator mounting

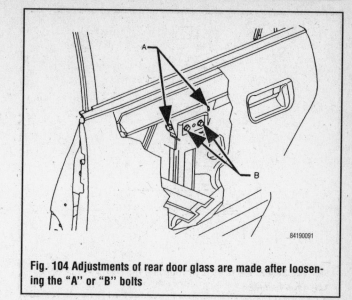

Fig. 104 Adjustments of rear door glass are made after loosening the "A" or "B" bolts

back to the front edge of the glass. Apply a light finger pressure on the guide and tighten the fasteners.

11. Once the glass has been properly adjusted and is operating smoothly, install the outer door panel assembly.

Electric Window Motor

REMOVAL & INSTALLATION

▶ **See Figures 105 and 106**

1. Remove the regulator from the door assembly. Refer to the procedure earlier in this section.

2. Except for the rear doors, drill a hole through the regulator sector gear and backplate. Install the bolt and nut to lock the sector gear in position. Be careful not to drill a hole closer than 7/16 in. (11mm) from the edge of the sector. Also, do not drill through the lift arm attaching portion of the sector gear, or the joint integrity will be jeopardized.

3. Drill out the ends of the motor attaching rivets using a 1/4 in. bit.

4. Remove the motor and the remaining portions of the rivets from the regulator. Except for the rear doors, one rivet will not be accessible until assembly.

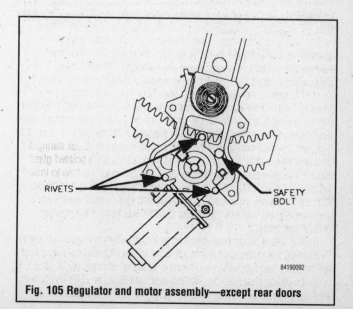

Fig. 105 Regulator and motor assembly—except rear doors

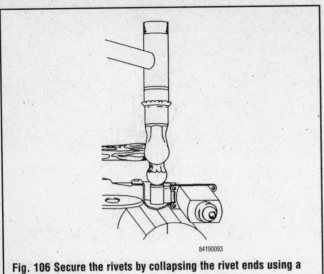

Fig. 106 Secure the rivets by collapsing the rivet ends using a ball peen hammer

To install:

5. Position the new motor to the regulator. With the aid of an assistant, install new rivets by collapsing or crushing the rivet ends using a ball peen hammer.

6. Except for the rear doors, once two of the three rivets have been installed, remove the bolt securing the sector gear to the backplate, then use an appropriate electrical source (such as the vehicle's window motor harness) to rotate the regulator, providing access to the remaining rivet.

7. Except for the rear doors, use a flat nosed rotary file to grind off one sector gear tooth from the side which does not contact the driven gear. This is necessary to reach the remaining rivet. Remove the old rivet, then install the remaining new rivet.

8. Install the regulator to the door assembly.

Windshield & Fixed Glass

REMOVAL & INSTALLATION

If your windshield, or other fixed window, is cracked or chipped, you may decide to replace it with a new one yourself. However, there are two main reasons why replacement windshields and other window glass should be installed only by a professional automotive glass technician: safety and cost.

The most important reason a professional should install automotive glass is for safety. The glass in the vehicle, especially the windshield, is designed with safety in mind in case of a collision. The windshield is specially manufactured from two panes of specially-tempered glass with a thin layer of transparent plastic between them. This construction allows the glass to "give" in the event that a part of your body hits the windshield during the collision, and prevents the glass from shattering, which could cause lacerations, blinding and other harm to passengers of the vehicle. The other fixed windows are designed to be tempered so that if they break during a collision, they shatter in such a way that there are no large pointed glass pieces. The professional automotive glass technician knows how to install the glass in a vehicle so that it will function optimally during a collision. Without the proper experience, knowledge and tools, installing a piece of automotive glass yourself could lead to additional harm if an accident should ever occur.

Cost is also a factor when deciding to install automotive glass yourself. Performing this could cost you much more than a professional may charge for the same job. Since the windshield is designed to break under stress, an often life saving characteristic, windshields tend to break VERY easily when an inexperienced person attempts to install one. Do-it-yourselfers buying

two, three or even four windshields from a salvage yard because they have broken them during installation are common stories. Also, since the automotive glass is designed to prevent the outside elements from entering your vehicle, improper installation can lead to water and air leaks. Annoying whining noises at highway speeds from air leaks or inside body panel rusting from water leaks can add to your stress level and subtract from your wallet. After buying two or three windshields, installing them and ending up with a leak that produces a noise while driving and water damage during rainstorms, the cost of having a professional do it correctly the first time may be much more alluring. We here at Chilton, therefore, advise that you have a professional automotive glass technician service any broken glass on your vehicle.

WINDSHIELD CHIP REPAIR

▶ See Figures 107 thru 121

➡Check with your state and local authorities on the laws for state safety inspection. Some states or municipalities may not allow chip repair as a viable option for correcting stone damage to your windshield.

Fig. 107 Small chips on your windshield can be fixed with an aftermarket repair kit, such as the one from Loctite®

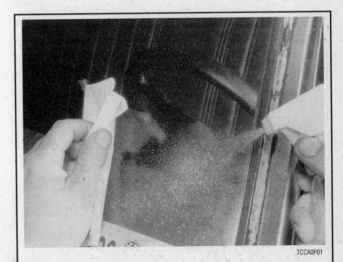

Fig. 108 To repair a chip, clean the windshield with glass cleaner and dry it completely

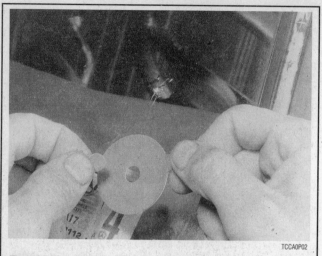

Fig. 109 Remove the center from the adhesive disc and peel off the backing from one side of the disc . . .

Fig. 112 Peel the backing off the exposed side of the adhesive disc . . .

Fig. 110 . . . then press it on the windshield so that the chip is centered in the hole

Fig. 113 . . . then position the plastic pedestal on the adhesive disc, ensuring that the tabs are aligned

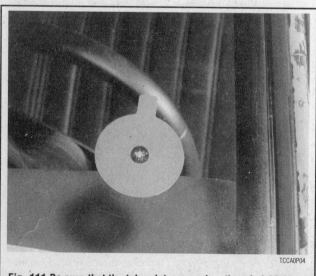

Fig. 111 Be sure that the tab points upward on the windshield

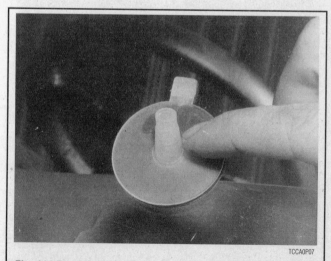

Fig. 114 Press the pedestal firmly on the adhesive disc to create an adequate seal . . .

Although severely cracked or damaged windshields must be replaced, there is something that you can do to prolong or even prevent the need for replacement of a chipped windshield. There are many companies which offer windshield chip repair products, such as Loctite's® Bullseye™ windshield repair kit. These kits usually consist of a syringe, pedestal and a sealing adhesive. The syringe is mounted on the pedestal and is used to create a vacuum which pulls the plastic layer against the glass. This helps make the chip transparent. The adhesive is then injected which seals the chip and helps to prevent further stress cracks from developing. Refer to the sequence of photos to get a general idea of what windshield chip repair involves.

➡**Always follow the specific manufacturer's instructions.**

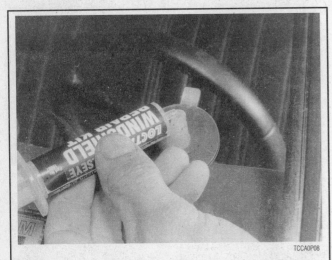

Fig. 115 . . . then install the applicator syringe nipple in the pedestal's hole

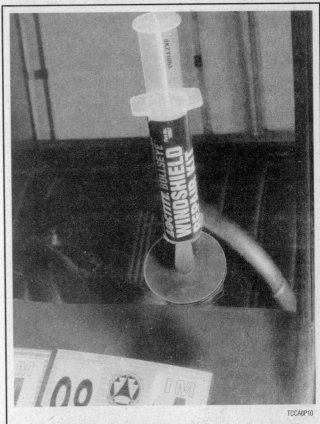

Fig. 117 After applying the solution, allow the entire assembly to sit until it has set completely

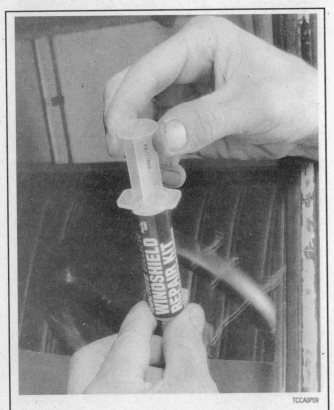

Fig. 116 Hold the syringe with one hand while pulling the plunger back with the other hand

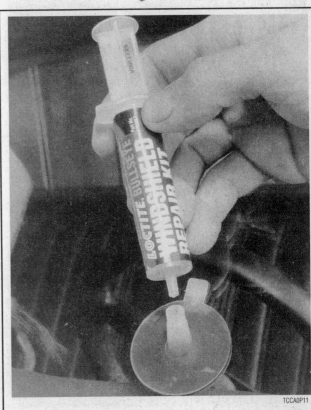

Fig. 118 After the solution has set, remove the syringe from the pedestal . . .

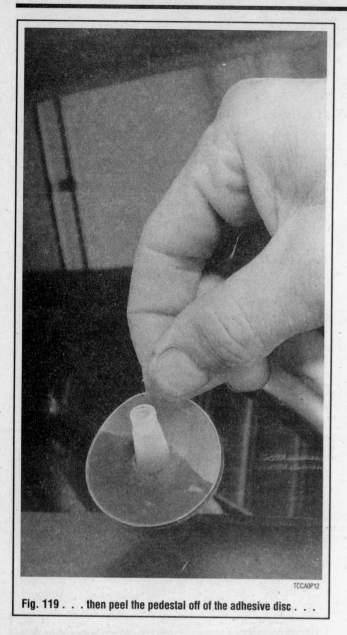

Fig. 119 . . . then peel the pedestal off of the adhesive disc . . .

TCCA0P12

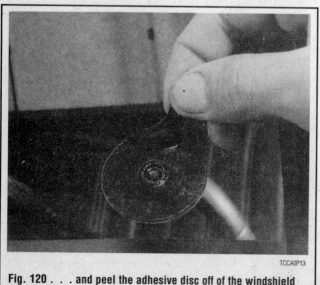

Fig. 120 . . . and peel the adhesive disc off of the windshield

TCCA0P13

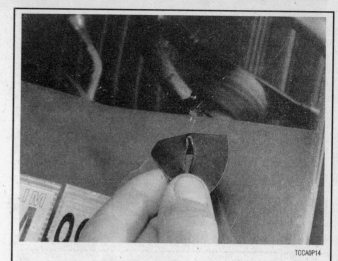

Fig. 121 The chip will still be slightly visible, but it should be filled with the hardened solution

TCCA0P14

Inside Rear View Mirror

The rear view mirror is attached to a support which is bonded to the windshield glass. If the support becomes unattached, it must be aligned and rebonded before the mirror can be mounted.

REPLACEMENT

▶ **See Figure 122**

1. Measure Dimension A, from the bottom of the encapsulated molding roof line to the bottom (flat side) or the mirror support. Dimension A is 6.4 in. (162mm) for the Sedan and Wagon or 7.9 in. (200mm) for the Coupe.

2. Mark the outside of the windshield glass using a wax pencil or crayon, then make a large diameter circle around the support location mark.

3. Clean the inside glass surface within the large marked circle using a glass cleaning solution or polishing compound. Rub the area until it is completely clean and dry.

4. Once dry, clean the area using an alcohol saturated paper towel to remove any traces of scouring powder or cleaning solution.

5. If still attached, use a Torx® bit to separate the mirror from the support, then use a piece of fine grit sandpaper to rough the bonding surface of

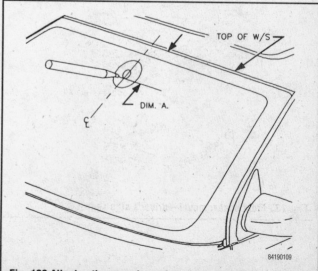

Fig. 122 Aligning the rear view mirror for installation

84190109

the mirror support. If the original support is being used, all traces of adhesive must be removed prior to installation.

6. Wipe the sanded mirror support with a clean alcohol saturated paper towel, then allow to dry.

7. Follow the directions on a manufacturer's kit (Loctite® Rearview Mirror adhesive or equivalent) and prepare the support for installation.

8. When ready, position the support to the marked location, making sure the rounded end is pointed upward. Press the support to the glass for 30–60 seconds using steady pressure. After about 5 minutes, any excess adhesive may be removed with an alcohol moistened paper towel or glass cleaning solution.

9. Allow additional time to cure, if necessary, per the adhesive manufacturer's instructions, then install the rear view mirror to the support.

Seats

REMOVAL & INSTALLATION

Front Seats

▶ **See Figure 123**

1. If removing the driver's seat, disconnect the negative battery cable, then remove the CPA device and unplug the seat buckle wire connector.

2. Push the lap belt through the seat guide loop, if equipped.

3. Remove the 4 seat adjuster-to-vehicle attaching bolts, then carefully remove the seat assembly from the vehicle. Place the seat on a clean, protected surface.

To install:

4. Position the seat in the vehicle while aligning it to the fastener mounting holes on the vehicle floor.

5. If installing the driver's seat, connect the wiring harness to the buckle connector, then install the CPA device.

6. Check that the seat adjusters are timed properly by making sure all adjuster mounting tabs are tight against their mounting surfaces and all bolt holes are aligned.

7. Apply a coat of Loctite® 242, or an equivalent threadlock, to the 4 seat adjuster bolts, then install the bolts and tighten to 26 ft. lbs. (35 Nm).

8. Install the lap belt through the seat guide loop, if equipped.

9. If applicable, connect the negative battery cable.

10. Check to be certain that all features of the seat mechanism operate properly.

Rear Seats

▶ **See Figures 124, 125 and 126**

1. To remove the cushion from the Coupe, remove the inserts and fasteners, then remove the rear seat center console. Disengage the seat cushion front floor clips and guide the cushion out of the rear attachment.

2. To remove the cushion from the Sedan and Wagon, push on the retainer tabs located at the bottom of the seat cushion assembly, then lift the front of the cushion from under the seat backs, guide the seat belt buckles through the cushion and remove the cushion assembly from the vehicle.

3. To remove the seat backs:
   a. Release the latch and tilt the seat forward.
   b. disengage the cargo floor carpet from the seat back.
   c. Remove the plastic clip from the outboard pivot pin.
   d. Slide the seat back toward the outside of the vehicle in order to position the pivot pin relief slot at the floor bracket, then lift the pivot pin out of the floor bracket.
   e. Lift the seat back out of the outboard floor bracket and remove the seat back from the vehicle.

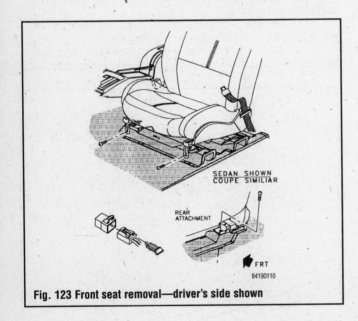

Fig. 123 Front seat removal—driver's side shown

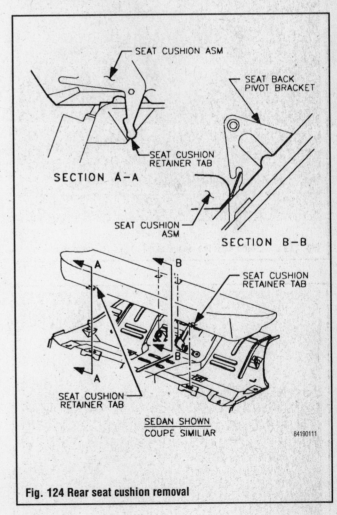

Fig. 124 Rear seat cushion removal

To install:

4. If removed, install the seat backs:

   a. Guide the seat back assembly onto the center pivot pin.

   b. Align the outer pivot pin relief with outer bracket, then guide the pin into the bracket.

   c. Slide the seat back inboard to engage the pivot pin into the bracket.

   d. Install the outboard pivot pin clip.

   e. Engage the cargo floor carpet with the seat backs.

5. Install the cushion in the Sedan or Wagon by positioning the cushion (rear edge first). Secure the wire frame correctly at the center buckle and pull the center lap belt through the seat cushion. Guide the seat cushion front floor clips into their floor pan holes, then push firmly on the front of the seat to engage the clips.

6. Install the cushion in the Coupe, by positioning the cushion with the rear edge first, then securing the wire frame at the center buckle. Install the rear center console, fasteners and inserts.

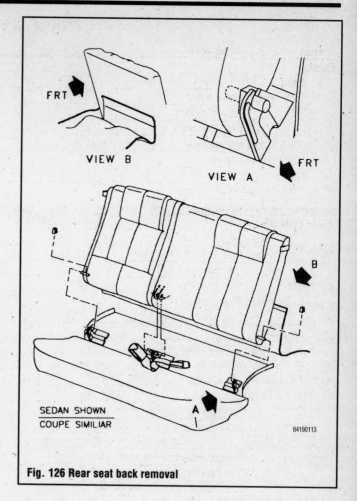

Fig. 126 Rear seat back removal

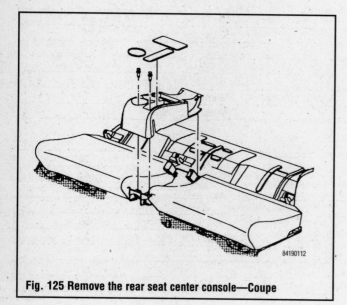

Fig. 125 Remove the rear seat center console—Coupe

TORQUE SPECIFICATIONS

System	Component	Ft. Lbs.	Nm
Exterior			
Doors			
	Hinge-to-door bolts		
	1991-95 models	26	35
	1996-98 models	22	30
	Check link retainers	12	16
	Striker retainers	18	25
Hood			
	Hinge-to-hood bolts	19	25
	Hood latch bolts	89 inch lbs.	10
Trunk lid			
	Hinge-to-trunk lid bolts	89 inch lbs.	10
	Lower panel-to-trunk lid screws		
	1991-96 coupe	53 inch lbs.	6
	1991-95 sedan	53 inch lbs.	6
	Package shelf retainers		
	Coupe and sedan	19	25
Liftgate			
	Hinge-to-liftgate bolts	89 inch lbs.	10
	Latch striker bolts	89 inch lbs.	10
Front bumper			
	Sedan and wagon		
	Bumper impact bar-to-vehicle frame	22	30
	Front fascia-to-front wheelhouse panel	31 inch lbs.	3.5
	Coupe		
	Bumper impact bar-to-vehicle frame	22	30
	Tow hook-to-impact bar	22	30
	Front fascia support-to-impact bar	31 inch lbs.	3.5
Rear bumper			
	Sedan and wagon		
	Bumper impact bar-to-vehicle	22	30
	Coupe		
	Bumper impact bar-to-vehicle	19	25
Outside mirrors			
	Mirror-to-door nuts		
	1991-95 models	53 inch lbs.	6
	1996-98 models	89 inch lbs.	10
	Mirror cable set screw	27 inch lbs.	3
Fenders			
	Fender-to-front body hinge pillar bolt	89 inch lbs.	10
	Top of fender-to-spaceframe bolts	89 inch lbs.	10
	Bottom rear fender bolts	89 inch lbs.	10
	Fender-to-fender reinforcement bracket bolts	89 inch lbs.	10
	Outer headlamp mounting bolt	61 inch lbs.	7
Power sunroof			
	Glass attaching screws	35-44 inch lbs.	4-5
	Motor screws	27-35 inch lbs.	3-4
	Guide assembly screws	35-44 inch lbs.	4-5
	Locator screws	35 inch lbs.	4
	Cable locator screws	35 inch lbs.	4

89560C01

TORQUE SPECIFICATIONS

System	Component	Ft. Lbs.	Nm
Interior			
Instrument Panel and Pad			
1991-94 models			
	Steering column assembly bolts	33	45
1995-98 models			
	Instrument panel assembly nuts and screws	89 inch lbs.	10
	Passenger side air bag module fasteners	89 inch lbs.	10
	Glove box mounting screws	20 inch lbs.	2.2
	H-bracket bolts	19	25
	Instrument panel junction block mounting screw	20 inch lbs.	2.2
	Junction block wiring connector retaining screw	5	6.5
	Ground wire-to-H bracket retaining screw	20 inch lbs.	2.2
	Instrument panel/HVAC outlet end caps	20 inch lbs.	2.2
Center console			
	Front console mounting screws	11 inch lbs.	1.2
	Rear console mounting screws	14 inch lbs.	1.6
	Parking brake lever cover	14 inch lbs.	1.6
	Center console armrest mounting screws	14 inch lbs.	1.6
Headliner			
	Shoulder belt mounting fastener	37	50
Interior door panels			
	Door pull cup attaching screw	22 inch lbs.	2.2
Door locks			
	Outer door panel mounting screws	53 inch lbs.	6
Seats			
	Seat retaining bolts	26	35

89560C02

GLOSSARY

AIR/FUEL RATIO: The ratio of air-to-gasoline by weight in the fuel mixture drawn into the engine.

AIR INJECTION: One method of reducing harmful exhaust emissions by injecting air into each of the exhaust ports of an engine. The fresh air entering the hot exhaust manifold causes any remaining fuel to be burned before it can exit the tailpipe.

ALTERNATOR: A device used for converting mechanical energy into electrical energy.

AMMETER: An instrument, calibrated in amperes, used to measure the flow of an electrical current in a circuit. Ammeters are always connected in series with the circuit being tested.

AMPERE: The rate of flow of electrical current present when one volt of electrical pressure is applied against one ohm of electrical resistance.

ANALOG COMPUTER: Any microprocessor that uses similar (analogous) electrical signals to make its calculations.

ARMATURE: A laminated, soft iron core wrapped by a wire that converts electrical energy to mechanical energy as in a motor or relay. When rotated in a magnetic field, it changes mechanical energy into electrical energy as in a generator.

ATMOSPHERIC PRESSURE: The pressure on the Earth's surface caused by the weight of the air in the atmosphere. At sea level, this pressure is 14.7 psi at 32°F (101 kPa at 0°C).

ATOMIZATION: The breaking down of a liquid into a fine mist that can be suspended in air.

AXIAL PLAY: Movement parallel to a shaft or bearing bore.

BACKFIRE: The sudden combustion of gases in the intake or exhaust system that results in a loud explosion.

BACKLASH: The clearance or play between two parts, such as meshed gears.

BACKPRESSURE: Restrictions in the exhaust system that slow the exit of exhaust gases from the combustion chamber.

BAKELITE: A heat resistant, plastic insulator material commonly used in printed circuit boards and transistorized components.

BALL BEARING: A bearing made up of hardened inner and outer races between which hardened steel balls roll.

BALLAST RESISTOR: A resistor in the primary ignition circuit that lowers voltage after the engine is started to reduce wear on ignition components.

BEARING: A friction reducing, supportive device usually located between a stationary part and a moving part.

BIMETAL TEMPERATURE SENSOR: Any sensor or switch made of two dissimilar types of metal that bend when heated or cooled due to the different expansion rates of the alloys. These types of sensors usually function as an on/off switch.

BLOWBY: Combustion gases, composed of water vapor and unburned fuel, that leak past the piston rings into the crankcase during normal engine operation. These gases are removed by the PCV system to prevent the buildup of harmful acids in the crankcase.

BRAKE PAD: A brake shoe and lining assembly used with disc brakes.

BRAKE SHOE: The backing for the brake lining. The term is, however, usually applied to the assembly of the brake backing and lining.

BUSHING: A liner, usually removable, for a bearing; an anti-friction liner used in place of a bearing.

CALIPER: A hydraulically activated device in a disc brake system, which is mounted straddling the brake rotor (disc). The caliper contains at least one piston and two brake pads. Hydraulic pressure on the piston(s) forces the pads against the rotor.

CAMSHAFT: A shaft in the engine on which are the lobes (cams) which operate the valves. The camshaft is driven by the crankshaft, via a belt, chain or gears, at one half the crankshaft speed.

CAPACITOR: A device which stores an electrical charge.

CARBON MONOXIDE (CO): A colorless, odorless gas given off as a normal byproduct of combustion. It is poisonous and extremely dangerous in confined areas, building up slowly to toxic levels without warning if adequate ventilation is not available.

CARBURETOR: A device, usually mounted on the intake manifold of an engine, which mixes the air and fuel in the proper proportion to allow even combustion.

CATALYTIC CONVERTER: A device installed in the exhaust system, like a muffler, that converts harmful byproducts of combustion into carbon dioxide and water vapor by means of a heat-producing chemical reaction.

CENTRIFUGAL ADVANCE: A mechanical method of advancing the spark timing by using flyweights in the distributor that react to centrifugal force generated by the distributor shaft rotation.

CHECK VALVE: Any one-way valve installed to permit the flow of air, fuel or vacuum in one direction only.

CHOKE: A device, usually a moveable valve, placed in the intake path of a carburetor to restrict the flow of air.

CIRCUIT: Any unbroken path through which an electrical current can flow. Also used to describe fuel flow in some instances.

CIRCUIT BREAKER: A switch which protects an electrical circuit from overload by opening the circuit when the current flow exceeds a predetermined level. Some circuit breakers must be reset manually, while most reset automatically.

COIL (IGNITION): A transformer in the ignition circuit which steps up the voltage provided to the spark plugs.

COMBINATION MANIFOLD: An assembly which includes both the intake and exhaust manifolds in one casting.

COMBINATION VALVE: A device used in some fuel systems that routes fuel vapors to a charcoal storage canister instead of venting them into the atmosphere. The valve relieves fuel tank pressure and allows fresh air into the tank as the fuel level drops to prevent a vapor lock situation.

COMPRESSION RATIO: The comparison of the total volume of the cylinder and combustion chamber with the piston at BDC and the piston at TDC.

CONDENSER: 1. An electrical device which acts to store an electrical charge, preventing voltage surges. 2. A radiator-like device in the air conditioning system in which refrigerant gas condenses into a liquid, giving off heat.

CONDUCTOR: Any material through which an electrical current can be transmitted easily.

CONTINUITY: Continuous or complete circuit. Can be checked with an ohmmeter.

COUNTERSHAFT: An intermediate shaft which is rotated by a mainshaft and transmits, in turn, that rotation to a working part.

CRANKCASE: The lower part of an engine in which the crankshaft and related parts operate.

CRANKSHAFT: The main driving shaft of an engine which receives reciprocating motion from the pistons and converts it to rotary motion.

CYLINDER: In an engine, the round hole in the engine block in which the piston(s) ride.

CYLINDER BLOCK: The main structural member of an engine in which is found the cylinders, crankshaft and other principal parts.

CYLINDER HEAD: The detachable portion of the engine, usually fastened to the top of the cylinder block and containing all or most of the combustion chambers. On overhead valve engines, it contains the valves and their operating parts. On overhead cam engines, it contains the camshaft as well.

DEAD CENTER: The extreme top or bottom of the piston stroke.

DETONATION: An unwanted explosion of the air/fuel mixture in the combustion chamber caused by excess heat and compression, advanced timing, or an overly lean mixture. Also referred to as "ping".

DIAPHRAGM: A thin, flexible wall separating two cavities, such as in a vacuum advance unit.

DIESELING: A condition in which hot spots in the combustion chamber cause the engine to run on after the key is turned off.

DIFFERENTIAL: A geared assembly which allows the transmission of motion between drive axles, giving one axle the ability to turn faster than the other.

DIODE: An electrical device that will allow current to flow in one direction only.

DISC BRAKE: A hydraulic braking assembly consisting of a brake disc, or rotor, mounted on an axle, and a caliper assembly containing, usually two brake pads which are activated by hydraulic pressure. The pads are forced against the sides of the disc, creating friction which slows the vehicle.

DISTRIBUTOR: A mechanically driven device on an engine which is responsible for electrically firing the spark plug at a predetermined point of the piston stroke.

DOWEL PIN: A pin, inserted in mating holes in two different parts allowing those parts to maintain a fixed relationship.

DRUM BRAKE: A braking system which consists of two brake shoes and one or two wheel cylinders, mounted on a fixed backing plate, and a brake drum, mounted on an axle, which revolves around the assembly.

DWELL: The rate, measured in degrees of shaft rotation, at which an electrical circuit cycles on and off.

ELECTRONIC CONTROL UNIT (ECU): Ignition module, module, amplifier or igniter. See Module for definition.

ELECTRONIC IGNITION: A system in which the timing and firing of the spark plugs is controlled by an electronic control unit, usually called a module. These systems have no points or condenser.

END-PLAY: The measured amount of axial movement in a shaft.

ENGINE: A device that converts heat into mechanical energy.

EXHAUST MANIFOLD: A set of cast passages or pipes which conduct exhaust gases from the engine.

FEELER GAUGE: A blade, usually metal, or precisely predetermined thickness, used to measure the clearance between two parts.

FIRING ORDER: The order in which combustion occurs in the cylinders of an engine. Also the order in which spark is distributed to the plugs by the distributor.

FLOODING: The presence of too much fuel in the intake manifold and combustion chamber which prevents the air/fuel mixture from firing, thereby causing a no-start situation.

FLYWHEEL: A disc shaped part bolted to the rear end of the crankshaft. Around the outer perimeter is affixed the ring gear. The starter drive engages the ring gear, turning the flywheel, which rotates the crankshaft, imparting the initial starting motion to the engine.

FOOT POUND (ft. lbs. or sometimes, ft.lb.): The amount of energy or work needed to raise an item weighing one pound, a distance of one foot.

FUSE: A protective device in a circuit which prevents circuit overload by breaking the circuit when a specific amperage is present. The device is constructed around a strip or wire of a lower amperage rating than the circuit it is designed to protect. When an amperage higher than that stamped on the fuse is present in the circuit, the strip or wire melts, opening the circuit.

GEAR RATIO: The ratio between the number of teeth on meshing gears.

GENERATOR: A device which converts mechanical energy into electrical energy.

HEAT RANGE: The measure of a spark plug's ability to dissipate heat from its firing end. The higher the heat range, the hotter the plug fires.

HUB: The center part of a wheel or gear.

HYDROCARBON (HC): Any chemical compound made up of hydrogen and carbon. A major pollutant formed by the engine as a byproduct of combustion.

HYDROMETER: An instrument used to measure the specific gravity of a solution.

INCH POUND (inch lbs.; sometimes in.lb. or in. lbs.): One twelfth of a foot pound.

INDUCTION: A means of transferring electrical energy in the form of a magnetic field. Principle used in the ignition coil to increase voltage.

INJECTOR: A device which receives metered fuel under relatively low pressure and is activated to inject the fuel into the engine under relatively high pressure at a predetermined time.

INPUT SHAFT: The shaft to which torque is applied, usually carrying the driving gear or gears.

INTAKE MANIFOLD: A casting of passages or pipes used to conduct air or a fuel/air mixture to the cylinders.

JOURNAL: The bearing surface within which a shaft operates.

KEY: A small block usually fitted in a notch between a shaft and a hub to prevent slippage of the two parts.

MANIFOLD: A casting of passages or set of pipes which connect the cylinders to an inlet or outlet source.

MANIFOLD VACUUM: Low pressure in an engine intake manifold formed just below the throttle plates. Manifold vacuum is highest at idle and drops under acceleration.

MASTER CYLINDER: The primary fluid pressurizing device in a hydraulic system. In automotive use, it is found in brake and hydraulic clutch systems and is pedal activated, either directly or, in a power brake system, through the power booster.

MODULE: Electronic control unit, amplifier or igniter of solid state or integrated design which controls the current flow in the ignition primary circuit based on input from the pick-up coil. When the module opens the primary circuit, high secondary voltage is induced in the coil.

NEEDLE BEARING: A bearing which consists of a number (usually a large number) of long, thin rollers.

OHM: (Ω) The unit used to measure the resistance of conductor-to-electrical flow. One ohm is the amount of resistance that limits current flow to one ampere in a circuit with one volt of pressure.

OHMMETER: An instrument used for measuring the resistance, in ohms, in an electrical circuit.

OUTPUT SHAFT: The shaft which transmits torque from a device, such as a transmission.

OVERDRIVE: A gear assembly which produces more shaft revolutions than that transmitted to it.

OVERHEAD CAMSHAFT (OHC): An engine configuration in which the camshaft is mounted on top of the cylinder head and operates the valve either directly or by means of rocker arms.

OVERHEAD VALVE (OHV): An engine configuration in which all of the valves are located in the cylinder head and the camshaft is located in the cylinder block. The camshaft operates the valves via lifters and pushrods.

OXIDES OF NITROGEN (NOx): Chemical compounds of nitrogen produced as a byproduct of combustion. They combine with hydrocarbons to produce smog.

OXYGEN SENSOR: Use with the feedback system to sense the presence of oxygen in the exhaust gas and signal the computer which can reference the voltage signal to an air/fuel ratio.

PINION: The smaller of two meshing gears.

PISTON RING: An open-ended ring with fits into a groove on the outer diameter of the piston. Its chief function is to form a seal between the piston and cylinder wall. Most automotive pistons have three rings: two for compression sealing; one for oil sealing.

PRELOAD: A predetermined load placed on a bearing during assembly or by adjustment.

PRIMARY CIRCUIT: the low voltage side of the ignition system which consists of the ignition switch, ballast resistor or resistance wire, bypass, coil, electronic control unit and pick-up coil as well as the connecting wires and harnesses.

PRESS FIT: The mating of two parts under pressure, due to the inner diameter of one being smaller than the outer diameter of the other, or vice versa; an interference fit.

RACE: The surface on the inner or outer ring of a bearing on which the balls, needles or rollers move.

REGULATOR: A device which maintains the amperage and/or voltage levels of a circuit at predetermined values.

RELAY: A switch which automatically opens and/or closes a circuit.

RESISTANCE: The opposition to the flow of current through a circuit or electrical device, and is measured in ohms. Resistance is equal to the voltage divided by the amperage.

RESISTOR: A device, usually made of wire, which offers a preset amount of resistance in an electrical circuit.

RING GEAR: The name given to a ring-shaped gear attached to a differential case, or affixed to a flywheel or as part of a planetary gear set.

ROLLER BEARING: A bearing made up of hardened inner and outer races between which hardened steel rollers move.

ROTOR: 1. The disc-shaped part of a disc brake assembly, upon which the brake pads bear; also called, brake disc. 2. The device mounted atop the distributor shaft, which passes current to the distributor cap tower contacts.

SECONDARY CIRCUIT: The high voltage side of the ignition system, usually above 20,000 volts. The secondary includes the ignition coil, coil wire, distributor cap and rotor, spark plug wires and spark plugs.

SENDING UNIT: A mechanical, electrical, hydraulic or electro-magnetic device which transmits information to a gauge.

SENSOR: Any device designed to measure engine operating conditions or ambient pressures and temperatures. Usually electronic in nature and designed to send a voltage signal to an on-board computer, some sensors may operate as a simple on/off switch or they may provide a variable voltage signal (like a potentiometer) as conditions or measured parameters change.

SHIM: Spacers of precise, predetermined thickness used between parts to establish a proper working relationship.

SLAVE CYLINDER: In automotive use, a device in the hydraulic clutch system which is activated by hydraulic force, disengaging the clutch.

SOLENOID: A coil used to produce a magnetic field, the effect of which is to produce work.

SPARK PLUG: A device screwed into the combustion chamber of a spark ignition engine. The basic construction is a conductive core inside of a ceramic insulator, mounted in an outer conductive base. An electrical charge from the spark plug wire travels along the conductive core and jumps a preset air gap to a grounding point or points at the end of the conductive base. The resultant spark ignites the fuel/air mixture in the combustion chamber.

SPLINES: Ridges machined or cast onto the outer diameter of a shaft or inner diameter of a bore to enable parts to mate without rotation.

TACHOMETER: A device used to measure the rotary speed of an engine, shaft, gear, etc., usually in rotations per minute.

THERMOSTAT: A valve, located in the cooling system of an engine, which is closed when cold and opens gradually in response to engine heating, controlling the temperature of the coolant and rate of coolant flow.

TOP DEAD CENTER (TDC): The point at which the piston reaches the top of its travel on the compression stroke.

TORQUE: The twisting force applied to an object.

TORQUE CONVERTER: A turbine used to transmit power from a driving member to a driven member via hydraulic action, providing changes in drive ratio and torque. In automotive use, it links the driveplate at the rear of the engine to the automatic transmission.

TRANSDUCER: A device used to change a force into an electrical signal.

TRANSISTOR: A semi-conductor component which can be actuated by a small voltage to perform an electrical switching function.

TUNE-UP: A regular maintenance function, usually associated with the replacement and adjustment of parts and components in the electrical and fuel systems of a vehicle for the purpose of attaining optimum performance.

TURBOCHARGER: An exhaust driven pump which compresses intake air and forces it into the combustion chambers at higher than atmospheric pressures. The increased air pressure allows more fuel to be burned and results in increased horsepower being produced.

VACUUM ADVANCE: A device which advances the ignition timing in response to increased engine vacuum.

VACUUM GAUGE: An instrument used to measure the presence of vacuum in a chamber.

VALVE: A device which control the pressure, direction of flow or rate of flow of a liquid or gas.

VALVE CLEARANCE: The measured gap between the end of the valve stem and the rocker arm, cam lobe or follower that activates the valve.

VISCOSITY: The rating of a liquid's internal resistance to flow.

VOLTMETER: An instrument used for measuring electrical force in units called volts. Voltmeters are always connected parallel with the circuit being tested.

WHEEL CYLINDER: Found in the automotive drum brake assembly, it is a device, actuated by hydraulic pressure, which, through internal pistons, pushes the brake shoes outward against the drums.

MASTER
INDEX